First International Tainan-Moscow Algebra Workshop

First International Tainan-Moscow Algebra Workshop

Proceedings of the International Conference held at
National Cheng Kung University
Tainan, Taiwan, Republic of China, July 23–August 22, 1994

Executive Editors

Y. Fong
U. Knauer
A.V. Mikhalev

Editorial Board

Walter de Gruyter · Berlin · New York 1996

Editors

Y. Fong
Department of Mathematics
National Cheng Kung University
700 Tainan, Taiwan
Republic of China

U. Knauer
Fachbereich Mathematik
Carl von Ossietzky Universität
D-26111 Oldenburg
Germany

A. V. Mikhalev
Department of Mech.-Math.
Moscow State University
Moscow, 117234
Russia

1991 Mathematics Subject Classification: 16-06, 17-06, 05-XX, 13-XX
Keywords: Ring, semigroup, Lie algebra, nearring, Minkowski algebra, Wiener index

♾ Printed on acid-free paper which falls within the guidelines of the ANSI to ensure permanence and durability.

Library of Congress Cataloging-in-Publication Data

International Tainan-Moscow Algebra Workshop (1st : 1994 : National Cheng Kung University)
First International Tainan-Moscow Algebra Workshop : proceedings of the international conference held at National Cheng Kung University, Tainan, Taiwan, Republic of China, July 23-August 22, 1994 / executive editors, Y. Fong, U. Knauer, A. V. Mikhalev.
p. cm.
ISBN 3-11-014864-1 (acid-free paper)
1. Algebra – Congresses. I. Fong, Yuen. II. Knauer, U., 1942– . III. Mikhalev, A. V. (Aleksandr Vasil'evich) IV. Title.
QA150.I585 1994
512'.02–dc20 96-43293
CIP

Die Deutsche Bibliothek – Cataloging-in-Publication Data

First International Tainan-Moscow Algebra Workshop: proceedings of the international conference held at National Cheng Kung University, Tainan, Taiwan, Republic of China, July 23 – August 22, 1994 / executive ed.: Y. Fong ... Ed. board K. I. Beidar ... – Berlin ; New York : de Gruyter, 1996
ISBN 3-11-014864-1
NE: Fong, Yuen [Hrsg.]; International Tainan-Moscow Algebra Workshop <1, 1994>; Guoli-Chenggong-Daxue <Tainan, Taiwan>; HST

Printed in Germany. Typeset using the authors' TeX files: I. Zimmermann, Freiburg.
Printing: A. Collignon GmbH, Berlin. Binding: Lüderitz & Bauer, Berlin.
Cover design: Thomas Bonnie, Hamburg.

Dedicated to my beloved teacher
Prof. I. H. Lin
who has supported me spiritually
through years

Fong Yuen

Foreword

This volume contains three invited survey articles and 22 other papers presented at the First International Tainan–Moscow Algebra Workshop on algebras and their related topics at the Mathematics Department of National Cheng Kung University, Tainan, Taiwan, Republic of China, during the period July 23–August 22, 1994.

All papers in this volume have been refereed by an international referee board. I would like to express my deepest thanks to all the referees who were so helpful and punctual in submitting their reports.

Thanks are also due to the Promotion and Research Center of the National Science Council of Republic of China, the Office of Research and Development of National Cheng Kung University, the National Kaohsiung Normal University, the National Taiwan University and the Eurasia Book Company for their financial support of this workshop.

The purpose of the workshop was to give a clear picture of the recent development and research in the fields of different kinds of algebras both in Taiwan and Russia. Further, we were hoping to enhance the possibility of future cooperation in research work between the algebraists of both nations. Here we would like to point out that this workshop was in fact the first international algebra workshop to be held in the southern part of this country and it will be continued every two years.

My special thanks go to Prof. W.-F. Ke and Prof. K. I. Beidar. They both spent many hours retyping some of the manuscripts and handling all the TEX related problems. To them I must say "Thanks, my Brothers!"

Those who made numerous, positive contributions to this meeting include Prof. Chen-Ku Hwang (President of National Kaohsiung Normal University), Prof. Shian Leou (Dean of Academic Affairs, NKNU), Prof. Long-Shung Shiao (Dean of College of Science, NKNU), Prof. Tai-Cheng Tso (Chairman of Graduate Institute of Mathematics Education, NKNU), Prof. Ngai-Ching Wong (National Sun Yat-Sun University), Prof. Cheng-I Weng (former Dean of Academic Affairs, NCKU), Prof. Yuh-Jia Lee (former Dean of Science Faculty, NCKU), Prof. Hung-Shan Weng (former Director General of the Office of Research and Development, NCKU), Prof. Yan-Kuin Su (Director General of the Office of Research and Development, NCKU) and James Liao (Managing Director of Eurasia Book Co.) To all of them I am truly grateful.

Here, I would like to express my gratitude Dr. M. Karbe of Walter de Gruyter & Co. for including this volume in the mathematics series. Special thanks must go to my dearly beloved brother Mr. Fong Ka-Wai for his spiritual and financial assistance through the last five years. Last but not least, I give my sincerest thanks to my wife Hsu Fang-Ai, my son Fong Tao and my daughters Fong Han, Fong Chun, Fong Yu and Fong Chi for their patience and endurance in living with an absent-minded fellow in the past years.

Fong Yuen
Tainan, Taiwan

Table of Contents

Foreword vii

Invited Addresses

P. M. Higgins
Ramsey's Theorem in algebraic semigroups 1

A. A. Mikhalev
Combinatorial aspects of the theory of Lie superalgebras 37

A. V. Mikhalev
Isomorphisms and anti-isomorphisms of endomorphism rings of modules 69

Contributed Papers

V. A. Artamonov
Algorithmic solution of quantum Serre conjecture 123

K. I. Beidar, Y. Fong and W.-F. Ke
On the simplicity of centralizer nearrings 139

K. I. Beidar, Y. Fong, U. Knauer, A. V. Mikhalev and K. P. Shum
Semigroups with generalized polynomial identities 147

A. Belov and T. Gateva-Ivanova
Radicals of monomial algebras 159

G. M. Brodskii
Lattice anti-isomorphisms of modules and the $AB5^*$ condition 171

M. A. Chebotar
On certain subrings and ideals of prime rings 177

B. Chen
The induced linear maps on Minkowski algebras of convex sets 181

S.-J. Cheng
Superconformal algebras and affine algebras with extended symmetry 199

M. I. Gekhtman and A. Stolin
Orbits of the coadjoint representation and Yang–Baxter equation 207

A. V. Grishin
On the finite basis property of T-spaces over a field of finite characteristic . . . 225

N. Lam
On the conjecture of the resolution of a finite set of points in projective space . . 229

V. N. Latyshev
Lie nilpotency: recognition and word problem 237

T.-K. Lee and T.-C. Lee
Annihilators of power values of a derivation 241

V. T. Markov
Modules with Krull and Gabriel dimensions over rings and algebras with polynomial identities 249

A. A. Mikhalev, U. U. Umirbaev and A. A. Zolotykh
A Lie algebra with cohomological dimension one over a field of prime characteristic is not necessarily free 257

V. A. Mushrub
Extended centroid of Ore extensions and injective ring endomorphisms 265

A. A. Nechaev
Polylinear recurring sequences over modules and quasi-Frobinus modules 283

L. S. Shiao and C. S. Wang
On semi-endomorphism near-rings of non-abelian groups of order pq 299

A. Slinko
Local structure of linearly compact algebras and local finiteness of coalgebras 307

G. Szeto and Y.-F. Wong
Azumaya algebras in skew group rings 317

A. A. Tuganbaev
On distributive rings and modules 323

B.-Y. Yang and Y.-N. Yeh
Chains of motley gems—and their Wiener indices 329

Ramsey's Theorem in algebraic semigroups

Peter M. Higgins

1. Introduction

In this series of talks I intend to give an introduction to several active areas in algebraic semigroup theory, some of which have been featured in the recent book [30]. The topics to be covered are quite distinct, but all involve finiteness in some form or other, which tends to give each a combinatorial flavour. My intention is to use this as my unifying theme but that in itself leaves such a broad selection of material on offer as to make the choice of starting point difficult, and the problem of finishing more difficult still. In order to draw the central thread more tightly therefore, all the subjects discussed will feature an application of Ramsey's Theorem.

Ramsey's Theorem has become a fundamental cornerstone of combinatorics. Its history seems to me (a non-expert) similar to that of Hall's Theorem on transversals—both results are relatively modern, are simple to state, and short proofs can be found which are both elementary and elegant. They each address basic combinatorial questions, and so, inevitably I suppose, their application has become widespread throughout mathematics. This in turn led to a growth in similar results and proof ideas, which shed light on earlier work. (I have in mind Menger's Theorem on vertex disjoint chains in a graph, which pre-dates and implies Hall's 'Marriage Theorem', which states that given any two collections, one consisting of n men, the other of women, each man may marry a woman that he knows if and only if each subset of k men collectively knows at least k women, $k = 1, 2, \ldots$. This can, of course, be formulated in terms of the existence of matchings within bipartite graphs, see [59].) Indeed Ramsey Theory has emerged as a burgeoning subject in its own right. However, although many books prove the Marriage Theorem, treatments of Ramsey's Theorem are less common. For this reason I record the proofs here. My source is essentially [21].

In this section we shall prove both the infinite and finite versions of Ramsey's Theorem, and give three immediate applications to finite semigroups. In particular, we shall prove a result of Simon that any finitely generated semigroup with only finitely many non-idempotents is itself finite. In Section 2 we shall identify which permutation identities have the *finite basis property*, meaning that any semigroup variety admitting such an identity necessarily has a finite set of defining identities. In Section 3 we move from varieties to so-called pseudovarieties of finite monoids and give a Ramsey style proof of Reiterman's Theorem, which says that any collection of finite semigroups closed under the taking of homomorphic images, subsemigroups, and finite direct products can be defined by a sequence of so-called pseudoidentities. The fourth section begins with a combinatorial proof of Isbell's Zigzag Theorem characterizing semigroup dominions, from which point we harness the Zigzag Theorem and Ramsey's Theorem in tandem to prove that any epimorphism from a finite commutative semigroup is necessarily surjective. In the final section we introduce Nambooripad's idea of the biordered set of idempotents of a semigroup, and use Ramsey to prove a result of Easdown that it is possible to decide, in an effective way, whether or not a given finite biordered set arises from a finite semigroup.

We shall write N to denote the set of positive integers, $N = \{1, 2, \ldots\}$, and denote the set $\{1, 2, \ldots, n\}$ by $[n]$. The cardinality of a set X will be written as $|X|$. By a *graph* G we shall mean a countable collection V of vertices together with a collection E of edges which is a set of unordered pairs, $\{u, v\}$ $\forall u, v \in V$. Our graphs will have no multiple edges, nor loops. A graph is *complete* if $\{u, v\} \in E$ for each pair of distinct vertices $u, v \in V$. The complete graph on $n \in N$ vertices will be denoted by K_n. By an *r-colouring* of a graph G, we mean an assignment of a label to each of the edges of G drawn from $[r]$; in other words a function $\chi : E \to [r]$. A graph is *monochromatic* if all of its edges have the same colour.

1.1. Ramsey's Theorem (Infinite Version). *Let G be an r-coloured complete graph on the vertex set N. Then G has a monochromatic subgraph isomorphic to G.*

Proof. Define a decreasing chain, $S_1, S_2, \ldots$, of infinite subsets of N, together with elements $x_i \in S_i$ as follows. Put $S_1 = N$ and choose $x_1 \in S_1$ arbitrarily. Given $x_i \in S_i$, let S_{i+1} be any infinite subset of S_i such that each of the edges $\{x_i, y\}$, $y \in S_{i+1}$, are all of the one colour - this is possible as there are infinitely many edges $\{x_i, y\}$, $y \in S_i$ and only finitely many colours by which they are labelled. The sequence $x_1, x_2, \ldots$ then has the property that for any fixed $i \in N$, $\chi\{x_i, x_j\} = \chi\{x_i, x_k\}$ for any $j, k > i$, as both x_j and x_k lie in S_{i+1}. We thus have a well-defined function χ^* whereby $\chi^*(x_i) = \chi(x_i, x_j), (j > i)$. Again, since there are only finitely many colours, we may extract an infinite subsequence $y_1, y_2, \ldots$ from the sequence (x_i) such that $\chi^*(y_i)$ is the same colour, j say, for all of the y_i. The complete subgraph of G with vertex set $\{y_1, y_2, \ldots\}$ is then a required monochromatic copy of G, as every edge is coloured j. □

We can give an immediate application to semigroups. Throughout we shall write $E = \mathrm{E}(S)$ for the set of idempotents of a semigroup S.

1.2. Lemma. *Let S be a finite semigroup and $s = a_1, a_2, \ldots$ an infinite sequence of elements drawn from S. Then there exists an infinite ascending sequence $i_1 < i_2 < \cdots$ of positive integers such that for any $j < k$, the product $a_{i_j} a_{i_j+1} \ldots a_{i_k-1} = e$, where $e \in \mathrm{E}(S)$ is some fixed idempotent of S. (The indices, and e, depend on s.)*

Proof. Take the complete graph G on vertex set N and label the edge from i to j $(i < j)$ by the product $a_i a_{i+1} \dots a_{j-1}$. Since S is finite, this in effect gives an r-colouring of G, where $r = |S|$. By Ramsey's Theorem, G contains a monochromatic copy G' of itself with each edge labelled by some $e \in S$. Take any 3 vertices, a_i, a_j, a_k of G' with $i < j < k$. We have $e = a_i a_{i+1} \dots a_{j-1} = a_j a_{j+1} \dots a_{k-1} = a_i a_{i+1} \dots a_{k-1}$, but since the product of the first two is equal to the third, we have $e^2 = e$, as required. □

For an infinite sequence $s = a_1, a_2, \dots$ of elements drawn from a semigroup S we shall call any product of the form $a_i a_{i+1} \dots a_j$ $(i \leq j)$ a *factor* of s, and denote it by A_{ij}. When $i = 1$ we shall just write A_{1j} as A_j and call A_j a *left factor* of s. Lemma 1.2 tells us more about factors in the case where S is finite. Of course, the $\mathcal{R}$-classes of the factors $A_1, A_2, \dots$ represent a descending chain which must stabilize when S is finite. The lemma tell us additionally that, for some indices $i \leq j$, the factor $A_{ij} \in \mathrm{E}(S)$ and that $A_{ik} \mathcal{R} A_{ij}$ for all $k \geq j$; it follows that, for sufficiently large i, all factors A_{ij} belong to the same regular $\mathcal{R}$-class.

Lemma 1.2 will be all the assistance from Ramsey needed in our work in Section 4. It does however have the obvious shortcoming that it provides no bounds whatever on the integers $i_1, i_2, \dots$ which, conceivably, may be forced to take on arbitrarily large values as we range over all possible sequences drawn from a fixed finite semigroup S. However, this is not the case, and we can gain some control over these indices through a finite version of Ramsey's Theorem which says that any sufficiently large r-coloured complete graph has a monochromatic complete subgraph of a prescribed order.

1.3. Ramsey's Theorem (Finite Version). *Let $r, m \geq 1$ be positive integers. Then there exists a positive integer $n = R(m, r)$ such that for every complete finite graph G having at least n vertices and for every r-colouring of the edges of G, there is a monochromatic subgraph of G isomorphic to K_m.*

As is often the case, infinite systems can be easier to handle than their finite analogues. In this instance the aspect of the infinite case that simplifies the proof is the characteristic property of infinite sets, first pointed out by Galileo, that they possess proper subsets of the same cardinality. The proof of the finite version is essentially the same as the infinite case, except that more care must be taken to ensure that the sets in our decreasing chain are large enough to guarantee that the construction does not run aground in the shallow waters of the empty set.

Proof of 1.3. Clearly $R(m, 1) = m$, so henceforth take $r \geq 2$. We show that a sufficiently large value of n is given by

$$n = 1 + r + r^2 + \dots + r^{r(m-1)} = (r^{r(m-1)+1} - 1)/(r - 1),$$

so let us suppose that $G = K_n$, with vertex set S and edge set E, and fix an arbitrary r-colouring $\chi : E \to [r]$ of E. We recursively build sets $S_1, S_2, \dots, S_{r(m-1)+1}$ and choose elements $x_i \in S_i$ as follows. Put $S_1 = S$, and choose $x_1 \in S_1$ arbitrarily. Having selected S_i and $x_i \in S_i$ we put $T_j = \{u \in S_i : \chi\{x_i, u\} = j\}$, $j = 1, 2, \ldots, r$. Set $S_{i+1} = T_j$, where T_j is chosen to have the largest order among the sets $T_1, T_2, \ldots, T_r$. Since $|T_1| + |T_2| + \cdots + |T_r| = |S_i| - 1$, it follows that $|S_{i+1}| \geq (|S_i| - 1)/r$. From this, together with the initial condition that $|S_1| = 1 + r + \cdots + r^{r(m-1)}$, it follows by induction that $|S_{i+1}| \geq 1 + r + \cdots + r^{r(m-1)-i}$, and in particular $|S_{r(m-1)+1}| \geq 1$. Hence the set $X = \{x_1, x_2, \ldots, x_{r(m-1)+1}\}$ can be formed by this procedure.

We next introduce a new colouring $\chi^* : X \to [r]$ by putting $\chi^*(x_i) = j$ where $j = \chi\{x_i, y\}$ $\forall y \in S_{i+1}$. Since this colouring partitions X into r sets and $|X| = r(m-1) + 1$ it follows that there exists a subset $Y = \{y_1, y_2, \ldots, y_m\}$ of X such that $\chi^*(y_i) = j$ for some particular colour j. The complete subgraph of G with vertex set Y is then the required monochromatic copy of K_m. □

The finite version of the Ramsey Theorem given above is about the existence of monochromatic colourings of sets of pairs of sufficiently large sets. The general Ramsey Theorem has as its subject colourings of k-sets of sufficiently large sets.

1.4. Ramsey's Theorem (Unabridged). *For any three positive integers m, r, k with $r \geq 1, m \geq k$ there exists a positive integer $n = R(m, r, k)$ such that for every finite set S having at least n members and for every colouring in r colours of the subsets of S having k elements there exist a subset F of E with m elements such that every subset of F of order k has the same colour.*

Our Theorem 1.3 expresses in the language of graph theory the special case of Theorem 1.4 in which $k = 2$. The case where $m = 2, k = 1$ is often called the Pigeon Hole Principle: if $r + 1$ letters are posted in r pigeon holes, then at least one pigeon hole contains 2 letters. This is equivalent to the statement that $R(2, r, 1) = r + 1$. The case where $k = 1$, and m is unrestricted, is thus a generalization of the Pigeon Hole Principle for it amounts to saying that if the members of an n-set ($n \geq R(m, r, 1)$) are given an r-colouring then there is a monochromatic m-subset. The value of $R(m, r, 1)$ is evidently $r(m-1) + 1$, a fact that was used in the proof of Theorem 1.3. Ramsey's Theorem can therefore be viewed as a generalization of the Pigeon Hole Principle, a fact which itself suggests that we should expect the theorem to arise frequently, for the Pigeon Hole Principle is used time and again throughout mathematics - as semigroupists we need look no further than the Pumping Lemma of automata theory, where we argue that if a word of length $r + 1$ is accepted by an automaton with r states then some state is visited twice, causing a loop in the corresponding path, and forcing a sequence of words of the form $uw^i v, i = 1, 2, \ldots$ to be accepted by the automaton. It would seem natural to expect therefore that more subtle Pumping Lemmas might be derived through application of the Ramsey Theorem, and this is implicitly done in [24].

The infinite version of the Pigeon Hole Principle is merely the observation that if we have an infinite number of letters to post into finitely many slots then at least one slot must contain infinitely many of them. This is a weak version of the infinite

Ramsey Theorem (in which edge colouring is replaced by vertex colouring) and it shall be used (countably many times) in our proof of Reiterman's Theorem. The unabridged Ramsey Theorem can also be put into a graph theoretic framework by passing from graphs, (the edge sets of which are pairs) to *hypergraphs* in which ordinary edges are replaced by k-sets (see [21]).

The general theorem is proved by induction on k (see Theorem 2, p. 7 of [21]), but involves no new ideas, and Theorem 1.3 is adequate for our purposes. Indeed we can now prove our bounded version of Lemma 1.2.

1.5. Lemma. *Let S be a finite semigroup of order r, m a positive integer, and $a_1, a_2, \ldots$ an infinite sequence of elements drawn from S. Then there exists a positive integer $n = n(S, m)$ such that there are indices $i_1 < i_2 < \cdots < i_{m+1} \leq n$, together with an idempotent $e \in S$, for which, for all $1 \leq j < k \leq m+1$, $a_{i_j} a_{i_j+1} \ldots a_{i_k-1} = e$. (Again, the indices and the idempotent, depend on s.)*

Lemmas 1.2 can be strengthened in another way. In the paper [12] T. C. Brown showed that the lengths $i_{j+1} - i_j$ of the factorizations of the idempotents can be bounded independently of m. Brown credits the first proof of this lemma to Justin [39], who discovered it as a consequence of the following result, also due to Brown [11], [12].

1.6. Lemma. *Let $X = \{x_1, x_2, \ldots, x_m\}$ be a finite set, and let $s = a_1, a_2, \ldots$ be an infinite sequence drawn from X. (We express this by writing $s = a_1 a_2 \ldots \in X^\omega$.) Then there exists a fixed integer k and $x \in X$ such that, for any n, there are indices $i_1 < i_2 < \cdots i_n$ (these depend on n) with $x = a_{i_1} = a_{i_2} = \cdots = a_{i_n}$ and $i_{j+1} - i_j \leq k$, $1 \leq j \leq n-1$.*

Proof. We proceed by induction on m. The result is immediate if $m = 1$, so let us assume that $m = r + 1 \geq 2$ and that the result holds for lesser values of m. If it is *not* true that a_{r+1} is absent from arbitrarily long factors of s, then we can take $x = a_{r+1}$ and we are finished. Otherwise, there is an infinite subset H of the free semigroup on $Y = \{a_1, \ldots, a_r\}$, each member of which occurs as a factor of s.

Suppose inductively that we have a finite product $B_n = b_1 b_2 \ldots b_n$ with $b_i \in Y$ which is a left factor of infinitely many words of H where we begin the induction at $n = 0$ by taking B_0 to be the empty word. Since there are infinitely many such words and only finitely many choices for the $(n+1)$st letter of each word, we may extend B_n to form a word B_{n+1} which is a left factor of infinitely many words of H, and thus define the sequence $t \in Y^\omega$, the left factors of which are precisely the B_n $n = 1, 2, \ldots$. By the inductive hypothesis, there exists $x \in Y$ and an integer k such that for any n there are integers $i_1 < i_2 < \cdots < i_n$ with $x = b_{i_1} = b_{i_2} = \cdots = b_{i_n}$ and $i_{j+1} - i_j \leq k$, $1 \leq j \leq n-1$. But every left factor of t is a left factor of words of H, which are in turn factors of s. It follows that every left factor of t, and hence every factor of t, is a factor of s also. Hence the factor $b_{i_1} \ldots b_{i_2} \ldots\ldots b_{i_n}$ is a required factor of s to complete the proof. □

The procedure for constructing the sequence t is a standard one, used a number of times in the Brown paper, and to save repetition, he calls it the construction of t from H *in the regular way*. The argument demonstrates the principle that, given an infinite set H, and a sequence of colourings $r_1, r_2, \ldots$ (meaning that each colouring r_i is a mapping of H onto some finite set), one can extract an infinite sequence $t = b_1, b_2, \ldots,$ from H such that, $r_i(b_j) = r_i(b_k)$ $\forall j, k \geq i$. We shall use this principle again in our proof of Reiterman's Theorem in Section 3. The formal enunciation of this dates back to Rado [49].

The proof of Lemma 1.6 has some subtlety. The argument does not gather any information about the value of the bound k, or of the first index i_1. Moreover, the result is similar to the Van der Waerden Theorem on arithmetic sequences which says that for any positive integer n and any m-colouring of N there exists an integer w such that $[w]$ contains a monochromatic arithmetic sequence of length n. In other words, Van der Waerden's Theorem allows us to replace the inequality in the previous lemma by equality, but at the cost of allowing the integer k to become dependent on the value of n. Van der Waerden's Theorem is difficult to prove, and the values for $w = w(n, m)$ are very large. In fact no 'reasonable' upper bounds for $w(n, 2)$ are known [21].

A recent paper of Fouche [19] on so-called complete factorizations of free semigroups interprets some classical results of Ramsey Theory, including Van der Waerden's Theorem, in this new context. Another recent contribution is that of Bergelson and Hindman [9] in which Van der Waerden's Theorem is generalized by placing the question in the context of products in semigroups. Another result in similar vein to Lemma 1.5 is to be found in [20]: if S is a finite semigroup then any product $A = a_1a_2 \ldots a_k$ where $k = |S| - |E| + 1$ contains a subword, (by which we mean a product formed from A by deletion of some of the a_i), which is idempotent.

1.7. Brown–Justin Lemma. *Let $s = a_1, a_2, \ldots$ be an infinite sequence drawn from a finite semigroup S. Then there is a fixed integer k and fixed idempotent $e \in \mathrm{E}(S)$ such that, for any n there exists indices $i_1 < i_2 < \cdots < i_{n+1}$ with $i_{j+1} - i_j \leq k$ $\forall 1 \leq j \leq n$ and all products $a_{i_j}a_{i_j+1} \ldots a_{i_{j+1}-1}$ equal e.*

Proof. The case where $|S| = 1$ is clear, so let us suppose that $|S| \geq 2$ and proceed by induction on $|S|$. For the given value of $|S|$ we proceed by induction on p, the cardinality of the generating set A of distinct generators a_i in s. If $p = 1$ we take k to be the least idempotent power of a, where a is the unique member of A. Let us assume the result for $p = r$ and suppose that $p = r + 1$. Let $a \in A$ be fixed. If a is absent from arbitrarily long factors of s, then, constructing in the regular way a new sequence from this set of factors, and arguing as in Lemma 1.6, the result follows by the inductive hypothesis on p. Otherwise we may assume that there is an integer m and integers $i_1, i_2, \ldots$ such that $a = a_{i_1} = a_{i_2} = \cdots$ and $0 < i_{j+1} - i_j \leq m$ for $j = 1, 2, \ldots$. There is no loss in assuming that $i_1 = 1$.

Next we take a new finite generating set $B = \{a_{i_j}a_{i_j+1} \ldots a_{i_{j+1}-1} \mid j = 1, 2, \ldots\}$ and regard s as a sequence in B^ω. If $|aS| < |S|$ we may apply the inductive hypothesis to infer the required conclusion of the lemma (if the length of the factorizations of the idempotents in the generating set B is bounded above by k, then mk is an upper bound for their lengths as products of the set A).

Thus we may assume that $|aS| = |S|$, and similarly that $|S| = |Sa|$, for all $a \in A$ (since the fixed element a was an arbitrary member of A). Since $|S|$ is finite, this amounts to saying that S is a (finite) group.

Consider the sequence $t = q_1, q_2, \ldots$ where $q_i = a_1 a_2 \ldots a_{i-1}$. Applying Lemma 1.6 yields an integer k such that for any n there are indices $i_1 < i_2 < \cdots < i_{n+1}$ with $q_{i_1} = q_{i_2} = \cdots = q_{i_{n+1}} = q$ say, and $i_{j+1} - i_j \leq k$ for $j = 1, 2, \ldots, n$. It follows from this that $a_{i_j} a_{i_j+1} \ldots a_{i_{j+1}-1}$ is the identity of the group S for each $j = 1, 2, \ldots n$, which is the conclusion we seek. □

As a consequence Brown proved the next theorem ([10], [12]) which he also noted holds if we replace local finiteness by local nilpotency.

1.8. Theorem. *Let $\phi : S \to T$ be a surjective homomorphism from a semigroup S onto a locally finite semigroup T such that $e\phi^{-1}$ is a locally finite subsemigroup of S for each $e \in \mathrm{E}(T)$. Then S is locally finite.*

Proof. First we show that it suffices to consider the case where T is finite. Suppose that the theorem holds in this case. Let X be a finite subset of S. We need to show that $\langle X \rangle$, the subsemigroup of S generated by X, is also finite. Since T is locally finite, $T' = \langle X\phi \rangle$ is a finite subsemigroup of T. Restricting ϕ to $T'\phi^{-1}$ yields a surjective homomorphism (with the required properties) of $T'\phi^{-1}$ onto the finite semigroup T'. By our assumption, $T'\phi^{-1}$ is locally finite, and so, since $X \subset T'\phi^{-1}$, $\langle X \rangle$ is finite, as required. □

Henceforth, we take T to be finite, and let $A = \{a_1, \ldots, a_m\}$ denote a finite subset of S, and $\langle A \rangle$ the subsemigroup of S generated by A. To show that $\langle A \rangle$ is finite, it is enough to show that any sufficiently long product in $\langle A \rangle$ is equal to some shorter product in $\langle A \rangle$. If this were false it would be possible to construct, in the regular way, a sequence $s = a_1 a_2 \ldots \in A^\omega$ no left factor of which could be represented as a member of $\langle A \rangle$ by a shorter product. Taking the sequence $s\phi =: a_1\phi a_2\phi \ldots$ and applying the previous lemma would yield a fixed bound k and a fixed idempotent $e \in \mathrm{E}(T)$ such that, for any chosen n, there exist n consecutive factors $e_1, e_2, \ldots, e_n$ of s with $e_i\phi = e$ for $i = 1, 2, \ldots, n$ and with $|e_i|$ bounded by k. The products e_i would then be members of the locally finite semigroup $e\phi^{-1}$. Since each e_i is a member of the finite set $A^{[k]}$ of products of k or fewer members of A, there are only finitely many possibilities for the elements $e_1, \ldots, e_n$, and so $e_1 \ldots e_n$ always belongs to the finite subsemigroup of $e\phi^{-1}$ generated by $A^{[k]} \cap e\phi^{-1}$ no matter how large is n. Thus if n is sufficiently large some factor $e_1 e_2 \ldots e_n$ can be contracted to a shorter product. This completes the proof.

One consequence of Theorem 1.8 is that any finitely generated band B is locally finite, as the greatest semilattice congruence η on B is such that $a\eta$ $(a \in B)$ is a rectangular band, and both the classes of semilattices, and rectangular bands are easily seen to be locally finite. This classic result of Green and Rees [22] is also a corollary of our closing application of Ramsey's Theorem in this section which involves the use of a theorem of Hotzel [34].

1.9. Theorem. *A finitely generated semigroup S is finite if and only if all subgroups of S are finite and S satisfies* $\min_R$, *the minimum condition on principal right ideals.*

In [55] Simon used this together with Ramsey to prove two other characterizations of the class of finite semigroups within the class of finitely generated semigroups. A semigroup S is *strongly n-periodic* if for every sequence $a_1, a_2, \ldots, a_n$ of n elements of S there exists an idempotent e and indices k, l with $1 \leq k \leq l \leq n$ such that $a_k a_{k+1} \ldots a_l = e$. We say that S is *strongly periodic* if S is strongly n-periodic for some positive integer n.

The values of k and of l in the preceding definition depend on the sequence under consideration, although the number n depends only on S. A weaker condition, which we shall here call *sequential periodicity* arises if we allows the value of n to also depend on the sequence in question: in other words, S is *sequentially periodic* if every sequence s drawn from S has some factor which is idempotent. An example of a sequentially periodic semigroup that is not strongly periodic is the following. Let $X = \{x_1, x_2, \ldots\}$ be an infinite countable set, let F_X be the corresponding free semigroup, and write $|w|$ for the length of a word $w \in F_X$. Let ρ be the congruence on F_X generated by the relations $\{x_n w = 0 \text{ where } |w| \geq n\}$. (Note that a relation of the type $v = 0$ means that $vy = yv = v$ $\forall y \in F_X$.) Put $S = F_X/\rho$. Then S is sequentially periodic, as for any sequence $a_1, a_2, \ldots$ in S we have $a_1 a_2 \ldots a_n = 0$, where x_n is the first letter of some word in F_X representing a_1 in S. However S is not strongly periodic because, given any positive integer n, no factor of the word $x_n x_{n+1} \ldots x_{2n-1}$ admits any non-trivial ρ-transition; in particular $x_k x_{k+1} \ldots x_l \notin \mathrm{E}(S)$ for $n \leq k \leq l \leq 2n-1$. As we shall see below, the conditions of strong periodicity and of sequential periodicity coincide for finitely generated semigroups.

1.10. Theorem. *Let S be a semigroup generated by some finite set X. The following are equivalent:*

(1) *S is finite;*

(2) *S has only a finite number of non-idempotents;*

(3) *S is strongly periodic;*

(4) *S is sequentially periodic.*

Proof. Certainly (1) implies (2), and (3) implies (4). The proof that (2) implies (3) uses Ramsey's Theorem, while the proof that (4) implies (1) uses Hotzel's Theorem.

(2) implies (3). Let $r = |S \backslash \mathrm{E}(S)|$, and put $n = R(3, r)$. Let $a_1, a_2, \ldots, a_n \in S$ and suppose that no product $a_k a_{k+1} \ldots a_l \in E(S)$ $(1 \leq k \leq l \leq n)$. Label the vertices of the complete graph K_n with $1, 2, \ldots, n$, and label the edge from i to j $(i < j)$ by $a_i a_{i+1} \ldots a_{j-1}$. This gives an r-colouring of the graph, and so there

exists a monochromatic triangle with vertices i, j, k say of K_n, yielding as before that $a_i a_{i+1} \dots a_{j-1} \in \mathrm{E}(S)$, a contradiction. It follows that S is strongly periodic.

(4) implies (1). We show that in the presence of condition (4) the hypotheses of Theorem 1.9 apply. Let G be a subgroup of S with identity e, and let $a_1, a_2, \dots$ be an infinite sequence drawn from G. Applying sequential periodicity to the sequence $a_1^{-1}a_2, a_2^{-1}a_3, \dots$ we infer that for some $i < j$, $a_i^{-1}a_{i+1}a_{i+1}^{-1}a_{i+2}\dots a_{j-1}^{-1}a_j = a_i^{-1}a_j = e$; that is to say, $a_i = a_j$. We conclude that it is impossible to write down a sequence consisting of infinitely many distinct members of G, whence G is finite.

Finally, suppose that we have a strictly descending chain of principal right ideals in S: $a_0S^1 \supset a_1S^1 \supset \cdots \supset a_tS^1 \supset \cdots$. (We use the symbols $\supset, \subset$ in the sense of strict containment.) Factorize each a_i as $a_{i-1}b_i$ where $b_i \in S$, and further, factor each b_i as $s_ix_it_i$ where $s_i, t_i \in S^1$, $x_i \in X$, and

$$a_{i-1}S^1 = a_{i-1}s_iS^1 \supset a_{i-1}s_ix_iS^1 \supseteq a_{i-1}s_ix_it_i = a_iS^1.$$

Then, since X is finite, some $x \in X$ occurs infinitely often in the preceding factorizations, yielding a sequence of indices, $1 \le i_1 < i_2 < \cdots$ such that $x = x_{i_1} = x_{i_2} = \cdots$.

Now put $u_0 = a_{i_1-1}s_{i_1}$ and $u_j = t_{i_j}b_{i_j+1}\dots b_{i_{j+1}-1}s_{i_{j+1}}$ for $j = 1, 2, \dots$. Then we have

$$u_0xu_1\dots xu_jS^1 \supset u_0xu_1\dots xu_jxS^1$$

for all $j \ge 1$. By sequential periodicity applied to the sequence $xu_1, xu_2, \dots$ there exists indices $k \le l$ such that $e = xu_kxu_{k+1}\dots xu_l = e^2$. But then

$$xu_k\dots xu_lS^1 = xu_k\dots xu_lxS^1,$$

which then yields the contradiction that $u_0xu_1\dots xu_lS^1 = u_0xu_1\dots xu_lxS^1$. This completes the proof. □

Remark. The original proof of Simon does not include condition (4) above. We could also show that (4) implies (3) by way of the contrapositive: if S were not strongly periodic we could construct, in the regular way, a sequence which violated sequential periodicity.

2. The finite basis property for semigroup identities

If X is a countable set and $p, q \in F_X$, the free semigroup on X, we say that a semigroup S satisfies the identity $p = q$ if $p\phi = q\phi$ for every homomorphism $\phi : F_X \to S$. Informally, we regard the members of X as variables, and we require that the words p and q give rise to equal members of S for every possible assignment of values of S to the variables. A class of semigroups is called a *variety*, or an *equational class*, if it consists of exactly all those semigroups satisfying some (possibly infinite) collection of identities $p_1 = q_1, p_2 = q_2, \dots$, and we call such a list a *basis* of the variety

$\mathcal{V}$ and write $\mathcal{V} = [p_1 = q_1, p_2 = q_2, \ldots]$. For example, the variety of rectangular bands is given by $[x = x^2, xyz = xz] = [x = xyx]$; that the equations of the first basis are a consequence of the second is seen as follows. Replacing y by x in $x = xyx$ yields $x = x^3$, which in turn implies $x^2 = x^4$. However, replacing y by x^2 gives $x = x^4$, whence we see that any semigroup satisfying $x = xyx$ is a band. But then, since $y = y(xz)y$ is a consequence of the identity $x = xyx$ we obtain $xz = (xyx)(zyz) = x(yxzy)z = xyz$, as claimed.

For a class K of semigroups, or indeed, of algebras in general, denote the classes of homomorphic images, subalgebras, and direct products of members of K by $\mathrm{H}(K)$, $\mathrm{S}(K)$, and $\mathrm{P}(K)$ respectively. By Birkhoff's Theorem, a class K forms a variety if and only if $K = \mathrm{HSP}(K)$.

A semigroup variety $\mathcal{V}$ is called *finitely based* if $\mathcal{V}$ admits some finite basis of identities, and we shall say that a semigroup identity has the *finite basis property* (FBP) if every variety admitting it is finitely based. The classic result in this topic is that the identity $xy = yx$ has the FBP [45, Theorem 9]. In this section we shall characterize those permutation identities, that is those identities of the form $x_1x_2\ldots x_n = x_{1\pi}x_{2\pi}\ldots x_{n\pi}$, where π is a permutation of $[n]$, which have the FBP.

Observe that any given basis P of a finitely based variety $\mathcal{V}$ must contain a finite basis Q, because each of the identities in any given finite basis F of $\mathcal{V}$ must each be derivable from a finite set of identities of the given basis P ; the subset of all identities of P used to derive all those of F will then be a required finite basis Q.

The *content* of a word $p \in F_X$ is the set of letters that occur in p. We write the number of occurrences of a letter x in p as $|p|_x$, and we write $|p|$ for the length of p. It is easy to see that any identity $p = q$ is a consequence of $xy = yx$ if and only if it is *balanced*, meaning that for every $x \in X$, $|p|_x = |q|_x$. By the *length* of a balanced identity $p = q$ we mean the common length of the words p and q. A variety $\mathcal{V}$ is called *uniformly periodic* if it admits an identity of the form $x^m = x^{m+n}$, for some $m, n \in N$. If $|p| \neq |q|$, then the identity $p = q$ has as a consequence the identity $x^{|p|} = x^{|q|}$, and so any variety admitting $p = q$ as one of its identities is uniformly periodic. Moreover, any identity $p = q$ which is not balanced implies uniform periodicity as, if $|p|_x \neq |q|_x$, we may replace x by x^2 throughout yielding a consequential identity $p' = q'$, say. Since at least one of the statements $|p| \neq |q|$ or $|p'| \neq |q'|$ is now true, it follows from our previous observation that the variety $[p = q]$ is uniformly periodic. In conclusion we have a dichotomy: a variety $\mathcal{V}$ is either *balanced*, meaning that $[xy = yx] \subseteq \mathcal{V}$, or $\mathcal{V}$ is *uniformly periodic*, meaning that every $S \in \mathcal{V}$ satisfies a single identity of the form $x^m = x^{m+n}$.

2.1. Example ([45], Theorem 2). The variety

$$\mathcal{V} = [xyzw = xzyw,\ yx^ky = xyx^{k-2}yx,\ k = 2, 3, \ldots]$$

is not finitely based.

Proof. We need only show that no initial sequence of these identities is a basis for $\mathcal{V}$. Consider such an initial sequence E_0, and let n be the greatest of all exponents appearing in E_0. Then the identity $yx^{n!+1}y = xyx^{n!-1}yx$ is not a consequence of the identities of E_0. (Here we use the fact that first appears in Euclid's proof of the infinitude of primes: none of the numbers $2, 3, \ldots, n$ divide $n! + 1$. Hence we cannot alter the word $yx^{n!+1}y$ through the use of any of the identities of E_0.) This completes the proof. □

It follows that a permutation identity has the FBP only if it is not a consequence of the *normal identity*, $xyzw = xzyw$, that is, only if it has the form $x_1x_2 \ldots x_n = x_{1\pi}x_{2\pi} \ldots x_{n\pi}$, where π is a permutation of $[n]$ such that either $1\pi \neq 1$ or $n\pi \neq n$. We shall prove that this necessary condition is also sufficient.

The next result allows us to restrict our attention to balanced varieties, while the succeeding one circumvents the difficulties that arise when dealing with arbitrary permutation identities.

2.2. Theorem ([45], Theorem 22). *All uniformly periodic varieties admitting a non-trivial permutation identity are finitely based.*

2.3. Theorem ([40], Proposition 3.1(ii)). *Every permutation identity of the form* $x_1x_2 \ldots x_n = x_{1\pi}x_{2\pi} \ldots x_{n\pi}$ *where* $n\pi \neq n$ *implies the identity* $x_1x_2 \ldots x_nxy = x_1x_2 \ldots x_nyx$.

Proof. We may write n as $(i-1)\pi$ for some $i \in \{2, 3, \ldots, n\}$. We then have:

$$\begin{aligned} x_1x_2 \ldots (x_nxy) &= x_{1\pi}x_{2\pi} \ldots (x_nxy)x_{i\pi} \ldots x_{n\pi} \\ &= x_{1\pi}x_{2\pi} \ldots (x_nx)(yx_{i\pi}) \ldots x_{n\pi} \\ &= x_1x_2 \ldots (yx_{i\pi}) \ldots (x_nx) = [x_1x_2 \ldots (yx_{i\pi}) \ldots x_n]x \\ &= [x_{1\pi}x_{2\pi} \ldots x_n(yx_{i\pi}) \ldots x_{n\pi}]x \\ &= [x_{1\pi}x_{2\pi} \ldots (x_ny)x_{i\pi} \ldots x_{n\pi}]x \\ &= [x_1x_2 \ldots (x_ny)]x = x_1x_2 \ldots x_nyx, \end{aligned}$$

as required. □

2.4. Theorem. *A permutation identity* $\phi : x_1x_2 \ldots x_n = x_{1\pi}x_{2\pi} \ldots x_{n\pi}$ *has the FBP if and only if* $1\pi \neq 1$ *or* $n\pi \neq n$.

Proof. It remains to prove the sufficiency of the condition and, by symmetry, it is enough to consider the case where $n\pi \neq n$.

Suppose to the contrary that there exists a permutation identity ϕ such that $n\pi \neq n$ yet ϕ does not possess the FBP. Then there exists an infinite sequence of balanced identities, $\phi, \phi_1, \phi_2, \ldots$, such that ϕ_k is not a consequence of its predecessors in the sequence ($k = 1, 2, \ldots$). By Theorem 2.3 we may replace ϕ by $\phi_0 : x_1x_2 \ldots x_nxy = x_1x_2 \ldots x_nyx$ and the previous statement remains true.

Since there are, up to the naming of variables, only finitely many distinct balanced identities of length less than or equal to n, we may, without loss, assume that $|\phi_i| > n$ for all $i = 0, 1, \ldots$, and so write ϕ_i in the form $\phi_i : p_i f_i = q_i g_i$, where $|p_i| = |q_i| = n$ and $|f_i| = |g_i| \geq 1$ $\forall i = 1, 2, \ldots$.

Let ϕ_i' be the identity $p_i f_i' = q_i g_i'$ where $f_i = f_i'$ and $g_i = g_i'$ are balanced identities. Then ϕ_i' is a consequence of the conjunction of ϕ_0 and ϕ_i. This allows us to replace ϕ_i by ϕ_i' in the list $\phi_0, \phi_1, \ldots$, and still maintain a sequence of balanced identities of length greater than n, no one of which is a consequence of its predecessors. In the course of the proof we shall rearrange the ϕ_i in this fashion without further comment, and without changing the name of the identity ϕ_i.

There are only finitely many possibilities for $|C(p_i)|$, $|C(q_i)|$, and $|C(p_i) \cap C(q_i)|$. Therefore, by passing to a suitable subsequence, we may assume that each of these three non-negative integers is independent of i, $i = 1, 2, \ldots$, and so we may take it that $C(p_i) = \{x_1, x_2, \ldots, x_l\}$, $C(q_i) = \{x_k, x_{k+1}, \ldots, x_t\}$ say, with $l \geq 1, k \leq l+1$, for all $i = 1, 2, \ldots$. Moreover, since there are only finitely many words of length n in a given set of variables, there are only finitely many possibilities for p_i and for q_i. Therefore, again passing to a suitable subsequence, we may assume that $p_i = p$, and $q_i = q$ say, for all $i = 1, 2, \ldots$.

We can now write the left hand side of ϕ_i as $px_1^{r_1}x_2^{r_2}\ldots x_t^{r_t}x_{t+1}^{s_1}x_{t+2}^{s_2}\ldots x_{t+m}^{s_m}$ where $r_j \geq 0, s_1 \geq s_2 \geq \cdots \geq s_m \geq 1$ ($r_1, \ldots, r_t, s_1, \ldots, s_m$ all depending on i). We associate a pair of vectors (u_i, v_i) with ϕ_i:

$$u_i = (r_1, r_2, \ldots, r_t), \quad v_i = (s_1, s_2, \ldots, s_m, 0, 0, \ldots),$$

and define a partial order on the $\{\phi_i\}_{i \in N}$ by $\phi_i \leq \phi_{i'}$ if $u_i \leq u_{i'}$ and $v_i \leq v_{i'}$, where the partial order on the u's and on the v's is that induced by the natural lineal order on the entries.

We now show that $\phi_i \leq \phi_{i'}$ implies that $\phi_{i'}$ is a consequence of the conjunction of ϕ_0 together with ϕ_i. Consider the left hand side of $\phi_{i'}$:

$$pf_{i'} = px_1^{r_1'}x_2^{r_2'}\ldots x_t^{r_t'}x_{t+1}^{s_1'}\ldots x_{t+m'}^{s_{m'}'}$$

and rewrite it in the form pf_ih where

$$h = x_1^{r_1'-r_1}x_2^{r_2'-r_2}\ldots x_t^{r_t'-r_t}x_{t+1}^{s_1'-s_1}\ldots x_{t+m'}^{s_{m'}'-s_{m'}},$$

$(s_{m+1} = s_{m+2} = \cdots = s_{m'} = 0)$. Applying the identity ϕ_i to pf_ih yields $qg_ih = qg_{i'}$, as ϕ_i and $\phi_{i'}$ are balanced.

Therefore, to complete the proof, it is enough to show that, given a sequence of pairs $(u_1, v_1), (u_2, v_2), \ldots, (u_r, v_r), \ldots$, where the u_i and the v_i are defined as before, there exists $i < i'$ such that $u_i \leq u_{i'}$ and $v_i \leq v_{i'}$. Since the u_i's are t-dimensional, it is easy to show by induction on t that the sequence of u_i's contains an infinite increasing chain. Therefore, without loss, we assume that $u_1 \leq u_2 \leq \cdots \leq u_r \leq \cdots$. It remains to show that $v_i \leq v_{i'}$ for some $i < i'$. This is a special case of the following application of the infinite version of the Ramsey Theorem in which we call an infinite sequence $(n_s)_{s\geq 0}$ over $N \cup \{0\}$ a *vector* over N^0. We call a vector over N^0 *decreasing* if $n_s \geq n_t$ for all $s \leq t$ $(s, t \in N)$. □

2.5. Lemma. *If $S_0 = (v_1, v_2, \ldots)$ is a sequence of decreasing vectors over N^0, then there exists $i < i'$ such that $v_i \leq v_{i'}$.*

Proof. First, let us assume that there is a greatest positive integer r which occurs among the entries of the vectors v_i, and write the vector v_i as $(v_{i1}, v_{i2}, \ldots)$. Suppose that, contrary to the assertion of the lemma, there is no $i < i'$ such that $v_i \leq v_{i'}$. Take the complete graph G on N and colour the edge from i to j $(i < j)$ by the positive integer k, where k is the value of the first entry in v_i which exceeds the corresponding entry of v_j. This assigns an r-colouring to the edges of G which, by the infinite version of the Ramsey Theorem, contains a copy of itself, all the edges of which are coloured k, say. This corresponds to an infinite subsequence $S_1 = (u_1, u_2, \ldots)$ of S_0 for which there exists a sequence of positive integers, $t_1, t_2, \ldots$, such that the $u_{it_i} = k > u_{(i+1)t_i}$. However, since the vectors are decreasing, we see that $t_1 > t_2 > t_3 > \cdots$, an impossibility for a sequence of positive integers. This contradiction completes the proof in the case where the upper bound r exists.

Suppose now that there is no upper bound on the entries of the vectors v_i. Let $m \geq 0$ be the minimum value of all the entries of S_0 and, by deleting some initial subsequence of S_0 if necessary, we may assume that m is the minimum entry of v_1, which first occurs at v_{1t} say. Let $i \geq 2$. Since $v_1 \not\leq v_i$ there exists $j \leq t-1$ such that $v_{1j} > v_{ij}$. Since there are only finitely many possibilities for j we may draw an infinite subsequence $S_1 = (u_1, u_2, \ldots)$ from S_0 such that for some fixed positive integer j, $u_1 = v_1$ and $u_{1j} > u_{ij}$ $\forall i \geq 2$. Let k denote the value u_{1j}. For each u_i we introduce $u_i' = (u_{i1}, u_{i2}, \ldots, u_{ij})$ and $u_i'' = (u_{i,j+1}, u_{i,j+2}, \ldots)$. By induction on j, one can construct an infinite subsequence $S_2 = (w_1, w_2, \ldots)$ of S_1 such that $w_1' \leq w_2' \leq \cdots$. The sequence $w_1'', w_2'', \ldots$ is an infinite sequence of decreasing vectors the entries of which are bounded above by k. Hence, by the first part of the proof, there exists $i < i'$ such that $w_i'' \leq w_{i'}''$. However, $w_i' \leq w_{i'}'$ and $w_i'' \leq w_{i'}''$ together are equivalent to $w_i \leq w_{i'}$, thus completing the proof. □

The above proof is essentially that in [26]. In [1] A. Aizenstat determined all identities with the FBP amongst the class of all identities ϕ for which each variable appears at most once on each side of ϕ. (The permutation identities are exactly those identities for which each variable appears exactly once on each side.) A great deal of research has gone into the general question of finding all identities with the FBP. The problem has not been completely solved: in particular the identity $xy = xy^2$ remains a stumbling block. References include Pollak [47], Pollak and Volkov [48], Sapir [51], and Sapir and Volkov [52].

3. Pseudovarieties and Reiterman's Theorem

The theory of pseudovarieties has its origins firmly rooted in automata and language theory. Although we shall present a purely algebraic development here, an historical digression, in which I will describe (without proofs) how this connection comes about, will serve to motivate the topic.

Let A^* be the free monoid on the finite *alphabet* $A = \{a_1, a_2, \ldots, a_n\}$. A subset L of A^* is called a *language*, and we say that L is *recognized* by a monoid M if there exists a homomorphism $\phi : A^* \to M$ and a subset P of M such that $L = P\phi^{-1}$. The image $M\phi$ of ϕ is a submonoid of M, (which also recognizes L), that is isomorphic to A^*/ρ, where $\rho = \phi \circ \phi^{-1}$. By definition, ρ *saturates* L, meaning that each ρ-class lies entirely within L or within $A^*\backslash L$. Hence $\rho \subseteq \sim$, where $\sim$ is the largest congruence that saturates L, known as the *syntactic congruence* of L: $u \sim v$ if and only if $(\forall x, y \in A^*)(xuy \in L \Leftrightarrow xvy \in L)$. The monoid $A^*/\sim$ is known as the *syntactic monoid* $M(L)$ of L, and the foregoing serves to show that $M(L)$ *divides* any monoid M that recognizes L, meaning that $M(L)$ is a homomorphic image of a submonoid of M. It is an easy exercise to show that, conversely, if $M(L)$ divides a monoid N, then N recognizes L. In particular, this is true of $M(L)$ itself. We say that L is *recognizable* if L can be recognized by some finite monoid.

The notion of recognizable language can equally be formulated in terms of automata. An automaton is a finite graph one of whose vertices (known as *states*) is designated the *initial state*, and some states are called *terminal* or *accepting* states. The letters of the alphabet A (and by a natural extension, the words of A^* also) act on the automaton, sending it from one state to another. The language, L, of an automaton, is the set of words in A^*, each of which sends the automaton from its initial state to some terminal state. We say that the automaton *recognizes* L. Each $a \in A$ defines a transformation of the state set S of the automaton. The monoid M generated by this collection of transformations is the *transition monoid* of the automaton. It can readily be verified that L can be recognized by its transition monoid (take $\phi : A^* \to M$ to be the natural homomorphism; then $L = P\phi^{-1}$ where $P = L\phi$). There is a unique *minimal automaton*, $\mathcal{A}$, for a recognizable language L, meaning that each state of $\mathcal{A}$ is *accessible* from the initial state, and that $\mathcal{A}$ can be constructed from any other accessible automaton $\mathcal{B}$ recognizing L by identifying two states p and q of $\mathcal{B}$ if the words which take p to any terminal state exactly match those which do the same for q. The transition monoid of the minimal automaton is the syntactic monoid, $M(L)$, defined above. Indeed $\mathcal{A}$ can be constructed in a manner akin to the construction of $M(L)$ as follows. Let ρ be the maximum right congruence saturating L, that is $(u, v) \in \rho$ if and only if $(\forall x \in A^*)((ux \in L) \Leftrightarrow (vx \in L))$. The states of $\mathcal{A}$ correspond to the ρ classes, with the initial state being the class of the empty word, 1, and the terminal states are the ρ-classes (wholly) contained in L. Define the action of $a \in A$ by $[u] \cdot a = [ua]$, where $[u]$ denotes the ρ-class of $u \in A^*$. This finite automaton is then the minimal automaton which recognizes L.

Another impetus to this theory comes via the celebrated theorem of Kleene, which states that the class of recognizable languages coincides with the class of *rational* languages, which is the smallest collection of languages that contains all the finite ones, and is closed under boolean operations (intersection, union, and complementation), set products, and the *star operation*: L^* is the submonoid of A^* generated by L. (References [36] or [41]). A third direction is provided by the theory of formal grammars, for recognizable languages are exactly the languages of the so-called *right linear grammars* [41] , and yet another comes from the new and exciting subject of automatic groups [18]. This all presents a rich theory, and a natural hope from an algebraic viewpoint would be a fruitful classification of recognizable languages through properties of their syntactic monoids. There now arises a question of approach. We cannot simply regard L and $M(L)$ as equivalent. After all, the transition monoid of an automaton 'forgets' about its designated initial and terminal states, so cannot expect to capture all information about L. Indeed, different languages may share the same syntactic monoid, and not every monoid arises as the syntactic monoid of a language (see, for instance, [41]). The most successful instrument of investigation so far has been provided by the Variety Theorem of Eilenberg and Schutzenberger, which establishes a natural bijection between certain classes of finite monoids, known as pseudovarieties, and so-called varieties (or streams) of rational languages. This theorem, published in 1976, provided a bridge linking algebraic semigroups and language theory which has carried a thriving traffic ever since.

A *variety* of rational languages is a collection of languages closed under boolean operations, inverse images of homomorphisms between free monoids, and left and right quotients (the *left quotient* $a^{-1}L = \{u \in A^* : au \in L\}$). The collection of all syntactic monoids of a variety of languages form a *pseudovariety* of monoids, that is a class K such that $\mathrm{H}(K)$, $\mathrm{S}(K)$, $\mathrm{P}_f(K) \subseteq K$, where $\mathrm{P}_f(K)$ denotes the class of all direct products of finitely many members of K. In other words, a pseudovariety of monoids is a class of finite monoids closed under division, and finite direct products. We have already seen how the division operation manifests itself, as the monoids recognizing a language L are exactly those which share $M(L)$ as a common divisor. Direct products arise because, for any two languages, L_1 and L_2, $M(L_1 \cap L_2)$ divides $M(L_1) \times M(L_2)$ (as does $M(L_1 \cup L_2)$, as the syntactic monoid of a language and its complement are easily seen to be equal, whence this follows from De Morgan's Law). Moreover, many interesting classes of finite monoids are pseudovarieties: groups, nilpotent monoids, $\mathcal{R}$-trivial monoids, $\mathcal{J}$-trivial monoids, and $\mathcal{H}$-trivial monoids, to name but a few, and to each corresponds a language variety.

A deep result of Simon [54] identifies the language variety of the $\mathcal{J}$-trivial monoids (monoids, the members of which can be distinguished by their principal ideals) as the class of all so-called *piecewise testable* languages, that is, languages, the word membership of which can be tested by inspection of subwords of length no greater than some fixed value n. (For an alternative algebraic proof, see [56].) A finite monoid has a trivial $\mathcal{H}$-relation if and only if it is *aperiodic*, meaning that all of its subgroups are trivial. A classic result in the field is that of Schutzenberger [53], which says that the pseudovariety of all aperiodic monoids corresponds to the variety of *star-free* languages, that is, those rational languages which can be constructed from the alphabet A without recourse to Kleene's star operation. A striking result of Ash [6] is the proof that the pseudovariety generated by inverse monoids is the pseudovariety of all monoids, the idempotents of which form a semilattice (containment in one direction is clear). The proof crucially invokes Ramsey's Theorem through Lemma 1.5 above.

Ash has also used Ramsey's Theorem in his proof of the Rhodes' Type II Conjecture [8], a deep question which had vexed leading researchers in the field for some years, the solution of which was the highlight of the conference in honour of Professor G. B. Preston held at Monash University in July 1990. The Proceedings of that conference have an outline of the proof by Ash [7], together with an article by Margolis [43] detailing its consequences, one of which is the earlier result of Ash on pseudovarieties. The problem which had been the obstacle to progress was the characterization of the elements of a finite monoid M that are related to the identity in every relational morphism with a finite group. (A *relational morphism* $\tau : S \to T$ between two semigroups S and T is a relation which satisfies (i) $a\tau \neq \emptyset\ \forall a \in S$ and (ii) $(a\tau)(b\tau) \subseteq (ab)\tau\ \ \forall a, b \in S$.) As conjectured by Rhodes, Ash proved that they form the smallest submonoid K of M closed under *weak conjugation*, meaning that if $yzy = y$ or $zyz = z$ then $yKz \subseteq K$. The application of Ramsey's Theorem employed in the proof is a corollary of Lemma 1.5 which can be formulated as follows. Let M be a finite monoid generated by a set $A = \{a_1, a_2, \ldots, a_n\}$. Then there exists an integer K such that any product $P = a_1a_2 \ldots a_t \in A^*$ has a factorization $u_1u_2 \ldots u_k$ in A^* such that $k \leq K$ and that, regarded as members of M, each $u_i \in A \cup \mathrm{E}(M)$. A suitable value of K is $R(3, r) - 1$, where $r = |A \cup \mathrm{E}(M)|$, for let P be as above and suppose that $u_1u_2 \ldots u_m$ is a factorization of P in A^* such that m is minimal with respect to the property that each $u_i \in A \cup \mathrm{E}(M)$. Then $m \leq K$ for otherwise, by Ramsey's Theorem, there would exist two successive factors of P with the same idempotent value in M, which could be combined to reduce the length m of the factorization to $m - 1$.

A unifying feature of the results of Ash and Simon on pseudovarieties is that each identifies the pseudovariety generated by a certain concrete class of monoids of relations: the class of symmetric inverse monoids in the case of Ash, (the members of which are partial one-to-one mappings), and the class of monoids of all reflexive relations on a finite set in the case of Simon's result or, alternatively, the class of monoids of all order-preserving and decreasing mappings on a finite chain (see [46]). The class of all $\mathcal{R}$-trivial monoids (monoids, the members of which can be distinguished by their principal right ideals) is generated as a pseudovariety by the monoids of all order-preserving and decreasing partial mappings on a chain [32]. There is however, as yet, no corresponding result for the pseudovariety of aperiodics.

Entire books have now been written on finite semigroups from the pseudovariety viewpoint (see [2], [46]). The introduction below however, following Ash [5], shows that the notion arises naturally from the concept of generalized variety, an idea which extends the ordinary notion of variety in universal algebra.

In the following discussion K will denote a class of algebras, all of the same finite similarity type (that is, the type involves only finitely many operations, each of which has finite arity). Although we shall be primarily interested in semigroups, and all examples will be taken from there, there is nothing to be gained by insisting that the symbol K stands for a class of semigroups, (although no harm is done by thinking in such terms).

In addition to the operations $\mathrm{H}(K)$, $\mathrm{S}(K)$, $\mathrm{P}(K)$, $\mathrm{P}_f(K)$ already introduced, we shall also use $\mathrm{Pow}(K)$, which will denote the classes of direct powers of members of K. Let E denote the set of all identities of the type of K, and for any $A \in K$, let $\mathrm{Id}(A)$ denote the set of identities satisfied by A. By a *generalized variety* we mean a class of algebras K closed under H, S, P_f, and Pow. A collection Γ of sets or of classes is *directed* if for all $A, B \in \Gamma$ there exists $C \in \Gamma$ such that $A \subseteq C$ and $B \subseteq C$. We write $\mathrm{Var}(K)$, [resp. $\mathrm{Gen}(K)$], for the variety, [resp. generalized variety], generated by K.

An example of a generalized variety that is not a variety is **Nil**, the class of all *nilpotent* semigroups, that is, semigroups S with a zero, 0, such that, for some positive integer n, $S^n = 0$. **Nil** is not a variety as it is not closed under arbitrary products: let S_n be the relatively free semigroup on a set X in the variety $[x_1x_2 \ldots x_n = 0]$. Then $S_1 \times S_2 \times \cdots \times S_n \times \cdots$ is not nilpotent. (Note that the given equation is equivalent to *two* identities, namely, $x_1x_2 \ldots x_n y = yx_1x_2 \ldots x_n = x_1x_2 \ldots x_n$, and so does define a variety.) Although not a variety, **Nil** can be characterized within the class of all semigroups by means of identities: **Nil** consists of all semigroups that satisfy *almost all* (by which we mean all but finitely many) of the identities: $x_1x_2 \ldots x_n = 0$, $(n = 1, 2, \ldots)$. The finite members of **Nil**, which form a pseudovariety of semigroups, are similarly defined within the class of finite semigroups. This behaviour is representative of generalized varieties, as we shall see.

All classes of algebras under consideration will be taken to be closed under isomorphism.

3.1. Theorem. *The following are equivalent for a class K of algebras.*

(1) *K is a generalized variety;*

(2) $K = \mathrm{HSP}_f\,\mathrm{Pow}(K)$*;*

(3) *K is the union of some directed family of varieties.*

Proof. (1) $\Rightarrow$ (2) is immediate.

(2) $\Rightarrow$ (3). For each finite collection $K_0 \subseteq K$, $\mathrm{P}(K_0) = \mathrm{P}_f\,\mathrm{Pow}(K_0)$ and so $\mathrm{Var}(K_0) = \mathrm{HSP}(K_0) = \mathrm{HSP}_f\,\mathrm{Pow}(K_0) \subseteq K$. Hence K is a union of the family $\Gamma = \{\mathrm{Var}(K_0) : K_0 \text{ is a finite subset of } K\}$. Also, since $\mathrm{Var}(K_1) \cup \mathrm{Var}(K_2) \subseteq \mathrm{Var}(K_1 \cup K_2)$, Γ is directed under $\subseteq$, as required.

(3) $\Rightarrow$ (1). Let K be the union of some directed family Γ of varieties, and let $A \in K$, so that $A \in V$ for some $V \in \Gamma$. Then $\mathrm{H}(A)$, $\mathrm{S}(A)$, $\mathrm{Pow}(A)$ are all in V, so that each of these lie in K. Let $A_1, A_2, \ldots, A_n \in K$ so that each A_i lies in some variety V_i of Γ. Since Γ is directed under $\subseteq$, it follows by induction on n that there exists $V \in \Gamma$ such that $V_1 \cup V_2 \cup \cdots \cup V_n \subseteq V$, whence $A_1 \times A_2 \times \cdots \times A_n \in V$

also, and we infer that K is closed under P_f, as required. □

We next define a *pseudovariety* to be any class of finite algebras closed under H, S, and P_f. Pseudovarieties are related to generalized varieties by the following theorem.

3.2. Theorem. *A class K of algebras is a pseudovariety if and only if it consists of the finite members of some generalized variety.*

Proof. If K is a generalized variety then its finite members certainly form a pseudovariety, since H, S, and P_f do not lead out of the class of finite algebras.

Conversely, if K_0 is a pseudovariety and $K = \mathrm{Gen}(K_0)$, we claim that every finite member of K is a member of K_0. Take A to be a finite algebra in K. Then there exist $A_1, A_2, \ldots, A_n \in K_0$, and sets $I_1, I_2, \ldots, I_n$, (not necessarily finite), a subalgebra B of $A_1^{I_1} \times \cdots \times A_n^{I_n}$, and a surjective homomorphism $\phi : B \to A$. For each $a \in A$, we may choose $b_a = (\xi_{1a}, \ldots, \xi_{na}) \in B$ with $\phi(b_a) = a$. This choice yields equivalence relations, $\equiv_k$ on each I_k, defined by $i \equiv_k j$ if, for all $a \in A$, $\xi_{ka}(i) = \xi_{ka}(j)$. Note that each $\equiv_k$ has no more than a finite number, $|A| \cdot |A_k|$, equivalence classes. Letting $J_k = I_k / \equiv_k$ we have a homomorphism from $A_1^{I_1} \times \cdots \times A_n^{I_n}$ to $A_1^{J_1} \times \cdots \times A_n^{J_n}$ $((\zeta_1, \ldots, \zeta_n) \mapsto (\zeta_1', \ldots, \zeta_n')$ where $\zeta_k'([i]_k) = \zeta_k(i)$, for each $\equiv_k$ class $[i]_k$, and where i is a fixed representative from $[i]_k$), which is one-to-one on the subalgebra of B generated by the b_a. Therefore, since each J_k is finite, we conclude that $A \in \mathrm{HSP}_f(K_0)$, and so that $A \in K_0$. □

The next theorem, a special case of [5, Theorem 3], is a generalization of the result of Eilenberg and Schutzenberger [17] characterizing pseudovarieties of monoids by sequences of identities. (Although the latter authors commented that their result held for universal algebras in general.)

3.3. Theorem. *Let K_0 denote a collection of finite algebras. Then the following are equivalent.*

(1) *K_0 is the pseudovariety.*

(2) *There is a chain $V_1 \subseteq V_2 \subseteq \cdots$ of varieties such that K_0 consists of all finite members of $V_1 \cup V_2 \cup \cdots$.*

(3) *There is a sequence of identities $I : \phi_1, \phi_2, \ldots,$ such that, for any finite algebra A, $A \in K_0 \Leftrightarrow \phi_n \in \mathrm{Id}(A)$ for almost all n.*

Proof. (1) implies (3). Suppose that K is a generalized variety with K_0 the collection of finite members of K. Let $K_0 = \{A_1, A_2, \ldots\}$ and write the set of finite algebras not in K as $B = \{B_1, B_2, \ldots\}$. For each j, $B_j \notin K$, and hence for each i, $B_j \notin \mathrm{Var}\{A_1, A_2, \ldots, A_i\}$, as this variety is contained in K by the proof of Theorem 3.1. Choose an identity $\phi_{ij} \in \mathrm{Id}\{A_1, \ldots, A_i\} - \mathrm{Id}(B_j)$. Let $I : \phi_1, \phi_2, \ldots$ be any listing of those ϕ_{ij} with $i > j$. Then, for each j, ϕ_n is of the form ϕ_{ij} for infinitely many n, and each $\phi_{ij} \notin \mathrm{Id}(B_j)$, so that each B_j fails to satisfy infinitely many of the identities of I. On the other hand, for each k, ϕ_n is some ϕ_{ij} with $i \geq k$ for almost all n, and for all $i \geq k$, $\phi_{ij} \in \mathrm{Id}(A_k)$, so that A_k satisfies almost all of the identities of I.

(3) implies (2). If $I : \phi_1, \phi_2, \ldots$ is such a sequence of identities, let V_n be the variety of all algebras defined by the identity sequence $I_n : \phi_n, \phi_{n+1}, \ldots$. Then $V_1 \subseteq V_2 \subseteq \ldots$, and a finite algebra A satisfies almost all the identities of I if and only if A satisfies all the identities of I_n, for some $n \geq 1$.

(2) implies (1). This is immediate from Theorem 3.1. □

For the remainder of this section all semigroups S under consideration will be taken as finite.

There is an alternative way of presenting pseudovarieties. Let us return to our example of the pseudovariety **Nil** of all nilpotent semigroups. Let S be a semigroup and put $k = |S| - |E| + 1$, where $E = \mathrm{E}(S)$ is the set of idempotents of S. Then $S^k = SES$, as, for any k elements of S, $a_1, a_2, \ldots, a_k$ say, either some product $A_i = a_1 a_2 \ldots a_i$ $(i \leq k)$ is idempotent, or, by the Pigeon Hole Principle, there exists i, j with $i < j$ such that $A_i = A_j$, whence $A_j = A_i(a_{i+1} \ldots a_j)^t$ $\forall t \geq 1$ whereupon, taking t sufficiently large so as to yield an idempotent power e we obtain $A_j = A_i e$. In either instance we conclude that $A_k \in SES$. It follows that a finite semigroup S with zero 0 is nilpotent if and only if ES) $= \{0\}$, or, equivalently, if and only if the (unique) idempotent power of every $a \in S$ is 0.

We introduce a unary (single variable) operation, $(\cdot)^\omega$ on the pseudovariety of all finite semigroups by defining a^ω as the idempotent of the monogenic semigroup generated by a in S. The pseudovariety **Nil** is then determined by the single equation $x^\omega = 0$. This is a clear improvement on the previous situation. Instead of determining membership of **Nil** by verifying that S 'ultimately' satisfies some infinite sequence of identities, the membership criterion has been expressed in a single equation which goes straight to the heart of the matter. Again we see this when considering the pseudovariety of all completely regular semigroups (those semigroups which are unions of groups). This class is defined by the single 'pseudoidentity' $x^{\omega+1} = x$, where, of course, $x^{\omega+1}$ means $x^\omega x$. A corresponding sequence of identities ultimately determining this class is: $[x = x^{n!+1}, n = 1, 2, \ldots]$ which is a less transparent definition, for we have been forced to introduce the factorial function into the equation list, which has no intrinsic connection with the pseudovariety in question.

As further examples, it is a straightforward exercise to check that the pseudovarieties $\mathcal{H}$, $\mathcal{R}$, and $\mathcal{J}$ of all aperiodic, $\mathcal{R}$-trivial, and $\mathcal{J}$-trivial semigroups respectively are similarly determined by the 'pseudoidentities' $[x^\omega = x^{\omega+1}]$, $[(xy)^\omega x = (xy)^\omega]$, and $[(xy)^\omega x = (xy)^\omega = y(xy)^\omega] = [(xy)^\omega = (yx)^\omega, x^\omega = x^{\omega+1}]$ respectively. Indeed, accepting for the moment that we shall clarify the idea of pseudoidentity and show that pseudovarieties will be determined by them, a presentation by means of pseudoidentities can be an efficient way of verifying that a class of algebras does form a pseudovariety. In addition to the examples just given, consider the pseudovariety of semigroups determined by $((xy)^\omega(yx)^\omega(xy)^\omega)^\omega = (xy)^\omega$. This pseudoidentity exactly corresponds to the condition that each regular $\mathcal{D}$-class of S is a (simple) semigroup. (To see this, use the fact that any two idempotents, e and f, of a finite semigroup S are $\mathcal{J}$-related, if and only if they are *conjugates*, meaning that there exist $u, v \in S$ such that $e = uv$ and $f = vu$.) Similarly, the class of all locally trivial semigroups, that is those semigroups such that $eSe = e$ for all $e \in E(S)$, is determined by $x^{\omega}yx^{\omega} = x^{\omega}$. It is not immediate that each of these classes K satisfies $K = \mathrm{HSP}_f(K)$ otherwise.

The key feature of our 'implicit' operation $(\cdot)^{\omega}$ is that it commutes with homomorphisms: that is, if $\alpha : S \to T$ is a homomorphism between finite semigroups then, for any $a \in S$, $\alpha(a^{\omega}) = (\alpha(a))^{\omega}$. In general, an n-ary *implicit operation* π on a class $\mathcal{C}$ of algebras associates to each $A \in \mathcal{C}$ an n-ary operation $\pi_A : A^n \to A$ in such a fashion that if $A, B \in \mathcal{C}$ and $\alpha : A \to B$ is a homomorphism then:

$$\pi_B(\alpha(a_1), \ldots, \alpha(a_n)) = \alpha(\pi_A(a_1, \ldots, a_n)) \ \forall a_1, \ldots, a_n \in A.$$

A basic source of implicit operations arises from the term functions within the class of algebras of a specific type, τ, under consideration—any term, $t(x_1, \ldots, x_n)$ of type τ in the variables $x_1, \ldots, x_n$ defines an n-ary implicit operation on $\mathcal{C}$ by letting t_A be the induced n-ary operation on $A \in \mathcal{C}$ obtained by substituting $a_i \in A$ for $x_i (i = 1, \ldots, n)$. As examples, consider the terms $x^2 + y^2$, and xy^{-1}. These define implicit operations on the classes of Rings, and of Inverse Semigroups respectively, as if α represents a typical homomorphism between members of the classes in question, $\alpha(a^2 + b^2) = (\alpha(a))^2 + (\alpha(b))^2$ is always true in the first case, while, $\alpha(ab^{-1}) = \alpha(a)(\alpha(b))^{-1}$ holds in the second. Implicit operations on a class of algebras defined by term functions are known as *explicit operations*. Every implicit operation π on a pseudovariety is the limit of a sequence of explicit operations (π_n) in the sense that the restrictions of π and of π_n to all algebras in the pseudovariety of cardinality no more that n are equal. This is a fundamental result of Reiterman [50], which reveals the topological nature of the subject. For a brief overview of the topic see Weil [58].

For $A \in \mathcal{C}$ and n-ary implicit operations π, ρ we write $A \models \pi = \rho$ to mean that $\pi_A = \rho_A$, and we call $\pi = \rho$ a pseudoidentity for $\mathcal{C}$ that is satisfied by A. If Σ is a set of pseudoidentities for $\mathcal{C}$, $A \models \Sigma$ means that A satisfies all the pseudoidentities in Σ, and $[\Sigma]_{\mathcal{C}}$ stands for the class of all $A \in \mathcal{C}$ such that $A \models \Sigma$.

3.4. Reiterman's Theorem. *Let V be a fixed pseudovariety of finite algebras and let $W \subseteq V$. Then W is a pseudovariety if and only if there is a set of pseudoidentities such that $W = [\Sigma]_V$.*

Proof. The reverse direction presents no difficulties since implicit operations commute with homomorphisms. To prove the forward implication, let W be a pseudovariety within V and let $W' = [\Sigma]_V$, where Σ is the set of all pseudoidentities each of which is satisfied by every member of W. Clearly $W \subseteq W'$. We prove the reverse inclusion by showing that if $A \notin W$ then $A \notin W'$.

Since W is a pseudovariety we have, by Theorem 3.3, a sequence of identities $I_0 : \phi_1, \phi_2, \ldots$ such that $B \in V$ is also in W if and only if B satisfies almost all the ϕ_i.

Since $A \notin W$, we can extract a subsequence I_1 of I_0 such that A satisfies no member of I_1.

Let $k \in N$. As there are, up to isomorphism, only finitely many $B \in W$ with $|B| \leq k$, it follows that all such B satisfy all the identities of I_1 after some identity ϕ that depends only on k. Thus we can draw a subsequence $I_2 : \phi_{2,1}, \phi_{2,2}, \ldots$ from I_1 with the property that all such B satisfy all the identities of I_2, except perhaps those preceding $\phi_{2,k}$.

Let n denote $|A|$. Form a new sequence of identities $I_3 : \phi_{3,1}, \phi_{3,2}, \ldots$ as follows. Suppose that $\phi_{2,k}$ has the form $f(x_1, x_2, \ldots, x_m) = g(x_1, x_2, \ldots, x_m)$. Since $A \not\models \phi_{2,k}$, there exists $a_1, a_2, \ldots, a_m \in A$ such that $f(a_1, a_2, \ldots, a_m) \neq g(a_1, a_2, \ldots, a_m)$. Since there are no more than n distinct members of A in the list, A does not satisfy some consequence of $\phi_{2,k}$ which involves at most n variables. This yields an identity in n variables, $\phi_{3,k}$, which is not satisfied by A, but which is satisfied by all $B \in W$ with $|B| \leq k$, thus forming $I_3 : \phi_{3,1}, \phi_{3,2}, \ldots$.

Define the k-equivalence of two identities, $f_1 = g_1$ and $f_2 = g_2$ to occur if the term functions defined by f_1 and f_2 coincide on all B with $|B| \leq k$, and the term functions g_1 and g_2 agree likewise. Since there are, up to isomorphism, only finitely many $B \in V$ with $|B| \leq k$ and, for any such B, only finitely many functions $f : B^n \to B$, it follows that for an infinite sequence of identities in no more than n variables, there is at least one infinite k-equivalence class.

Extract our final sequence $I : \psi_1, \psi_2, \ldots$ from I_3 as follows. Put $\psi_1 = \phi_{3,1}$. Suppose inductively that we have chosen $\psi_1, \ldots, \psi_{k-1}$ and have deleted identities from I_3 in such a way that the remaining sequence is still infinite, begins $\psi_1, \ldots, \psi_{k-1}$, and all the identities after ψ_j are j-equivalent ($j = 1, 2, \ldots, k-1$). (The inductive hypothesis clearly holds if $k = 2$.) Choose ψ_k from the remaining identities such that infinitely many of the remaining identities are k-equivalent to ψ_k, and discard all those which are not. The identity ψ_k then satisfies the inductive requirements. This process yields a sequence of identities $I : \psi_1, \psi_2, \ldots$ in n variables, none of which are satisfied by A, and such that, for each $B \in V$ with $|B| \leq k$, the term functions defined by the left hand sides of $\psi_k, \psi_{k+1}, \ldots$ all coincide on B, as do those defined by the right hand sides. Moreover, if $B \in W$, then B satisfies all of $\psi_k, \psi_{k+1}, \ldots$.

We use I to define a pseudoidentity $\pi = \rho$ as follows. Let $B \in V$ be algebra with $|B| = k$, and take $\psi_k : f_k(x_1, \ldots, x_n) = g_k(x_1, \ldots, x_n)$. Define $\pi_B : B^n \to B$ by $\pi_B(b_1, \ldots, b_n) = f_k(b_1, \ldots, b_n)$, and similarly define ρ using g_k. To show that π is an n-ary implicit operation, let $\alpha : B \to C$ be a homomorphism. Then $|\alpha(B)| = l \leq k$. We obtain:

$$\begin{aligned}\alpha(\pi_B(b_1, \ldots, b_n)) &= \alpha(f_k(b_1, \ldots, b_n)) \\ &= f_k(\alpha(b_1), \ldots, \alpha(b_n)) = f_l(\alpha(b_1), \ldots, \alpha(b_n))\end{aligned}$$

as f_k and f_l are l-equivalent, which then equals $\pi_C(\alpha(b_1), \ldots, \alpha(b_n))$. Similarly ρ is an n-ary implicit operation, and so $\pi = \rho$ is a pseudoidentity on V.

Now for any $B \in W$ with $|B| = k$ say, $\pi_B = f_k = g_k = \rho_B$, and thus $(\pi = \rho) \in \Sigma$. Nevertheless, $A \not\models \pi = \rho$. To see this recall that A does not satisfy $\psi_n : f_n = g_n$, and so there exist $a_1, \ldots, a_n \in A$ such that:

$$\pi_A(a_1, \ldots, a_n) = f_n(a_1, \ldots, a_n) \neq g_n(a_1, \ldots, a_n) = \rho_A(a_1, \ldots, a_n).$$

Thus $A \notin W'$, which completes the proof of Reiterman's Theorem. □

The preceding proof, taken from [29], is free from topology, and has the advantage of not invoking the Axiom of Choice. (In the original proof, the axiom is implicitly employed when the Tychonoff Theorem is invoked to ensure compactness of a certain direct product of compact spaces.) We can also replace the phrase 'set of pseudoidentities' Σ, by a 'sequence of pseudoidentities': the proof constructs for each finite algebra A not in the given pseudovariety a pseudoidentity admitted by the pseudovariety W which serves to disqualify A. The sequence so formed will then be a defining set of pseudoidentities for W.

Pseudoidentities can be used to capture varietal properties and so assist in the search for the calculation of joins in the lattice of psuedovarieties in a technique analagous to that used in lattices of varieties in general. For example, it is straightforward to verify that all $\mathcal{R}$-trivial and all $\mathcal{L}$-trivial semigroups satisfy the pseudoidentity: $(xy)^\omega x(zx)^\omega = (xy)^\omega(zx)^\omega$, and so the pseudovariety $\mathcal{V}$ defined by this single pseudoidentity is an upper bound for the join, $\mathcal{R} \vee \mathcal{L}$, in the lattice of all pseudovarieties of finite semigroups. We then can consider $\mathcal{V}$ as a possible candidate for $\mathcal{R} \vee \mathcal{L}$, and indeed in [4] Almeida and Azevedo proved that $\mathcal{V} = \mathcal{R} \vee \mathcal{L}$, and thus showed that the membership problem for $\mathcal{R} \vee \mathcal{L}$ is decidable. The proof makes use of Reiterman's Theorem and a precise description of all the pseudoidentites of $\mathcal{V}$. An interesting open question is whether or not the pseudovariety of all finite semigroups is join irreducible. One would expect it to be, but, surprisingly perhaps, it was proved by Almeida in [3], that there exist two pseudovarieties $\mathcal{J}_1, \mathcal{J}_2$, lying below $\mathcal{J}$, the pseudovariety of all $\mathcal{J}$-trivial semigroups, such that $\mathcal{J} = \mathcal{J}_1 \vee \mathcal{J}_2$. The proof again uses the approach outlined above to show that $\mathcal{R} \vee \mathcal{L} = \mathcal{V}$, along with the completeness theorem from equational logic.

4. Epimorphisms, dominions, and the Zigzag Theorem

The semigroup $U = (Z, \cdot)$, the integers under multiplication, is a 'large' subsemigroup of $S = (Q, \cdot)$, the rationals under multiplication in that U generates S as a group. A consequence of this is that U *dominates* S in the sense that, for any two semigroup homomorphisms $\alpha, \beta : S \to T$, if $\alpha|U = \beta|U$ then $\alpha = \beta$. To verify this, we need only check that, in these circumstances, $\alpha(u^{-1}) = \beta(u^{-1})$ $\forall u \in U$. Indeed, let us be more general and take U to be a *full* subsemigroup of an inverse semigroup S, meaning that $\mathrm{E}(U) = \mathrm{E}(S)$, and suppose that U generates S as an inverse semigroup. The following meandering calculation proves the point: $\alpha(u^{-1}) = \alpha(u^{-1}uu^{-1}) = \alpha(u^{-1}u)\alpha(u^{-1}) = \beta(u^{-1}u)\alpha(u^{-1})$ (since

$u^{-1}u \in U) = \beta(u^{-1})\beta(u)\alpha(u^{-1}) = \beta(u^{-1})\alpha(u)\alpha(u^{-1}) = \beta(u^{-1})\alpha(uu^{-1}) = \beta(u^{-1})\beta(uu^{-1}) = \beta(u^{-1}uu^{-1}) = \beta(u^{-1})$. This gives:

4.1. Proposition. *If U is a full subsemigroup of an inverse semigroup S which generates S as an inverse semigroup, then U dominates S.*

We can now give an example of a finite semigroup S with a dominating subsemigroup U. Let S be the 5-element combinatorial Brandt semigroup $M^0[1; I, I; \Delta]$, where $|I| = 2$, 1 denotes the trivial group, and Δ denotes the 2×2 identity matrix. By the first proposition, S is dominated by its subsemigroup $U = \{(1; 1, 1), (1; 1, 2), (1; 2, 2), 0\}$. Our two examples so far show that it is possible for commutative semigroups, and for finite semigroups, to possess proper dominating subsemigroups. Our goal in this section will be to show, through an application of Ramsey's Theorem, that this is not possible for the finite commutative semigroups. For a more extensive introduction to this topic, see Chapter 4 of [30].

First we look at the connection between epimorphisms and dominions. We shall henceforth write our mappings from left to right.

We begin by noting that if a product of two homomorphisms $\alpha\beta$ is an epimorphism (that is, is pre-cancellative, or left cancellative) then β is an epimorphism, and that the composition of two epimorphisms is itself an epimorphism. From this it follows that a homomorphism $\alpha : S \to T$ is epi if and only if the same is true of the corresponding inclusion $i : S\alpha \to T$. If this is the case we say that $S\alpha$ is *dense* or is *epimorphically embedded* in T. Therefore the study of epimorphisms is equivalent to the study of epimorphic embeddings.

We have seen a class of examples above in which a semigroup S is dominated by a subsemigroup U. For an arbitrary subsemigroup U of a semigroup S however, the domination of U may only be partial. In general, we define the *dominion* of U in S, denoted by $\mathrm{Dom}(U, S)$, to consist of all elements $d \in S$ which are *dominated* by U in the sense that whenever $\alpha, \beta : S \to T$ are homomorphisms such that $\alpha|U = \beta|U$ then $d\alpha = d\beta$. It is easy to see that $\mathrm{Dom}(U, S)$ is a subsemigroup of S containing U; indeed, it is routine to verify that that $\mathrm{Dom}(\cdot, S)$ defines a closure operator on the subsemigroups of S in that $U \subseteq \mathrm{Dom}(U, S)$, and if $U \subseteq V \subseteq S$ then $\mathrm{Dom}(U, S) \subseteq \mathrm{Dom}(V, S)$, and $\mathrm{Dom}(\mathrm{Dom}(U, S), S) = \mathrm{Dom}(U, S)$. We call a semigroup U *closed* in S if $\mathrm{Dom}(U, S) = U$, and U is *absolutely closed* if it is closed in every containing semigroup S. At the other extreme, U is dense in S if $\mathrm{Dom}(U, S) = S$. A weaker condition on U than that of being absolutely closed is that of being *saturated*, which means that U cannot be properly epimorphically embedded in another semigroup. All these definitions can of course be applied equally well to other algebras, although for some they do not lead to an interesting theory, for instance, all objects are absolutely closed in the category of groups and group morphisms. This is because to say that an algebra A is absolutely closed is equivalent to saying that A is a so-called special amalgamation base. Every group is an amalgamation base of any group amalgam in which it is embedded, as every group amalgam can be embbeded in a group without collapse (Schreier's Theorem). However, finding an elementary proof that every epimorphism from one group to another is necessarily surjective is an interesting exercise (see [42]).

There are several points to note. First, it follows from these definitions that to say that every epi from a semigroup U is onto is equivalent to saying that every homomorphic image of U is saturated. (There does exist a saturated semigroup with a morphic image which is not; [27, Example 2]). Also, there is no reason to expect that a subsemigroup is dense in its own dominion: for a counterexample see Ex. 4.1.5 of [30]. Finally, the notion of dominion could be introduced more generally for subsets U of S, but in that case we see that $\mathrm{Dom}(U, S) = \mathrm{Dom}(\bar{U}, S)$, where $\bar{U}$ is the subsemigroup of S generated by U. Semigroup dominions are described completely by the following theorem.

4.2. Isbell's Zigzag Theorem. *Let U be a subsemigroup of S. Then $d \in \mathrm{Dom}\,(U, S)$ if and only if $d \in U$ or there is a series of factorizations of d as follows:*

$$d = u_0y_1 = x_1u_1y_1 = x_1u_2y_2 = x_2u_3y_2 = \cdots = x_mu_{2m-1}y_m = x_mu_{2m},$$

where $u_i \in U$, $x_i, y_i \in S$, $u_0 = x_1u_1$, $u_{2i-1}y_i = u_{2i}y_{i+1}$, $x_iu_{2i} = x_{i+1}u_{2i+1}$ $(1 \le i \le m-1)$ and $u_{2m-1}y_m = u_{2m}$.

Such equations are called a *zigzag in S over U* with *value d*, *length m*, and *spine* $u_0, u_1, \ldots, u_{2m}$ (in that order). To familiarize himself with the zigzag equations, the reader may wish to pause to write down appropriate (length one) zigzags for the element u^{-1} of our first example, and the element $(1; 2, 1)$ of our second.

The proof below is not the original, due to Isbell [38], nor that based on tensor products to be found in Howie [35]. It is a combinatorial proof that uses a construction which owes its origins to the theory of HNN extensions in group theory. The proof is however self-contained and short, originally appearing in [31].

Proof. The reverse implication follows from a straightforward manipulation of the zigzag which is a direct generalization of the calculation of our first example. Suppose that d is the value of a length m zigzag Z, as given in the statement of the theorem. Let α, β be two homomorphisms from S to a semigroup T which agree on U, and extend α and β to homomorphisms from S^1 to T^1 by defining $1\alpha = 1 = 1\beta$ if $1 \notin S$. Taking $x_0 = y_{m+1} = 1$, we show inductively that $d\alpha = (x_iu_{2i})\beta y_{i+1}\alpha$ for $i = 0, 1, \ldots, m$. In particular, this gives the required $d\alpha = (x_mu_{2m})\beta = d\beta$. Now,

$$d\alpha = (u_0y_1)\alpha = u_0\alpha y_1\alpha = u_0\beta y_1\alpha = (x_0u_0)\beta y_1\alpha.$$

Suppose inductively that $d\alpha = (x_iu_{2i})\beta y_{i+1}\alpha$ $(0 \le i \le m-1)$. Then using the zigzag equations we obtain:

$$\begin{aligned} d\alpha &= (x_{i+1}u_{2i+1})\beta y_{i+1}\alpha = x_{i+1}\beta u_{2i+1}\beta y_{i+1}\alpha \\ &= x_{i+1}\beta u_{2i+1}\alpha y_{i+1}\alpha = x_{i+1}\beta(u_{2i+1}y_{i+1})\alpha \\ &= x_{i+1}\beta(u_{2i+2}y_{i+2})\alpha = x_{i+1}\beta u_{2i+2}\alpha y_{i+2}\alpha \\ &= x_{i+1}\beta u_{2i+2}\beta y_{i+2}\alpha = (x_{i+1}u_{2i+2})\beta y_{i+2}\alpha, \end{aligned}$$

thus completing the proof in this direction.

Conversely, suppose that $d \in \mathrm{Dom}(U, S)\backslash U$. Form a semigroup H by adjoining a new element t to S subject to the relations $t^2 = 1, tu = ut$, and $tut = u$ for all $u \in U$ (the latter relation is a consequence of the others, but it is convenient in the word argument to follow to regard it as an elementary relation). Consider the homomorphisms $\phi_1, \phi_2 : S \to H$ whereby $s\phi_1 = s$, and $s\phi_2 = tst$. (In fact both these mappings are embeddings, but we do not use this here.) Clearly $\phi_1|U = \phi_2|U$ so that $tdt = d$, or what is the same, $td = dt$ in H. We prove that the latter equation implies that d is the value of some zigzag in S over U.

Since $td = dt$, there is a sequence of transitions of minimal length $I : td \to \cdots \to dt$, where each transition $pwq \to pw'q$ $(p, w, w', q \in H)$ it either a *t-transition*, that is involves a relation in which t occurs, or is a *refactorization*, that is $w = w'$ in S. We claim that no transition in I involves any of the relations $t^2 = 1$ or $tut = u$ $(u \in U)$. Suppose to the contrary that I contains a transition $\alpha : pq \to pt^2q$ $(p, q \in H)$. Clearly α is not the final transition of I, so consider the subsequent transition $\beta : pt^2q \to$. The right hand side of β has one of the forms (i) pq; (ii)$p't^2q$; (iii) pt^2q'; (iv) the product p has the form $p'u$ or $p'tu$ $(u \in U)$; or (v) a similar remark applies to q. In case (iv) β could then have the form $\beta : p'ut^2q \to p'tutq$ or $\beta : p'tut^2q \to p'utq$, as the case may be. Dual remarks apply to case (v). Therefore cases (i), (iv), and (v) contradict our minimum length assumption on I. In cases (ii) and (iii) however, the two transitions α and β can be performed in the opposite order without changing the net effect. It follows that all transitions of the form $pq \to pt^2q$ can be taken to appear together at the end of I, and thus there are none.

Next suppose that α has the form $puq \to ptutq$, and once again consider the next transition β. If p has the form $p'v$, $p't$, or $p'tv$ $(v \in U)$ then β could have the form $p'vtutq \to p'tvutq$, $p'ttutq \to p'utq$, or $p'tvtutq \to p'vutq$ respectively. In each case the pair α, β could be replaced by single transitions: $p'vuq \to p'tvutq$, $p'tuq \to p'utq$, or $p'tvuq \to p'vutq$ respectively. Again, dual remarks apply if q has one of the corresponding forms vq', tq', or vtq'. Another possibility for β is $ptutq \to put^2q$, or $ptutq \to pt^2q$, but here again, α and β could be replaced by a single transition. Again, for β to cancel α would contradict the minimality of the length of I, leaving as the remaining possibilities that β involves only p, or only q, or has the form $ptutq \to pt^2ut^2q$. Arguing similarly to above, we conclude that all the transitions of the form $puq \to ptutq$ can be taken to occur together at the end of I, and so there are none.

Call a t-transition of the form $putq \to ptuq$ [resp. $ptuq \to putq$] a *left* [resp. *right*] *transition*, so that our sequence I consists entirely of refactorizations and of left and right transitions, with exactly one occurrence of the letter t in each word of I. Suppose that ptq is a product occurring in I, and that the next t-transition occurring in the sequence is a left transition. We may assume that this transition occurs immediately, or is preceded by just one refactorization of the form $ptq \to p'utq$, for it is clear that any refactorization of p can be performed in one step, while any refactorization of q can be postponed until after the left transition. Next, suppose that I contains two left transitions with no intervening right transition, which we may assume to have the form:

$$putq \to ptuq \to p'vtuq \to p'tvuq,$$

or simply

$$p'vutq \to p'vtuq \to p'tvuq \ (u, v \in U).$$

In the latter case the pair of transitions can be replaced by a single left transition, while the three transitions of the first case may be replaced by two:

$$putq \to p'vutq \to p'tvuq.$$

Coupling all this with the dual arguments for right transitions allows us to conclude that I consists of alternate left and right transitions, separated by a single refactorization; furthermore, both the first and last t-transitions are right. The sequence I then yields equalitites in H of the form:

$$\begin{aligned} td &= tu_0y_1 = u_0ty_1 = x_1u_1ty_1 = x_1tu_1y_1 = x_1tu_2y_2 = x_1u_2ty_2 \\ &= x_2u_3ty_2 = \cdots = x_{m-1}u_{2m-2}ty_m = x_mu_{2m-1}ty_m \\ &= x_mtu_{2m-1}y_m = x_mtu_{2m} = x_mu_{2m}t = dt, \end{aligned}$$

for some $m \geq 1, u_i \in U \ (0 \leq i \leq 2m)$, $x_i, y_i \in S^1$, and $u_0 = x_1u_1, u_{2i-1}y_i = u_{2i}y_{i+1}, x_iu_{2i} = x_{i+1}u_{2i+1} \ (1 \leq i \leq m-1)$, and $u_{2m-1}y_m = u_{2m}$. In fact, $x_i, y_i \in S$, for if $x_i = 1$, then in S we have:

$$d = u_0y_1 = x_1u_1y_1 = x_1u_2y_2 = \cdots = u_{2i}y_{i+1} = \cdots = x_mu_{2m};$$

and so that I could be shortened by beginning with $td \to tu_{2i}y_{i+1}$, with a dual remark applying if some $y_i = 1$. In other words, d is the value of the zigzag defined by deleting all instances of the letter t in the above equations, thus completing the proof. □

Since epimorphisms are a generalization of surjective homomorphisms, they might be expected to share some structure preserving qualities, and indeed the epimorphic image of a commutative semigroups is commutative, a result that we shall need in our main application of the Zigzag Theorem. We show a stronger result, the proof of which is curiously reminiscent of the technique of proof that we have just employed. What is more, the proof clearly demonstrates the efficacy of the Zigzag Theorem—I know of no way of proving this result without recourse to its use.

4.3. Theorem ([37]). *Let U be a commutative subsemigroup of a semigroup S. Then* $\mathrm{Dom}(U, S)$ *is commutative.*

Proof. Let Z be a zigzag for some $d \in \mathrm{Dom}(U, S)$ and let $u \in U$. Then

$$\begin{aligned} ud &= uu_0y_1 = u_0uy_1 = x_1u_1uy_1 = x_1uu_1y_1 = x_1uu_2y_2 = \cdots \\ &= x_mu_{2m-1}uy_m = x_muu_{2m-1}y_m = x_muu_{2m} = x_mu_{2m}u = du, \end{aligned}$$

from which it follows that U is central in its dominion. It remains to show that any two members $d, t \in \mathrm{Dom}(U, S)$ also commute. Since t commutes with all members of the spine of Z, (by the previous argument), we may repeat the argument with u replaced by t to conclude that $td = dt$. □

4.4. Theorem. *Any epimorphism $\alpha : S \to T$ from a finite commutative semigroup S to another semigroup T is surjective.*

Proof. Suppose that α is an epimorphism. The inclusion $i : S\alpha \to T$ is then an epimorphism from the commutative semigroup $S\alpha$ into T, which is itself commutative by Theorem 4.3. It follows that it is enough for us to prove that any epimorphic embedding of a finite commutative semigroup U into a (commutative) semigroup S is necessarily onto.

First we show that there exists $d \in S\backslash U$ which has a zigzag of minimum length in S over U, the first spine member of which is an idempotent. Take any $d_0 \in S\backslash U$ and, using the Zigzag Theorem, factorize $d_0 = a_1 d_1 = a_1 a_2 d_2 = \cdots = a_1 a_2 \ldots a_i d_i = \cdots$, where $d_j = a_{j+1} d_{j+1}$ is the first line of a zigzag with value d_j of minimum length amongst all zigzags in S over U with value d_j ($j = 0, 1, 2, \ldots$). Consider the sequence $(a_1, a_2, \ldots)$ in U and invoke Ramsey's Theorem through Lemma 1.2 to infer that $e = a_m a_{m+1} \ldots a_n \in \mathrm{E}(U)$ for some indices m and n. Put $d = d_{m-1}$; then, of all zigzags in S over U with value d there is one of minimum length which begins $d = a_m d_m = x_1 b d_m = x_1 c y_2$ say, where $a_m = x_1 b$, $bd_m = cy_2$. We replace the first two lines of this zigzag by $d = ed_n = x_1 b a_{m+1} a_{m+2} \ldots a_n d_n = x_1 c y_2$ with $e = a_m a_{m+1} \ldots a_n = x_1 b a_{m+1} a_{m+2} \ldots a_n$ and $b a_{m+1} a_{m+2} \ldots a_n d_n = b d_m = c y_2$ to yield a zigzag of the same minimum length for d the first spine member of which is an idempotent.

We now derive a contradiction by showing that any zigzag in S over U of length $m > 1$, the first spine member of which is an idempotent, can be replaced by a zigzag of length $m - 1$ withe the same value. (A slight modification of the argument shows that in the $m = 1$ case the value of the zigzag is a member of U.)

Take $d \in S\backslash U$ with $d = a_0 y_1 = x_1 a_1 y_1 = \cdots = x_m a_{2m-1} y_m = x_m a_{2m}$, a zigzag in S over U with value d such that $a_0 \in \mathrm{E}(U)$. Since U is finite it satisfies some identity $x^k = x^l$ for some $k < l$. A required zigzag is as follows:

$$\begin{aligned}
d &= a_0 a_1^{l-k-1} a_2 y_2 \\
&= x_2 a_3 a_1^{l-k} y_2 \\
&= x_2 a_1^{l-k} a_4 y_3 \\
&= x_3 a_5 a_1^{l-k} y_3 \\
&\;\;\vdots \\
&= x_i a_{2i-1} a_1^{l-k} y_i \\
&= x_i a_1^{l-k} a_{2i} y_{i+1}
\end{aligned}$$

$$= x_{i+1}a_{2i+1}a_1^{l-k}y_{i+1}$$

$$\vdots$$

$$= x_{m-1}a_1^{l-k}a_{2m-2}y_m$$
$$= x_m a_{2m-1}a_1^{l-k}y_m$$
$$= x_m a_1^{l-k}a_{2m}$$

with

$$a_0a_1^{l-k-1}a_2 = x_2a_3a_1^{l-k},\ x_ia_1^{l-k}a_{2i} = x_{i+1}a_{2i+1}a_1^{l-k}, \quad i = 2, 3, \ldots, m-1,$$
$$a_{2i-1}a_1^{l-k}y_i = a_1^{l-k}a_{2i}y_{i+1}, \quad i = 2, 3, \ldots, m-1,$$
$$\text{and} \quad a_{2m-1}a_1^{l-k}y_m = a_1^{l-k}a_{2m}.$$

We next verify the required zigzag equalities. First, the value of the new zigzag is d as:

$$a_0a_1^{l-k-1}a_2y_2 = a_0^ka_1^{l-k-1}a_2y_2 = (x_1a_1)^ka_1^{l-k-1}a_2y_2 = x_1^ka_1^{l-1}a_2y_2$$
$$= x_1^ka_1^ly_1 = x_1^ka_1^ky_1 = (x_1a_1)^ky_1 = a_0^ky_1 = a_0y_1 = d.$$

Next, $a_0a_1^{l-k-1}a_2 = x_1a_1^{l-k}a_2 = x_1a_2a_1^{l-k} = x_2a_3a_1^{l-k}$, and in general,

$$x_ia_1^{l-k}a_{2i} = x_ia_{2i}a_1^{l-k} = x_{i+1}a_{2i+1}a_1^{l-k}, \quad i = 2, 3, \ldots, m-1$$
$$a_{2i-1}a_1^{l-k}y_i = a_1^{l-k}a_{2i-1}y_i = a_1^{l-k}a_{2i}y_{i+1}, \quad i = 1, 2, \ldots, m,$$

where we take $y_{m+1} = 1$, as required. □

The argument of this proof is amplified in [25] where it is shown that any epimorphism from a finite semigroup S is surjective if SES, the ideal generated by the idempotents of S, satisfies some non-trivial permutation identity. In this case it follows from a theorem of Khan [40] that SES in fact satisfies the *normality identity* $xyzw = xzyw$, and so the condition is effective, and indeed often gives the best practical way of showing that certain small semigroups are saturated. That all finite commutative semigroups are saturated was first shown in [37], where it was proved that a sufficient condition for saturation of a commutative semigroup is the descending chain condition on principal ideals. Other classes of saturated semigroups are the classes of nilpotent semigroups, generalized inverse semigroups (orthodox semigroups the idempotents of which form a normal band), and the class of completely semisimple semigroups with no infinite chain of $\mathcal{J}$-classes (which includes all finite regular semigroups, and all completely 0-simple semigroups). Recently I have proved that any epimorphism from a locally inverse semigroup is surjective, [33]. (A regular semigroup S is *locally inverse* if each local subsemigroup eSe is inverse for each $e \in E(S)$. This extends the above quoted result on generalized inverse semigroups which are exactly the orthodox locally inverse semigroups [60].) On the other hand, non-saturated classes include bands, commutative cancellative semigroups (as seen in our first example), commutative periodic semigroups, and finite idempotent-generated semigroups.

The following classes of semigroups are absolutely closed: right simple semigroups, full (and partial) transformation semigroups, inverse semigroups, (which, more generally, have the strong amalgamation property), finite monogenic semigroups, and the semigroup of all endomorphisms of a vector space over a division ring. Semigroups which lack this property include finite null semigroups, finite right normal bands, and finite rectangular bands.

These results, most of which can be found in [30], all depend for their proof on the Zigzag Theorem. A similar theorem arises in Ring Theory and is used to establish a similar theory, (see, for example, [28]).

Although much is known, the following questions, I believe, remain open:

(1) Is the class of absolutely closed semigroups closed under the taking of (a) homomorphic images; or (b) ideals?
(2) Is the class of saturated semigroups closed under the taking of (a) direct products; (b) ideals; and (c) adjunction of an identity?

The class of absolutely closed semigroups is not closed under the adjunction of an identity: the Baer-Levi semigroup U on the base set N (the semigroup of all injective mappings whose ranges have infinite complements) is absolutely closed because it is right simple, however $\mathrm{Dom}(U^1, \mathcal{T}_N) = V$, where V is the submonoid of all injective mappings of the full transformation semigroup $\mathcal{T}_N$ [27, Example 7].

Zigzags fully describe dominions in the categories of Semigroups, and Commutative Semigroups [37], and the proof of the Zigzag Theorem in [35] (making use of tensor products) adapts to show that the same is true in the category of finite semigroups. Since every (finite) semigroup can be embedded in a (finite) regular semigroup, it follows that the Zigzag Theorem also holds in the categories of Finite Semigroups and Finite Regular Semigroups. A long-standing open question is:

(3) Does the Zigzag Theorem hold in the category of Bands?

A category of semigroups in which zigzags do not fully describe dominions is the variety $[x^2y^2 = y^2x^2]$. This example, due to T. E. Hall [23], is also Ex. 4.2.3 of [30].

5. Biordered sets

The *biordered set* of a semigroup S consists of its set of idempotents $E = \mathrm{E}(S)$, together with a partial multiplication inherited from S with domain $D_E = \{(e, f) \in E \times E : ef \text{ or } fe \in \{e, f\}\}$. (Note that if $ef \in \{e, f\}$, let us say $ef = f$, then $fe \in E$.) This type of algebra, which has been axiomatized by Nambooripad [44], can be regarded as the 'skeleton' of the semigroup, which influences its overall shape. The degree of this influence is variable: for groups, and for idempotent-free semigroups, the notion is irrelevant, but, for semibands (idempotent-generated semigroups) the opposite is true, and, in particular, any fundamental regular semiband is determined by its biordered set (*fundamental* meaning that the largest congruence contained in Green's $\mathcal{H}$ relation is trivial).

A basic question addressed in the Ph.D. thesis of David Easdown was whether or not a given biordered set could arise from a semigroup of a particular class, or whether it is somehow forbidden. This was motivated by a theorem of Nambooripad which says that a biordered set can arise from a regular semigroup if and only if every so-called 'sandwich set', $S(e, f)$, of a pair of idempotents is not empty. Easdown introduced a fundamental representation theorem for biordered sets which allowed him to recover Nambooripad's result and to give similar characterizations for the classes of bands, unions of groups, E-solid semigroups, finite semigroups, periodic semigroups, group bound semigroups, and eventually regular semigroups.

The situation in which one is 'given a biordered set' needs to be clarified. We might begin with a semigroup and consider its biordered set, but this severely limits our approach. In practice, one might begin with some 'arrangement' of idempotents and wish to decide, say, whether such an arrangement might represent the biordered set of a finite semigroup. Before proceeding, we would want to know whether the given arrangement could represent the biordered set of any semigroup at all, and to answer such a question requires a system of axioms which formally captures the essential properties of a biordered set, and two such (equivalent) systems were invented by Nambooripad [44]. These axioms take some time to become familiar! Their statement is complicated, (but the Easdown arrow notation can facilitate their use [see 30]). For instance, Axiom B4 states that if $e, f, g \in E$ are such that $f = ef, g = eg$ and $(fe)(ge) = fe$, then there exists $f' \in E$ such that $ef' = f' = f'g$ and $f'e = fe$. In a semigroup, this is satisfied by taking $f' = fg$ for then $ef' = f' = f'g$ and

$$f'^2 = fgfg = f(eg)(ef)g = (fe)(ge)fg = (fe)(fg) = f(efg) = f^2g = f'$$

and

$$f'e = fge = f(eg)e = (fe)(ge) = fe.$$

That these axioms are appropriate was proved by Easdown [15] by showing that any abstract biordered set is the biordered set of some semigroup. The proof strategy is to begin with the free semigroup F_E, where E is a given biordered set with domain D_E, and show that E is the biordered set of F/ρ, where ρ is the congruence on F generated by $\rho_0 = \{(z, xy) : (x, y) \in D_E \text{ and } xy = z\}$. The proof then involves the demonstration of three conditions:

(i) $e, f \in E$ and $e\rho f \Rightarrow e = f$ (no collapse);
(ii) $ef = e \Leftrightarrow ef\rho e$ and $fe = e \Leftrightarrow fe\rho e$ (no new biordered set relations);
(iii) $w \in F$ and $w^2\rho w \Leftrightarrow w\rho e$ for some $e \in E$ (no new idempotents).

The proof of (iii) involved a particularly difficult word argument. If we satisfy ourselves with verifying just conditions (i) and (ii), we can still infer the weaker result that E is a biordered subset of F/ρ, an idea which we now explain. In general, for a biordered set E we call $F \subset E$ a *biordered subset* of E if F is a *full partial subalgebra* of E, in the sense that $D_F = D_E \cap (F \times F)$ and that F satisfies all the biordered set axioms (which we have not stated here, but are not all automatically inherited by F: Axiom B4, and its dual, $B4^*$, are implications which require, in specified circumstances, the existence of an element of F with certain properties). If E and F are biordered sets and $\theta : E \to F$ is a mapping, then θ is a *biordered set morphism* if θ satisfies:

$$(e, f) \in D_E \Rightarrow (e\theta, f\theta) \in D_F \quad \text{and} \quad (ef)\theta = e\theta f\theta.$$

We call θ a *biordered set isomorphism* if it is an isomorphism of partial algebras; that is a bijection such that both θ and θ^{-1} are morphisms. The requirement that θ^{-1} also be a morphism is not automatically satisfied by partial algebras: without it, the possibility arises that F might have products for which there are no counterparts in E.

Returning to our conditions (i) and (ii) above, given that they hold, it follows that the mapping whereby $e \mapsto e\rho$ is an injective biordered set morphism, and indeed it defines a biordered set isomorphism onto a biordered subset of the biordered set of F/ρ. Thus the satisfaction of (i) and (ii) proves that every biordered set is a biordered subset of some semigroup. This fact was first established by Easdown [14] using a different construction which has proved to very fruitful, and is akin to the classical idea of the Schutzenberger representation of a semigroup by action on $\mathcal{R}$- and $\mathcal{L}$-classes.

For a biordered set E define the equivalence $\mathcal{L}'$ by the rule that $e\mathcal{L}'f$ if and only if $e = ef$ and $f = fe$. Dually, we define the $\mathcal{R}'$-classes of E. Denote the set of $\mathcal{L}'$- and of $\mathcal{R}'$- classes of E by $E/\mathcal{L}'$ and of $E/\mathcal{R}'$ respectively. An arbitrary member of $E/\mathcal{L}'$ [resp. $E/\mathcal{R}'$] will be denoted by L [resp. R] and the $\mathcal{L}'$-class [resp. $\mathcal{R}'$-class] containing a particular $e \in E$ will be written as L_e [resp. R_e]; the notation is chosen in recognition of the fact that biordered sets model biordered sets of semigroups, in which case the $\mathcal{L}'$ and $\mathcal{R}'$ relations are just the corresponding Green's relations restricted to the idempotents.

For convenience, we shall write the full transformation semigroup on a set X by $T(X)$, while its left-right dual will be written as $T^*(X)$.

Our representation of E is parallel to the representation of an eventually regular semigroup S (see Section 1.4 of [30]: a semigroup is *eventually regular* if each of its members has a regular power) the kernel of which is the maximum idempotent-separating congruence of S. Suppose that $\infty \notin E/\mathcal{L}' \cup E/\mathcal{R}'$, and put $X = E/\mathcal{L}' \cup \{\infty\}$ and $Y = E/\mathcal{R}' \cup \{\infty\}$.

5.1. Definition.

(i) $\rho : E \to T(X)$, where

$$\rho_e : L \mapsto \begin{cases} L_{xe}, & \text{if there exists } x \in L \text{ such that } ex = x; \\ \infty, & \text{otherwise;} \end{cases}$$

(ii) $\lambda : E \to T(Y)$, where

$$\lambda_e : R \mapsto \begin{cases} R_{ex}, & \text{if there exists } x \in R \text{ such that } xe = x; \\ \infty, & \text{otherwise,} \end{cases}$$

with ∞ a fixed point of both ρ_e and λ_e.

(iii) $\phi : E \to T(X) \times T^*(Y)$, where $e \mapsto \phi_e = (\rho_e, \lambda_e)$.

The Easdown Representation Theorem [14] is then

5.2. Theorem. *The mapping ϕ is an injective biordered set morphism from E into* $\mathrm{E}T(X) \times T^*(Y))$*. Also $E\phi$ is a biordered subset of* $\mathrm{E}T(X) \times T^*(Y))$ *and E and $E\phi$ are isomorphic as biordered sets.*

As a consequence of this theorem, it follows that any biordered set can at least be realised as a biordered subset of a biordered set of some semigroup in a manner that preserves finiteness. This representation was the main tool used by Easdown in gaining his other characterizations. Call a semigroup S *idempotent-consistent* if, for each congruence ρ on S, every idempotent congruence class necessarily contains an idempotent: in other words, S 'satisfies Lallement's Lemma'. The *core* of S is $\langle E \rangle$, the subsemigroup of S generated by its idempotents. Easdown and Hall [16] went on to prove

5.3. Theorem. *Let C be a class of idempotent-consistent semigroups closed under the taking of homomorphic images and cores. Let E be any biordered set. Then* $E = \mathrm{E}(S)$ *for some $S \in C$ if and only if* $E\phi = \mathrm{E}(\langle E\phi \rangle)$ *and* $\langle E\phi \rangle \in C$.

This is a powerful result, as it reduces the question of whether a biordered set E arises in a certain class of semigroup to a study of the idempotents of one particular semigroup which is constructed from E. Moreover, the theorem is applicable to many interesting types of semigroup. For instance, C could be any of the following classes: finite semigroups, periodic semigroups, bands, regular semigroups, completely regular semigroups, group bound semigroups, and eventually regular semigroups. In particular, a finite biordered set E is the biordered set of some finite semigroup if and only if no idempotents are generated in $\langle E\phi \rangle$ other than those originally given by E; that is to say, if and only if $E\phi = \mathrm{E}\langle E\phi \rangle)$. It is not necessary to check arbitrarily long products of idempotents in order to verify this.

5.4. Theorem. *A finite biordered set E is the biordered set of some finite semigroup S if and only if* $\mathrm{E}\langle E^n\phi \rangle) = E\phi$ *where* $n = R(3, r) - 1$*, and* $r = |\langle E\phi \rangle|$.

Proof. By Theorem 5.3, the given condition is certainly necessary. Suppose then that $\mathrm{E}(E\phi)^n) = E\phi$ and let $x = e_1\phi e_2\phi \dots e_m\phi \in (E\phi)^m$ $(e_i \in E)$ for some $m \geq n+1$ and suppose that $x \in \mathrm{E}(\langle E\phi \rangle)$. We proceed by induction on m to show that $x \in E\phi$. Consider the product $x_{n+1} = e_1\phi e_2\phi \dots e_{n+1}\phi$. By Lemma 1.5, and its proof, x_{n+1}

has two successive consecutive products which are equal to some idempotent of $\langle E\phi \rangle$, and so, by the given condition, to some $e\phi \in E\phi$. Replacing these two successive products by the single idempotent $e\phi$ then yields a shorter product of idempotents from $\langle E\phi \rangle$ which is then a member of $E\phi$ by the inductive hypothesis. □

5.5. Corollary. *A finite biordered set E is the biordered set of some finite semigroup if and only if* $\mathrm{E}(\langle E^n \rangle \phi) \subseteq E\phi$ *for* $n = (r+2)(r-1)/2$.

Proof. This is immediate from the previous result and a standard upper bound for Ramsey numbers:

$$R(m, r) \leq C(m + r - 2, m - 1)$$

where $C(p, q)$ denotes the binomial coefficent. In the case in point the value of m is 3. This upper bound is itself a consequence of the inequality (shown in [21] to be strict):

$$R(m, r) \leq R(m, r - 1) + R(m - 1, r)$$

which is a byproduct of proving Ramsey's Theorem using a double induction on m and r. The proof given in Section 1 was preferred as the argument was applicable to both the finite and infinite cases. The inductive proof does however furnish a better upper bound, and is straightforward. (See [21, p. 3].) Unfortunately, we have only an exponential bound on the value of r, for it is bounded above by $(t+1)^t(s+1)^s$, where t and s are the respective number of $\mathcal{L}'$- and of $\mathcal{R}'$-classes E. A lower (but still exponential) bound could be obtained from the standard method of constructing a subsemigroup of the full transformation semigroup from a given set of generators, as appears in [46, p. 21]. □

The investigation of biordered sets of certain types was also undertaken contemporaneously with Easdown's work by Tret'yakova [57].

The study of finite semigroups has blossomed over the past decade mainly through the connection with language theory and pseudovarieties. This in turn has opened the door to the development of combinatorial techniques within our subject. The particular significance of Ramsey's Theorem has been dramatically underlined by the two breakthroughs of Ash ([6], [7]). We can expect therefore that future problems will yield to applications of other deep aspects of Ramsey Theory. At least this seems a hopeful direction.

References

[1] Aizenstat, A., Permutation identities, Modern Algebra, No.3 (Leningrad Gos. Ped. Inst., Leningrad) (1975), 3–12.

[2] Almeida, J., Finite semigroups and universal algebra, World Scientific Publishers, 1994.

[3] Almeida, J., On direct product decompositions of finite $\mathcal{J}$-trivial semigroups, Internat. J. Algebra Comput. 1 (3) (1991), 329–337.

[4] Almeida, J., and A. Azevedo, The join of the pseudovarieties of $\mathcal{R}$-trivial and $\mathcal{L}$-trivial monoids, J. Pure Appl. Algebra 60 (1989), 129–137.

[5] Ash, C. J., Pseudovarieties, generalized varieties, and similarly described classes, J. Algebra 92 (1) (1985), 104–115.

[6] Ash, C. J., Finite semigroups with commuting idempotents, J. Austral. Math. Soc. Ser. A 43 (1987).

[7] Ash, C. J., Inevitable sequences and a proof of the "Type II Conjecture", in: Proceedings of the Monash Conference on Semigroup Theory, Hall et. al. (eds.), World Scientific Publishers 1991, pp. 31–42.

[8] Ash, C. J., Inevitable graphs: A proof of the type II conjecture and some related decision procedures, Internat. J. Algebra Comput. 1 (1) (1991), 127–146.

[9] Bergelson, V., and N. Hindman, Ramsey Theory in Noncommutative Semigroups, Trans. Amer. Math. Soc. 330 (2) (1992), 433–446.

[10] Brown, T. C., On locally finite semigroups, Ukrainian Math. J. 20 (6) (1968), 732–738 (in Russian).

[11] Brown, T. C., Van der Waerden's Theorem on arithmetic progressions, Notices Amer. Math. Soc. 16 (1), (1969), 245.

[12] Brown, T. C., An interesting combinatorial method in the theory of finite semigroups, Pacific, J. Math. 36 (2) (1971), 285–89.

[13] Easdown, D., Biordered sets of eventually regular semigroups, Proc. London Math. Soc. 49 (3) (1984), 485–503.

[14] Easdown, D., Biordered sets are biordered subsets of semigroups, J. Austral. Math. Soc. 37 (1984), 258–268.

[15] Easdown, D., Biorderd sets come from semigroups, J. Algebra 96 (1985), 581–591.

[16] Easdown, D., and T. E. Hall, Reconstructing some idempotent-generated semigroups from their biordered sets, Semigroup Forum 29 (1984), 207–216.

[17] Eilenberg, S., and M. P. Schutzenberger, On pseudovarieties of monoids, Adv. Math. 19 (3) (1976), 413–448.

[18] Epstein, D. B. A., et. al, Word Processing in Groups, Jones and Barlett, London 1992.

[19] Fouche, W. L., Reflective properties of free semigroups, Quart. J. Math. Oxford (2) 44 (1993), 435–449.

[20] Gilliam, D. W. H., Hall, T. E., and N. H. Williams, On finite semigroups and idempotents, Bull. London Math. Soc. 4 (1972), 143–44.

[21] Graham, R. L., Rothschild, B. L., and J. H. Spencer, Ramsey Theory, Wiley and Sons, (1980)

[22] Green, J. A., and Rees, D., On semigroups in which $x^r = x$, Proc. Cambridge Phil. Soc. 48 (1952), 35–40.

[23] Hall, T. E., Epimorphisms and dominions, Semigroup Forum 24 (1982), 271–83.

[24] Hashiguchi, K., Notes on finitely generated semigroups and pumping conditions for regular languages, Theoret. Comput. Sci. 46 (1986), 53–66.

[25] Higgins, P. M., Epimorphisms, permutation identities and finite semigroups, Semigroup Forum 29 (1984), 87–97.

[26] Higgins, P. M., A new determination of the permutation identities which ensure that a semigroup variety is finitely based, J. Pure Appl. Algebra 38 (1985), 65–69.

[27] Higgins, P. M., Epimorphisms, dominions and semigroups, Algebra Universalis 21 (1985), 225–33.

[28] Higgins, P. M., Epimorphisms and amalgams, Colloq. Math. LVI (1988), 1–17.

[29] Higgins, P. M., An algebraic proof that pseudovarieties are defined by pseudoidentities, Algebra Universalis 27 (1990), 597–99.

[30] Higgins, P. M., Techniques of Semigroup Theory, Oxford University Press, Oxford 1992.

[31] Higgins, P. M., A short proof of Isbell's Zigzag Theorem, Pacific J. Math. 144 (1) (1990), 47–50.

[32] Higgins, P. M., Divisors of semigroups of order-preserving mappings on a finite chain, University of Essex Mathematics Rpt. 94-4, to appear in Internat. J. Algebra Comput.

[33] Higgins, P. M., Epis from locally inverse semigroups are onto, University of Essex Mathematics Rpt. 94-13.

[34] Hotzel, E., On finiteness conditions in semigroups, J. Algebra 60 (2) (1979), 352–370.

[35] Howie, J. M., An Introduction to Semigroup Theory, Academic Press, London 1976.

[36] Howie, J. M., Automata and Languages, Oxford University Press, Oxford 1991.

[37] Howie, J. M., and J. R. Isbell, Epimorphisms and dominions II, J. Algebra 6 (1967), 7–21.

[38] Isbell, J. R., Epimorphisms and dominions, in: Proceedings of the Conference on Categorical Algebra, La Jolla (1965), Lange and Springer-Verlag, Berlin, pp. 232–46.

[39] Justin, J., Generalisation du théorème de Van der Waerden sur les semi-groupes répétitifs, J. Combin Theory (1972).

[40] Khan, N. M., On saturated permutative varieties and consequences of permutation identities, J. Austral. Math. Soc. Ser A 38 (2) (1985), 186–97.

[41] Lallement, G., Semigroups and Combinatorial Applications, Wiley and Sons, New York 1979.

[42] Linderholm, G. E., A group epimorphism is surjective, Amer. Math. Monthly 77 (1970), 176–77.

[43] Margolis, S. W., Consequences of Ash's proof of the Rhodes Type II Conjecture in: Monash Conference on Semigroup Theory, Hall et. al. (eds.), World Scientific Publishers, 1991, pp. 180–205.

[44] Nambooripad, K. S. S., The structure of regular semigroups I, Mem. Amer. Math. Soc. 22 No. 224 (1979), vii + 119pp.

[45] Perkins, P., Bases for equational theories of semigroups, J. Algebra 11 (1968), 298–314.

[46] Pin, J. E., Varieties of Formal Languages, Plenum Press, London 1986.

[47] Pollak, G., Some sufficient conditions for heriditarily finitely based varieties of semigroups, Acta. Sci. Math. (Szeged) 50 (1986), 299–330.

[48] Pollak, G., and M. Volkov, On almost simple semigroup identities, in: Semigroups, North Holland Publishers, Amsterdam 1985, pp. 287–323.

[49] Rado, R., Axiomatic treatment of rank in infinite sets, Canadian J. Math. 1 (1949), 337–343.

[50] Reiterman, J., The Birkhoff Theorem for finite algebras, Algebra Universalis 14 (1982), 1–10.

[51] Sapir, M., Burnside type problems and finite basis property in semigroup varieties Izv. Akad. Nauk SSSR Ser. Mat. 51 (1987), 319–40 (in Russian); English translation: Math. USSR-Izv. 30 (1988), 295–314.

[52] Sapir, M., and M. Volkov, The HFB property and the structure of semigroups, in: Contributions to General Algebra, Wien 6 (1988), pp. 303–310.

[53] Schutzenberger, M. P., On finite monoids having only trivial subgroups, Inform. and Control 8 (1965), 190–194.

[54] Simon, I., Piecewise testable events, Proc. 2nd GI Conference, Lecture Notes in Comput. Sci. 33, Springer-Verlag, Berlin 1975, pp. 214–222.

[55] Simon, I., Conditions de finitude pour des semi-groupes, C.R. Acad. Sci. Paris Ser. A 290 (1980), 1081–82.

[56] Straubing, H., and D. Therien, Partially ordered finite monoids and a theorem of I. Simon, J. Algebra 119 (1988), 393–99.

[57] Tret'yakova, E. G., Biordered sets of idempotents of semigroups of certain types, Izv. Vyssh. Uchebn. Zaved. Mat. 3 (1986), 77–80.

[58] Weil, P., Implicit operations on pseudo-varieties: An introduction, in: Monoids and Semigroups with Applications, J. Rhodes (ed.), World Scientific Publishing, 1991.

[59] Wilson, R. J., Introduction to Graph Theory, Oliver and Boyd, Edinburgh 1972.

[60] Yamada, M., Regular semigroups whose idempotents satisfy permutation identities, Pacific J. Math. 21 (1967), 371–92.

Department of Mathematics
University of Essex
Wivenhoe Park
Colchester CO4 3SQ, England

Combinatorial aspects of the theory of Lie superalgebras

*Alexander. A. Mikhalev**

Introduction

In the present article we consider some questions of the combinatorial theory of Lie superalgebras in the spirit of infinite dimensional Lie algebras. The need of such consideration is motivated by the importance of Lie superalgebras which prove to be useful in different areas of mathematics and physics (see for example [Bah1], [BD], [BZ], [Ber], [BP], [CNS], [Fer], [GD], [GM], [Ka2], [Kap], [KM], [Kem], [Le1], [Le2], [Man1], [Man2], [Man3], [Man4], [Mar], [Mas], [Mi2], [Mi7], [Mi13], [MZ3], [Mos], [Nee], [NR], [Ni], [Pe], [O], [RW], [Roe], [Ros1], [Ros2], [Sch1], [Sch4], [She], [Stan], [VdL], [VOW], [Va], [Wais], [Zai], [Ze1], [Ze2]). There have appeared several books and surveys on Lie superalgebras [Kac1],[Kac2], [Sch2], [Ber], [Le3], [BMPZ].

Free Lie superalgebras and their Z-graded subalgebras appeared in a natural way in algebraic topology and homological algebra (see [Av], [Bab], [Hain], [Lem], [MM]). At the same time the theory of free Lie algebras is now one of the well developed branches of Lie algebra theory, see [Bah2], [Bo2], [BKu], [CFL], [Co1], [Co2], [Bo5], [Du], [DT], [MHall1], [Ku1], [Ku2], [Ku3], [Ku4], [Ku6], [Lab], [MR], [Mic], [Reu1], [Reu2], [Schü1], [Shi1], [Shi2], [Shi3], [Shi4], [Shi5], [Shi6], [Shp1], [Shp3], [Um1], [Um2], [Um4], [Ün], [Vi], [We1], [We2], [Wit1], [Wit2], [Wit3], and has many important connections with group theory, see [Bah2], [Baum], [CFL], [Co1], [PHall], [MHall1], [MHall2], [LS], [Ma1], [Ma2], [Ma3], [Ma4], [MKS], [She], [We1], [Wit1]. The theorem about freeness of subalgebras in free Lie algebras was proved by A. I. Shirshov [Shi1] and in free Lie rings and p-algebras by E. Witt [Wit3]. The theorem on the finiteness of the rank for the intersection of two subalgebras of finite rank of free Lie p-algebras is due to G. P. Kukin [Ku3].

*Partially supported by the Russian Foundation for Fundamental Research, grant 93-011-1543, and by the International Science Foundation, grant M22000

From the point of view of Computer Algebra and Symbolic Calculations it is important to have the theory of canonical forms (so-called Gröbner bases) in context of finitely presented algebraic systems, see for example [BSt], [Berg], [Bo3], [BLo], [Da], [Gol1], [Gre], [Gro], [K-RW], [Lat], [LC], [Mi4], [Mi5], [Mi9], [MR], [New], [PP], [Rob], [Uf1]. Computation in algebra using Gröbner bases is now one of intensive developed branches of Computer Algebra, canonical forms of elements give possibilities to make effective calculations in algebras, for example, to compute dimensions, Hilbert and Poincaré series, to check some conjectures, to find algorithms for solving some algorithmic problems, see for example [BKu], [Comp], [Da], [DK1], [GAKR], [GL], [Gra], [Hu], [KV]. The Composition Lemma for Lie algebras was suggested by A. I. Shirshov in [Shi3]. It is becoming to be a rather useful tool in Ring Theory for solving a number of algorithmic problems. In associative case see also L. A. Bokut' [Bo3], G. M. Bergman [Berg], D. J. Anick [An], E. S. Golod [Gol1], [Gol2], V. N. Latyshev [Lat]; in group theory see J. R. J. Groves [Gro]; in Lie superalgebras see A. A. Mikhalev [Mi4], [Mi5], [Mi9], [Mi16].

Shuffle products were introduced by R. Ree in [Re1], [Re2] (he also considered some properties of free colour Lie superalgebras). It gives the dual Hopf algebra structure on the free associative algebra (see [Reu2]). The structure of free shuffle algebras was considered by R. Radford [Ra]. Applications of shuffle algebras to free Lie algebras can be found in the book [Reu2] by Ch. Reutenauer.

Free differential calculus for free groups and for free group rings was introduced by R. Fox in [Fox]. Now it is a very useful tool in group theory and in knot theory to solve a number of problems, see for example the book [Gu] by N. Gupta. J. Birman in [Bi] gave a matrix criterion for an endomorphism of free group to be an automorphism. In [Um2] U. U. Umirbaev proved an analog of this result for free Lie algebras (Ch. Reutenauer in [Reu1] also obtained such criterion; more general case was considered by V. Shpilrain in [Shp1] and by U. U. Umirbaev in [Um5]; endomorphisms of free Lie algebras over commutative rings were studied by A. A. Mikhalev and A. A. Zolotykh in [MZ8]). The general Jacobian problem for varieties of algebras was considered by A. V. Yagdgev, he solved it in [Y] for the varieties of all algebras and of anticommutative algebras. As for famous Jacobian problem for polynomial rings see for example the articles [Kel] by O. H. Keller and [BCW] by H. Bass, E. H. Connell, and D. Wright. Jacobian problem for free associative algebras were solved by W. Dicks and J. Lewin in [DL] and by A. H. Schofield in [Scho] (as for free associative algebras over commutative rings see the article [MZ7] by A. A. Mikhalev and A. A. Zolotykh). Universal derivations in associative algebras were considered by G. M. Bergman and W. Dicks in [BD1], [BD2]. O. G. Kharlampovich in [Kha] used Fox differential calculus in connection with Lyndon condition for Lie algebras. In [Shp3] V. Shpilrain introduced the rank of an element of a free Lie algebra and gave an algorithm for calculation the rank in the case of homogeneous relative to the length element.

Differential Lie algebras are due to the papers on derivations of R. Baer [Baer] and N. Jacobson [J]. V. K. Kharchenko in [Kh1], [Kh2] considered them in connection with differential identities of associative rings. K. I. Beidar, W. S. Martindale and

A. V. Mikhalev in [BMM] touched differential Lie algebras from the point of view of generalized polynomial identities.

Some aspects of combinatorial theory of Lie superalgebras were exposed with full details in the book [BMPZ] by Y. A. Bahturin, A. A. Mikhalev, V. M. Petrogradsky and M. V. Zaicev. Now in preparation in CRC Press is the new book Combinatorial Aspects of Lie Superalgebras by A. A. Mikhalev and A. A. Zolotykh devoted to combinatorial aspects and algorithms in infinite dimensional Lie superalgebras (with a package of programs).

The article is organized as follows. In Section 1 we study linear structure of free Lie superalgebras (bases, Witt formulas, associative support, images of inner derivations, bases of free products with the amalgamated subalgebra, identical relations among normed monomials, Gröbner bases, shuffle products). In Section 2 we collect results on subalgebras of free Lie superalgebras and their properties. In Section 3 we use Fox differential calculus in free Lie superalgebras to study endomorphisms and automorphisms of free Lie superalgebras and orbits of elements under the action of automorphism groups of free Lie superalgebras. In Section 4 we introduce differential Lie superalgebras and formulate Poincaré–Birkhoff–Witt theorems for their universal enveloping algebras.

1. Linear structure of free Lie superalgebras

1.1. Colour Lie (p-) superalgebras

Let G be an abelian group, K a commutative associative ring with identity, char $K \neq 2$, $U(K)$ the group of invertible elements of K, $\varepsilon : G \times G \to U(K)$ a skew symmetric bilinear form,

$$\varepsilon(g_1 + g_2, h) = \varepsilon(g_1, h)\varepsilon(g_2, h), \quad \varepsilon(g, h_1 + h_2) = \varepsilon(g, h_1)\varepsilon(g, h_2),$$
$$\varepsilon(g, h)\varepsilon(h, g) = 1, \quad \varepsilon(g, g) = \pm 1$$

for all $g, g_1, g_2, h, h_1, h_2 \in G$,

$$G_- = \{g \in G \mid \varepsilon(g, g) = -1\}, \quad G_+ = \{g \in G \mid \varepsilon(g, g) = +1\}.$$

We say that a G-graded algebra $R = \bigoplus_{g \in G} R_g$ over K is *a colour Lie superalgebra* if

$$[x, y] = -\varepsilon(d(x), d(y))[y, x], \quad [u, u] = 0 \quad \text{with} \quad d(u) \in G_+,$$
$$[x, [y, z]] = [[x, y], z] + \varepsilon(d(x), d(y))[y, [x, z]], \quad [v, [v, v]] = 0$$

with $d(v) \in G_-$ for homogeneous elements $x, y, z, u, v \in R$ where $d(a) = g$ for $a \in R_g$.

Let char $K = p$, p is a prime number. We say that a colour Lie superalgebra R over

K is a colour Lie p-superalgebra if on homogeneous components $R_g, g \in G_+$, we have a mapping $x \to x^{[p]}, d(x^{[p]}) = pd(x)$, such that for all $z \in R, \alpha \in K, x, y \in R, d(x) = d(y) \in G_+$, the following conditions are satisfied:

$$(\alpha x)^{[p]} = \alpha^p x^{[p]}, \quad (\mathrm{ad}(x^{[p]}))(z) = [x^{[p]}, z] = (\mathrm{ad}\, x)^p(z),$$
$$(x+y)^{[p]} = x^{[p]} + y^{[p]} + \sum s_i(x, y),$$

where $is_i(x, y)$ is the coefficient on t^{i-1} in the polynomial $(\mathrm{ad}(tx+y))^{p-1}(x)$ (here $(\mathrm{ad}\, a)(b) = [a, b]$).

Let $R_+ = \bigoplus_{g \in G_+} R_g$, $R_- = \bigoplus_{g \in G_-} R_g$. For convenience we use the notation $\varepsilon(a, b)$ instead of $\varepsilon(d(a), d(b))$.

If Q is a G-graded associative algebra over K then $[Q]$ denotes the colour Lie superalgebra with the operation $[,]$ where $[a, b] = ab - \varepsilon(d(a), d(b))ba$ for homogeneous elements $a, b \in Q$.

If $\mathrm{char}\, K = p > 0$, p is a prime number, and $x^{[p]} = x^p$ for all $x \in Q$ with $d(x) \in G_+$, then $[Q]$ with the operation $[p]$ is a colour Lie p-superalgebra denoted by $[Q]^p$.

1.2. Bases of free colour Lie (p)-superalgebras

Let $X = \cup_{g \in G} X_g$ be a G-graded set, i.e., $X_g \cap X_f = \emptyset$ for $g \neq f$, $d(x) = g$ for $x \in X_g$, let also $\Gamma(X)$ be the groupoid of nonassociative monomials in the alphabet X, $u \circ v = (u)(v)$ for $u, v \in \Gamma(X)$, and $S(X)$ be the free semigroup of associative words with the bracket removing homomorphism $\bar{\ } : \Gamma(X) \to S(X)$; $[u]$ being an arrangement of brackets on the word $u \in S(X)$. For $u = x_1 \ldots x_n \in S(X)$, $x_i \in X$ we consider the word length $l(u) = n$, $d(u) = \sum_{i=1}^n d(x_i) \in G$. For $z \in \Gamma(X)$ we set $l(z) = l(\bar{z})$ and $d(z) = d(\bar{z})$.

Let K be a commutative associative ring with 1, $A(X)$ and $F(X)$ the free associative and nonassociative K-algebras, respectively, $S(X)_g = \{u \in S(X), d(u) = g\}$, $\Gamma(X)_g = \{v \in \Gamma(X), d(v) = g\}$, let $A(X)_g$ and $F(X)_g$ be the K-linear spans of the subsets $S(X)_g$ and $\Gamma(X)_g$ respectively, $A(X) = \oplus_{g \in G} A(X)_g$ and $F(X) = \oplus_{g \in G} F(X)_g$ the free G-graded associative and nonassociative algebras respectively. Let now I be the G-graded ideal of $F(X)$ generated by the homogeneous elements of the form

$$a \circ b + \varepsilon(a, b) b \circ a$$

and

$$(a \circ b) \circ c - a \circ (b \circ c) + \varepsilon(a, b) b \circ (a \circ c), \qquad e \circ e, \quad f \circ (f \circ f)$$

where $a, b, c, e, f \in \Gamma(X)$, $d(e) \in G_+$, $d(f) \in G_-$; then $L(X)$ is *the free colour Lie K-superalgebra* (i.e., every G-map φ of degree zero from X into any colour Lie K-superalgebra R with the same group G and form ε $(d(\varphi(x)) = d(x),\ x \in X)$ can be uniquely extended to a colour Lie superalgebra homomorphism $\bar{\varphi} : L(X) \to R$).

It is clear that this universal property defines the free colour Lie superalgebra $L(X)$ uniquely up to a colour Lie superalgebra isomorphism. For $u \in F(X)$ we set $\widetilde{u} = u + I \in L(X)$.

Suppose that the set $X = \cup_{g \in G} X_g$ is totally ordered and the set $S(X)$ is ordered lexicographically, i.e., for $u = x_1 \dots x_r$ and $v = y_1 \dots y_m$ where $x_i, y_j \in X$ we have $u < v$ if either $x_i = y_i$ for $i = 0, 1, \dots, t-1$ and $x_t < y_t$ or $x_i = y_i$ for $i = 1, 2, \dots, m$ and $r > m$. Then $S(X)$ is linearly ordered.

For the element $a \in A(X)$ we denote by $\hat{a}$ *the leading term of* a (i.e., the monomial with a coefficient which is the oldest firstly by the length and then lexicographically).

A word $u \in S(X)$ is said to be *regular* if $u \neq 1$ ann it follows from $u = ab$, $a, b \in S(X)$, $a, b \neq 1$, that $u = ab > ba$ (this condition is equivalent to the condition $u = ab > b$). A word $w \in S(X)$ is said to be s*-regular* if either w is a regular word, or $w = uu$ with u a regular word, $d(u) \in G_-$.

A monomial $u \in \Gamma(X)$ is said to be *regular* if either $u \in X$ or:

a) from $u = u_1 \circ u_2$ if follows that u_1, u_2 are regular monomials with $\bar{u}_1 > \bar{u}_2$;

b) from $u = (u_1 \circ u_2) \circ u_3$ it follows that $\bar{u}_2 \leq \bar{u}_3$.

Observe that if $x \in X$, v is a regular monomial and $x > \bar{v}$ then $x(v)$ is a regular monomial.

A monomial $u \in \Gamma(X)$ is said to be s*-regular* if either u is a regular monomial or $u = (v)(v)$ with v a regular monomial and $d(v) \in G_-$.

If u is an s-regular monomial, then $\bar{u}$ is an s-regular word. Conversely, if u is an s-regular word, then there is a unique arrangement of brackets $[u]$ on u such that $[u]$ is an s-regular monomial. Consider the subalgebra $[X]$ of $[A(X)]$ generated by X and the homomorphism $\pi : L(X) \to [X]$ such that $\pi(\tilde{x}) = x$ for all $x \in X$. Then for regular monomial $z \in L(X)$ we have $\widehat{\pi(z)} = \bar{z}$; for s-regular monomial $w = (z)(z)$, $d(z) \in G_-$, we have $\widehat{\pi(w)} = 2\bar{w}$. If u is a regular word, then considering the regular monomial $[u]$ we see that $u = (u_1)(u_2)$, where u_2 is the longest regular proper subword of u from the right side.

Theorem 1.1 ([Mi1]). *Let K be a commutative ring with 1; $2 \in U(K)$, and let $L(X) = F(X)/I$ be a free colour Lie superalgebra. Then the cosets in $F(X)/I$ whose representatives are s-regular monomials give us a basis of $L(X)$ as a free G-graded K-module. Moreover, $\pi : L(X) \to [X]$ is an isomorphism.*

Remark. The assertion of this theorem also takes place if instead of s-regular monomials we consider another arrangements of brackets $[u]'$ on regular words u such that $\widehat{[u]'} = u$ for each regular word u. It gives us an analog of Shirshov basis (in the case when K is a field and $L(X)$ is the free Lie superalgebra see also [Sht]). On the base of this theorem we have an algorithm to rewriting any element of $L(X)$ to the canonical form. In the same way one can construct Hall, Lazard–Viennot and other bases for free colour Lie superalgebras, for the details see [Me] and [DDKM] (this types of bases of free Lie algebra could be found for example in [Vi]). As in the case of free Lie algebras, it is also interesting to construct a basis of free Lie superalgebras which consists from left normed monomials (see also [BL]).

Corollary 1.1 (an analog of Witt's formula). *Let* $2 \in U(K)$, $X = \cup_{g\in G} X_g$, $X_+ = \{x_1, \dots, x_t\}$, $X_- = \{x_{t+1}, \dots, x_{t+s}\}$, *and let* $L(X)$ *be the free colour Lie* K*-superalgebra,* $\mu(l)$ *the Möbius function,* $W(\alpha_1, \dots, \alpha_k)$ *the rank of the free module of elements of multidegree* $\alpha = (\alpha_1, \dots, \alpha_k)$ *in the free Lie algebra of rank* k,

$$W(\alpha_1, \dots, \alpha_k) = \frac{1}{|\alpha|} \sum_{e|\alpha_i} \mu(e) \frac{(|\alpha|/e)!}{(\alpha_1/e)! \dots (\alpha_k/e)!},$$

where $|\alpha| = \sum_{i=1}^{k} \alpha_i$ *(Witt's formula). Let* $SW(\alpha_1, \dots, \alpha_{t+s})$ *be the rank of the free module of elements of multidegree* $\alpha = (\alpha_1, \dots, \alpha_{t+s})$ *in the free colour Lie superalgebra* $L(X)$ *of rank* $t+s$. *Then*

$$SW(\alpha_1, \dots, \alpha_{t+s}) = W(\alpha_1, \dots, \alpha_{t+s}) + \beta W(\frac{\alpha_1}{2}, \dots, \frac{\alpha_{t+s}}{2}),$$

where

$$\beta = \begin{cases} 0, & \text{if there exists an } i \text{ such that} \\ & \quad \alpha_i \text{ is odd, or if } \frac{1}{2}\sum_{i=t+1}^{t+s} \alpha_i \cdot d(\alpha_i) \in G_-; \\ 1, & \text{otherwise.} \end{cases}$$

Some other analogs of Witt's formula can be found in papers [Re2], [Bab], [Ka1] and [MT].

For $w \in L(X)$ we define *the length* $l_X(w)$ as the greatest length of s-regular monomials in the presentation of w as a linear combination of s-regular monomials. If $\pi(\widetilde{w}) = \sum_{i=1}^{k} \alpha_i u_i$ where $0 \neq \alpha_i \in K$, $u_i \in S(X)$, $u_l \neq u_j$ for $i \neq j$, then $l_X(w) = \max_{1\le i\le k}\{l_X(u_i)\}$. Therefore $l_X(w)$ does not depend on the ordering of X.

Let now w_0 denote *the leading part of* $w \in L(X)$ (i.e., the sum of summands whose length is equal to $l_X(w)$ in the s-regular presentation of w). If $w = w_0$ then we say that w is *l-homogeneous* . By analogy, we define the length $l_x(w)$, the leading part and property of being homogeneous relative to $x \in X$. We say that an element $w \in L(X)$ is *multihomogeneous* if w is homogeneous relative to all $x \in X$.

The cardinality $|X|$ is called *the rank* of the free colour Lie superalgebra $L(X)$ (notation: rank $L(X)$). It is easy to see that

$$\operatorname{rank} L(X) = \dim L(X)/[L(X), L(X)], \quad |X_g| = \dim(L(X)/[L(X), L(X)])_g,$$

and therefore rank $L(X)$ and $|X_g|$ do not depend on the choice of a set of free generators in $L(X)$.

Consider now bases of free colour Lie p-superalgebras. Let K be a field with char $K = p > 2$, $X = \bigcup_{g\in G} X_g$ a G-graded set and $A(X)$ the free G-graded associative algebra. We say that an element $w \in A(X)$ is an *as-regular monomial* relative to the alphabet X if $w = \pi(\tilde{u})$ where u is an s-regular monomial relative to X, and, respectively, a *ps-regular monomial*, if either w is an *as*-regular monomial or $w = \underbrace{(\dots((\pi(\tilde{v}))^p)^p \dots)^p}_{k \text{ times}} = (\pi(\tilde{v}))^{p^k}$, where $k \in \mathbb{N}$, and v is an s-regular monomial with $d(v) \in G_+$. By definition, *ps*-regular monomials are just the associative presentations of s-regular monomials and their p-th powers for even s-regular monomials.

Let $L^p(X)$ be the subalgebra of $[A(X)]^p$ generated by X. The set of ps-regular monomials give us a linear basis in the colour Lie p-superalgebra $L^p(X)$; $L^p(X)$ is a free colour Lie p-superalgebra with the set $X = \bigcup_{g \in G} X_g$ of free generators (i.e., X generates the algebra $L^p(X)$ and any mapping $\varphi : X \to R$ where R is any colour Lie p-superalgebra with the same G and ε, can be extended to a homomorphism $\bar{\varphi} : L^p(X) \to R$ of colour Lie p-superalgebras).

The number $\operatorname{rank} L^p(X) = |X|$ is called *the rank* of the free colour Lie p-superalgebra. It is clear that the rank of $L^p(X)$ does not depend on the choice of a set of free generators.

1.3. Associative support of the free colour Lie superalgebra

If $M \subset A(X)$ then under its associative support we understand the set of all associative words which have nonzero coefficients with elements from M. For a word $u = x_1 \dots x_n$ we set $\ddot{u} = x_n \dots x_1$. For a word $a \in S(X)$ we denote by $m(a)$ its multidegree.

Theorem 1.2 ([Mi14]). 1) *Let K be either a field,* char $K \neq 2$, *or* $K = \mathbb{Z}$, *L(X) a free colour Lie superalgebra over K,*

$$Y_1 = \{a^k \mid a \in S(X),\ k \geq 3 \text{ with } d(a) \in G_-,\ k \geq 2 \text{ otherwise}\},$$

$$Y_2 = \{a \mid a \in S(X),\ a = u\ddot{u} \text{ with } u \in S(X),\ d(u) \in G_+,$$
$$\text{or } u = v^l,\ v \in S(X),\ l \in \mathbb{N}\}.$$

The set $S(X) \setminus (Y_1 \cup Y_2)$ is the associative support of L(X).

2) *Let K be a field,* char $K = p > 2$, *and $L^p(X)$ a free colour Lie p-superalgebra over K. Suppose that Y_1 and Y_2 are as in* 1). *Let also*

$$Z_1 = \{a^{p^t} \mid a \in S(X) \setminus (Y_1 \cup Y_2),\ t \in \mathbb{N},\ d(a) \in G_+\};$$

$$Z_2 = \{u_1 \dots u_{(p^t-1)/2} v \ddot{u}_{(p^t-1)/2} \dots \ddot{u}_1 \mid u_i, v \in S(X) \setminus (Y_1 \cup Y_2),$$
$$m(u_i) = m(v)\ \forall i,\ v = \ddot{v}\}.$$

The set $(S(X) \setminus (Y_1 \cup Y_2)) \cup Z_1 \cup Z_2$ is the associative support of $L^p(X)$.

For free Lie algebras such description of the associative support was given by G. Duchamp and J. Thibon in [DT].

1.4. Images of inner derivations

We use the notation $\operatorname{Im}(\operatorname{ad} u) = \{a \in L^{(p)}(X) \mid a = [u, v], v \in L^{(p)}(X)\}$.

Theorem 1.3 ([Mi12]). *Let K be a commutative ring,* char $K \neq 2$, $G = G_+$, $L(X)$ *a free colour Lie algebra, $u \in L(X)$, $\alpha\hat{u} \in S(X)$ for some $\alpha \in U(K)$. Then the K-submodule* $\operatorname{Im}(\operatorname{ad} u)$ *is free and a direct summand in the free K-module $L(X)$.*

In the case when $K = \mathbb{Z}$ and $u \in X$ this theorem was obtained by D. Doković in [Do].

Theorem 1.4 ([Mi12]). *Let* $2 \in U(K)$, $L(X)$ *be the free colour Lie superalgebra over* K. *If a* G*-homogeneous element* u *belongs to* $L(X)$ ($u \in L^p(X)$), $\alpha\hat{u} \in S(X)$ *for some* $\alpha \in U(K)$, $\alpha\hat{u}$ *is a regular word, then the* K*-submodule* $\mathrm{Im}(\mathrm{ad}\, u)$ *is free and a direct summand of the free* K*-module* $L(X)$ (*of* $L^p(X)$ *respectively*).

Theorem 1.5 ([Mi12]). 1) *Let* $2 \in U(K)$, $L(X)$ *be the free colour Lie superalgebra over* K, $v \in L(X)$, $d(v) \in G_-$, $u = [v, v]$, $\alpha\hat{u} \in S(X)$ *for some* $\alpha \in U(K)$. *Then the* K*-module* $\mathrm{Im}(\mathrm{ad}\, u)$ *is free and a direct summand of the free* K*-module* $L(X)$.

2) *Let* $2 \in U(K)$, $\mathrm{char}\, K = p$, *and* $L^p(X)$ *be the free colour Lie* p*-superalgebra over* K. *If* $v \in L^p(X)$, $d(v) \in G_+$, $u = v^{p^k}$, $k \in \mathbb{N}$, $\alpha\hat{u} \in S(X)$ *for some* $\alpha \in U(K)$, *then the* K*-module* $\mathrm{Im}(\mathrm{ad}\, u)$ *is free and a direct summand of the free* K*-module* $L^p(X)$.

1.5. Bases of free metabelian colour Lie superalgebras

If in a colour Lie superalgebra L any two commutators commute, then we say that L is a *metabelian colour Lie superalgebra.*

The free metabelian colour Lie superalgebra $M(X)$ with the set X of free generators is the algebra $L(X)/L^{(2)}(X)$ where $L(X)$ is the free colour Lie superalgebra, $L^{(2)} = [L^{(1)}(X), L^{(1)}(X)]$, $L^{(1)}(X) = [L(X), L(X)]$.

Theorem 1.6 (Bahturin and Drensky [BD]). *Let* $X = \cup_{g \in G} X_g = X_+ \cup X_-$ *be a* G*-graded set,* $M(X)$ *the free metabelian colour Lie superalgebra in* X *over a field* K, $\mathrm{char}\, K \neq 2, 3$. *Then a basis in* $M(X)$ *is formed by all normed monomials of the form*

$$x_i \in X_+, \ y_j \in X_-,$$

$$[x_{i_1}, x_{i_2}, \ldots, x_{i_p}, y_{j_1}, \ldots, y_{j_q}], \ i_1 > i_2 \leq \cdots \leq i_p, \ j_1 < j_2 < \cdots < j_q, \ p > 1,$$

$$[y_{j_1}, y_{j_2}, \ldots, y_{j_q}], \ j_1 \geq j_2 < \cdots < j_q, \ q > 1,$$

$$[y_{j_1}, x_{i_1}, \ldots, x_{i_p}, y_{j_2}, \ldots, y_{j_q}], \ i_1 \leq \cdots \leq i_p, \ j_2 < \cdots j_q, \ p, q \geq 1.$$

Remark. It is interesting to construct bases for free solvable, free polynilpotent Lie superalgebras (for Lie algebras case see [Bo1]).

1.6. Free products with amalgamated subalgebra

Let K be a field, T a set, and H^0, $H_\alpha, \alpha \in T$, colour Lie K-superalgebras (with the same group G and skew symmetric bilinear form ε), H^0_α be G-homogeneous subalgebra of H_α, $\delta_\alpha : H^0 \to H^0_\alpha$ an isomorphism of colour Lie superalgebras. We say that a colour Lie K-superalgebra Q (with the group G and the form ε) is *the free* *product of colour Lie superalgebras* $H_\alpha, \alpha \in T$, *with amalgamated homogeneous subalgebra* H^0 (notation: $Q = \prod^*_{H^0, \alpha\in T} H_\alpha$) if the following conditions are satisfied:

1. there are homogeneous subalgebras H'_α in Q, isomorphisms $\sigma_\alpha : H_\alpha \to H'_\alpha$ of colour Lie superalgebras such that $\sigma_\alpha\delta_\alpha(h) = \sigma_\beta\delta_\beta(h)$ for all $\alpha, \beta \in T$ and $h \in H^0$;
2. if M is a colour Lie K-superalgebra (with the same group G and form ε), $\gamma_\alpha : H_\alpha \to M$ is a homomorphism of colour Lie superalgebras, $\gamma_\alpha\delta_\alpha(h) = \gamma_\beta\delta_\beta(h)$ for all $\alpha, \beta \in T$ and $h \in H^0$, then there is one and only one homomorphism $\psi : \prod^*_{H^0, \eta\in T} H_\eta \to M$ of colour Lie superalgebras such that $\psi\delta_\alpha = \gamma_\alpha$ for all $\alpha \in T$.

It is clear that the colour Lie superalgebra $\prod^*_{H^0, \alpha\in T} H_\alpha$ is defined uniquely (up to an isomorphism of colour Lie superalgebras).

Let $B^0 = \{f_\beta, \beta \in T^0\}$ be a G-homogeneous basis in H^0, $B^0_\alpha = \{e_{\alpha\beta}, e_{\alpha\beta} = \delta_\alpha(f_\beta), \alpha \in T, \beta \in T^0\}$, and let B_α be a homogeneous basis in H_α, $B^0_\alpha \subset B_\alpha$ and $B_\alpha = \{e_{\alpha\beta}, \beta \in T_\alpha\}$ for all $\alpha \in T$. Then the multiplication in H_α for all $\alpha \in T$ can be given using structure constants $p^\tau_{\alpha\beta\gamma} \in K : [e_{\alpha\beta}, e_{\alpha\gamma}] = \sum_\tau p^\tau_{\alpha\beta\gamma} e_{\alpha\tau}, \beta, \gamma \in T_\alpha$. For any $\alpha \in T$ consider a G-graded set $X_\alpha = \{x_{\alpha\beta}, \beta \in T_\alpha, d(x_{\alpha\beta}) = d(e_{\alpha\beta})\}$.

Let X be a set obtained from $\bigcup_{\alpha\in T} X_\alpha$ by the identification $x_{\alpha\gamma} = x_{\beta\gamma}$ for $\gamma \in T^0$. Consider T^0 with a total order and a total order extension of T^0 to T_α for all $\alpha \in T$ such that $\beta > \gamma$ for $\beta, \gamma \in T_\alpha, \beta \notin T^0, \gamma \in T^0$. Let J be an ideal of the free colour Lie superalgebra $L(X)$ generated by the following subset of G-homogeneous elements $D = \{d_{\alpha\beta\gamma}\} = [x_{\alpha\beta}, x_{\alpha\gamma}] - \sum_\tau p^\tau_{\alpha\beta\gamma} x_{\alpha\tau}, \alpha \in T, \beta, \gamma \in T_\alpha$, where $\beta > \gamma$ if $x_{\alpha\beta} \in X_+$ and $\beta \geq \gamma$ if $x_{\alpha\beta} \in X_-\}$.

It is easy to see that the quotient-algebra $L(X)/J$ is isomorphic to the algebra $\prod^*_{H_0, \alpha\in T} H_\alpha$.

Now we order X by the following way: for $x_{\alpha\beta} \neq x_{\gamma\delta}$ we set $x_{\alpha\beta} > x_{\gamma\delta}$ if $\alpha > \gamma$ or $\alpha = \gamma$ and $\beta > \delta$.

We say that a regular monomial $w \in \Gamma(X)$ is *special* if the word $\bar{w}$ does not contain subwords $x_{\alpha\beta}x_{\alpha\gamma}, \beta > \gamma$, and also subwords $x_{\alpha\beta}x_{\alpha\gamma}$, where $\beta \geq \gamma$ for $x_{\alpha\beta} \in X_-$. We say that a monomial $w \in \Gamma(X)$ is called *s-special* if either w is a special monomial or $w = [v, v]$ where v is a special monomial, $d(v) \in G_-, l_X(v) > 1$.

Theorem 1.7 ([Mi2]). *Let K be a field,* char $K \neq 2$. *Then the cosets whose representatives are s-special monomials form a linear basis of the algebra* $L(X)/J = \prod^*_{H^0, \alpha\in T} H_\alpha$.

In the case of Lie algebras a basis of free product with amalgamated subalgebra was constructed by A. I. Shirshov in [Shi4].

If, in the definition of the free product of colour Lie superalgebras with amalgamated subalgebra, in place of colour Lie superalgebras we consider colour Lie p-superalgebras then we have the definition of *the free product of colour Lie p-superalgebras with amalgamated homogeneous subalgebra.* It is clear that such product is defined uniquely (up to an isomorphism of colour Lie p-superalgebras). Let J be an ideal of the free colour Lie p-superalgebra $L^p(X)$ generated by elements $d_{\alpha\beta\gamma}$ and $h_{\alpha\beta} = x_{\alpha\beta}^p - \sum_\tau q_{\alpha\beta}^\tau x_{\alpha\tau}$ where $\alpha \in T$, $\beta, \tau \in T_\alpha$, $q_{\alpha\beta}^\tau \in K$, $x_{\alpha\beta} \in X_+$ and $e_{\alpha\beta}^p = \sum_\tau q_{\alpha\beta}^\tau e_{\alpha\tau}$ in H_α. It is clear that the algebra $L^p(X)/J$ is the free product of colour Lie p-superalgebras H_α with amalgamated homogeneous subalgebra H^0.

We say that a regular monomial $w \in \Gamma(X)$ is *p-special* if the word $\bar{w}$ does not contain subwords of the form $x_{\alpha\beta}x_{\alpha\gamma}$, $\beta > \gamma$; $x_{\alpha\beta}x_{\alpha\gamma}$, $\beta \geq \gamma$ with $x_{\alpha\beta} \in X_-$; or $x_{\alpha\beta}^p$ with $x_{\alpha\beta} \in X_+$. A monomial $w \in \Gamma(X)$ is called *ps-special* if either w is a p-special monomial or $w = [u, u]^{p^r}$ where u is a p-special monomial, $d(u) \in G_-$, $l_X(u) > 1$, $r \in \mathbb{N}$, or $w = v^{p^t}$ where $t \in \mathbb{N}$, v is a p-special monomial, $d(v) \in G_+$, $l_X(v) > 1$.

Theorem 1.8 ([Mi5], [BMPZ]). *Let K be a field,* char $K = p > 2$. *Then the cosets, whose representatives are ps-special monomials, form a linear basis of $L^p(X)/J$.*

In the case of Lie p-algebras a basis of free product with amalgamated subalgebra was constructed by G. P. Kukin in [Ku4].

1.7. Identical relations among left normed monomials

Let δ, $\delta' : A(X) \to L(X)$ be linear mappings such that for all $x_1, \ldots, x_n \in X$ $\delta(x_1 \ldots x_n) = [x_1 \ldots x_n]$ is the left normed and $\delta'(x_1 \ldots x_n) = [x_1 \ldots x_n]'$ is the right normed monomial respectively (δ and δ' are derivations of the algebra $L(X)$). It is clear that any element of $L(X)$ can be written as a linear combination of left normed (right normed) monomials.

Theorem 1.9 ([Mi8]). *Let either K be a field,* char $K \neq 2$, *or $K = \mathbb{Z}$, $a \in A(X)$. Then*

1) $\delta(a) = 0$ *if and only if* $a = \sum(a_i\delta(b_i) + \varepsilon(d(a_i), d(b_i)b_i\delta(a_i))v_i$;

2) $\delta'(a) = 0$ *if and only if* $a = \sum w_i(\delta'(c_i)e_i + \varepsilon(d(c_i), d(e_i)\delta'(e_i)c_i)$,

where $i \in \mathbb{N}$, a_i, b_i, c_i, e_i are G-homogeneous elements from $A(X)$, $v_i, w_i \in A(X)$.

For Lie algebras such criterion was obtained by P. M. Cohn in [Co1].

1.8. Canonical bases in Lie superalgebras

In all what follows in this subsection K is a field, char $K \neq 2$.

Consider the weight function $\varphi : X \to \mathbb{N}$, let $\varphi(x_1 \dots x_n) = \sum_{i=1}^{n} \varphi(x_i)$ for $x_1, \dots, x_n \in X$ and let $\hat{a}$ be the leading term in $a \in A(X)$ first relative to the weight φ and then lexicographically.

Lemma 1.1 ([Mi4], [Mi9]). *Let u, v be regular words, $a, b \in S(X)$.*

1. *If $u = avb$ then on u there is an arrangement of brackets $[u] = [a[v]b]$ such that $[v]$ is a regular monomial, $\widehat{[u]} = u$ in $L(X)$.*
2. *Let $u = avvb, d(v) \in G_-, a, b \in S(X)$. Then on u there is an arrangement of brackets $[u]$ such that $[u] = [a[v, v]b]$ where $[v, v]$ is an s-regular monomial, $\widehat{[u]} = 2u$ in $L(X)$.*
3. *Let $u = av^{p^t}b, d(v) \in G_+$, then there exists $[u] = [a[v]^{p^t}b]$ where $[v]$ is a regular monomial, $\widehat{[u]} = u$ in $L^p(X)$.*
4. *If $u = a(vv)^{p^t}b, d(v) \in G_-$, then there exists $[u] = [a[v, v]^{p^t}b]$ where $[v, v]$ is an s-regular monomial, $\widehat{[u]} = 2^{p^t}u$ in $L^p(X)$.*
5. *If $u \neq v, k, l \in \mathbb{N}$, and the word v^k is a subword of the word u^l, then the word v^k is a subword of the word u.*

Theorem 1.10 ([Mi6]). *Any Lie superalgebra (Lie p-superalgebra) over a field with countable set of generators can be isomorphically embedded in a Lie superalgebra (in a Lie p-superalgebra) with two generators over the same field.*

For Lie algebras this statement was proved by A. I. Shirshov in [Shi2] (in the case of solvable Lie algebras see [Ku5]).

Let a, b be some G-homogeneous elements in $L(X)$ (in $L^p(X)$). Assume that their leading terms a_0 and b_0 in the $(p)s$-regular presentation (first relative to the weight φ and then lexicographically) have coefficients which are equal to 1 and $\bar{a_0} = e_1e_2, \bar{b_0} = e_2e_3$ where $e_1, e_2, e_3 \in S(X)$. Then the word $e_1e_2e_3$ is regular. Now we define the element $(a, b) \in L(X)$ which is called *the composition for elements $a, b \in L(X)$ (relative to a subword e_2)*:

1. in the case where $a_0 \neq b_0$ we set $(a, b) = [(a)e_3] - \alpha[e_1(b)]$ where $[(a)e_3]$ is obtained from the monomial $[[a_0]e_3]$ with the arrangement of brackets described in Lemma 1.1 by replacing a_0 by a and analogously $[e_1(b)]$ is obtained from $[e_1[b_0]]$ by replacing b_0 by b; and α is such that $\widehat{[(a)e_3]} = \alpha\widehat{[e_1(b)]}$;
2. for $a = b, \bar{a_0} = uu$, where u is a regular word and $d(u) \in G_-$, we set $(a, a) = [a, [u]] = [a - [u, u], [u]]$ where $[u]$ is a regular monomial;
3. for $a = b, \bar{a_0} = u^{p^t}$, where u is a regular word with $d(u) \in G_+$, we set $(a, a) = [a, [u]]$ where $[u]$ is a regular monomial;

4. for $a = b, \bar{a_0} = (vv)^{p^t}$, where v is a regular word with $d(u) \in G_-$, we set $(a, a) = [a, [v]]$ where $[v]$ is a regular monomial.

Let $\bar{a_0} = e_1\bar{b_0}e_2, e_1, e_2 \in S(X)$. We define the element $\langle a, b\rangle \in L(X)(L^p(X))$ which is called *the composition relatively subwords for elements* a, b: we set $\langle a, b\rangle = a - \alpha[e_1(b)e_2]$ where $[e_1(b)e_2]$ is obtained from the monomial $[e_1[b_0]e_2]$ with the arrangements of brackets described in Lemma 1.1 by replacing b_0 by b, and $\alpha \in K$ is such that $\hat{a} = \alpha[\widehat{e_1(b)e_2}]$.

A G-homogeneous subset $S \subset L(X)$ is called *stable* if the following conditions are satisfied:

a) the coefficients of the leading monomials in s-regular presentation for all elements from S are equal to 1;

b) all words $\hat{s}$, where $s \in S$, are distinct;

c) if $a, b \in S$ and the composition (a, b) does exist then either $(a, b) \in S$ (up to a coefficient from K) or after expressing the elements associatively we have

$$(a, b) = \sum \alpha_i A_i s_i B_i,$$

where $\alpha_i \in K, s_i \in S, A_i, B_i \in S(X)$,

$$\alpha_1 A_1 \hat{s}_1 B_1 = \widehat{(a, b)}, \quad A_i \hat{s}_i B_i < A_1 \hat{s}_1 B_1 (i \neq 1) \quad \text{for all } i;$$

d) if for $a, b \in S$ the composition relatively subwords $\langle a, b\rangle$ does exists then the element $\langle a, b\rangle$ satisfies the same conditions as (a, b) in c).

Theorem 1.11 (Composition Lemma, [Mi4], [Mi9] and [Mi16]). *Let K be a field,* char $K \neq 2$, *$L(X)$ a free colour Lie superalgebra (in the case when* char $K = p$, *$L^p(X)$ a free colour Lie p-superalgebra), S a stable set,* id(S) *an ideal in $L(X)$ (in $L^p(X)$) generated by S. Then for any element $a \in$* id(S) *the word $\hat{a}$ contains a subword $\hat{b}$, $b \in S$. Furthermore, $(p)s$-regular monomials u such that $\hat{u}$ does not contain subwords $\hat{b}$, $b \in S$, form a linear basis of $L(X)/$*id(S) *(of $L^p(X)/$*id(S)*).*

As corollaries one can derive a number of algorithms to solve algorithmic problems in Lie superalgebras, for example, an algorithm for solving the equality problem in colour Lie p-superalgebras with stable or φ-homogeneous finite sets of defining relations (see [BMPZ]). Using the Composition Lemma it is possible to introduce Gröbner bases for colour Lie superalgebras and for colour Lie p-superalgebras. For any ideal I of the free colour Lie (p)-superalgebra there exist a stable set of defining relations. We say that such set is a Gröbner basis of I. Moreover, there exist the unique (up to coefficients) minimal Gröbner basis of I which can be called the reduced Gröbner basis of I. In the recent article [Bo5] L. A. Bokut' proved that a subset S of the free Lie algebra $L(X)$ is stable if and only if it is a stable subset in the free associative algebra $A(X)$, for necessary definitions see [Bo3], [Berg], [Lat]. Note that it is possible also to prove this fact using H-presentations (see Theorem 2.11, point 2, page 22 of [Lat]). The same assertion takes place for free colour lie superalgebras and for free colour Lie p-superalgebras.

1.9. Shuffle algebras

Let V and Z be G-graded sets, $V \cap Z = \emptyset$, $v_1, \ldots, v_k$ pairwise distinct elements of $V, z_1, \ldots, z_l$ pairwise distinct elements of Z. We say that a word $w \in S(V \cup Z) \setminus 1$ is a shuffle word of the words $v_1 \ldots v_k$ and $z_1 \ldots z_l$ if w has the multidegree $(1)m(w) = v_1 + \cdots + v_k + z_1 \ldots z_l$ and

$$w|_{z_1=1,\ldots,z_l=1} = v_1 \ldots v_k; \quad w|_{v_1=1,\ldots,v_k=1} = z_1 \ldots z_l.$$

The parity $\sigma(w)$ of a shuffle word w of the words $v_1 \ldots v_k$ and $z_1 \ldots z_l$ is the sum of all $\varepsilon(z_i, v_j)$ such that z_i is situated before v_j in w.

Let X be a G-graded set. For $x_{i_1}, \ldots, x_{i_k}, x_{j_1}, \ldots, x_{j_l} \in X$ we define the shuffle product $(x_{i_1} \ldots x_{i_k}) * (x_{j_1} \ldots x_{j_l})$ as the following linear combination of words of multidegree $x_{i_1} + \cdots + x_{i_k} + x_{j_1} \cdots + x_{j_l}$:

$$(x_{i_1} \ldots x_{i_k}) * (x_{j_1} \ldots x_{j_l}) = \sum_w \sigma(w) w|_{v_s = x_{i_s}, s=1,\ldots,k; z_t = x_{j_t}, t=1,\ldots,l}$$

where w is running through all shuffle words of the words $v_1 \ldots v_k$ and $z_1 \ldots z_l$ with $d(v_s) = d(x_{i_s}), d(z_t) = d(x_{j_t})$. Taking $1 * u = u * 1 = u$ for all $u \in S(X) \setminus 1$ and extending $*$ on the free G-graded associative algebra $A(X)$ on X over a commutative associative ring K with the identity element by linearity, we define the shuffle product $*$ on $A(X)$. Then $A(X)$ with this product is an ε-commutative and associative algebra.

One can define the shuffle product $*$ on $A(X)$ in the following way:

$$1 * u = u * 1 = u; \qquad (xu) * (yv) = x(u * (yv)) + \varepsilon(y, x)\varepsilon(y, u)y((xu) * v)$$

for all $x, y \in X$, $u, v \in S(X) \setminus 1$ (with the extension $*$ on $A(X)$ by linearity).

If we consider $A(X)$ as the universal enveloping algebra of the free colour Lie superalgebra $L(X)$, then $*$ is the adjoint of coproduct δ of $A(X)$.

D. Radford in [Ra] proved that in the case of trivial grading group the free associative algebra as a shuffle algebra is the free commutative associative algebra with a set of free generators consisting of Lyndon words. In the general case we have the following theorem.

Theorem 1.12 ([MZ6]). *If K is a Q-algebra, then $A(X)$ with the shuffle product $*$ is the free ε-commutative algebra with a set of free generators consisting of s-regular words.*

For applications of shuffle algebras to free Lie algebras see for example [Reu2].

2. Subalgebras of free Lie superalgebras

2.1. The freeness of subalgebras

We say that a subset M of G-homogeneous elements of $L(X)$ is *independent* if M is a set of free generators of the subalgebra, generated by M in $L(X)$.

A subset $M = \{a_i\}$ of G-homogeneous elements in $L(X)$ is called *reduced* if for any i the leading part a_i^0 of the element a_i does not belong to the subalgebra generated by the set $\{a_j^0, j \neq i\}$.

Let $S = \{s_\alpha, \alpha \in I\}$ be a G-homogeneous subset of $L(X)$. We say that a mapping $\omega\ S \to L(X)$ is *an elementary transformation* if either $\omega(s_\alpha) = s_\alpha$ for all $\alpha \in I \setminus \beta$, $\omega(s_\beta) = \lambda s_\beta + \omega(s_{\alpha_1}, \ldots, s_{\alpha_t})$, where

$$0 \neq \lambda \in K,\ \alpha_1, \ldots, \alpha_t \neq \beta,\ d(\omega(s_{\alpha_1}, \ldots, s_{\alpha_t})) = d(s_\beta),$$

$$\omega(y_{\alpha_1}, \ldots, y_{\alpha_t}) \in L(y_1, y_2, \ldots),\ d(y_i) = d(s_i),$$

or elements of S and $\omega(S)$ are linear combinations of each other.

Such elementary transformation is called *triangular* transformation if $l_X(\omega(s_{\alpha_1}, \ldots, s_{\alpha_t})) \leq l_X(s_\beta)$. It is clear that elementary transformations of the set X of free generators induce elementary automorphisms of the free colour Lie superalgebra $L(X)$.

Theorem 2.1 ([Mi1]). *Let K be a field,* char $K \neq 2$. *Then any reduced subset of a free colour Lie superalgebra $L(X)$ is independent. Thus any G-homogeneous subalgebra of $L(X)$ is free.*

This theorem is rather useful for solving algorithmic problems. For example, as a corollary we have an algorithm to decide if an element of $L(X)$ is in a given finite generated subalgebra, or not.

Theorem 2.2 ([Mi8]). *Let K be a commutative domain over which all projective modules are free,* char $K \neq 2$, *B an isolated subalgebra of the free colour Lie (p-)superalgebra $L = L(X)$ ($L = L^p(X)$) (i.e., $\{a \in L \mid \exists\alpha \in K, \alpha \neq 0, \alpha a \in B\} = B$) which is a direct summand of the K-module L. Then B is a free colour Lie (p-)superalgebra.*

Now we pass to the corollaries of the theorem about the freeness of subalgebras.

Theorem 2.3. (an analog of P. M. Cohn's theorem on generators of the automorphism group of a free Lie algebra of finite rank, [Mi5], [BMPZ]). *The set of elementary automorphisms generates the automorphism group of a free colour Lie superalgebra of finite rank.*

From these theorems we derive an algorithm to decide whether some finite set of G-homogeneous elements of a free colour Lie superalgebra is an independent set (it is necessary to transform these elements using triangular elementary transformations to a reduced set) and algorithm to solve the entry problem for finite generated subalgebras of free colour Lie superalgebra, see [BMPZ].

Generators of the derived algebra. *Let X be a G-graded set, $|X| \geq 2$. Suppose that X is totally ordered in such way that from $x \in X_+$, $y \in X_-$ it follows that $y > x$. The derived algebra $L^2(X) = [L(X), L(X)]$ is a free colour Lie superalgebra, and, as a free set of generators, we can choose left-normed monomials*

$$[x_1x_2 \ldots x_{s-1}x_s] = [[x_1x_2 \ldots x_{s-1}], x_s]$$

where $x_i \in X, s \geq 2, x_1 > x_2 \leq x_3 \leq \cdots \leq x_s$ and $x_i \neq x_{i+1}$ for $x_i \in X_-$, and also if $x_1, \ldots, x_s \in X_-$ then $x_1 \geq x_2 < x_3 < \cdots < x_s$.

Now we pass to subalgebras of the free colour Lie p-superalgebra $L^p(X)$. As in the case of $L(X)$, we have

Theorem 2.4 ([Mi3]). *Let K be a field, $p = \operatorname{char} K > 2$, $L^p(X)$ the free colour Lie p-superalgebra. Then any reduced subset of $L^p(X)$ is independent. Thus any G-homogeneous subalgebra H of $L^p(X)$ is a free colour Lie p-superalgebra.*

As in the case of $L(X)$, the set of elementary automorphisms generates the automorphism group of a free colour Lie p-superalgebra of finite rank.

The following theorem shows that any subalgebra of finite rank of $L(X)$ is completely defined by its (nontrivial) even component.

Theorem 2.5 ([Mi5]). *Let K be a field, $\operatorname{char} K \neq 2$, $|X_-| \geq 1$, and let A, B be G-homogeneous subalgebras of finite rank in the free colour Lie superalgebra $L(X)$, $A = A_+ \bigoplus A_-$, $B = B_+ \bigoplus B_-$. If $A_- = B_- \neq \{0\}$, then $A = B$.*

Now we give some examples concerning this theorem.

1. Let $X_+ = \{x\}$, $X_- = \{y\}$, $A = L(X)$, $B = L(Z(x))$ where $Z(x) = \{y, \operatorname{ad}^k x(y) \mid k \in \mathbb{N}\}$. Then since $x \in A_+$ and $x \notin B_+$, we see that $A_+ \neq B_+$; but at the same time $A_- = B_-$ (here rank $B = \infty$).
2. In general, from $A_+ = B_+$ it does not follow that $A = B$. Indeed, let $X_+ = \{x\}$, $X_- = \{y\}$, $A = L(X)$, $B = L(W(y))$, where $W(y) = \{x, [x, y], [y, y]\}$. Then rank A, rank $B < \infty$ and $A_+ = B_+$. Since $y \in A_-$ and $y \notin B_-$, we have at the same time $A_- \neq B_-$.

At the end of this subsection we remark that the algebra $L = L(X)(L = L^p(X))$ is finitely separable:

Theorem 2.6 ([Mi11]). *If a be a G-homogeneous element of L, B a G-homogeneous finitely generated subalgebra of L, $a \notin B$, then there exist a finite dimensional colour Lie $(p$-$)$superalgebra H and an epimorphism $\varphi : L \to H$ such that $\varphi(a) \notin \varphi(B)$.*

In the case when L is a free Lie algebra this result is due to U. U. Umirbaev [Um1].

2.2. The cardinality of ranks of subalgebras

Theorem 2.7 (an analog of O. Schreier's formula, [Mi1]). *Let K be a field,* $\operatorname{char} K \neq 2$, $L = L(X) = L_+ \bigoplus L_-$ *the free colour Lie superalgebra,* $\operatorname{rank} L = |X| = N < \infty$, *and let M be a G-homogeneous subalgebra,* $M = M_+ \bigoplus M_-$, $M_+ = L_+$, $\dim L_-/M_- = s < \infty$. *Then* $\operatorname{rank} M = 2^s(N-1)+1$.

Corollary 2.1. *Let K be a field,* $\operatorname{char} K \neq 2, 3$, $L = L_+ \bigoplus L_-$ *the finitely generated colour Lie superalgebra, and let* $M = M_+ \bigoplus M_-$ *be a G-homogeneous subalgebra of L such that* $M_+ = L_+$ *and* $\dim L_-/M_- < \infty$. *Then M is a finitely generated colour Lie superalgebra.*

Theorem 2.8 ([Mi5], [BMPZ]). *Let* $|X_-| \geq 1$, $L(X) = L_+ \bigoplus L_-$ *be a free colour Lie superalgebra,* $M = M_+ \bigoplus M_-$ *a G-homogeneous subalgebra in $L(X)$ such that* $M_- = L_-$ *and* $\dim L_+/M_+ = s < \infty$. *Then* $\operatorname{rank} M = \infty$, $M = L(Y)$ *and* $|Y_-| = \infty$.

Proposition 2.1. *Any nonzero ideal* $H = H_+ \bigoplus H_-$ *of the free colour Lie superalgebra* $L(X) = L_+ \bigoplus L_-$, $|X| > 1$, *such that* $H \neq 0$ *and* $H_+ \neq L_+$, *is a free colour Lie superalgebra of infinite rank.*

Theorem 2.9 ([Mi3]). *Let K be a field,* $p = \operatorname{char} K > 2$, $|X| = N < \infty$. *Let also* $H = H_+ \bigoplus H_-$ *be a G-homogeneous subalgebra of the free colour Lie p-superalgebra* $L^p = L^p(X)$. *Suppose that* $\dim L^p_+/H_+ = t < \infty$ *and* $\dim L^p_-/H_- = s < \infty$. *Then* $\operatorname{rank} H = 2^s p^t (N-1)+1$.

Now we consider the intersection of two finitely generated subalgebras of $L(X)$ (of $L^p(X)$) over the field K with $\operatorname{char} K > 2$.

Theorem 2.10 ([Mi3]). *Let K be a field,* $p = \operatorname{char} K > 2$, $L^p(X)$ *the free colour Lie p-superalgebra, and let A and B be some G-homogeneous finitely generated subalgebras of $L^p(X)$. Then the subalgebra $A \cap B$ is also finitely generated.*

Theorem 2.11 ([Mi3]). *Let K be a field,* $p = \operatorname{char} K > 2$, $L(X)$ *a free colour Lie superalgebra, and A and B be some G-homogeneous finitely generated subalgebras of $L(X)$. Then their intersection $A \cap B$ is also finitely generated.*

Now we consider the action of a finite group of linear automorphisms on the free colour Lie superalgebra $L(X)$. In the Lie algebra case this problem was considered by R. Bryant in [Br]. Let V_g be the free K-module spanned by the set X_g, $GL(V_g)$ the group of all automorphisms of the K-module V_g, $F = \prod_{g \in G} GL(V_g)$, H a finite subgroup in F. Then $L(X)$ is a KH-module, moreover, any element of H acts as an automorphism of the algebra $L(X)$. Denote by L^H the subalgebra of all fixed points in $L(X)$ relatively the action of the group H.

Theorem 2.12 ([Mi10]). *Let K be a principal ideal domain,* char $K \neq 2, 2 \leq |X| < \infty$ *and the group H be finite and nontrivial. Then L^H is a free colour Lie superalgebra of infinite rank.*

As a corollary we have that in the case where $|X| > 1, |X_-| \geq 1$, the even component L_+ of $L(X) = L_+ \bigoplus L_-$ has infinite rank; in [Mi5], [BMPZ] it is possible to find a construction of a set of free generators of L_+.

3. Fox differential calculus in free Lie superalgebras

In all what follows in this section K is a field, char $K \neq 2$.

3.1. Jacobian matrices and automorphisms

For any element a of the free associative algebra $A(X)$, $X = \{x_1, \dots, x_n\}$, we have the unique presentation in the form $a = x_1a_1 + x_2a_2 + \cdots + x_na_n + \alpha \cdot 1$, where $a_i \in A(X), \alpha \in K$. We call the element a_i the right Fox partial derivative of the element a and denote $a_i = \partial a/\partial x_i, i = 1, \dots, n$.

Theorem 3.1 ([Mi11]). *Let $X = \{x_1, \dots, x_n\}$, $L(X)$ be the free colour Lie superalgebra (in the case when* char $K = p$ *$L^p(X)$ the free colour Lie p-superalgebra). Let also $f_1, \dots, f_n$ be G-homogeneous elements of $L(X)$ (of $L^p(X)$), $d(f_i) = d(x_i), i = 1, \dots, n$. Then the endomorphism $\varphi : L(X) \to L(X)$ ($\varphi : L^p(X) \to L^p(X)$) where $\varphi(x_i) = f_i, i = 1, \dots, n$, is an automorphism if and only if the matrix $(\partial f_i/\partial x_j), 1 \leq i, j \leq n$, is invertible over $A(X)$.*

3.2. The rank and primitivity of elements

We say that an element $a \in A(X)$ *is depending on* $x_i \in X$ if in the associative presentation of a we have with nonzero coefficient a monomial which contains x_i.

We say that an element a of the free colour Lie superalgebra $L(X)$ has *the rank* k if k is the least number of free generators from $X = \{x_1, \dots, x_n\}$ on which $\varphi(a)$ depends on (where φ is an automorphism of $L(X)$). We denote $k = \text{rank}(a)$. By analogy we define the rank for an element of the free colour Lie p-superalgebra $L^p(X)$.

The free associative algebra $A(X)$ over a field K is a free ideal ring (P. M. Cohn's theorem, see his book [Co3]), therefore each left ideal of $A(X)$ is a free $A(X)$-module of the unique rank.

Theorem 3.2 (Mikhalev and Zolotykh [MZ2], [MZ4]). 1. *Let* char $K = 0$, $h \in L(X)$. *Then* rank(h) *is equal to the rank of the left ideal of $A(X)$ generated by elements $\{\partial h/\partial x, x \in X\}$ (as a free left $A(X)$-module).*

2. *In the case when* char $K = p > 2$ *the assertion* 1 *takes place for an element* h *of* $L^p(X)$.

As a corollary, we have an algorithm for finding the rank of an element h which is reduced to the construction of finite Gröbner basis of the left ideal of $A(X)$.

We say that an element $h \in L(X)$ ($h \in L^p(X)$) is *primitive* if it can be embedded in a set of free generators of $L(X)$ (respectively, of $L^p(X)$).

Theorem 3.3 (Mikhalev and Zolotykh [MZ2], [MZ4]). 1. *Let* char $K = 0$, $h \in L(X)$. *Then the following conditions are equivalent:*

a) *the element h is primitive;*

b) *there are elements* $m_1, \ldots, m_n \in A(X)$ *such that*

$$\sum_{i=1}^{n} m_i(\partial h/\partial x_i) = 1.$$

2. *In the case when* char $K = p > 2$ *the assertion* 1 *takes place for an element* $h \in L^p(X)$.

Remark. The assertions of these theorems do not take place in the case when char $K = p > 2$ and $L(X)$ is the free Lie algebra. Indeed, let $X = \{x, y, z\}$, $h = x + [y, z] + (\operatorname{ad} x)^p(z)$. Then h is not a primitive element but the rank of above mentioned left ideal is equal to 1 (h is a primitive element of $L^p(X)$, but not of $L(X)$).

Using these theorems one can define the rank of a system of elements and give a criterion for a system of elements to bee primitive (see [MZ2], [MZ4]). In [Ku1] G. P. Kukin gave a criterion for an element of the free Lie algebra L to be primitive (a is a primitive element of L if and only if $L/\operatorname{id}(a)$ is a free Lie algebra where $\operatorname{id}(a)$ is the ideal of L generated by a). In [Shp3] V. Shpilrain gave another description of the rank of a homogeneous (relative to the length) element of the free Lie algebra.

Theorem 3.4 (Mikhalev, Umirbaev and Zolotykh [MUZ1]). *Let K be a field with* char $K = p > 2$, $X = \{x, y, z\}$, $L(X)$ *the free Lie algebra over* K,

$$h = x + [y, z] + (\operatorname{ad} x)^p(z),$$

$\operatorname{id}(h)$ *the ideal of* $L(X)$ *generated by* h. *Then the algebra* $L(X)/\operatorname{id}(h)$ *is not a free Lie algebra but the cohomology dimension of* $L(X)/\operatorname{id}(h)$ *is equal to one.*

Remark. The algebra $L(X)/\operatorname{id}(h)$ is not a free Lie algebra, but its universal enveloping algebra is the free associative algebra of rank two. Note that G. L. Feldman in [Fel] proved that any two generated Lie algebra of cohomological dimension one is free.

3.3. Automorphisms and primitivity of elements

Theorem 3.5 (Mikhalev and Zolotykh [MZ1], [MZ5]). 1. *Let K be a field,* char $K = 0$, $|X| < \infty$, *and let φ be an endomorphism of the free colour Lie superalgebra $L(X)$ such that for any primitive element a of $L(X)$ the element $\varphi(a)$ is also a primitive element of $L(X)$. Then φ is an automorphism of $L(X)$.*

2. *In the case when* char $K = p > 2$ *the assertion* 1 *takes place for the free colour Lie $(p$-$)$superalgebra $L^p(X)$.*

Remark. This theorem gives a positive answer for a question of V. Shpilrain [Shp2] in the case of free colour Lie $(p$-$)$superalgebras.

4. Differential Lie superalgebras and their universal enveloping algebras

4.1. Differential Lie superalgebras

Let G be an abelian group, $R = \bigoplus_{g\in G} R_g$ a G-graded associative ring with 1, $C(R)$ the center of the ring R, Φ a subring in R_0, $1 \in \Phi$, $\Phi \subset C(R)$, $U(\Phi)$ the group of invertible elements in Φ, $\varepsilon : G \times G \to U(\Phi)$ a skew symmetric bilinear form. We say that Φ-linear map $\mu : R \to R$, $\mu(\Phi) = 0$, is *ε-derivation of weight $d(\mu) \in G$* if $d(\mu(x)) = d(\mu) + d(x)$, $\mu(xy) = \mu(x)y + \varepsilon(\mu, x)x\mu(y)$ for all homogeneous elements $x, y \in R$. Denote the Φ-linear span of all ε-derivations of R by Der R. Then Der R is a G-graded associative Φ-algebra, [Der R] is a colour Lie superalgebra; in the case of prime characteristic $[\mathrm{Der}\, R]^p$ is a colour Lie p-superalgebra over Φ.

Let $C = Z(R)$ be the colour center of R, i.e., a G-graded subring of R such that it is a Φ-linear span of such G-homogeneous elements $x \in R_+$ that $xa = \varepsilon(x, a)ax$ for all homogeneous $a \in R$. It is clear that $\Phi \subset C_0$. Then Der R is a left C-module: for $c \in C$, $\mu \in \mathrm{Der}\, R$, $x \in R$ we denote $(c\mu)(x) = c\mu(x)$. The following identities are fulfilled in the C-module Der R:

$$[c\mu, \nu] = c[\mu, \nu] - \varepsilon(c, \nu)\varepsilon(\mu, \nu)\nu(c)\mu$$

for all homogeneous elements $c \in C$, $\mu, \nu \in \mathrm{Der}\, R$; in the case of char $R = p$, p is a prime number,

$$(c\delta)^p = T_{p,\delta}(c) + (\varepsilon(\delta, c))^{p(p-1)/2} c^p \delta^p$$

$$\text{where} \quad T_{1,\delta}(c) = c, \; T_{k+1,\delta}(c) = c\delta(T_{k,\delta}(c))$$

for all homogeneous $c \in C$, $\delta \in (\mathrm{Der}\, R)_+$.

Let $R = \bigoplus_{g\in G} R_g$ be a colour Lie $(p$-$)$superalgebra over Φ, $F = \bigoplus_{g\in G} F_g$ a G-graded supercommutative ring with 1 (i.e., $F = F_+, ab = \varepsilon(a, b)ba$ for all homogeneous elements $a, b \in F$), $\Phi \subset F_0$, Der F the colour Lie (p)-superalgebra over Φ of all ε-derivations of F (Der $F|_\Phi = 0$).

The algebra R is called *a (restricted) differential Lie F-superalgebra* if

a) R is a G-graded left F-module;

b) there is a homomorphism$\tilde{}$: $R \to$ Der F of colour Lie (p-)superalgebras over Φ such that $d(\tilde{x}) = d(x)$ for all homogeneous elements $x \in R$,$\tilde{}$is a homomorphism of left F-modules and the following identities are fulfilled:

$$[c\mu, \nu] = c[\mu, \nu] - \varepsilon(c, \nu)\varepsilon(\mu, \nu)\tilde{\nu}(c)\mu$$

for all homogeneous elements $c \in F$, $\mu, \nu \in R$;

in the case when char $R = p$, p is a prime number,

$$(c\delta)^p = T_{p,\delta}(c) + (\varepsilon(\delta, c))^{p(p-1)/2}c^p\delta^p$$

where

$$T_{1,\delta}(c) = c, \quad T_{k+1,\delta}(c) = c\tilde{\delta}(T_{k,\delta}(c))$$

for all homogeneous $c \in F$, $\delta \in R_+$.

4.2. Universal enveloping algebras of differential Lie superalgebras and Poincaré–Birkhoff–Witt theorem

Let A be a G-graded associative algebra with 1 over commutative ring Φ with 1. Suppose that A is a faithful left module over G-graded supercommutative ring F ($\Phi \subset F_0$). Let also L be a left F-submodule in A such that L is a colour Lie (p-)subsuperalgebra over Φ in $[A]^{(p)}$ and

$$l(c) = l(c \cdot 1) - \varepsilon(l, c)cl \in F \cdot 1$$

for all homogeneous $c \in F$, $l \in L$.

Then L is a (restricted) differential Lie F-superalgebra.

We say that a G-graded associative algebra A over Φ is *a (restricted) associative envelope of (restricted) differential Lie F-superalgebra* D if there are a (restricted) differential Lie F-superalgebra $L \subset A$ and a F-linear homomorphism of colour Lie (p-)superalgebras over Φ $\xi_A : D \to L$ such that $\tilde{f} = \xi_A(f) \in$ Der F for all $f \in D$ and the set $\xi_A(D)$ generate A as a G-graded associative algebra with 1 over Φ. A (restricted) associative envelope U of D is called *a (restricted) universal enveloping algebra of a (restricted) differential Lie F-superalgebra* D if for any (restricted) associative envelope A of D there is a F-linear homomorphism of G-graded associative Φ-algebras with 1 $\psi : U \to A$ such that $\psi\xi_U = \xi_A$.

It is clear that there exists the (restricted) universal enveloping algebra and it is uniquely determined (up to isomorphisms of G-graded associative algebras with 1).

Let a (restricted) differential Lie F-superalgebra D be a free G-graded left F-module and let F be a free G-graded Φ-module. Denote by $\bar{D}$ the algebra D as G-graded set. Let $\Phi\langle\bar{D}\rangle$ be the free G-graded associative algebra with 1 over Φ on $\bar{D}$. Consider the free product $P = F * \Phi\langle\bar{D}\rangle$ over Φ of G-graded associative algebras with 1. Let I be the ideal in P generated by all elements of the form

$$\overline{\delta+\mu}-\bar{\delta}-\bar{\mu};\quad \overline{c\delta}-c\bar{\delta};\quad \overline{[\delta,\mu]}-[\bar{\delta},\bar{\mu}];\ [\bar{\delta},c]-\bar{\delta}(c)$$

for all homogeneous $\delta, \mu \in D$, $c \in F$;

if D is a restricted differential Lie F-superalgebra, then we add to the generating set of I the elements $\overline{\delta^{[p]}} - (\bar{\delta})^p$ for all homogeneous $\delta \in D_+$.

One can easily see that the algebra P/I is the (restricted) universal enveloping algebra of the (restricted) differential Lie F-superalgebra D.

Theorem 4.1 (PBW-Theorem, [Mi13]). *Let $E = E_+ \cup E_-$ be a total ordered G-graded basis of the free F-module D, $x < y$ for all $x \in E_+$ and $y \in E_-$, F a free Φ-module, $\xi : D \to U(D)$ the (restricted) universal enveloping algebra of the (restricted) differential Lie F-superalgebra D. Then $U(D)$ is a free F-module with a G-homogeneous basis consisting of* 1 *and all monomials of the form*

$$(\xi(e_1))^{\alpha_1}(\xi(e_2))^{\alpha_2}\dots(\xi(e_m))^{\alpha_m}(\xi(f_1))^{\beta_1}\dots(\xi(f_n))^{\beta_n}$$

with $e_i \in E_+$, $f_j \in E_-$, $e_1 < e_2 < \cdots < e_m$, $f_1 < f_2 < \cdots < f_n$, $\alpha_l \in \mathbb{N} \cup \{0\}$, $\beta_j \in \{0, 1\}$ (in the case when D is a restricted differential Lie F-superalgebra, additionally we suppose that $\alpha_i < p$).

PBW-theorems for (restricted) universal enveloping algebras of differential Lie algebras can be found in [BMM]. As for PBW-theorems for (restricted) universal enveloping algebras of colour Lie (p-)superalgebras see, for example, [BMPZ]. Note that universal enveloping algebras of colour Lie (p-)superalgebras have the natural structure of G-graded Hopf algebras (see [Mi7], [BMPZ]). It would be interesting to introduce a structure of "differential" G-graded Hopf algebra on (restricted) universal enveloping algebras of (restricted) differential Lie superalgebras.

References

[An] D. J. Anick, On the cohomology of associative algebras, Trans. Amer. Math. Soc. 296 (1986), 641–659.

[Av] L. L. Avramov, Free Lie subalgebras of the homology of local rings, Trans. Amer. Math. Soc. 270 (1982), 589–608.

[Bab] I. K. Babenko, Analytic properties of Poincaré series of a loop space, Mat. Zametki 27 (1980), 751–765.

[Bah1] Yu. A. Bahturin, Identities in the universal enveloping algebra for a Lie superalgebra, Mat. Sb. 127 (1985), 384–397; English translation: Math. USSR-Sb. 55 (1986), 383–396.

[Bah2] Yu. A. Bahturin, Identical Relations in Lie Algebras, VNU Science Press, Utrecht 1987.

[BD] Yu. A. Bahturin and V. S. Drensky, Identities of the soluble colour Lie superalgebras, Algebra i Logika 26 (1987), 403–418; English translation: Algebra and Logic 26 (1987), 229–240.

[BMPZ] Yu. A. Bahturin, A. A. Mikhalev, M. V. Zaitsev, and V. M. Petrogradsky, Infinite Dimensional Lie Superalgebras, de Gruyter Exp. Math. 7, Walter de Gruyter, Berlin–New York 1992.

[BZ] Yu. A. Bahturin and M. V. Zaicev, Residual finiteness of colour Lie superalgebras, Trans. Amer. Math. Soc. 337 (1993), 159–180.

[Baer] R. Baer, Algebraische Theorie der differentiierbaren Funktionenkörper. I, Sitzungsberichte, Heidelberger Akademie 1927, 15–32.

[BCW] H. Bass, E. H. Connell, and D. Wright, The Jacobian conjecture: reduction of degree and formal expansion of the inverse, Bull. Amer. Math. Soc. 7 (1982), 287–330.

[Baum] G. Baumslag, Free Lie algebras and free groups, J. London Math. Soc. 4 (1972), 523–532.

[BSt] P. Beckman and J. Stückrad, The concept of Gröbner algebras, J. Symbolic Comput. 10 (1990), 465–479.

[BMM] K. I. Beidar, W. S. Martindale 3-rd, and A. V. Mikhalev, Rings with generalized identities , Marcel Dekker, to appear.

[Ber] F. A. Berezin, Introduction in Algebra and Analysis with Anticommuting Variables, Izdat MGU, Moscow, 1983; English translation: Introduction to superanalysis, Reidel, 1987.

[BP] J. Bergen and D. S. Passman, Delta methods in enveloping algebras of Lie superalgebras, Trans. Amer. Math. Soc. 334 (1992), 259–280.

[Berg] G. M. Bergman, The diamond lemma for ring theory, Adv. Math. 29 (1978), 178–218.

[BD1] G. M. Bergman and W. Dicks, On universal derivations, J. Algebra 36 (1975), 193–211.

[BD2] G. M. Bergman and W. Dicks, Universal derivations and universal ring constructions, Pacific J. of Math. 79 (1978), 293–337.

[Bi] J. S. Birman, An inverse function theorem for free groups, Proc. Amer. Math. Soc. 41 (1973), 634–638.

[BL] D. Blessenohl and H. Laue, A basis construction for free Lie algebras, Preprint del Dipartimento di Matematica, Università degli studii di Lecce, no. 16, 1989.

[Bo1] L. A. Bokut', A base of free polynilpotent Lie algebras, Algebra i Logika, 2 (1962), no. 4, 13–19.

[Bo2] L. A. Bokut', Unsolvability of the equality problem and subalgebras of finitely presented Lie algebras, Izv. Akad. Nauk SSSR 36 (1972), no. 6, 1173–1219; English translation: Math. USSR Izvestija 6 (1972), no. 6, 1153–1199.

[Bo3] L. A. Bokut', Embeddings into simple associative algebras, Algebra i Logika 15 (1976), 117–142; English translation: Algebra and Logic 15 (1976), 73–90.

[Bo4] L. A. Bokut', On algebraically closed and simple Lie Algebras, Trudy MIAN 148 (1978), 30–41.

[Bo5] L. A. Bokut', Lie algebra and its universal enveloping algebra have the same Gröbner–Shirshov basis, Internat. J. Algebra Comput., to appear.

[BKl] L. A. Bokut' and A. A. Klein, Serre relations and Gröbner–Shirshov bases for simple Lie algebras I, II, in press.

[BKu] L. A. Bokut' and G. P. Kukin, Algorithmic and Combinatorial Algebra, Kluwer Academic Publ. Dordrecht, 1994.

[BM] L. A. Bokut' and P. Malcolmson, Gröbner–Shirshov bases for quantum enveloping algebras, in press.

[Br] R. M. Bryant, On the fixed points of a finite group acting on a free Lie algebra, J. London Math. Soc. (2) 43 (1991), 215–224.

[BLo] B. Buchberger and R. Loos, Algebraic Simplification, Computing Suppl. 4 (1982), 11–43; Computer Algebra, Symbolic and Algebraic Computation, 2-nd ed., Springer-Verlag, Wien–New York 1983.

[CFL] K. T. Chen, R. H. Fox, and R. C. Lyndon, Free differential calculus, IV. The quotient groups of the lower central series, Ann. Math. 68 (1958), 81–95.

[Co1] P. M. Cohn, Integral modules, Lie rings and free groups, Ph.D. Thesis, Cambridge University, 1951.

[Co2] P. M. Cohn, Subalgebras of free associative algebras, Proc. London Math. Soc. 14 (1964), 618–632.

[Co3] P. M. Cohn, Free rings and their relations, 2-nd ed., Academic Press, London–New York 1985.

[Comp] Computers in Nonassociative Rings and Algebras, ed. R. E. Beck and B. Kolman, Academic Press, New York 1977.

[CNS] L. Corwin, Y. Ne'eman, and S. Sternberg, Graded Lie algebras in mathematics and physics (Bose-Fermi symmetry), Rev. Modern Phys. 47 (1975), 573–603.

[Da] J. H. Davenport, Algebraic Computations and Structures, Computer Algebra, Marcel Dekker, 1989.

[DDKM] J. Desarmenien, G. Duchamp, D. Krob, and G. Melancon, Quelques remarques sur les super-algèbres de Lie libres, C.R. Acad. Sci. Paris 318 (1994), 419–424.

[DL] W. Dicks and J. Lewin, A Jacobian conjecture for free associative algebras, Comm. Algebra 10 (1982), 1285–1306.

[Do] D. Doković, On some inner derivations of free Lie algebras over commutative rings, J. Algebra 119 (1988), 233–245.

[Du] G. Duchamp, Orthogonal projection onto the free Lie algebra, Theoret. Comput. Sci. 79 (1991), 227–239.

[DK1] G. Duchamp and D. Krob, Computing with P.B.W. in enveloping algebras, Lecture Notes in Control and Inform. Sci. 165 (1991), 223–240.

[DT] G. Duchamp and J. Y. Thibon, Le support de l'algèbre de Lie libre, Discrete Math. 76 (1989), 123–129.

[Fel] G. L. Feldman, Ends of Lie algebras, Uspekhi Mat. Nauk 38 (1983), no. 1, 199–200.

[Fer] S. Ferrara, Supersymmetry, North-Holland, Amsterdam 1987.

[Fox] R. H. Fox, Free differential calculus. I. Derivations in free group rings, Ann. Math. 57 (1953), 547–560.

[GD] I. M. Gelfand and Y. L. Daleckii, Some Formal Differential Structures Connected with Lie Superalgebras, preprint no. 85, Inst. Prikl. Mat. AN SSSR, 1994.

[GM] S. I. Gelfand and Yu. I. Manin, Methods of Homologigal Algebra, I, Nauka, Moscow, 1988.

[GAKR] V. P. Gerdt, I. R. Akselrod, V. E. Kovtun, and V. N. Robuk, Construction of a Lie algebra by a subset of generators and commutation relations, Comp. Alg. in Phys. Research. World Scientific, Singapore 1991, 306–312.

[GL] V. P. Gerdt and W. Lassner, Isomorphism verification for complex and real Lie algebras by Groebner basis technique, in: Modern Group Analysis, Adv. Anal. and Comp. Math. in Math. Phys, Kluwer Academic Publ., Amsterdam 1993, 245–254.

[Gol1] E. S. Golod, Standard bases and homology, Lecture Notes in Math. 1352, Springer-Verlag, 1988, 88–95.

[Gol2] E. S. Golod, Standard bases and homology, II, Trudy MIAN, in press.

[Gra] P. K. H. Gragert, Lie algebra computations, Acta Appl. Math. 16 (1989), 231–242.

[Gre] E. L. Green, An introduction into noncommutative Gröbner bases, Computational Algebra, Marcel Dekker, 1993.

[Gro] J. R. J. Groves, Rewriting systems and homology of groups, Lecture Notes in Math. 1456 Springer-Verlag, 1990, 114–141.

[Gu] N. Gupta, Free Group Rings, Contemp. Math. 66, Amer. Math. Soc., Providence, RI 1987.

[Hain] R. M. Hain, Iterated Integrals and Homotopy Periods, Amer. Math. Soc., Providence, RI 1984.

[MHall1] M. Hall, A basis for free Lie algebras and higher commutators in free groups, Proc. Amer. Math. Soc. 1 (1950), 575–581.

[MHall2] M. Hall, The Theory of Groups, Macmillan, New York 1959.

[PHall] P. Hall, A contribution to the theory of groups of a prime power order, Proc. London Math. Soc. 36 (1934), 29–95.

[Hu] D. I. Hughes, Symbolic computation with fermions, J. Symbolic Comput. 10 (1990), 657–664.

[J] N. Jacobson, Abstract derivations and Lie algebras, Trans. Amer. Math. Soc. 42 (1937), no. 2, 206–224.

[Kac1] V. G. Kac, Lie superalgebras, Adv. Math. 26 (1977), 8–96.

[Kac2] V. G. Kac, A sketch of Lie superalgebras theory, Comm. Math. Phys. 53 (1977), 31–64.

[Kac3] V. G. Kac, Infinite Dimensional Lie Algebras, Birkhäuser, Boston 1983.

[K-RW] A. Kandri-Rody and V. Weispfenning, Non-commutative Groebner bases in algebras of solvable type, J. Symbolic Comput. 9 (1990), 1–26.

[Ka1] I. L. Kantor, An Analogue of E. Witt's Formula for the Dimensions of Homogeneous Components of Free Lie Superalgebras, deposited VINITI, 17th April 1984, no. 2384-84 Dep.

[Ka2] I. L. Kantor, Jordan and Lie superalgebras determined by a Poisson algebra, Amer. Math. Soc. Transl. (2) 151 (1992), 55–80.

[Kap] I. Kaplansky, Superalgebras, Pacific J. Math. 86 (1980), 93–98.

[KM] M. V. Karasev and V. P. Maslov, Nonlinear Poison Brackets. Geometry and Quantization, Nauka, Moscow, 1991.

[Kel] O. H. Keller, Ganze Cremona Transformationen, Monatshefte Math. Phys. 47 (1939), 299–306.

[Kem] A. R. Kemer, Varieties and Z_2-graded algebras, Izv. AN SSSR Ser. Mat. 48 (1984), no. 5, 1042–1059.

[Kh1] V. K. Kharchenko, Differential identities of semiprime rings, Algebra i Logika 18 (1979), no. 1, 86–119; English translation: Algebra and Logic 18 (1979), no. 1, 58–80.

[Kh2] V. K. Kharchenko, Automorphisms and Derivations of Associative Rings, Kluwer Academic Publishers, Dordrecht–Boston–London 1991.

[Kha] O. G. Kharlampovich, Lyndon condition for solvable Lie algebras, Izvestia Uuzov. Mat. 1984, no. 9, 50–59.

[KV] A. V. Kondrat'ev and A. B. Verevkin, Package of Programs for Calculation of Gröbner Basis, Canonical Elements, Hilbert Series, and to Define the Rationality of the Hilbert Series of Graded Associative Algebras, Ul'yanovsk Branch of Moscow State University.

[Ku1] G. P. Kukin, Primitive elements of free Lie algebras, Algebra i logika, 9 (1970), no. 4, 458–472.

[Ku2] G. P. Kukin, Subalgebras of free sums of Lie algebras with the amalgamated subalgebra, Algebra i Logika 11 (1972), no. 1, 59–86; English translation: Algebra and Logic 11 (1970), 33–50.

[Ku3] G. P. Kukin, On subalgebras of free Lie p-algebras, Algebra i Logika 11 (1972), 535–550; English translation: Algebra and Logic 11 (1972), 294–303.

[Ku4] G. P. Kukin, On free products of restricted Lie algebras, Mat. Sb. 95 (1974), no. 1, 53–83; English translation: Math. USSR-Sb. 24 (1974), 49–78.

[Ku5] G. P. Kukin, Embeddings of solvable Lie algebras of countable rank into Lie algebras with two generators, Algebra i Logika 14 (1975), no. 4, 414–421.

[Ku6] G. P. Kukin, On derivations of free Lie p-algebras, Algebra i Logika 20 (1981), no. 5, 555–562.

[Lab] J. P. Labute, Free Lie algebras as modules over their universal enveloping algebras, Proc. Amer. Math. Soc. 68 (1978), 135–139.

[Lat] V. N. Latyshev, Combinatorial Theory of Rings, Standard bases, Moscow State University, 1988.

[LC] Ph. Le Chenadec, Canonical Forms in Finitely Presented Algebras, Pitman, London–New York 1986.

[Le1] D. A. Leites, Introduction to the theory of supermanifolds, Uspekhi Mat. Nauk 35 (1980), 3–57.

[Le2] D. A. Leites, The Theory of Supervarieties, Petrozavodsk, 1983.

[Le3] D. A. Leites, Lie Superalgebras, VINITI, Itogi Nauki i Tekhniki, Modern Problems of Math. Recent Progress 25 (1984), 3–50; English translation: J. Soviet Math. 30 (1985), 2481–2512.

[Lem] J. M. Lemaire, Algèbres connèxes et homologie des éspaces de lacets, Lecture Notes in Math. 422), Springer-Verlag, Berlin 1974.

[LS] R. C. Lyndon and P. E. Schupp, Combinatorial Ring Theory. Springer-Verlag, Berlin 1977.

[Ma1] W. Magnus, Beziehungen zwischen Gruppen und Idealen in einem speziellen Ring, Math. Ann. 111 (1935), 259–280.

[Ma2] W. Magnus, Über Beziehungen zwischen höheren Kommutatoren, J. Reine Angew. Math. 177 (1937), 105–115.

[Ma3] W. Magnus, Über Gruppen und zugeordnete Liesche Ringe, J. Reine Angew. Math. 182 (1940), 142–149.

[Ma4] W. Magnus, A connection between the Baker–Hausdorff formula and a problem of Burnside, Ann. Math. 52 (1950), 11–126.

[Ma5] W. Magnus, Algebraic Aspects in the Theory of Systems of Linear Differential Equations, New York University Research Report BR-3, 1953.

[MKS] W. Magnus, A. Karras, and D. Solitar, Combinatorial Group Theory, 2-nd ed, Dover, New York 1976.

[Man1] Yu. I. Manin, Gauge Fields and Complex Geometry, Nauka, Moscow 1984.

[Man2] Yu. I. Manin, Quantum Groups and Non-Commutative Geometry, Preprint CRM-1561, Montreal 1988.

[Man3] Yu. I. Manin, Multiplicative matrices in Hopf superalgebras, Vestnik Moskov. Univ. Ser. I Mat. Meh. 1989, no. 2, 13–17.

[Man4] Yu. I. Manin, Topics in Non-Commutative Geometry, Princeton Univ. Press, 1991.

[Mar] W. Marcinek, Generalized Lie Algebras and Related Topics, I, II, Acta Univ. Wratislav. Mat. Fiz. Astronom. LV, no. 1170, Wroclaw, 3–52.

[Mas] I. M. Masson, Enveloping algebras of Lie superalgebras: a survey, Contemp. Math. 124 (1992), 141–149.

[Me] G. Melancon, Réécritures dans l'algèbre de Lie libre, le groupe libre et l'algèbre associative libre , Publications du LACIM, 8 (1991), Montréal.

[MR] G. Melancon and Ch. Reutenauer, Une présentation combinatoire de la superalgèbre de Lie libre, C.R. Acad. Sci. Paris 315 (1992), 1215–1220.

[Mic] J. Michel, Bases de algèbres de Lie libres et série de Hausdorff, Séminaire Dubreil (algèbre) 27 (1974), no. 6, Institut Henri Poincaré, Paris.

[Mi1] A. A. Mikhalev, Subalgebras of free colour Lie superalgebras, Mat. Zametki 37 (1985), 653–661; English translation: Math. Notes, 37 (1985), 356–360.

[Mi2] A. A. Mikhalev, Free colour Lie superalgebras, Dokl. Akad. Nauk SSSR 286 (1986), 551–554; English translation: Soviet Math. Dokl. 33 (1986), 136–139.

[Mi3] A. A. Mikhalev, Subalgebras of free Lie p-superalgebras, Mat. Zametki 43 (1988), 178–191; English translation: Math. Notes 43 (1988), 99–106.

[Mi4] A. A. Mikhalev, The junction lemma and the equality problem for colour Lie superalgebras, Vestnik Moskov. Univ. Ser.1. Mat. Mekh., 1989, no. 5, 88–91; English translation: Moscow Univ. Math. Bull. 44 (1989), 87–90.

[Mi5] A. A. Mikhalev, Free Lie superalgebras, Ph.D. Thesis, Moscow State University, Moscow 1989.

[Mi6] A. A. Mikhalev, Embedding of Lie superalgebras of countable rank into a two generated Lie superalgebras, Uspekhi Mat. Nauk SSSR 45 (1990), 139–140; English translation: Russian Math. Surveys 45 (1990), 162–163.

[Mi7] A. A. Mikhalev, Ado–Iwasawa theorem, graded Hopf algebras and residual finiteness of color Lie (p-)superalgebras and their universal enveloping algebras, Vestnik. Moscov. Univ. Ser. 1 Mat. Mekh. 1991, no. 5, 72–74.

[Mi8] A. A. Mikhalev, Free colour Lie super-rings, Izvestia Vuzov. Mat., 1991, no. 10, 46–49.

[Mi9] A. A. Mikhalev, The composition lemma for colour Lie superalgebras and for Lie p-superalgebras, Contemp. Math. 131 (1992), 91–104.

[Mi10] A. A. Mikhalev, On fixed points of a free colour Lie superalgebra under the action of a finite group of linear automorphisms, Uspekhi Mat. Nauk 47 (1992), no. 4, 205–207. English translation: Russian Math. Surveys 47 (1992), no. 4, 220–221.

[Mi11] A. A. Mikhalev, On right ideals of the free associative algebra generated by free colour Lie (p-)superalgebras, Uspekhi Mat. Nauk 47 (1992), no. 5, 187–188. English translation: Russian Math. Surveys 47 (1992), no. 5, 196–197.

[Mi12] A. A. Mikhalev, Images of inner derivations of elements of free Lie algebras and superalgebras, Vestnik Moskov. Univ. Ser. 1 Mat. Mekh. 1993, no. 1, 35–38.

[Mi13] A. A. Mikhalev, Differential Lie superalgebras, Uspekhi Mat. Nauk 48 (1993), no. 5, 179–180. English translation: Russian Math. Surveys 48 (1993), no. 5, 179–180.

[Mi14] A. A. Mikhalev, On some properties of super Lie elements in the free associative algebra, Contemp. Math. 184 (1995).

[Mi15] A. A. Mikhalev, Some aspects of combinatorial theory of Lie superalgebras, in: Proc. of the Intern. School "Algebra and Analysis", Baikal Lake, 1992, to appear.

[Mi16] A. A. Mikhalev, A. I. Shirshov composition techniques in Lie superalgebras (non commutative Gröbner bases), Trudy Sem. Petrovsk. 18 (1995).

[MSZ] A. A. Mikhalev, V. E. Shpilrain, and A. A. Zolotykh, Subalgebras of free algebras, preprint 1995.

[MUZ1] A. A. Mikhalev, U. U. Umirbaev, and A. A. Zolotykh, An example of a nonfree Lie algebra of cohomological dimension one, Uspekhi Mat. Nauk 49 (1994), no. 1, 203–204.

[MUZ2] A. A. Mikhalev, U. U. Umirbaev, and A. A. Zolotykh, A Lie algebra with cohomological dimension one over a field of prime characteristic is not necessary free, in: First International Workshop Tainan–Moscow, Proceedings of the First International Conference held at National Cheng Kung University, Tainan, Taiwan, R.O.C., 1994, Walter de Gruyter, Berlin 1996, 257–264.

[MZ1] A. A. Mikhalev and A. A. Zolotykh, Endomorphisms of the free Lie algebra preserving primitivity of elements are automorphisms, Uspekhi Mat. Nauk 48 (1993), no. 6, 149–150. English translation: Russian Math. Surveys 48 (1993), no. 6, 189–190.

[MZ2] A. A. Mikhalev and A. A. Zolotykh, Rank of an element of the free Lie (p-)superalgebra, Doklady Akad. Nauk 334 (1994), no. 6, 690–693. English translation: Russian Acad. Sci. Dokl. Math. 49 (1994), no. 1, 189–193.

[MZ3] A. A. Mikhalev and A. A. Zolotykh, Applications of Fox differential calculus to free Lie superalgebras, Proc. of the III Internat. Conference on non associative algebras, Oviedo 1993, to appear.

[MZ4] A. A. Mikhalev and A. A. Zolotykh, Rank and primitivity of elements of free colour Lie (p-)superalgebras, Internat. J. Algebra Comput. 4 (1994), 617–655.

[MZ5] A. A. Mikhalev and A. A. Zolotykh, Automorphisms and primitive elements of free Lie superalgebras, Commun. Algebra 22 (1994), 5889–5901.

[MZ6] A. A. Mikhalev and A. A. Zolotykh, Bases of free super shuffle algebras, Uspekhi Mat. Nauk 50 (1995), no. 1, 199–200.

[MZ7] A. A. Mikhalev and A. A. Zolotykh, Endomorphisms of free associative algebras over commutative rings and their Jacobian matrices, Fundamental'naya i Prikladnaya Matematika 1 (1995), 177–190.

[MZ8] A. A. Mikhalev and A. A. Zolotykh, An inverse function theorem for free Lie algebras over commutative rings, to appear in Algebra Colloq.

[MZ9] A. A. Mikhalev and A. A. Zolotykh, Ranks of subalgebras of free Lie superalgebras, to appear in Vestnik Moskov. Univ. Ser. 1 Mat. Mekh.

[MZ10] A. A. Mikhalev and A. A. Zolotykh, Test elements for monomorphisms of free Lie algebras and superalgebras, to appear in Commun. Algebra.

[MM] J. W. Milnor and J. C. Moore, On the structure of Hopf algebras, Ann. of Math. 81 (1965), 211–264.

[MT] A. I. Molev and L. M. Tsalenko, Representation of the symmetric group in the free Lie (super)algebra and in the space of harmonic polynomials, Funktsional. Anal. i Prilozhen. 20 (1986), 76–77; English translation: Funktional Anal. Appl. 20 (1986), 150–152.

[MR] A. Montorsi and M. Rasetti, Symbolic manipulation techniques for (Z_2-graded) Lie, Hopf and Hecke algebras, Rend. Sem. Fac. Sci. Univ. Calgliari 61 (1991), no 1, 107–110.

[Mos] M. V. Mosolova, Functions on non commuting operators that generate a graded Lie algebra, Mat. Zametki 29 (1981), 35–44; English translation: Math. Notes 29 (1981), 17–22.

[Nee] Y. Neeman, The application of graded Lie algebras to invariance considerations in particle physics, Lecture Notes in Math. 570, Springer-Verlag, 1977, 109–144.

[New] M. H. A. Newman, On theories with a combinatorial definition of "equivalence", Ann. Math. 43/2 (1942), 223–243.

[NR] A. Nijenhuis and R. W. Richardson, Cohomology and deformations in graded Lie algebras, Bull. Amer. Math. Soc. 72 (1966), no. 1, Pt. 1, 1–29.

[Ni] K. Nishiyama, Oscillator representations for orthosympletic algebras, J. Algebra 129 (1990), 231–262.

[O] A. L. Onishchik, Lie superalgebras of vector fields, Amer. Math. Soc. Transl. (2) 148 (1991), 87–97.

[PP] F. Pauer and M. Pfeifholder, The theory of Grbner bases, L'Enseigment Math. 2nd Ser. 34 (1988), 215–232.

[Pe] V. M. Petrogradsky, Identities in the enveloping algebras for modular Lie superalgebras, J. Algebra 145 (1992), 1–21.

[Ra] D. E. Radford, A natural ring basis for the shuffle algebra and an application to group schemes, J. Algebra 58 (1979), 432–454.

[Re1] R. Ree, Lie elements and an algebra associated with shuffles, Annals of Math. 68 (1958), 210–220.

[Re2] R. Ree, Generalized Lie elements, Canadian J. Math. 12 (1960), 493–502.

[Reu1] Ch. Reutenauer, Applications of a noncommutative Jacobian matrix, J. Pure Appl. Algebra 77 (1992), 169–181.

[Reu2] Ch. Reutenauer, Free Lie Algebras, Clarendon Press, Oxford 1993.

[RW] V. Rittenberg and D. Wyler, Generalized superalgebras, J. Nucl. Phys. B 139 (1978), 189–202.

[Rob] L. Robbiano, Introduction to the theory of Gröbner bases, Queen's papers in Pure and Appl. Math., no. 80, V, (1988).

[Roe] M. Roefols, Prolongation Structures of Supersymmetric Systems, Ph. D. Thesis, Univ. Twente 1993.

[Ros1] L. E. Roos, On Representations and Cohomology of Graded Lie Algebras, Ph. D. Thesis, Berkeley Univ. 1964.

[Ros2] L. E. Ross, Representations of graded Lie algebras, Trans. Amer. Math. Soc. 120 (1965), no. 1, 17–23.

[Sch1] M. Scheunert, Generalized Lie algebras, J. Math. Phys. 20 (1979), 712–720.

[Sch2] M. Scheunert, The Theory of Lie Superalgebras. An Introduction, Lecture Notes in Math. 716, Springer-Verlag, 1979.

[Sch4] M. Scheunert, Graded tensor calculus, J. Math. Phys. 24 (1983), 2658–2670.

[Scho] A. H. Schofield, Representations of Rings over Skew Fields, London Math. Soc. Lecture Notes Series 92, Cambridge University Press, Cambridge 1985.

[Schü1] M. P. Schützenberger, Sur une propriéte combinatoire des algèbres de Lie libres pouvant etre utilisée dans un problème de Mathématiques appliqués, Sem. Dubreil-Pisot, 1958–59, I.H.P. Paris.

[She] I. P. Shestakov, Quantization of Poison superalgebras and speciality of Jordan superalgebras of Poison type, Dokl. Akad. Nauk 334 (1994), no. 1, 29–32.

[Shi1] A. I. Shirshov, Subalgebras of free Lie algebras, Mat. Sb. 33 (1953), 441–452.

[Shi2] A. I. Shirshov, On free Lie rings, Mat. Sb. 45 (1958), 113–122.

[Shi3] A. I. Shirshov, Some algorithmic problems about Lie algebras, Sib. Mat. Zh. 3 (1962), 292–296.

[Shi4] A. I. Shirshov, About one conjecture in the theory of Lie algebras, Sibirsk. Mat. Zh. 3 (1962), 297–301.

[Shi5] A. I. Shirshov, On bases of free Lie algebras, Algebra i Logika 1 (1962), 14–19.

[Shi6] A. I. Shirshov, Collected Works. Rings and Algebras, Nauka, Moscow 1984.

[Shp1] V. Shpilrain, On generators of L/R^2 Lie algebras, Proc. Amer. Math. Soc. 119 (1993), 1039–1043.

[Shp2] V. Shpilrain, Recognizing automorphisms of the free groups, Arch. Math. (Basel) 62 (1994), 385–392.

[Shp3] V. Shpilrain, On the rank of an element of a free Lie algebra, Proc. Amer. Math. Soc. 123 (1995), 1303–1307.

[Shp4] V. Shpilrain, Test elements for endomorphisms of free groups and algebras, to appear in Israel J. Math..

[Sht] A. S. Shtern, Free Lie superalgebras, Sibirsk. Mat. Zh. 27 (1986), 170–174.

[She] W. Specht, Die linearen Beziehungen zwischen höheren Kommutatoren, Math. Z. 51 (1948), 367–376.

[St] J. R. Stallings, On torsion-free groups with infinitely many ends, Ann. Math. 88 (1968), no. 2, 312–334.

[Stan] R. P. Stanley, Unimodality and Lie superalgebras, Stud. Appl. Math. 72 (1985), no. 3, 263–281.

[Sw] R. G. Swan, Groups of cohomological dimension one, J. Algebra 12 (1969), 585–610.

[Uf1] V. A. Ufnarovsky, Combinatorial and asymptotic methods in algebra, VINITI, Itogi Nauki i Tehn. Sovr. Probl. Mat. Fund. Napravl. 57 (1990), 5–177.

[Um1] U. U. Umirbaev, On the approximation of free Lie algebras with respect to entry, Tatu Riikl. Ül. Toimetised = Sci. Notes of Tartu Univ. 878 (1990), 147–152.

[Um2] U. U. Umirbaev, On Jacobian matrices of Lie algebras, The VI All-Union School on Varieties of Alg. Systems, Abstracts of reports, Magnitogorsk 1990, 32–33.

[Um3] U. U. Umirbaev, A note on primitive elements of free groups, Proc. of the 5-th Siberian School "Algebra and Analysis", to appear.

[Um4] U. U. Umirbaev, On Jacobian Matrices of Lie Algebras, preprint.

[Um5] U. U. Umirbaev, Partial derivatives and endomorphisms of some relatively free Lie algebras, Sibirsk. Mat. Zh. 34 (1993), no. 6, 179–188.

[Um6] U. U. Umirbaev, Primitive elements of free groups, Uspekhi Mat. Nauk 49 (1994), no. 2, 175–176.

[Um7] U. U. Umirbaev, The entry problem and subalgebras of free algebras, D. Sci. Thesis, Almaty, Novosibirsk, 1995.

[Um8] U. U. Umirbaev, On ranks of elements of free groups, Uspekhi Mat. Nauk 50 (1995).

[Um9] U. U. Umirbaev, Universal derivations and subalgebras of free algebras, Proc. of the III Intern. Conf. in Algebra, Krasnoyarsk, 1993, to appear.

[Ün] Y. Ünlü, Graded subalgebras associated with a subalgebra of a graded free Lie algebra, J. Pure Appl. Algebra 37 (1985), 103–110.

[VdL] J. Van der Jeugt, Unimodal polynomials arising from Lie superalgebras, Stud. Appl. Math. 77 (1987), 47–59.

[VOW] F. Van Oystaeyen and L. Willaert, Quantum Sections of Colour Lie Superalgebras, preprint.

[Va] M. A. Vasiliev, De Sitter supergravity with positive cosmological constant and generalized Lie superalgebras, Class. and Quantum Gravity 2 (1985), 645–652.

[Vi] G. Viennot, Algèbres de Lie Libres et Monoides Libres, Lecture Notes in Math. 691, Springer-Verlag, Berlin 1978.

[Wais] A. Wais, On special varieties of Lie algebras, Algebra i Logika 28 (1988), 29–40.

[We1] F. Wever, Operatoren in Lieschen Ringen, J. Reine Angew. Math. 189 (1947), 44–55.

[We2] F. Wever, Über Invarianten in Lieschen Ringen, Math. Ann. 120 (1949), 563–580.

[Wit1] E. Witt, Treue Darstellung Lieschen Ringe, J. Reine Angew. Math. 177 (1937), 152–160.

[Wit2] E. Witt, Über freie Liesche Ringe und ihre Unterringe, Math. Zeitschr. 58 (1953), 113–114.

[Wit3] E. Witt, Die Unterringe der freien Lieschen Ringe, Math. Z. 64 (1956), 195–216.

[Y] A. V. Yagzhev, On endomorphisms of free algebras, Sibirsk. Mat. Zh. 21 (1980), no. 1, 181–192.

[Zai] M. V. Zaicev, Locally representable varieties of Lie superalgebras, Mat. Zametki 52 (1992), no. 5, 33–41.

[Ze1] E. I. Zelmanov, On the nilpotency of nil algebras, Lecture Notes in Math. 1352, Springer-Verlag, 1988, 227–240.

[Ze2] E. I. Zelmanov, Superalgebras and identities, Amer. Math. Soc. Transl. (2) 148 (1991), 39–46.

Department of Mechanics and Mathematics
Chair of Higher Algebra
Moscow State University
Moscow 119899, Russia
e-mail: aamikh@cnit.math.msu.su

Isomorphisms and anti-isomorphisms of endomorphism rings of modules

*A. V. Mikhalev**

Abstract. Let A and B be rings and M_A and N_B a right A-module and a right B-module respectively. Let

$$\alpha\colon \operatorname{End}(M_A) \to \operatorname{End}(N_B)$$

be either an isomorphism or an anti-isomorphism of their endomorphism rings. The Wedderburn–Baer–Kaplansky problem is to characterize all such α. The problem has a long history. The aim of these lectures is to consider main results in this direction for modules M_A and N_B which are close to free ones (in the epicenter we put strong generators, i.e., modules with a cyclic free direct summand). The problems under consideration are connected with conditions for inducing of isomorphisms of endomorphism rings by semilinear transformations, by Morita equivalences or by full embeddings. We consider also multiplicative isomorphisms of endomorphism semigroups of modules. We touch some related topics. First of all, we describe recent results on isomorphisms of general linear groups over arbitrary associative rings (Schreier – van der Waerden problem), because of a nice reduction of this group problem to the ring problem of classification of isomorphisms and anti-isomorphisms of matrix rings. We consider Malcev type theorems on elementary equivalences of linear groups and rings of matrices. At the same time we deal with the characterization problem of endomorphism rings of modules from different classes (in particular, of matrix rings). Some comments are related to operator rings and semigroups of topological and ordered linear spaces and to Frobenius type results on A-linear mappings on matrix rings A_n.

Contents

1. Introduction 70

1.1. Isomorphisms of endomorphism rings 72

1.2. Anti-isomorphisms of endomorphism rings 78

1.3. Some comments on endomorphism rings and semigroups and automorphism groups of Abelian groups 79

*Partially supported by the Russian Foundation for Fundamental Research, grant 93-11-1544, by the Russian Government and the International Science Foundation, grant JBX100, and by INTAS.

1.4. Characterization of endomorphism rings 80
1.5. Books and surveys 82
1.6. Acknowledgments 82
2. Strong generators 83
3. Isomorphisms of endomorphism rings of strong generators 88
4. Anti-isomorphisms of endomorphism rings of strong generators 93
5. Endomorphism semigroups of modules 99
6. Isomorphisms of the general linear group $\mathrm{GL}_n(A)$ over an associative ring A (Schreier–van der Waerden problem) 101
7. On Malcev type theorems on elementary equivalences of linear groups and rings of matrices 103
8. Related topics 104
8.1. Representation of rings as full matrix rings (hidden matrices) 104
8.2. Isomorphisms and automorphisms of triangular and structural matrix rings 105
8.3. Isomorphisms and automorphisms of incidence rings and algebras 105
8.4. Isomorphisms and anti-isomorphisms of operator rings and semigroups of Banach and topological linear spaces 106
8.5. Automorphisms and anti-automorphisms of endomorphism semigroups of partially ordered linear spaces 107
8.6. Frobenius type results 108
References 108

1. Introduction

This is the summary of survey-lectures I gave at the International Tainan–Moscow Algebra Workshop in August 1994. Expositions of the results in Sections 3–7 is based mostly on the papers [199, 200, 203, 24, 23, 21, 22, 20, 107, 108, 109, 110, 111, 103, 104, 105, 106].

Let A and B be associative rings with unit elements, and M_A and N_B right modules over A and B, respectively. The Wedderburn–Baer–Kaplansky problem is to classify all isomorphisms and anti-isomorphisms

$$\alpha\colon \mathrm{End}(M_A) \to \mathrm{End}(N_B)$$

of their endomorphism rings.

The classical Wedderburn–Artin structure theorem (see, for example, [7, 64, 125, 128, 275]), characterizing the ring $\mathrm{End}(M_A)$ of linear operators on a finite dimensional linear space M_A over a skew field A as a simple right Artinian ring, in its uniqueness part, is one of the first examples of solution of the WBK-problem for finite dimensional linear spaces over skew fields. Wolfson [276] and Baer [13, p. 183] showed that every ring isomorphism $\alpha: \mathrm{End}(M_A) \to \mathrm{End}(N_B)$ for the arbitrary linear spaces M_A and N_B over skew fields A and B is induced by a semilinear transformation (see also [10, 54]). We mention also the uniqueness part of the density theorem for primitive rings (see Jacobson [127, 128]). Note that for operator rings of Banach spaces this problem was considered by Eidelheit [63], Mackey [172, 173], Arnold [9], Rickart [231]; see more details on this topic in our Subsection 8.4.

The well known Noether–Skolem theorem says that if R is a simple Artinian algebra, finite dimensional over its center F, then all F-automorphisms of R are inner. As a corollary of it we get that every automorphism of the matrix algebra A_n over a field A is inner. In the case of a commutative ring A this result is not far from being true (even for the class of central separable algebras, see Rosenberg and Zelinsky [234]). In particular, if $\mathrm{Aut}_A(A_n)$ is the group of all A-algebra automorphisms of A_n and $\mathrm{Inn}_A(A_n)$ is the normal subgroup of all inner automorphisms of A_n, then the factor group (which is called *the Skolem–Noether group of* A)

$$\mathrm{SN}_n(A) = \mathrm{Aut}_A(A_n)/\,\mathrm{Inn}_A(A_n)$$

may be viewed as a measure of the extent to which the Skolem–Noether theorem fails for $(n \times n)$-matrices over A. An exposition of the case of matrix algebras over commutative rings using the most elementary technique is presented by Isaacs in [126]. The group $\mathrm{SN}_n(A)$ is in fact under control:

Theorem 1.1. *Let A be any commutative ring with* 1 *and $n > 0$. Then:*

a) $\mathrm{SN}_n(A)$ *is an Abelian group;*

b) $x^n = 1$ *for all* $x \in \mathrm{SN}_n(A)$.

Theorem 1.2.

a) *If A is a unique factorization domain, then* $\mathrm{SN}_n(A) = 1$.

b) *If A is a Dedekind domain, then* $\mathrm{SN}_n(A)$ *is isomorphic to the subgroup of the class group of A, consisting of all the elements of order dividing n.*

Note that Guralnik and Montgomery in [116] have proved a slightly weaker version for an affine semiprime Noetherian PI-ring A.

The next very important step was made in Baer–Kaplansky isomorphism theorem (Kaplansky in 1952 announced in [134] his generalization of a theorem of Baer [12] on endomorphism rings of bounded torsion groups; see also Kaplansky [135], Fuchs [79, Theorem 108.1]):

Theorem 1.3. *If M and N are torsion Abelian groups, then every isomorphism of* $\mathrm{End}(M)$ *with* $\mathrm{End}(N)$ *is induced by an isomorphism of M with N.*

In his book [135] Kaplansky in fact considered the isomorphism problem for modules over a complete discrete valuation ring.

Note that analysing the technique which was created in the proofs of isomorphism problems it is possible to recognize the importance of strong generators M_A, i.e., modules with cyclic free direct summands (see our Section 2). If $e^2 = e \in \mathrm{End}(M_A)$ is a projection on a cyclic free direct summand, then there are nice relations between the two derived Morita contexts induced by M_A and by e (see Amitsur [6]).

1.1. Isomorphisms of endomorphism rings

1.1.1. Isomorphisms induced by semilinear transformations. We start with the following problem which was raised by Kaplansky: if A and B are rings with unit element and $A_n \cong B_n$ for some $n > 1$, is it true that $A \cong B$? Note that by Morita theory the centers of the rings A and B are isomorphic.

In such generality, this question has been answered in the negative: examples of rings without IBN-property (invariant basis number; see Cohn's book [45]). In particular:

Theorem 1.4 (Leavitt [156], see Corner [47, 50], Cohn [44]). *Let K be a commutative ring with 1, and q a positive integer. Then there exists a K-algebra A such that for positive integers m and n the free modules A_A^m and A_A^n are isomorphic if and only if $m \equiv n \pmod q$. In particular, there exists a (noncommutative) domain A with $A_A^{nq} \cong A_A^n$.*

Taking endomorphism rings of these modules we obtain $(A_q)_n \cong A_n$, but $A_q \ncong A$, since A is a domain and A_q has zero divisors.

We continue the list of examples of non-isomorphic rings $A \ncong B$ with isomorphic rings of matrices (see Lam [155] for more information):

a) A and B are 2-sided Noetherian rings (Knus [147, Ex. 4.1]).

b) In [130, p. 44] and [131] Jategaonkar for every $n > 1$ constructed non-isomorphic right hereditary right Noetherian right subcommutative domains A and B such that $A_k \cong B_k$ precisely when $n|k$.

c) In [283, Cor. 2.3] Wolfson has constructed two non-isomorphic domains A and B such that $A_n \cong B_n$ and A_n, B_n are prime principal left ideal rings, $n = 2, 4, 6, \ldots$; thus this answers in the negative a conjecture of Jategaonkar [130, p. 45, 137, Problem 4] on the uniqueness part in the third Goldi's theorem (by resemblance with Wedderburn–Artin theorem).

d) B is the first Weyl algebra, P_B is a nonzero right ideal of B, $A = \mathrm{End}(P_B)$, $n \geq 2$ (Webber [273], Smith [257], Stafford [260]).

e) Noetherian rings A and B, $n = 2$ (Chatters [34, 35], Lam [155, § 5]).

f) In [155, Definition 1.6] by Lam two rings A and B are called to be *matrix-isomorphic* if $A_n \cong B_n$ for all $n \geq 2$. The notion to be matrix-isomorphic for A and B is a refinement of $A_n \cong B_n$ for some n, which is in its turn a refinement of the Morita-equivalence of A and B. Note that it is quite possible that $A_k \ncong B_k$

for all $k < n$, but $A_n \cong B_n$. Indeed, Goodearl in [114, Ex. 6.13] constructed two von Neumann regular rings A and B such that for some $n \geq 2$ the matrix rings A_n and B_n are isomorphic, but A_k is not isomorphic to B_k for $k = 1, 2, \dots, n-1$. As it was shown in [20] by Beidar, Knauer, and Mikhalev, it gives an example of rings A and B such that their multiplicative monoids $(A, \cdot)$ and $(B, \cdot)$ are Morita equivalent but non-isomorphic.

g) Chatters shows in [38] that for any positive integer n there exists a division algebra D of generalized rational quaternions and n matrix-isomorphic pairwise non-isomorphic maximal $\mathbb{Z}$-orders in D.

h) An example of two Morita-equivalent rings A and B (i.e., $B \cong eA_ne, e^2 = e \in A_n$, $A_neA_n = A_n$) with A_m not isomorphic to B_n for any integers m and n can be found in [32] by Camillo.

All these examples show that Kaplansky problem should be considered for various classes of rings. We list as illustrations some results in positive direction, when under some additional assumptions we have that $A \cong B$ follows from $A_n \cong B_n$, $n \geq 2$ (see Jategaonkar [130] and Lam [155] for more details):

1. A is commutative (see Kaplansky, Jategaonkar [131, p. 24], Stephenson [262], Isaacs [126], Lam [153, (17, 18)]).
2. A is a right Artinian (see Lam [153, (17, 19)]).
3. A is a 2-sided self injective (see Gentile [90]).
4. A is a (semi) principal right ideal domain (see Robson [232], Jategaonkar [130]).
5. A and B are regular rings whose primitive factor-rings are Artinian (see [114, Cor. 6.12] by Goodearl).
6. A and B are directly finite regular rings satisfying general comparability (see [114, Cor. 8.17] by Goodearl).
7. A and B are regular right self-injective rings (see [114, Cor. 10.35] by Goodearl).

Now we mention some results on isomorphisms of endomorphism algebras. Wolfson used in [279] Kaplansky's method to prove the following

Theorem 1.5. *If M_A and N_A are torsion-free modules over the complete discrete valuation ring A, then every A-algebra isomorphism of* $\mathrm{End}(M_A)$ *with* $\mathrm{End}(N_A)$ *is induced by an isomorphism of M_A with N_A.*

The development of isomorphism theorems even for classes of mixed Abelian groups was hampered by the nonavailability, in general, of indecomposable summands that could be mapped to torsion-free elements. Developing Kaplansky's technique, May and Toubassi [188] studied mixed groups of torsion-free rank one having simply presented torsion. See also the survey of May [182] on endomorphism rings of mixed Abelian groups. Sufficient conditions for isomorphisms of endomorphism algebras of reduced modules over complete discrete valuation rings to be induced by isomorphisms are given in [183] by May. In [184] May gives conditions on a module M_A over an incomplete discrete valuation ring to guarantee that there exist many nonisomorphic modules with endomorphism algebras isomorphic to $\mathrm{End}(M_A)$.

Next, we come to results on isomorphisms of endomorphism rings which are induced by semilinear transformations.

In [211, Theorem 1.3] Morita has proved that any isomorphism of (strongly dense sub)rings of faithful modules over self-basic quasi-Frobenius rings is induced by a semilinear transformation (in [211] this theorem is a corollary of a more general result on isomorphisms of endomorphism rings of faithful modules over quasi-Frobenius rings induced by category equivalences, which have been earlier classified in [210]).

In [280] Wolfson showed that any isomorphism of endomorphism rings of free modules over principal left ideal domains is induced by a semilinear transformation (in fact, Wolfson modified Baer's technique from [13]).

Recall that a module M_A is called *locally free* if each finite set of elements of M_A is contained in a (finitely generated) free direct summand of M (a locally free Abelian group is just a homogeneous separable group of null type).

In [281] Wolfson generalized his previous theorem from [280] to a class of locally free modules over principal left ideal domains.

In [199, Prop. 1, Theorems 1–3] and [200, Prop. 2, Cor. 1] Mikhalev considered some sufficient conditions on isomorphisms of endomorphism rings of strong generators to be induced by a semilinear transformations (later in [203] it is shown that in some sense these conditions are necessary). Gluskin in [97] has touched the isomorphism problem for endomorphism rings of free modules with indecomposable cyclic direct summands. Note that Mikhalev [199, 200] and Gluskin [97] started also to consider multiplicative isomorphisms of endomorphism submonoids (see our Section 5).

Stephenson [262, Th. 3], [264] gives some equivalent conditions on rings A and B to be Morita equivalent in the terms of isomorphisms of different infinite matrix rings and he shows that if $A_n \cong B_n$ for regular rings A and B then injective hulls of A and B are isomorphic (note that Camillo in [32] proved the theorem that rings A and B are Morita equivalent if and only if $\mathrm{End}(A_A^{(\mathbb{N})}) \cong \mathrm{End}(B_B^{(\mathbb{N})})$,conjectured by Stephenson in [262]).

In [130, Th. 1.6] and [131] Jategaonkar has proved equivalence of the following conditions:

1. A is a PF_n*-ring* (i.e., for every right A-module P with $(P_A)^{(n)} \cong (A_A)^{(n)}$ we have $P_A \cong A_A$).
2. If B is a ring with an isomorphism
$$\alpha\colon \mathrm{End}(B_B^{(n)}) \to \mathrm{End}(A_A^{(n)}),$$
then α is induced by a semilinear transformation.

Hauptfleisch in [124] solved, in particular, the isomorphism problem for endomorphism rings of locally free modules over subrings of rationals.

McDonald in [193, Section 4] also suggests some sufficient conditions for isomorphisms of endomorphism (sub)rings of infinitely generated projective strong generators to be induced by semilinear transformations (they are very close to Mikhalev [199, 200]; note also that Abrams [4], Garsia and Simon [89] constructed some counterexamples to McDonald's results and made necessary corrections).

A nice exposition of some results on automorphisms of matrix algebras over commutative rings can be found in [126] by Isaacs.

In [74] Franzsen and Schultz identified a large class of rings over which locally free modules can be determined by their endomorphism rings (note that the example of the ring A with $A \cong \mathrm{End}(A_A^{(\mathbb{N})})$ in [74, Section 6] is invalid, as it was shown by Abrams in [3]).

Hacque in [119] studies isomorphisms and automorphisms of almost simple rings (i.e., rings with left isotype and left faithful socle; these rings can be realized as endomorphism rings of linear systems (in the sense of Mackey [173]) over skew fields).

Bolla in [27] and [28] gives a characterization of isomorphisms between endomorphism rings of progenerators and infinitely generated free modules over arbitrary rings in terms of Morita equivalences; in addition in [28] he indicates in some cases the interrelationship between this categorical description and the classical description which uses semilinear transformations.

Since a prime (left) Goldie ring R, in which each finitely generated left ideal is principal, is the endomorphism ring $\mathrm{End}({}_AM)$ of a free module ${}_AM$, of finite rank, over a (left) Ore domain A (see [101], [130], [232], [125, Ch. 2]), Wolfson in [282, 283] has found necessary and sufficient conditions to insure that whenever

$$R \cong \mathrm{End}(M_A) \cong \mathrm{End}(N_B)$$

(M_A, N_B are free and finitely generated over domains A and B) each isomorphism $\alpha\colon \mathrm{End}(M_A) \to \mathrm{End}(N_B)$ is induced by a semilinear isomorphism of M_A upon N_B.

In [203, Theorems 2.3, 2.7, 2.10] among three criteria, there is the third one for an isomorphism between endomorphism rings of strong generators to be induced by a semilinear transformation (in [203] it is deduced from the more general first criterion to be induced by a faithful embedding, see our Section 3).

Note that some related results to this topic can be found in Subsection 1.1.4.

1.1.2. Isomorphisms induced by Morita equivalences. Morita [211] remarked that the classical assertion that any two faithful modules over a simple ring are isomorphic if their endomorphism rings are isomorphic, however, cannot be extended to the case of modules over a semisimple ring (even using semilinear isomorphisms). Indeed, let $A = I \oplus J$ be a semisimple ring such that I and J are isomorphic respectively to the full matrix rings K_2 and K_3 over a field K, X and Y the simple right ideals of I and J respectively,

$$M_A = X \oplus X \oplus Y \quad \text{and} \quad N_A = X \oplus Y \oplus Y.$$

Then the endomorphism rings $\mathrm{End}(M_A)$ and $\mathrm{End}(N_A)$ are isomorphic, but there is no semilinear isomorphism of M_A onto N_A. By this reason Morita in [211] suggested to use category equivalences (just classified in [210]) instead of semilinear transformations as a new more general solution for description of isomorphisms between endomorphism rings. In fact, in [211, Theorem 1.2] Morita has proved that any isomorphism between endomorphism rings $\mathrm{End}(M_A)$ and $\mathrm{End}(N_B)$ for faithful modules M_A and N_B over quasi-Frobenius rings is induced by a category equivalence from Mod-A onto Mod-B.

As we have remarked in our Subsection 1.1.1, Bolla in [27] has proved the following

Theorem 1.6. *If M_A and N_B are progenerators (i.e., projective finitely generated generators) over arbitrary rings A and B, then any isomorphism between endomorphism rings* $\mathrm{End}(M_A)$ *and* $\mathrm{End}(N_B)$ *is induced by a Morita equivalence from* Mod-A *onto* Mod-B.

Corollary 1.7. *For arbitrary rings A and B with* 1 *any isomorphism between matrix rings A_m and B_n is induced by a Morita equivalence.*

Camillo in [32] has proved that rings A and B are Morita equivalent if and only if endomorphism rings $\mathrm{End}(A_A^{(\mathbb{N})})$ and $\mathrm{End}(B_B^{(\mathbb{N})})$ of countably generated free modules are isomorphic (this theorem was conjectured in [262]). In [28] Bolla gave a characterization of ring isomorphisms between endomorphism rings of infinitely generated free modules over arbitrary rings in terms of Morita equivalences.

In [203, Theorem 2.7] Mikhalev characterized isomorphisms between endomorphism rings $\mathrm{End}(M_A)$ and $\mathrm{End}(N_B)$ of strong generators M_A and N_B over arbitrary rings A and B, which are induced by Morita equivalences (among corollaries there is Bolla's theorem [27] on endomorphism rings of progenerators which is deduced from this by the canonical matrix reduction from a progenerator to a strong generator; see our Section 3).

We note that in [88, Section 4] Garsia and Simon give some sufficient conditions for an isomorphism of endomorphism rings of locally free modules to be induced by Morita equivalences.

See also our Subsection 1.1.4 on related topics on Morita equivalences and infinite matrix rings and note that in [119] and in [268] it is possible to find some close technique with Morita contexts and idempotents.

1.1.3. Isomorphisms induced by full embeddings. In [203, Theorem 2.3] Mikhalev has proved a criterion for an isomorphism between endomorphism rings $\mathrm{End}(M_A)$ and $\mathrm{End}(N_B)$ of strong generators M_A and N_B to be induced by a full embedding from Mod-A into Mod-B (i.e., by a functor of the form $h^U = \mathrm{Hom}(U_A, -)$, where ${}_BU_A$ is a B-A-bimodule, U_A is a generator and $B = \mathrm{End}(U_A)$). As a corollary two other criteria are deduced from this for an isomorphism between $\mathrm{End}(M_A)$ and $\mathrm{End}(N_B)$ to be induced by a Morita equivalence from Mod-A onto Mod-B or by a semilinear transformation. In [87] Garsia and Simon give examples of two modules M_A and N_B

in the class of all countably generated locally free projective modules such that rings $\mathrm{End}(M_A)$ and $\mathrm{End}(N_B)$ are isomorphic, but rings A and B are not Morita equivalent; another example shows that there exist a non-finitely generated projective and locally free module M_A and an automorphism of its endomorphism ring $\mathrm{End}(M_A)$ such that it is not induced by any Morita auto-equivalence. Remark that in [88, Section 3] Garsia and Simon give also an example of two rings A and B and two non-finitely generated locally free modules M_A and N_B such that $\mathrm{End}(M_A) \cong \mathrm{End}(N_B)$, but rings A and B are not Morita equivalent, and they also suggest to consider some generalization of Morita equivalence as a possible form of solution in the isomorphism problem.

1.1.4. Infinite matrix rings. Denote the ring of row-finite $(\mathbb{N} \times \mathbb{N})$-matrices over the ring A (i.e., the endomorphism ring of the $\mathbb{N}$-free A-module) by $\mathrm{RFM}_{\mathbb{N}}(A)$. Let $\mathrm{FM}_{\mathbb{N}}(A)$ be the subring of $\mathrm{RFM}_{\mathbb{N}}(A)$ consisting of those matrices with a finite number of nonzero elements. As we have mentioned above, Stephenson [262, 264] has shown that A and B are Morita equivalent if and only if $\mathrm{FM}_{\mathbb{N}}(A) \cong \mathrm{FM}_{\mathbb{N}}(B)$, and it was shown by Camillo in [32] that it is also equivalent to $\mathrm{RFM}_{\mathbb{N}}(A) \cong \mathrm{RFM}_{\mathbb{N}}(B)$. In this connection Abrams [2] posed the following question: *Do intermediate matrix rings*

$$\mathrm{FM}_{\mathbb{N}}(A) \subseteq T \subseteq \mathrm{RFM}_{\mathbb{N}}(A)$$

also determine the Morita equivalence of rings? He showed that in general there are many such "good" intermediate rings.

Let now $\mathrm{FC}_{\mathbb{N}}(A)$ be the intermediate subring consisting of all elements of $\mathrm{RFM}_{\mathbb{N}}(A)$ which have at most a finite number of nonzero columns. Garsia [83] showed that the Morita equivalence can be determined by $\mathrm{FC}_{\mathbb{N}}(A)$; Xu [293, 294] extended results of Camillo and Garsia on infinite matrices over an arbitrary indexed set Γ. Some general form of these results was presented in [296] by Xu, Shum and Turner-Smith, [295] by Xu, Shum and Fong, [255] by Simon.

As we have seen earlier there are many examples of rings A (without IBN property) such that $A \cong A_n$ for $n > 1$. In [3] Abrams constructed examples of rings A with the property $A \cong \mathrm{RFM}(A) = \mathrm{End}(A_A^{(\mathbb{N})})$ (it was noted in [3] that the example from [74] by Franzsen and Schultz of such a ring is invalid). Garsia in [84] proves some general relation between $\mathrm{FC}(A)$ and $\mathrm{RFM}(A)$ in the spirit of the paper of Wolfson [286].

In [4] Abrams examines the automorphisms of certain rings T with

$$\mathrm{FM}_{\mathbb{N}}(A) \subseteq T \subseteq \mathrm{RFM}_{\mathbb{N}}(A),$$

in particular, he gives characterizing conditions for A which ensure that such a ring T exists with the property that some automorphism of T does not preserve finite rank endomorphisms. This result contrasted with McDonald [193, Section 4]. Indeed, in [193] McDonald assumed the following assertion to hold:

(∗) If Σ is a unitial subring of $\mathrm{RFM}(A)$, containing $\mathrm{FC}(A)$ and α is a ring automorphism of Σ, then

$$\mathrm{End}_A\left(e^{\alpha} A^{(\mathbb{N})}\right) \cong A$$

(here e is the upper left corner matrix unit of $\mathrm{RFM}(A)$). As we have remarked, Abrams [4] pointed out that it seems unlikely that (*) holds. In fact [4, Prop. 1.4] he showed that (*) would not be true if one replaces the condition "$\mathrm{FC}(A) \subseteq \Sigma$" by the weaker one "*$\Sigma$ contains all the matrices of* $\mathrm{RFM}(A)$ *with a finite number of nonzero entries.*" In [89, Prop. 3.1] Garsia and Simon show that, indeed, assertion (*) of MsDonald is not true, by considering a class of conterexamples.

At last note that Bonami in [30] considers the following two classes of rings:

$$\mathcal{A} = \{A \mid \forall B,\ B \overset{M}{\sim} A \Rightarrow B \cong A\}$$

(i.e., rings which are uniquely determined by their module category);

$$\mathcal{B} = \{A \mid \mathrm{End}(A_A^{(I)}) \in \mathcal{A} \text{ for all infinite sets } I\}$$

(i.e., endomorphism rings of infinite free modules over A which are uniquely determined by their module category. In particular, commutative Noetherian rings, semisimple rings, local rings and semifirs belong to $\mathcal{B}$.

1.2. Anti-isomorphisms of endomorphism rings

If M_A and N_B are linear spaces over skew fields A and B, then Baer in [13] showed that $\mathrm{End}(M_A)$ and $\mathrm{End}(N_B)$ are anti-isomorphic if and only if M_A and N_B are finite dimensional and mutually the duals of each other. Gewirtzman in [92] extended this result to free modules over principal left ideal domains. In his next paper [93] he considered locally free modules over principal left ideal domains and torsion-free modules over complete discrete valuation rings, but Lemma 1.4 of [93] is false (in the set theoretical arguments to compare the ranks of M_A and ${}_AM^*$), as are Theorems 1.1, 1.2 and 2.3 which depend on it. Avoiding these problems, in [284] Wolfson considers locally free modules M_A and N_B over a domain A and over a ring B, finitely generated left ideals in which is free. Then $\mathrm{End}(M_A)$ is anti-isomorphic to $\mathrm{End}(N_B)$ if and only if M_A and N_B are reflexive and mutually dual (the anti-isomorphism can be identified, using the adjoint mapping of $\mathrm{End}(M_A)$ into $\mathrm{End}({}_AM^*)$ (see [284, Theorem 2.3]). In fact, in [284, Theorem 2.1], for a torsionless strong generator M_A (with $e^2 = e \in \mathrm{End}(M_A)$, $r(e) = 1$) it is shown that the anti-isomorphism $\alpha\colon \mathrm{End}(M_A) \to \mathrm{End}(N_B)$ is induced by an anti-semilinear isomorphism of ${}_AM^*$ onto N_B if and only if $e^\alpha N_B$ is cyclic free; in this case both M_A and N_B are reflexive modules, and M_A is anti-semilinear isomorphic to ${}_BN^*$ (compare with our Corollary 4.9).

In [24, Theorem 4], [23] (see exposition in Section 4 of our paper) Beidar and Mikhalev have proven the criterion for an anti-isomorphism of endomorphism rings of (torsionless) strong generators to be induced by a Morita anti-equivalence. In addition it is indicated the relationship between this categorical description and Wolfson's characterization with anti-semilinear mappings. Note also that in [24, Theorem 1] it is shown that Morita equivalences preserve reflexive modules. As a corollary of these results, it is shown that any anti-isomorphism of matrix rings (and also of endomorphism rings of progenerators) is induced by a Morita anti-equivalence.

1.3. Some comments on endomorphism rings and semigroups and automorphism groups of Abelian groups

A well known classical result of Baer [12] and Kaplansky [134, 135] on definability of a torsion Abelian group by its endomorphism ring opened an extremely interesting direction of isomorphism theorems in the theory of Abelian groups and modules over rings.

As for endomorphism rings and semigroups of Abelian groups, in addition to [135, § 19] and [78, Ch. XV], see recent surveys of May [182], Mikhalev and Mishina [204, § 7], Sebeldin [249], and also a new book of Krylov and Mikhalev [149, Ch. 5] which is now in preparation.

Note that even for torsion-free Abelian groups the abundance of the indecomposable ones makes it impossible to prove that isomorphism of the endomorphism rings implies isomorphism of the underlying groups. For example (see [281] by Wolfson), for any cardinal number c there exist homogeneous completely decomposable groups G and H of rank c which are not isomorphic but have isomorphic rings of endomorphisms. For p-local Abelian groups see also [184] by May.

In addition we give only a selective additional bibliography on endomorphism rings of Abelian groups: Bazzoni and Metelli [19], Dugas and Göbel [59], Göbel and May [99], Krylov [148], May [181, 182, 183, 184], May and Toubassi [185, 186, 187, 188, 189], Metelli and Salce [197], Mikhalev [203], Puusemp [229], Sebeldin [239, 238, 240, 241, 242, 243, 244, 248], Wolfson [279].

We mention also some articles on endomorphism groups: Greenspon [115], Sebeldin [240, 241, 243].

Note that Puusemp [228] and Sebeldin [245, 246, 247] considered the problem of definability of Abelian groups by their multiplicative endomorphism semigroups.

As for automorphism groups of Abelian groups we note that Leptin ([157], [79, Ch. 16]) and Liebert [169, 170] have proved that any Abelian p-group ($p \geq 3$) is definable by its automorphism group, and even more, in [169, 170] it is possible to find a description of isomorphisms between automorphism groups of primary Abelian groups.

As for additional information, see Nikiforov [217], Dixon and Evans [55], and Solonina [258].

The book of Bekker and Kozuhov [25] is devoted to automorphisms of torsion free Abelian groups.

1.4. Characterization of endomorphism rings

In the books of Baer [13], Faith [64, 65], Fuchs [77, 78, 79], Goodearl [114], Jacobson [128], Jategaonkar [130], Petrich [225, 223], Rowen [236], Wisbauer [275] it is possible to find exposition of some characterization theorems for endomorphism rings of modules from different classes; see also recent surveys of Göbel [98], May [182], Mikhalev and Mishina [204, Section 8].

Recall that the Wedderburn–Artin theorem gives a characterization of complete matrix ring over a skew field as a simple right Artinian ring, the density theorem describes dense rings of linear transformations of (integral) infinite dimensional linear spaces over skew fields as primitive rings (see for example [7], [64], [128], [236]). Wolfson in [276] determines necessary and sufficient conditions for an abstract ring to be isomorphic to the ring $\mathrm{End}(M_A)$ of all linear transformations of the linear space M_A over a skew field A (he also characterizes the ring of all linear transformations of finite rank).

The characterization of a complete blocked triangular matrix ring over a division ring in [161] was given by Li and Zelmanovitz using conditions analogous to the Wedderburn–Artin characterization as a prime left Artinian ring.

The characterization of an indecomposable nonsingular generalized ring is given in [102] by Goldie and in [212] by Murase as a complete blocked triangular matrix ring over a division ring. Note that the first example of a simple Noetherian ring not isomorphic to a matrix ring over a domain was constructed by Zalesskii and Neroslavskii [216] (see also [112]). In [113] Goodearl constructs a family of simple Noetherian rings, of arbitrary large uniform (Goldie) dimension, that are not isomorphic to matrix rings over integral domains.

K. C. Smith and van Wyk in [256] introduced the concept of matrix units in a ring associated with a quasi-order relation and used it to provide an internal characterization of structural matrix rings. Note that the characterization problem for incidence algebras was considered in [72] by Feinberg.

Some necessary and sufficient conditions for a ring from some class of rings to be isomorphic to the endomorphism ring of an Abelian groups can be found in [78, Chapter 15], for example: a skew field is the endomorphism ring of some Abelian group if and only if it is a simple field ([78, Th. 111.1]); a simple ring is isomorphic to the endomorphism ring of some Abelian group if and only if it is the complete matrix ring over a simple field ([78, Th. 111.2]).

The isomorphism problem for Abelian groups is closely related to realization theorems which state that rings from a given class can be realized as endomorphism rings of Abelian groups from a given class of groups (see [77, Pr. 44]). The bulk of research has been devoted to the torsion and torsion-free classes. We start with a number of results of Corner [46, 48, 49, 50, 51] which have many followers (see abovementioned references), it is beginning with the theorem ([46], see also [78, Th. 110.1, 110.2]) that every countable ring whose additive group is reduced and torsion-free is isomorphic to $\mathrm{End}(M_{\mathbb{Z}})$ for some countable, reduced torsion-free Abelian group ($M_{\mathbb{Z}}$).

In fact, there are $2^{\aleph_0}$ nonisomorphic groups $M_{\mathbb{Z}}$ with the same endomorphism ring, this shows how difficult are the isomorphism problems for endomorphism rings of countable torsion-free Abelian groups.

In [48] Corner characterizes topological rings which are topologically isomorphic to finitely topologized endomorphism rings of countable reduced torsion-free Abelian groups. At about the same time Liebert (see [162, 163, 164]) found a characterization of the endomorphism rings of separable p-primary groups as topological rings in the finite topology.

Corner [49] has proved so called "a split realization theorem" for endomorphisms of separable p-groups: if R is a ring whose additive group $(R, +)$ is the completion of a free p-adic module of at most countable rank, then there exist a separable p-group $M_{\mathbb{Z}}$ such that

$$\mathrm{End}(M_{\mathbb{Z}}) \cong R \oplus \mathrm{End}_s(M_{\mathbb{Z}}),$$

where $\mathrm{End}_s(M_{\mathbb{Z}})$ is the ideal of small endomorphisms of $M_{\mathbb{Z}}$.

In [165, 166, 167] Liebert has received another topological description of the endomorphism algebras of reduced and divisible torsion-free modules over complete discrete valuation rings.

Liebert in [168] found a set of algebraic and topological conditions that are necessary and sufficient for an abstract ring R to be isomorphic to the endomorphism ring $\mathrm{End}(M_A)$ of a free module M_A over a principal left ideal domain A (remark that the main result in [168] refers to "principal ideal domains", however, as Wolfson [286, Footnote] has remarked, the statement and proof remain correct if "principal ideal domain" is replaced whenever it occurs by "principal left ideal domain"). Some modification of this result was suggested by Wolfson [286] (in particular, the topology on R can be defined by taking finite intersections of maximal right annihilators as a neighborhood basis of zero).

Metelli and Salce [197] give a solution for the characterization problem for endomorphism rings of locally free modules over subrings of rationals. Using technique due to [197] and [81], Franzsen and Schultz in [74] established a characterization theorem for endomorphism rings of locally free modules.

In [287] Wolfson studies rings which generalize prime principal left ideal rings in that they are endomorphism rings $\mathrm{End}(M_A)$ of locally free modules M_A (of possibly infinite rank) over a left Noetherian left hereditary domain A. He gives necessary and sufficient conditions for an abstract ring R to be isomorphic to such an endomorphism ring (which reduces to a prime principal left ideal ring in case when M_A is of finite rank) and characterizes the endomorphism ring of a locally free module over a principal left ideal domain (generalizing previous results in [168] and [286] on endomorphism rings of free modules, see also [74]).

Garsia and Simon in [88] consider the characterization problem for $\mathrm{End}(M_A)$ for M_A from the class of locally free modules (they use techniques of non-commutative localization and equivalences of categories; in particular, in their approach localization methods replace the topological techniques from the preceding papers).

In a series of papers [58, 59, 60] Dugas and Göbel using some ideas of Shelah consider cotorsion-free Abelian groups and modules (see also [250] by Shelah); in particular, they show that a ring R is isomorphic to the endomorphism ring of a cotorsion-free Abelian group M if and only if the additive group $(R, +)$ of R is cotorsion-free.

The paper [61] is devoted to endomorphism rings of separable torsion-free Abelian groups (some corrections to [61] are included in [100]). See also the paper [62] by Dugas and Irwin on countable rings as endomorphism rings.

A survey-lecture of Göbel on endomorphism rings of Abelian groups can be found in [98].

1.5. Books and surveys

We mention only some books in which these topics are discussed: Anderson and Fuller [7], Artin [10], Baer [13], Bass [18], Bekker and Kozuhov [25], Chatters and Hajarnavis [39], Cohn [45], Diedonné [54], Faith [64, 65], Fuchs [77, 78, 79], Goodearl [114], Hahn and O'Meara [123], Herstein [125], Jacobson [128], Jategaonkar [130], Kaplansky [135, 136], Krylov and Mikhalev [149], Lam [151, 152, 153], McConnell and Robson [190], McDonald [191, 196], Mitchell [208], Petrich [225, 226], Plotkin [227], Rowen [236], Suprunenko [265], Wehrfritz [274], Wisbauer [275].

We give also a list of recent surveys: Göbel [98], Hahn [121], Isaacs [126], Lam [154], Marcus [176, 177], Markov, Mikhalev, Skornyakov, and Tuganbaev [180], May [182], Mikhalev [201], Mikhalev and Mishina [204], O'Meara [218, 220], Sebeldin [249], Zalesskii [297].

1.6. Acknowledgments

My deep appreciation goes to all participants of the Moscow–Tainan Algebra Workshop for their personal interest and for the inspiration, that came from all discussions, and especially to Prof. Y. Fong and Prof. K. I. Beidar, who more than anyone else favoured the success of the Workshop, thank you very much for your constant interest in our project and the support and encouragement that you gave it.

I am most grateful to L. I. Mikhaleva and A. A. Mikhalev for their professionalism and patience with the bringing of the text to final form and for their efforts in bringing the work to fruition.

2. Strong generators

Definition 2.1. Let A be a ring, M_A a right A-module, $\mathrm{End}(M_A)$ its endomorphism ring. We say that:

a) the module M_A is *a strong generator*, if $M_A = P \oplus H$ where $P \cong A_A$ (i.e., M_A contains a free cyclic direct summand P);

b) an idempotent $e^2 = e \in \mathrm{End}(M_A)$ is called *an idempotent endomorphism of rank 1* (notation: $r(e) = 1$) if eM is a free right A-module of rank 1 (i.e., $M_A = eM \oplus H$, $eM \cong A_A$; in this case M_A is a strong generator).

Remark 2.2.

1. Any strong generator is a generator (recall that M_A is *a generator* in the category Mod-A of right A-modules if

$$\underbrace{M_A \oplus \cdots \oplus M_A}_{n} \cong A_A \oplus H_A,$$

i.e., it is a strong generator, for some $n \in \mathbb{N}$).

2. It is clear that any free module (and any locally free module) is a strong generator.

3. Kaplansky [135] remarked that if A is a complete discrete valuation ring and M is a torsion-free R-module which is not divisible then it is a strong generator.

4. Any faithful module over a selfbasic QF-ring is also a strong generator (see [215], [66]).

5. If R is a ring and $e^2 = e \in R$ is an idempotent of R, then the right eRe-module Re_{eRe} is a strong generator. Indeed,

$$Re_{eRe} = eRe_{eRe} \oplus (1-e)Re_{eRe}$$

(in connection with this example see Lemma 2.9).

6. Let A be a simple Noetherian ring of Krull dimension 1. Then every large enough finitely generated torsion-free A-module is a strong generator (see Chatters and Hajarnavis [39, Section 16]).

7. Rather than attacking the endomorphism ring $\tilde{R} = \mathrm{End}(\widetilde{M_A})$ Jategaonkar in [132] explores the underlying idea in the proofs to prove the results for $R = \mathrm{End}(\widetilde{M_A} \oplus A_A)$ and to transfer the required information from R to the corner subring $\tilde{R}$. The reason why it is any easier to look at R rather than $\tilde{R}$ is that $M_A = \widetilde{M_A} \oplus A_A$ is a strong generator in Mod-A and a Morita type transfer of properties from A to R is available.

Lemma 2.3. *Let M_A be a right A-module, $E = \mathrm{End}(M_A)$ its endomorphism ring, R a subring of E, $e^2 = e \in R$, $r(e) = 1$ (i.e., $M_A = eM_A \oplus H_A$, $eM_A \cong A_A$). Then the following conditions are equivalent:*

1. *there is a basis $\{x\}$ of eM_A such that $Rx = M$;*
2. *$Ry = M_A$ for any basis $\{y\}$ of eM_A;*
3. *Re is a left ideal of the ring $E = \mathrm{End}(M_A)$.*

Proof. Implication 2) $\Rightarrow$ 1) is evident. We show that 1) $\Rightarrow$ 3). Let $d \in E$ and $r \in R$.

Since $x \in eM$, $ex = x$ and $Rex = Rx = M$. Therefore

$$drex \in dRex = dM \subseteq M = Rex,$$

i.e., $drex = bex$ for some $b \in R$. If $m \in M$ and $em = xa$ for $a \in A$, then

$$(dre)m = (dre)(em) = (dre)(xa) = (drex)a = (bex)a = (be)(xa) = (be)m.$$

Thus, $dre = be$, i.e., $ERe \subseteq Re$.

3) $\Rightarrow$ 2). Let Re be a left ideal of the ring E and $\{y\}$ a basis of eM_A. Since $ERe = Re$, $y = ey$, $Ey = M$, we get

$$M \supseteq Ry = R(ey) = (Re)y = (ERe)y \supseteq Eey = Ey = M.$$

Therefore, $Ry = M$. □

Definition 2.4. We say that a subring of the endomorphism ring $\mathrm{End}(M_A)$ for a strong generator M_A is *abundant* if it satisfies one of the equivalent conditions 1)–3) in Lemma 2.3.

Remark 2.5.

1. The endomorphism ring $E = \mathrm{End}(M_A)$ of any strong generator M_A is abundant.
2. Let R be a primitive ring with nonzero socle, M a minimal left ideal of R, $A = \mathrm{End}({}_RM)$. Then R is an abundant subring of $E = \mathrm{End}(M_A)$.

Definition 2.6. Let ${}_RM_A$ be an R-A-bimodule, and let ${}_SN_B$ be an S-B-bimodule. A triple (α, β, γ), where $\alpha: R \to S$ and $\gamma: A \to B$ are isomorphisms of rings and $\beta: M \to N$ is an isomorphism of the Abelian groups with $(rma)^\beta = r^\alpha m^\beta a^\gamma$ for all $r \in R, m \in M, a \in A$, is called *a semilinear isomorphism of the bimodules* ${}_RM_A$ and ${}_SN_B$.

Lemma 2.7. *Let M_A and N_B be right modules, $R = \mathrm{End}(M_A)$ and $S = \mathrm{End}(N_B)$ their endomorphism rings, (β, γ) a semilinear isomorphism of the modules M_A and N_B. Then there exists (and unique) an isomorphism of rings $\alpha: R \to S$ such that (α, β, γ) is a semilinear isomorphism of the bimodules ${}_RM_A$ and ${}_SN_B$.*

Proof. Define the mapping $\alpha: R \to S$ setting

$$r^\alpha(n) = [r(n^{\beta^{-1}})]^\beta$$

for all $n \in N$, $r \in R$. It is clear that α is an isomorphism of the rings R and S, and moreover

$$r^\alpha m^\beta a^\gamma = [r((m^\beta a^\gamma)^{\beta^{-1}})]^\beta = (rma)^\beta.$$

Suppose now that for some α the triple (α, β, γ) is a semilinear isomorphism of the bimodules ${}_RM_A$ and ${}_SN_B$. Then for every $n = \beta(m) \in N$ (i.e., $m = n^{\beta^{-1}}$) we have

$$r^\alpha(n) = r^\alpha m^\beta = (rm)^\beta = [r(n^{\beta^{-1}})]^\beta,$$

thus we have proved uniqueness of α. □

Notation 2.8. Let A, B, R, and S be rings, ${}_SN_B$ a bimodule, $\alpha\colon R \to S$ and $\gamma\colon A \to B$ isomorphisms of rings. We use the notation $[\alpha, N, \gamma]$ for the R-A-bimodule N, in which

$$r * n = r^{\alpha} n \quad \text{and} \quad n * a = n a^{\gamma} \quad \text{for all} \quad r \in R, n \in N, a \in A.$$

In particular, if $\alpha = \mathrm{id}_R$ we get an A-module $[N, \gamma]$.

Lemma 2.9. *Let R be an abundant subring of the endomorphism ring $E = \mathrm{End}(M_A)$ of a strong generator M_A with $e^2 = e \in R$, $r(e) = 1$, a basis $\{x\}$ of eM_A, the left ideal Re of E (see Lemma 2.3 and Definition 2.4). Then:*

1. *the mapping $\beta\colon {}_ERe \to {}_EM$, where $(ye)^{\beta} = yex$ for $y \in R$, is an isomorphism of the left E-modules;*
2. $\mathrm{End}({}_RM) = \mathrm{End}({}_EM) \cong A$;
3. *the mapping $\gamma\colon eRe \to A$, where $m(eye)^{\gamma} = (m^{\beta^{-1}} eye)^{\beta}$ for $m \in M$ and $y \in R$, is an isomorphism of rings.*

Moreover, $(\mathrm{id}_R, \beta, \gamma)$ is a semilinear isomorphism of the bimodules ${}_RRe_{eRe}$ and ${}_RM_A$.

Proof. 1) It is clear that β is a homomorphism of Abelian groups. If $(ye)^{\beta} = 0$ for some $y \in R$, then $yex = o$, and therefore

$$(ye)M = (ye)(eM) = (ye)(xA) = 0,$$

i.e., $ye = 0$ and β is a monomorphism. Since $x = ex$ and $Rx = M$, $M = Rx = Rex$ and β is surjective. Furthermore, for $d \in E$ we have

$$(dre)^{\beta} = (dre)x = d(rex) = d(re)^{\beta},$$

thus β is an isomorphism of the left E-modules.

2) It is clear that $\mathrm{End}({}_EM) \subseteq \mathrm{End}({}_RM)$. Let $\varphi \in \mathrm{End}({}_RM)$, $m \in M$ and $d \in E$. Since $ex = x$ and $Rex = Rx = M$, we get $m = rex$ for some $x \in R$. It follows from $ERe = Re$ that $dre \in Re$. Therefore

$$(dm)\varphi = (drex)\varphi = [(dre)x]\varphi = (dre)(x\varphi) = d\,[(rex)\varphi] = d(m\varphi).$$

It means that $\varphi \in \mathrm{End}({}_EM)$. Thus, $\mathrm{End}({}_EM) = \mathrm{End}({}_RM)$.

By Remark 2.2 our strong generator $M_A \cong A_A \oplus H$ is a generator, and therefore (see, for example [64, Prop. 4.1]) $\mathrm{End}({}_EM) \cong A$ (under the canonical mapping $E \to \mathrm{End}({}_EM)$).

3) By 1) the pair (id_R, β) is a semilinear isomorphism of the left R-modules ${}_RRe$ and ${}_RM$, and by Lemma 2.7 the mapping γ is an isomorphism of rings such that the triple $(\mathrm{id}_R, \beta, \gamma)$ is a semilinear transformation of the bimodules ${}_RRe_{eRe}$ and ${}_RM_A$. □

Lemma 2.10. *Let M_A be a right A-module, $u^2 = u \in R = \mathrm{End}(M_A)$ and $v^2 = v \in R = \mathrm{End}(M_A)$. Then uRu-vRv-bimodules*

$$uRv \equiv u(\mathrm{End}(M_A))v \cong \mathrm{Hom}(vM_A, uM_A)$$

are canonically isomorphic. In particular:

if $u = 1$, then R-vRv-bimodules

$$Rv \equiv (\mathrm{End}(M_A))v \cong \mathrm{Hom}(vM_A, M_A)$$

are canonically isomorphic;

if $v = 1$, then uRu-R-bimodules

$$uR \equiv u\,(\mathrm{End}(M_A)) \cong \mathrm{Hom}(M_A, uM_A)$$

are canonically isomorphic.

Lemma 2.11. *Let $e^2 = e \in R = \mathrm{End}(M_A)$, $r(e) = 1$, $0 \neq w^2 = w \in R$. Then the following conditions are equivalent:*

1. *wM_A is a finitely generated projective right A-module;*
2. *$(wRe)_{eRe}$ is a finitely generated projective right eRe-module;*
3. *$wReRw = wRw$.*

Proof. At the beginning we show that 1) $\Leftrightarrow$ 2). Taking into account that $R = \mathrm{End}(M_A)$ is an abundant subring and by Lemma 2.9, we see that the triple $(\mathrm{id}_R, \beta, \gamma)$ is a semilinear isomorphism of the bimodules ${}_R Re_{eRe}$ and ${}_R M_A$. Let ${}_R Re_A = [\mathrm{id}_R, Re, \gamma^{-1}]$ (see Notation 2.8). Then for $y \in R$, we have

$$[(ye) * a]^\beta = (yea^{\gamma^{-1}})^\beta = (ye)^\beta a,$$

it means that β is an isomorphism of the R-A-bimodules ${}_R Re_A$ and ${}_R M_A$, and moreover, $(wRe)^\beta = w(Re)^\beta = wM$.

Since $\gamma\colon eRe \to A$ is an isomorphism of rings, the right eRe-module wRe is finitely generated and projective if and only if wM_A is a finitely generated and projective A-module.

2) $\Rightarrow$ 3). Suppose we have 2). By Lemma 2.9, point 3, the triple

$$(\mathrm{id}_R, \beta, \gamma)\colon {}_R Re_{eRe} \to {}_R M_A$$

is a semilinear isomorphism. Therefore,

$$R \equiv \mathrm{End}(M_A) \cong \mathrm{End}(Re_{eRe}).$$

By Lemmas 2.9 and 2.10:

$$wRw \cong \mathrm{End}(wM_A) \cong \mathrm{End}(wRe_{eRe});$$

$$eRw \cong \mathrm{Hom}(wM_A, eM_A) \cong \mathrm{Hom}(wRe_{eRe}, eRe_{eRe}).$$

By our assumption, wRe_{eRe} is a finitely generated projective right eRe-module, and therefore there exists a finite dual basis of wRe_{eRe} (see, for example [64, Lemma 3.23])

$$\{x_i \in wRe;\ y_i \in eRw \mid 1 \le i \le n\}$$

for which

$$\sum_{i=1}^{n} x_i(y_i z) = z \quad \text{for all} \quad z \in wRe.$$

It follows that

$$\sum_{i=1}^{n}(x_i y_i)z = z \quad \text{and} \quad (w - \sum_{i=1}^{n} x_i y_i)z = 0 \quad \text{for all} \quad z \in wRe.$$

Furthermore,

$$w - \sum_{i=1}^{n} x_i y_i \in wRw = \mathrm{End}(wRe_{eRe}).$$

Therefore, $w = \sum_{i=1}^{n} x_i y_i \in wReRw$. Thus

$$wReRw \supseteq (wRw)(wReRw)(wRw) \supseteq (wRw)w(wRw) = wRw,$$

it means that $wReRw = wRw$.

3) $\to$ 2). Suppose we have 3). Then

$$w = \sum_{i=1}^{n} x_i y_i \quad \text{for some} \quad x_1, \dots x_n \in Re,\ y_1, \dots y_n \in Rw.$$

It is clear that $y_i z \in eRe$ for all $z \in wRe$, $1 \le i \le n$, and $z = \sum_{i=1}^{n} x_i(y_i z)$. As we have remarked, $eRw \cong \mathrm{Hom}(wRe_{eRe}, eRe_{eRe})$. Thus,

$$\{x_i \in wRe;\ y_i \in eRw \mid 1 \le i \le n\}$$

is a finite dual basis of the right eRe-module wRe. It shows that wRe is a finitely generated projective right eRe-module. □

Corollary 2.12. *Let M_A be a strong generator, $0 \ne w^2 = w \in R = \mathrm{End}(M_A)$. Then the following conditions are equivalent:*

1) *wM_A is a finitely generated projective right A-module;*
2) *there exists an idempotent $e^2 = e \in R$, $r(e) = 1$, for which wRe_{eRe} is a finitely generated projective right eRe-module;*
2′) *for any idempotent $e^2 = e \in R$ such that $r(e) = 1$ the right eRe-module wRe_{eRe} is finitely generated and projective;*
3) *there exists an idempotent $e^2 = e \in R$, $r(e) = 1$, for which $wReRw = wRw$;*
3′) *$wReRw = wRw$ for all $e^2 = e \in R$ with $r(e) = 1$.*

3. Isomorphisms of endomorphism rings of strong generators

Definition 3.1. Let Mod-A and Mod-B be the categories of right modules over the rings A and B respectively. A covariant additive functor h: Mod-$A \to$ Mod-B is called *a full embedding* if for any right A-modules X_A and Y_A the canonical mapping

$$h: \mathrm{Hom}(X_A, Y_A) \to \mathrm{Hom}(h(X)_B, h(Y)_B)$$

is an isomorphism of Abelian groups (see [64, Ch.2]).

Let ${}_BU_A$ be a B-A-bimodule and $h^U = \mathrm{Hom}(U_A, -)$. If U_A is a generator and $B = \mathrm{End}(U_A)$, then h^U: Mod-$A \to$ Mod-B is a full embedding (see, for example [64, Theorem 4.24]). If U_A is a finitely generated projective generator, then h^U is a Morita equivalence (see for example [64, Theorem 4.29]). For $M_A \in$ Mod-A we set

$$\mathrm{tr}(M_A) = \sum_{f \in \mathrm{Hom}(M_A, A_A)} f(M)$$

(*the trace of the module* M_A, see [64, Ch.3]).

Definition 3.2. Let A and B be rings with 1, M_A and N_B right modules, $R = \mathrm{End}(M_A)$ and $S = \mathrm{End}(N_B)$ their endomorphism rings respectively, $\alpha: R \to S$ an isomorphism of rings. We say that α is induced by a generator U_A if:

1. $B = \mathrm{End}(U_A)$;
2. there exists an isomorphism $\tau: h^U(M_A) \to N_B$ of B-modules such that $[h^U(r)x]^\tau = r^\alpha x^\tau$ for all $x \in h^U(M), r \in R$ (i.e., $\alpha = \tau h^U \tau^{-1}$).

Theorem 3.3 ([203, Theorem 2.3]). *Let A and B be rings with* 1, *modules M_A and N_B strong generators, $R = \mathrm{End}(M_A)$ and $S = \mathrm{End}(N_B)$ their endomorphism rings, $\alpha: R \to S$ an isomorphism of the endomorphism rings. Then the following conditions are equivalent:*

1. *for any idempotent endomorphism $e^2 = e \in R$ of rank* 1 *the B-module $e^\alpha N_B$ is finitely generated and projective;*
2. *for any idempotent endomorphisms $e^2 = e \in R$, $u^2 = u \in S$, $r(e) = 1$, $r(u) = 1$, we have $e^\alpha SuSe^\alpha = e^\alpha Se^\alpha$;*
3. *for any idempotent $u^2 = u \in S$ of rank* 1 *the right A-module $u^{\alpha^{-1}}M$ is a generator in* Mod-A;
4. *there exists an idempotent $e^2 = e \in R$ of rank* 1 *such that $e^\alpha N_B$ is a finitely generated projective right B-module;*
5. *there exist idempotents $e^2 = e \in R$, $u^2 = u \in S$, $r(e) = 1$, $r(u) = 1$, such that $e^\alpha SuSe^\alpha = e^\alpha Se^\alpha$;*
6. *there exists an idempotent $u^2 = u \in S$ of rank* 1 *such that the right A-module $u^{\alpha^{-1}}M_A$ is a generator in* Mod-A;

7. *the isomorphism α is induced by h^U for some generator U_A in* Mod-A.

Proof. By Lemma 2.11 we see that 1) $\Leftrightarrow$ 2).

We show that 2) $\Leftrightarrow$ 3). Let now $e^2 = e \in R$, $r(e) = 1$, $u^2 = u \in S$, $r(u) = 1$, $v = u^{\alpha^{-1}}$. Consider the semilinear isomorphism $(\mathrm{id}_R, \beta, \gamma)$ of bimodules ${}_R Re_{eRe}$ and ${}_R M_A$ from Lemma 2.9, which induces isomorphism of the right A-modules $[vRe, \gamma^{-1}]_A$ and vM_A. Thus, vM_A is a generator in Mod-A if and only if $[vRe, \gamma^{-1}]_A$ is a generator in Mod-A. Since γ is an isomorphism of the rings eRe and A, $[vRe, \gamma^{-1}]_A$ is a generator in Mod-A if and only if vRe_{eRe} is a generator in Mod-eRe, what means in its turn that $eRvRe = \mathrm{tr}_{eRe}(vRe) = eRe$ (see [64, Prop. 3.26], Lemmas 2.10 and 2.11). Taking into account that α is an isomorphism and $u = v^\alpha$, we see that equalities $eRvRe = eRe$ and $e^\alpha SuSe^\alpha = e^\alpha Se^\alpha$ are equivalent.

Analogously we check that 5) $\Leftrightarrow$ 6).

The implication 1) $\Rightarrow$ 4) is evident.

6) $\Leftrightarrow$ 7). Suppose that 6) is fulfilled and $v = u^{\alpha^{-1}}$. Then vM_A is a generator in Mod-A. Let $(\mathrm{id}_S, \varepsilon, \sigma)$ be the semilinear isomorphism of bimodules ${}_S Su_{uSu}$ and ${}_S N_B$ from Lemma 2.9 and $\xi = \sigma^{-1}\alpha^{-1}$. Then $\xi: B \to vRv$ is an isomorphism of rings. Let ${}_B U_A = [\xi, vM]_A$. By Lemma 2.10 we see that $\mathrm{End}(vM_A) = vRv$. Therefore $\mathrm{End}(U_A) = B$. It is clear that U_A is a generator in Mod-A and A-modules U_A and vM_A coincide. By Lemma 2.10, $\mathrm{Hom}(vM_A, M_A) = Rv$ and thus $h^U(M)_B = [Rv, \xi]_B$.

Let τ be the composition of isomorphisms of Abelian groups

$$Rv \xrightarrow{\alpha} Su \xrightarrow{\varepsilon} N.$$

We show that $\tau = \alpha\varepsilon$ is an isomorphism of B-modules $[Rv, \xi]_B$ and N_B. Indeed, if $x \in Rv$ and $b \in B$, then

$$x * b = x(b^{\sigma^{-1}})^{\alpha^{-1}}, \quad b^{\sigma^{-1}} \in uSu \subseteq S, \quad (b^{\sigma^{-1}})^{\alpha^{-1}} \in vRv \subseteq R.$$

Thus

$$(x * b)^\tau = [(x * b)^\alpha]^\varepsilon = [(x(b^{\sigma^{-1}})^{\alpha^{-1}})^\alpha]^\varepsilon = [x^\alpha b^{\sigma^{-1}}]^\varepsilon.$$

Since $(\mathrm{id}_S, \varepsilon, \sigma)$ is a semilinear isomorphism of the bimodules ${}_S Su_{uSu}$ and ${}_S N_B$,

$$[x^\alpha b^{\sigma^{-1}}]^\varepsilon = (x^\alpha)^\varepsilon (b^{\sigma^{-1}})^\sigma = x^\tau b.$$

Thus, $(x * b)^\tau = x^\tau b$ for all $x \in Rv$, $b \in B$.

By Lemma 2.10, the canonical embedding $Rv \to \mathrm{Hom}(vM_A, M_A)$ is an injection of R-modules. Therefore, $h^U(r)x = rx$ for all $x \in RV$, $r \in R$. Hence

$$\left[h^U(r)x\right]^\tau = [(rx)^\alpha]^\varepsilon = \left[r^\alpha x^\alpha\right]^\varepsilon = \left(r^\alpha\right)^{\mathrm{id}_S} \left(x^\alpha\right)^\varepsilon = r^\alpha x^\tau.$$

Thus, the isomorphism α is induced by the generator U_A in Mod-A.

We prove that 7) $\Rightarrow$ 1). Suppose that 7) takes place and $e^2 = e \in R$, $r(e) = 1$, U_A is a generator in Mod-A which induces the isomorphism α and $\tau: h^U(M)_B \to N_B$ is such an isomorphism of B-modules that $(h^U(r)x)^\tau = r^\alpha x^\tau$ for all $r \in R$, $h \in h^U(M)$.

Since eM_A is a free A-module of rank 1, $\mathrm{Hom}(U_A, eM_A)$ is a finitely generated right B-module ([64, Corollary 4.2.1]). Therefore,

$$\left(e^\alpha N\right)^{\tau^{-1}} = h^U(e)h^U(M) = h^U(e)\,\mathrm{Hom}(U_A, M_A) = \mathrm{Hom}(U_A, eM_A).$$

Thus, $e^\alpha N$ is a finitely generated projective right B-module. □

Corollary 3.4. *Let M_A be a generator in* Mod-A, *and let N_B be a progenerator in* Mod-B *(i.e., a finitely generated projective generator in* Mod-B*). Suppose that $\alpha\colon \mathrm{End}(M_A) \to \mathrm{End}(N_B)$ is an isomorphism of their endomorphism rings. Then α is induced by h^U for some generator U_A in* Mod-A.

Proof. Since $\widetilde{M_A} = \oplus_{i=1}^r M_i$, $\widetilde{N_B} = \oplus_{i=1}^r N_i$ where $M_i \cong M_A$, $N_i \cong N_B$, are strong generators for a sufficiently large number r, we may consider the isomorphism $\tilde{\alpha}\colon \tilde{R} \to \tilde{S}$ of their endomorphism rings $\tilde{R} = \mathrm{End}(\widetilde{M_A}) = R_r$ and $\tilde{S} = \mathrm{End}(\widetilde{N_B}) = S_r$ for which $\tilde{\alpha}(E_{ii}) = E_{ii}$ (here $\{E_{ij}\}$ is the set of matrix units of the ring of $(r\times r)$-matrices). Now by Theorem 3.3 (point 4, for the finitely generated projective B-module $e^{\tilde{\alpha}}\widetilde{N_B}$) there exists a generator U_A for which the functor h^U induces the isomorphism $\tilde{\alpha}$ (it means that $B = \mathrm{End}(U_A)$, and there exists an isomorphism $\tau\colon h^U(\widetilde{M_A}) \to \widetilde{N_B}$ with $\tau h^U \tau^{-1} = \tilde{\alpha}$. Then

$$\begin{aligned} h^U(M_A) &= h^U(E_{11}\widetilde{M_A}) = h^U(E_{11})h^U(\widetilde{M_A}) = \tau^{-1}\tilde{\alpha}(E_{11})\tau h^U(\widetilde{M_A}) \\ &= \tau^{-1}E_{11}\widetilde{N_B} = \tau^{-1}N_B, \end{aligned}$$

i.e., $\tau\colon h^U(M_A) \to N_B$ is an isomorphism and for all $r \in R = \mathrm{End}(M_A)$, $x \in h^U(M) = h^U(E_{11}\widetilde{M_A})$ we have

$$\left[h^U(r)x\right]^\tau = \left[h^U(rE_{11})x\right]^\tau = (rE_{11})^{\tilde{\alpha}}x^\tau = r^\alpha E_{11}x^\tau = r^\alpha x^\tau.$$

Thus, α is induced by the generator U_A in Mod-A. □

Definition 3.5. A ring A is called *a right GI-ring* if each nonzero right ideal of A is a generator in Mod-A.

Corollary 3.6. *Let A be a right GI-ring, M_A a torsionless module (in the sense of H. Bass, i.e., $0 \to M_A \to M_A^{**}$, see* [64, Ch.3]*), N_B a generator in* Mod-B, *$\alpha\colon \mathrm{End}(M_A) \to \mathrm{End}(N_B)$ an isomorphism of their endomorphism rings. Then α is induced by a full emdedding h^U for some generator U_A in* Mod-A.

Theorem 3.7 ([203, Theorem 2.7]). *Let M_A and N_B be strong generators in* Mod-A *and* Mod-B *respectively, $\alpha\colon \mathrm{End}(M_A) \to \mathrm{End}(N_B)$ an isomorphism of their endomorphism rings. Then the following conditions are equivalent:*

1. *if $e^2 = e \in \mathrm{End}(M_A)$, $r(e) = 1$, then $e^\alpha N_B$ is a progenerator in* Mod-B;
2. *if $e^2 = e \in \mathrm{End}(M_A)$, $r(e) = 1$, and $u^2 = u \in \mathrm{End}(N_B)$, $r(u) = 1$, then $e^\alpha N_B$ and $u^{\alpha^{-1}} M_A$ are finitely generated projective right modules;*

3. *if* $e^2 = e \in \mathrm{End}(M_A)$, $r(e) = 1$, *and* $u^2 = u \in \mathrm{End}(N_B)$, $r(u) = 1$, *then* $e^\alpha N_B$ *and* $u^{\alpha^{-1}} M_A$ *are generators in* Mod-B *and* Mod-A, *respectively;*
4. *there exists an idempotent* $e^2 = e \in \mathrm{End}(M_A)$ *such that* $r(e) = 1$ *and* $e^\alpha N_B$ *is a progenerator in* Mod-B*;*
5. *there exist idempotents* $e^2 = e \in \mathrm{End}(M_A)$ *and* $u^2 = u \in \mathrm{End}(N_B)$ *such that* $r(e) = 1$, $r(u) = 1$, *and* $e^\alpha N_B$, $u^{\alpha^{-1}} M_A$ *are finitely generated projective modules;*
6. *there exist idempotents* $e^2 = e \in \mathrm{End}(M_A)$ *and* $u^2 = u \in \mathrm{End}(N_B)$ *such that* $r(e) = 1$, $r(u) = 1$, *and* $e^\alpha N_B$, $u^{\alpha^{-1}} M_A$ *are projective generators in* Mod-B *and* Mod-A, *respectively;*
7. *there exists a Morita equivalence* $h\colon \text{Mod-}A \to \text{Mod-}B$ *such that* $h(M_A) = N_B$ *and* $h(r)x = r^\alpha x$ *for all* $x \in N_B$ *and* $r \in R$ *(i.e.,* $r^\alpha = h(r)$*, in other words* α *is induced by a Morita equivalence* h*).*

Proof. 1) $\Leftrightarrow$ 2). The condition 1) that $e^\alpha N_B$ is a finitely generated projective generator in Mod-B for all $e^2 = e \in R$, $r(e) = 1$, by Theorem 3.3 (for α^{-1}, 3) $\Leftrightarrow$ 1)) is equivalent to the fact that $u^{\alpha^{-1}} M_A$ is a finitely generated projective A-module for all $u^2 = u \in S$, $r(u) = 1$, i.e., to the condition 2).

1) $\Leftrightarrow$ 3). Again by Theorem 3.3 (1) $\Rightarrow$ 3)) the condition 1) is equivalent to the fact that the module $u^{\alpha^{-1}} M_A$ is a generator in Mod-A for all $u^2 = u \in S$ with $r(u) = 1$, i.e., to the condition 3).

Analogously, we prove 4) $\Leftrightarrow$ 5) and 4) $\Leftrightarrow$ 6), changing quantifiers $\forall$ and $\exists$.

3) $\Leftrightarrow$ 6). The condition that the module $e^\alpha N_B$ is a generator for all $e^2 = e \in R$ with $r(e) = 1$ is equivalent (by Theorem 3.3 for α^{-1}, 3) $\Leftrightarrow$ 6)) to existence of $e^2 = e \in R$, with $r(e) = 1$, for which $e^\alpha N_B$ is a generator. By Theorem 3.3 (for α, 3) $\Leftrightarrow$ 6)), $u^{\alpha^{-1}} M_A$ is a generator in Mod-A for all $u^2 = u \in S$ with $r(u) = 1$ if and only if there exists $u^2 = u \in S$ with $r(u) = 1$ for which $u^{\alpha^{-1}} M_A$ is a generator. Thus we have shown that 3) $\Leftrightarrow$ 6).

3) $\Rightarrow$ 7). Since for all $e^2 = e \in R$ with $r(e) = 1$ the module $e^\alpha N_B$ is a generator in Mod-B, by Theorem 3.3 (for α^{-1}, 3) $\Rightarrow$ 1)) for all $u^2 = u \in S$ with $r(u) = 1$ the module $u^{\alpha^{-1}} M_A$ is finitely generated projective, and by 3) it is a generator. By Theorem 3.3, point 7), α is induced by the functor $h = h^U$, $U_A = u^{\alpha^{-1}} M_A$, which is a Morita equivalence.

7) $\Rightarrow$ 1). Let α is induced by h^U, where U_A ia a finitely generated projective generator in Mod-A,

$$\tau\colon h^U(M_A) = \mathrm{Hom}(U_A, M_A) \to N_B$$

is an isomorphism of B-modules and $[h^U(r)x]^\tau = r^\alpha x^\tau$ for all $x \in h^U(M_A)$, $r \in R$, i.e., $\tau h^U \tau^{-1} = \alpha$. Then for any $e^2 = e \in R$ with $r(e) = 1$ the A-module eM_A is a finitely generated projective generator in Mod-A, and therefore

$$e^{\alpha}N_B \cong (e^{\alpha}N_B)^{\tau^{-1}} = \mathrm{Hom}(U_A, eM_A) = h(eM_A)$$

is a finitely generated projective generator, i.e., we get 1). □

Corollary 3.8 (Bolla's Theorem [27, Theorem 2.1]). *Let M_A and N_B be two right progenerators,*

$$\alpha: R = \mathrm{End}(M_A) \to \mathrm{End}(N_B) = S$$

an isomorphism of their endomorphism rings. Then there exists a Morita equivalence h: Mod-A $\to$ Mod-B such that $h(M_A) = N_B$ and $h(r)x = r^{\alpha}x$ for all $x \in N_A$, $r \in R$, i.e., $r^{\alpha} = h(r)$.

Proof. By Corollary 3.4, the isomorphism α is induced by the generator U_A which is a direct summand $u^{\alpha^{-1}}\tilde{M}$ of the module $\widetilde{M_A} = M \oplus \cdots \oplus M$, and therefore U_A is a finitely generated projective A-module, it means that $h = h^U$ is a Morita equivalence inducing α. □

Corollary 3.9. *Let A and B be right GI-rings, M_A and N_B torsionless modules, α: $\mathrm{End}(M_A) \to \mathrm{End}(N_B)$ an isomorphism of the endomorphism rings. Then there exists a Morita equivalence h: Mod-A $\to$ Mod-B such that $h(M_A)= N_B$ and $h(r)x = r^{\alpha}x$ for all $x \in N_B$ and $r \in R$, i.e., $r^{\alpha} = h(x)$ and α is induced by h.*

Theorem 3.10 (Mikhalev [199, 200], [203, Theorem 2.10]). *Suppose that M_A and N_B are strong generators in* Mod-A *and* Mod-B*, respectively, and let*

$$\alpha: \mathrm{End}(M_A) \to \mathrm{End}(N_B)$$

be an isomorphism of their endomorphism rings. Then the following conditions are equivalent:

1. *if $e^2 = e \in \mathrm{End}(M_A)$ and $r(e) = 1$, then $r(e^{\alpha}) = 1$, i.e., $e^{\alpha}N_B \cong B_B$;*
2. *there exists an idempotent $e^2 = e \in \mathrm{End}(M_A)$ such that $r(e) = 1$ and $r(e^{\alpha}) = 1$;*
3. *α is induced by a semilinear isomorphism of the modules M_A and N_B.*

Proof. It is clear that 1) $\Rightarrow$ 2). By Lemma 2.9, 2) $\Rightarrow$ 3) (see also [199]).

3) $\Rightarrow$ 1). If the isomorphism $\alpha: R = \mathrm{End}(M_A) \to \mathrm{End}(N_B) = S$ is induced by a semilinear isomorphism $(\beta, \gamma): M_A \to N_B$, i.e., $(rma)^{\beta} = r^{\alpha}m^{\beta}a^{\gamma}$ for all $r \in R, m \in M, a \in A$ (see Lemma 2.7), then it follows from $M_A = eM_A \oplus H_A$, where $eM_A = xA$ is a free cyclic module with a basis $\{x\}$, that $e^{\alpha}N_B = (eM_A)^{\beta} = (xA)^{\beta} = x^{\beta}B$ and $0 = xa, a \in A$, if and only if $0 = (xa)^{\beta} = x^{\beta}a^{\gamma}$. Thus $r(e^{\alpha}) = 1$. □

Corollary 3.11. *Let B be a right fir (i.e., a ring of free right ideals), A a ring without nontrivial idempotents, M_A a generator in* Mod-A*, N_B a torsionless module,*

α: End(M_A) $\to$ End(N_B) an isomorphism of their endomorphism rings. Then α is induced by a semilinear isomorphism of the modules M_A and N_B.

Corollary 3.12. *Let $\mathbb{Z}$ be the ring of integers, M a subgroup of a direct power of $\mathbb{Z}$, $R = \operatorname{End}(M_{\mathbb{Z}})$. Then every automorphism of the ring R is inner.*

Remark 3.13. There exists a torsionless Abelian group which is not locally free (compare with [74]).

4. Anti-isomorphisms of endomorphism rings of strong generators

Definition 4.1. For $M_A \in$ Mod-A we consider the canonical homomorphism

$$\delta_M : M_A \to M^{**} = \operatorname{Hom}_A(\operatorname{Hom}_A(M_A, A_A), {}_AA)$$

with $\delta_M(m)[f] = f(m)$ for any $m \in M_A$ and $f \in {}_AM^* = \operatorname{Hom}_A(M_A, A_A)$. Recall that the module M_A is said to be *torsionless* (respectively, *reflexive*) if the canonical mapping $\delta_M : M_A \to M^{**}$ is injective (bijective). Analogously we define torsionless and reflexive left modules.

Lemma 4.2 (Wolfson [284, Lemma 1.1]). *Let M_A be a torsionless right module. Then there is a unique monomorphism $\eta \to \eta^*$ of* End(M_A) *into* End(${}_AM^*$) *satisfying $[f, \eta m] = [f\eta^*, m]$ for all $m \in M_A$ and $f \in {}_AM^*$. If M_A is reflexive, then the mapping is an isomorphism of* End(M_A) *onto* End(${}_AM^*$). *Conversely, if the mapping is onto* End(${}_AM^*$) *and M_A is a strong generator, then M_A is reflexive.*

Theorem 4.3 ([23], [24]). *Let ${}_SM$ be a reflexive S-module and let*

$$T: S\text{-Mod} \to R\text{-Mod}$$

be a Morita equivalence of the categories of left modules. Then $T(M)$ is a reflexive R-module.

Proof. By the Morita Theorem (see Definition 3.1) there is a bimodule ${}_SU_R$ such that ${}_SU$ is a progenerator, $R = \operatorname{End}(S_U)$ and the functors T and $h = \operatorname{Hom}_S({}_SU, -)$ are naturally equivalent. By [275, 46.4], the mapping

$$\alpha: U \otimes_R \operatorname{Hom}({}_SU, {}_SM) \to {}_SM, \tag{1}$$

for which $\alpha(u \otimes f) = f(u)$ for all $u \in U$, $f \in \operatorname{Hom}({}_SU, {}_SM)$, is an isomorphism of S-modules.

Taking into account [275, 12.12], we see that the mapping

$$\beta: \operatorname{Hom}_R\Big(h(M), \operatorname{Hom}_S(U_S, U_S)\Big) \to \operatorname{Hom}_S\Big(U \otimes_R h(M), U_S\Big) \tag{2}$$

with $\beta(f)(u \otimes x) = f(x)(u)$ for all $f \in \mathrm{Hom}_R(h(M), \mathrm{Hom}_S(U_S, U_S))$, $u \in U$, $x \in h(M)$ is an isomorphism of right R-modules.

Since U_S is a finitely generated projective S-module, the mapping

$$\gamma\colon \mathrm{Hom}_S\left(U_S, M_S\right) \to \mathrm{Hom}_R\left(\mathrm{Hom}_S(M_S, U_S), R\right) \tag{3}$$

with $\gamma(f)(g) = f \cdot g$ for all $f \in \mathrm{Hom}_S(U_S, M_S)$, $g \in \mathrm{Hom}_S(M_S, U_S)$, is an isomorphism of left R-modules.

The statement (3) for $U_S = S_S$ (and therefore for $R = \mathrm{End}(U_S) = S$) follows from reflexivity of M_A. In general, it follows from additivity of the functor $\mathrm{Hom}(-,-)$.

Consider now the following diagram:

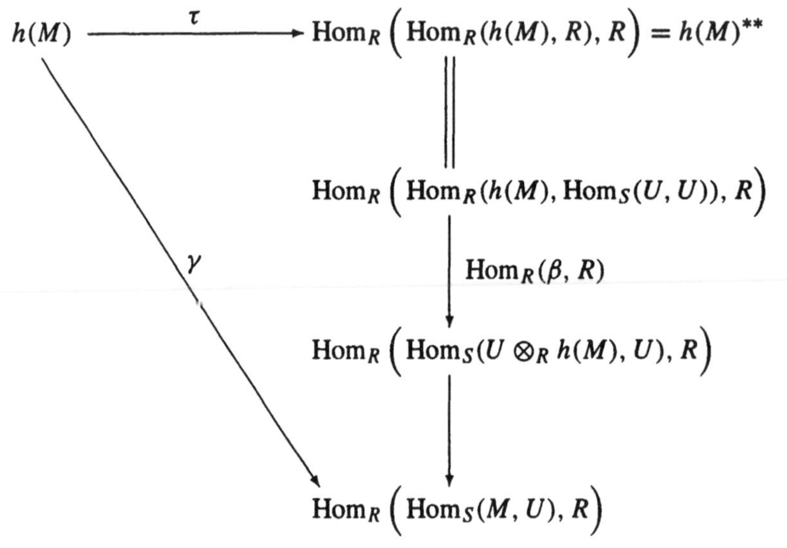

where τ is the canonical homomorphism. It is easy to check that this diagram is commutative. Therefore, τ is an isomorphism of R-modules and $T(M)$ is a reflexive R-module. □

Remark 4.4. Let $A_1 = A_1(F)$ be the first Weyl algebra over a field F of zero characteristic which is defined by two generators x and y and one defining relation $xy - yx = 1$. Then A_1 is a simple hereditary (non Artinian) Noetherian domain (see [190, Example 5.2.11]) for which the category of A_1-modules of finite length is "wild" (see [142]), i.e., for any F-algebra R generated by less than $\aleph_i$ elements (here $\aleph_i$ is the first uncountable inaccessible cardinal) there exists an exact full embedding

$$\mathcal{F}_R\colon \text{Mod-}R \to \text{Mod-}A_1$$

such that $\mathcal{F}_R(M)$ has socle height 2 whenever $M \neq 0$ (moreover, if R is a finitely generated F-algebra, then $\mathcal{F}_R$ takes finite dimensional R-modules into A_1-modules of finite length). It is clear that R_R is a reflexive module for which $\mathcal{F}_R(R_R)$ has socle height 2, therefore, $\mathrm{Hom}_{A_1}(\mathcal{F}_R(R_R), A_1) = 0$, i.e., $\mathcal{F}_R(R_R)$ is not a reflexive module. Thus, images of reflexive modules under full (exact) embeddings may not be reflexive.

Let now M_A be a right A-module, $M^* = \mathrm{Hom}_A(M_A, A_A)$ and $R = \mathrm{End}(M_A)$. It is clear that M^* is a right R-module with $(fr)m = f(rm)$ for all $f \in M^*$, $m \in M$ and $r \in R$.

Definition 4.5. We say that the anti-isomorphism

$$\alpha\colon \mathrm{End}(M_A) \to \mathrm{End}(N_B)$$

of the endomorphism rings *is induced by a Morita anti-equivalence* if there exists an equivalence $H\colon \text{Mod-}A^{\mathrm{op}} \to \text{Mod-}B$ of the categories of modules for which the canonical homomorphism $R \to \mathrm{End}({}_AM^*)$ is an isomorphism, $H(M^*) = N$ and $H(r)n = r^\alpha n$ for all $r \in R^{\mathrm{op}}$, $n \in N$.

Theorem 4.6. *Let M_A and N_B be strong generators and*

$$\alpha\colon \mathrm{End}(M_A) \to \mathrm{End}(N_B)$$

an anti-isomorphism of their endomorphism rings. Suppose that M_A is a torsionless module (i.e., the canonical homomorphism $M_A \to M_A^{**}$ is a monomorphism). Then the following conditions are equivalent:

1. *if $e^2 = e \in \mathrm{End}(M_A)$, $r(e) = 1$, then $e^\alpha N_B$ is a progenerator in* Mod-B;
2. *if $e^2 = e \in \mathrm{End}(M_A)$, $r(e) = 1$, and $u^2 = u \in \mathrm{End}(N_B)$, $r(u) = 1$, then $e^\alpha N_B$ and $u^{\alpha^{-1}} M_A$ are finitely generated projective modules;*
3. *if $e^2 = e \in \mathrm{End}(M_A)$, $r(e) = 1$, and $u^2 = u \in \mathrm{End}(N_B)$, $r(u) = 1$, then $e^\alpha N_B$ and $u^{\alpha^{-1}} M_A$ are generators;*
4. *there exists an idempotent $e^2 = e \in \mathrm{End}(M_A)$ such that $r(e) = 1$ and $e^\alpha N_B$ is a progenerator in* Mod-B;
5. *there exist idempotents $e^2 = e \in \mathrm{End}(M_A)$, $r(e) = 1$, and $u^2 = u \in \mathrm{End}(N_B)$, $r(u) = 1$, such that $e^\alpha N_B$ and $u^{\alpha^{-1}} M_A$ are finitely generated projective modules;*
6. *there exist idempotents $e^2 = e \in \mathrm{End}(M_A)$, $r(e) = 1$, and $u^2 = u \in \mathrm{End}(N_B)$, $r(u) = 1$, such that $e^\alpha N_B$ and $u^{\alpha^{-1}} M_A$ are generators;*
7. *the anti-isomorphism α is induced by a Morita anti-equivalence*

$$H\colon \text{Mod-}A^{\mathrm{op}} \to \text{Mod-}B.$$

Under any of the conditions 1)–7) *we have that M_A and N_B are in fact reflexive modules.*

Proof. Let $R = \mathrm{End}(M_A)$ and $S = \mathrm{End}(N_B)$. Since M_A and N_B are strong generators, there exist idempotents $w^2 = w \in R$, $r(w) = 1$, and $v^2 = v \in S$, $r(v) = 1$ (see Section 2). By Lemma 2.9, without loss of generality we may suppose that $A = wRw$, $B = vSv$, $M_A = Rw_{wRw}$, $N_B = Sv_{vSv}$.

2) $\Rightarrow$ 1). Let $e^2 = e \in R$ with $r(e) = 1$, as in 1). Applying 2) for $e \in R$ and $u = v \in S$, $r(v) = 1$, we have that $e^\alpha N_B = e^\alpha Sv_{vSv}$ and $v^{\alpha^{-1}} M_A = v^{\alpha^{-1}} Rw_{wRw}$ are finitely generated projective modules. From Lemma 2.11 (for e and $w = v^{\alpha^{-1}}$ in R, 2) $\Rightarrow$ 3)) it follows that

$$v^{\alpha^{-1}} ReRv^{\alpha^{-1}} = v^{\alpha^{-1}} Rv^{\alpha^{-1}}.$$

Applying the anti-isomorphism α, we get

$$vSe^\alpha Sv = vSv. \tag{4}$$

By Lemma 2.10,

$$vSe^\alpha = \mathrm{Hom}_{vSv}(e^\alpha Sv, vSv).$$

Taking into account (4) we see that $e^\alpha Sv$ is a generator in Mod-vSv. Therefore, $e^\alpha Sv_{vSv}$ is a progenerator in Mod-vSv.

3) $\Rightarrow$ 1). Let again $e^2 = e \in R$ with $r(e) = 1$ as in 1). Applying 3) for $e \in R$ and $u = v \in S$, $r(v) = 1$, we have that $e^\alpha N_B = e^\alpha Sv_{vSv}$ and $v^{\alpha^{-1}} M_A = v^{\alpha^{-1}} Rw_{wRw} = v^{\alpha^{-1}} Re_{eRe} =$ are generators.

By Lemma 2.10,

$$eRv^{\alpha-1} = \mathrm{Hom}_{eRe}(v^{\alpha^{-1}} Re, eRe).$$

Since $v^{\alpha^{-1}} Re_{eRe}$ is a generator,

$$eRv^{\alpha^{-1}} Re = \mathrm{Hom}_{eRe}(v^{\alpha^{-1}} Re, eRe)v^{\alpha^{-1}} Re = eRe.$$

Thus, $eRv^{\alpha^{-1}} Re = eRe$. Applying α we get $e^\alpha SvSe^\alpha = e^\alpha Se^\alpha$. By Lemma 2.11 (for v, $r(v) = 1$, and e^α in S, 3) $\Rightarrow$ 2)), $e^\alpha Sv_{vSv}$ is a right finitely generated projective vSv-module. Since we have known that $e^\alpha Sv_{vSv}$ is a generator, it is in fact a progenerator.

1) $\Rightarrow$ 2) and 1) $\Rightarrow$ 3). Suppose now that we have 1), and let $e^2 = e \in \mathrm{End}(M_A) = R$, $r(e) = 1$, and and $u^2 = u \in \mathrm{End}(N_B) = S$, $r(u) = 1$, as in 2) and 3). By 1), we know that $e^\alpha N_B = e^\alpha Sv_{vSv} = e^\alpha Su_{uSu}$ is a progenerator in Mod-B.

Since $e^\alpha N_B = e^\alpha Su_{uSu}$ is finitely generated and projective, by Lemma 2.11 (for u, $r(u) = 1$, and e^α in S, 2) $\Rightarrow$ 3)) we have

$$e^\alpha SuSe^\alpha = e^\alpha Se^\alpha. \tag{5}$$

By Lemma 2.10,

$$\mathrm{Hom}_{uSu}(e^\alpha Su, uSu) = uSe^\alpha.$$

Since $e^\alpha N_B = e^\alpha Su_{uSu}$ is a generator, we see that

$$uSe^\alpha Su = uSu \tag{6}$$

Applying the anti-isomorphism α^{-1} to (5) and (6) we get

$$eRu^{\alpha^{-1}}Re = eRe, \quad \text{and} \quad u^{\alpha^{-1}}ReRu^{\alpha^{-1}} = u^{\alpha^{-1}}Ru^{\alpha^{-1}}.$$

By Lemma 2.11 (for e, $r(e) = 1$, and $u^{\alpha^{-1}}$ in R, 3) $\Rightarrow$ 2)), it follows from the second equality that $u^{\alpha^{-1}}M_A = u^{\alpha^{-1}}Re_{eRe}$ is a finitely generated projective A-module. Since, by Lemma 2.10,

$$eRu^{\alpha^{-1}} = \mathrm{Hom}_{eRe}(u^{\alpha^{-1}}Re, eRe),$$

it follows from the first equality that $u^{\alpha^{-1}}M_A = u^{\alpha^{-1}}Re_{eRe}$ is a generator.

Therefore, 1) $\Rightarrow$ 2) and 1) $\Rightarrow$ 3). Thus, 1) $\Leftrightarrow$ 2) $\Leftrightarrow$ 3). Equivalences 4) $\Leftrightarrow$ 5) $\Leftrightarrow$ 6) are proved analogously. It is clear that 1) $\Rightarrow$ 4).

Now we show that 4) $\Rightarrow$ 7). Without loss of generality we may suppose that by 4) there exists $e^2 = e \in \mathrm{End}(M_A) = R$ such that $r(e) = 1$ and $e^\alpha N_B = e^\alpha Sv_{vSv}$ is a progenerator in Mod-B. By Lemma 2.10,

$$_{eRe}eR_R = \mathrm{Hom}_{eRe}(Re, eRe). \tag{7}$$

Let $f \in \mathrm{Hom}_{eRe}(Re, eRe)$. Consider the mapping $f^\alpha: e^\alpha S \to e^\alpha Se^\alpha$, setting $xf^\alpha = (fx^{\alpha^{-1}})^\alpha$ for $x \in e^\alpha S$. It is clear that f^α is a homomorphism of left $e^\alpha Se^\alpha$-modules. Since α is an anti-isomorphism of rings R and S, the mapping $f \to f^\alpha$ induces an isomorphism of Abelian groups $\mathrm{Hom}_{eRe}(Re, eRe)$ and $\mathrm{Hom}_{e^\alpha Se^\alpha}(e^\alpha S, e^\alpha Se^\alpha)$. From (7) it follows that

$$Se^\alpha = \mathrm{Hom}_{e^\alpha Se^\alpha}(e^\alpha S, e^\alpha Se^\alpha). \tag{8}$$

By Lemma 2.10,

$$Se^\alpha = \mathrm{Hom}_{vSv}(e^\alpha Sv, Sv). \tag{9}$$

Since $e^\alpha Sv_{vSv}$ is a progenerator, by Morita theorem (see, for example, [275, Theorem 46.4]) and by (9)

$$S = \mathrm{End}_{vSv}(Sv) = \mathrm{End}_{e^\alpha Se^\alpha}(Se^\alpha). \tag{10}$$

From Lemma 2.10 and (10) we get now that

$$e^\alpha S = \mathrm{Hom}_{e^\alpha Se^\alpha}(Se^\alpha, e^\alpha Se^\alpha). \tag{11}$$

At the same time from (11) and (8) it follows that Se^α is a reflexive $e^\alpha Se^\alpha$-module. By Theorem 4.3, reflexivity is preserved by equivalences of module categories, and therefore from (9) and [275, Theorem 46.4] it follows that Sv is a reflexive vSv-module.

Taking into account that we have proved 4) $\Rightarrow$ 5) and 4) $\Rightarrow$ 6) and applying it for $u = v$, we see that $v^{\alpha^{-1}}M_A = v^{\alpha^{-1}}Re_{eRe}$ is a progenerator in Mod-eRe. Applying these results to the anti-isomorphism α^{-1} we get that Re is a reflexive eRe-

module. From Lemma 4.2 we have that $R = \mathrm{End}_{eRe}(eR)$. It is clear that eR is a strong generator in Mod-$(eRe)^{\mathrm{op}}$ and $R^{\mathrm{op}} = \mathrm{End}_{(eRe)^{\mathrm{op}}}(eR)$. Furthermore, e is an idempotent of rank 1 in R^{op}, and the anti-isomorphism α induces a ring isomorphism $\beta: R^{\mathrm{op}} \to S$. Since $e^{\beta} = e^{\alpha}$ and $e^{\alpha}N_B = e^{\alpha}Sv_{vSv}$ is a progenerator, by Theorem 3.3 (4) $\Rightarrow$ 7)) there exists an equivalence H: Mod-$(eRe)^{\mathrm{op}} \to$ Mod-vSv such that $H(eR) = Sv$ and $H(r)x = r^{\alpha}x$ for all $x \in Sv, r \in R^{\mathrm{op}}$. Thus, 4) $\Rightarrow$ 7).

7) $\Rightarrow$ 1). Let H: Mod-$(wRw)^{\mathrm{op}} \to$ Mod-vSv be a Morita equivalence such that $H(r)x = r^{\alpha}x$ for all $x \in N_B = Sv_{vSv}$ and $r \in R^{\mathrm{op}}$. By Lemma 2.10, $wR = \mathrm{Hom}_{wRw}(Rw, wRw)$. Taking into account our condition,

$$R = \mathrm{End}_{wRw}((Rw)^*) = \mathrm{End}_{wRw}(wR) \quad \text{and} \quad H(wR) = Sv.$$

Let now $e^2 = e \in \mathrm{End}(M_A) = R$ be any idempotent with $r(e) = 1$, as in 1). By Lemma 2.9, $R = \mathrm{End}_{eRe}(Re)$. Therefore, the rings $\mathrm{End}_{eRe}(Re)$ and $\mathrm{End}_{wRw}(Rw)$ are identically isomorphic. By Theorem 3.7 (7) $\Leftrightarrow$ 1)) ${}_{wRw}wRe$ is a progenerator in wRw-Mod. Applying H to

$$wR \xrightarrow{e} wRe,$$

we get

$$Sv \xrightarrow{e^{\alpha}} e^{\alpha}Sv = H(wRe).$$

Thus, $e^{\alpha}N_B = e^{\alpha}Sv_{vSv}$ is a progenerator, i.e., we have 1). □

Corollary 4.7. *Let M_A and N_B be progenerators and let*

$$\alpha: \mathrm{End}(M_A) \to \mathrm{End}(N_B)$$

be an anti-isomorphism of their endomorphism rings. Then α is induced by some Morita anti-equivalence, i.e., by an equivalence

$$H: \text{Mod-}A^{\mathrm{op}} \to \text{Mod-}B.$$

In particular, any anti-isomorphism of the ring of matrices $\alpha: A_m \to B_n$ is induced by a Morita anti-equivalence.

Corollary 4.8. *Let A and B be right GIR (i.e., rings in which every right ideal is a generator) and M_A, N_B torsionless modules. If $\alpha: \mathrm{End}(M_A) \to \mathrm{End}(N_B)$ is an anti-isomorphism of rings, then α is induced by a Morita equivalence, $M^{**} = M$ and $N^{**} = N$.*

Corollary 4.9 (c.f. Theorem 3.10 and [284, Theorem 2.1]). *Under the conditions of Theorem 4.6, the following statements are equivalent:*

1. *if $e^2 = e \in \mathrm{End}(M_A)$ and $r(e) = 1$, then $r(e^{\alpha}) = 1$, i.e., $e^{\alpha}N_B \cong B_B$;*
2. *there exists an idempotent $e^2 = e \in \mathrm{End}(M_A)$ such that $r(e) = 1$ and $r(e^{\alpha}) = 1$;*
3. *α is induced by a anti-semilinear module isomorphism of ${}_AM^*$ onto N_B.*

In this case both M_A and N_B are reflexive modules.

5. Endomorphism semigroups of modules

In many situations endomorphism rings of modules are rings with unique addition (UA-rings), and therefore, in this case multiplicative isomorphisms of endomorphism rings are automatically additive, thus they are ring isomorphisms, and it gives us possibility to apply results on isomorphisms of endomorphism rings of modules. Note that the beginning of the theory of UA-rings and additional bibliography on this topic can be found, for example, in [202] and [263].

If we consider isomorphisms of endomorphism semigroups of modules, then sometimes it is possible to apply some rather general results on additive extensions of multiplicative isomorphisms of multiplicative subsemigroups of rings (see [202, §3, Theorem 3.1]).

In this section we present only one direct version of multiplicative analog of our criterion in Theorem 3.10. We note that investigation of multiplicative isomorphisms of endomorphism semigroups of modules was started in [199], [97]. The problem of multiplicative isomorphisms of operator rings of Banach spaces was posed by Eidelheit [63], see more details in our Subsection 8.4. Later Gluskin [96] studied isomorphisms of multiplicative endomorphism semigroups of linear spaces over skew fields, in particular, he showed that linear spaces are definable by their endomorphism semigroups. See also [225, 226] by Petrich. Note that in [1] it was noted that the ring of triangular endomorphisms of countable dimensional linear space is a UA-ring (see also [263]), and as a corollary it was given a description of automorphisms of multiplicative semigroups of triangular endomorphisms.

As for nonadditive analogs of these results for strong generator acts over monoids and bibliography see [144]. Note that definability of a free act by its endomorphisms semigroup was shown by Knauer and Mikhalev [143], and by Fleischer [73].

Definition 5.1. A right module M_A is called *a 2-strong generator,* if $M_A = P_1 \oplus P_2 \oplus T$, where $P_1 \cong P_2 \cong A_A$.

Notation 5.2. By $D_0(A, M)$ we denote the semigroup of all $\delta \in R = \mathrm{End}(M_A)$ such that $\mathrm{Im}\,\delta = a_\delta I$, where $a_\delta \in M_A$, I is a right ideal of the ring A. We write $[f, m] = f(m)$ for $f \in {}_AM^* = \mathrm{Hom}_A(M_A, A_A)$ and $m \in M_A$ and say that M_A and ${}_AM^*$ give us *a dual pair* if $[M^*, m] = 0$ for $m \in M$ implies that $m = 0$.

Proposition 5.3 ([199, Lemma 1]). *If $M_A = P \oplus H$, $P \cong A_A$, is a strong generator with $e^2 = e \in R$, $r(e) = 1$, $\{x\}$ is a basis of $eM_A = P_A$ (see Section 2) and D is a multiplicative subsemigroup in $R = \mathrm{End}(M_A)$ containing $D_0(A, M)$, then:*

1. *the mapping $\beta\colon De \to M$, where $(ye)^\beta = yex$ for $y \in D$, is a bijection;*
2. *the mapping $\gamma\colon eDe \to A$, where $m(eye)^\gamma = (m^{\beta^{-1}}eye)^\beta$ for $m \in M$ and $y \in D$, is an isomorphism of semigroups;*
3. *the mapping $\delta\colon eD \to M^*$, where $[(ey)^\delta, m]x = (ey)m$ for $m \in M$ and $y \in D$, is a bijection.*

Moreover, the triple $(\mathrm{id}_D, \beta, \gamma)$ *is a semilinear bi-isomorphism of biacts* ${}_{eDe}eD_D$ *and* ${}_AM^*D$, *for which*

$$\left[ey_1, y_2e\right]^{\gamma} = \left[(ey_1)^{\delta}, (y_2)^{\beta}\right]$$

with $y_1, y_2 \in D$.

Proof. As in Lemma 2.9. □

Lemma 5.4 (see [199, Lemma 3]; c.f. also [96, Lemma 3.2]). *Let* $M_A = P_1 \oplus P_2 \oplus T$, *where* $P_1 \cong P_2 \cong A_A$, *be 2-strong generators. Suppose that* N_B *and* ${}_BN^*$ *are a dual pair,* $\beta\colon M_A \to N_B$ *and* $\delta\colon {}_AM^* \to {}_BN^*$ *are bijections,* $\gamma\colon A \to B$ *is an isomorphism of the rings A and B such that*

$$[f^{\delta}, m^{\beta}] = [f, m]^{\gamma} \quad \text{for all} \quad f \in {}_AM^*,\ m \in M,$$

and $(P_1 + P_2)^{\beta}$ *is a submodule. Then* γ *is an isomorphism of rings,* (β, γ) *and* (γ, δ) *are semilinear transformations of modules.*

Proposition 5.5. *Let* $M_A = P_1 \oplus P_2 \oplus T$, *where* $P_1 \cong P_2 \cong A_A$, *be a 2-strong generator with the projection e from M onto* P_1 *along* $P_2 \oplus T$ *and the projection f from M onto* $P_1 \oplus P_2$ *along T. Suppose that* N_B *and* ${}_BN^*$ *are a dual pair,* $\alpha\colon D_1 \to D_2$ *is an isomorphism of endomorphism semigroups of the modules* M_A *and* N_B *respectively,* $D_1 \supseteq D_0(A, M)$ *and* $D_2 \supseteq D_0(B, N)$.

If $f \in D_1$ *and* $e^{\alpha}N_B \cong B_B$, *then there exists a semilinear transformation* $(\beta, \gamma)\colon M_A \to N_B$ *which induces an isomorphism* α *of the semigroups* D_1 *and* D_2.

Proof. The proof immediately follows from Proposition 5.3 and Lemma 5.4. □

Theorem 5.6. *Let* M_A *be a 2-strong generator,* N_B *a strong generator,*

$$\alpha\colon \mathrm{End}(M_A) \to \mathrm{End}(N_B)$$

an isomorphism of their multiplicative endomorphism semigroups. Then the following conditions are equivalent:

1. *if* $e^2 = e \in \mathrm{End}(M_A)$, $r(e) = 1$, *then* $r(e^{\alpha}) = 1$ *(i.e.,* $e^{\alpha}N_B \cong B_B$*);*
2. *there exists* $e^2 = e \in \mathrm{End}(M_A)$ *such that* $r(e) = 1$ *and* $r(e^{\alpha}) = 1$*;*
3. α *is induced by a semilinear isomorphism of modules* M_A *and* N_B.

Proof. The implication 1) ⇒ 2) is evident. By Proposition 5.5 we obtain 2) ⇒ 3). As for the implication 3) ⇒ 1), see Theorem 3.10. □

6. Isomorphisms of the general linear group $GL_n(A)$ over an associative ring A (Schreier–van der Waerden problem)

The classification problem of automorphisms of classical linear groups goes back to Schreier and van der Waerden [237], where they described automorphisms of $PSL_n(A)$, $n \geq 3$, over a field A. Diedonné [53] and Rickart [231], developing ideas of Mackey [172], invented involution method and described automorphisms of $GL_n(A)$, $n \geq 3$, over a skew field A. The series of monographs of Baer [13], Diedonné [54], Artin [10], Plotkin [227], Wehrfritz [274], Suprunenko [265] give excellent and complete exposition of classical results in this field.

As for the next period in this theory connected with automorphisms of linear groups over different classes of rings, see the recent superbly organized book of Hahn and O'Meara [123] (we mention also two books of McDonald [191, 196], lectures by O'Meara [218], and some recent surveys [220, 122, 297, 121]).

The aim of this subsection is to discuss a reduction from the group problem of description of isomorphisms of linear groups to the ring problems of description of isomorphisms and anti-isomorphisms of matrix rings where we can apply correspondingly Bolla Theorem ([27], see Section 3) and Beidar-Mikhalev Theorem ([24, 23], see Section 4). As for general definition of an isomorphism of "standard" type, which is critical to this study, we may mention the papers of Hahn [120] and Bolla [29] (see also [123, § 3.3] and [171]).

Let now $GL_n(R)$ be the general linear group over an associative ring R, and let $E_n(R)$ be the elementary subgroup of $GL_n(R)$ generated by elementary matrices of the form $1 + ae_{ij}$, $a \in A$, $1 \leq i \neq j \leq n$, $n \geq 2$ (by Suslin [266], $E_n(R) \triangleleft GL_n(R)$ over a commutative ring). Consider the factor-group $PGL_n(A)$ of $GL_n(A)$ by its center with the canonical epimorphism

$$\pi_A : GL_n(A) \to PGL_n(A).$$

In [270] Waterhouse, developing results of McDonald [192, 194], proved that if A is commutative ring in which 2 is invertible, then all automorphisms of $GL_n(3)$ for $n \geq 3$ are of standard type.

In [108] Golubchik and Mikhalev proved the following

Theorem 6.1. *Let A and B be associative rings in which 2 is invertible, $m \geq 3$, $n \geq 3$,*

$$\varphi : GL_m(A) \to GL_n(B)$$

an isomorphism of linear groups. Then there exists central idempotents $e^2 = e \in A_m$ and $f^2 = f \in B_n$ in matrix rings together with a ring isomorphism

$$\theta_1 : (eA)_m = eA_m \to fB_n = (fB)_n$$

and with a ring anti-isomorphism

$$\theta_2 : ((1-e)A)_m = (1-e)A_m \to (1-f)B_n = ((1-f)B)_n,$$

so that

$$\varphi(X) = \theta_1(eX) + \theta_2((1-e)X^{-1})$$

for all $X \in E_m(A)$.

Using a slightly different approach Zelmanov in [298] has proved this theorem for $n \geq 2$.

In [223] Petechuk proves that any automorphism of groups $\mathrm{GL}_n(A)$, $\mathrm{SL}_n(A)$, $E_n(A)$ for $n \geq 4$ over any commutative ring is standard. Note, that if $n = 3$ and $\frac{1}{2} \notin A$, then one can find a nonstandard isomorphism of $\mathrm{GL}_3(A)$ (over a commutative local ring A see Petechuk [222]; another example over a commutative ring A see in [160]).

In [103, 105] Golubchik generalized Theorem 6.1 for any associative rings A and B with unity, $m \geq 4, n \geq 2$.

In [107] Golubchik and Mikhalev considered epimorphisms of projective groups over associative rings, and in [104] Golubchik has proved the following generalization of Theorem 6.1 (compare with pessimistic remark in [123, p. 137] by Hahn and O'Meara to apply this method to $\mathrm{PGL}_n(A)$ and $\mathrm{PE}_n(A)$):

Theorem 6.2. *Let A and B be associative rings in which 2 is invertible, $m \geq 3, n \geq 2$,*

$$E_m(A) \subseteq G \subseteq \mathrm{GL}_m(A), \quad E_n(A) \subseteq H \subseteq \mathrm{GL}_n(B),$$

H a normal subgroup of $\mathrm{GL}_n(B)$, $\varphi\colon \mathrm{PG} \to \mathrm{PH}$ an isomorphism of projective groups. Then there exists central idempotents $e^2 = e \in A_m$ and $f^2 = f \in B_n$ in matrix rings together with a ring isomorphism

$$\theta_1\colon (eA)_m = eA_m \to fB_n = (fB)_n$$

and with a ring anti-isomorphism

$$\theta_2\colon ((1-e)A)_m = (1-e)A_m \to (1-f)B_n = ((1-f)B)_n,$$

so that

$$\varphi\pi_A(X) = \pi_B(\theta_1(eX) + \theta_2(1-e)X^{-1}))$$

for all $X \in E_m(A)$.

Let $X \in \mathrm{GL}_m(A)$ and

$$\psi\pi_A(X) = \varphi^{-1}\pi_B(\theta_1(eX) + \theta_2((1-e)X^{-1})).$$

Then φ is an automorphism of the projective group PG which is identical on $\mathrm{PE}_n(A)$. Gerasimov [91] has shown existence of an associative ring A for which there exists a nonidentical automorphism of $\mathrm{PGL}_n(A)$ with identical restriction on $\mathrm{PE}_n(A)$. Anyway, for many good rings the situation is better.

In [110] Golubchik and Mikhalev proved the following

Theorem 6.3. *Let A be a PI-ring with unit element, $n \geq 3$, H a subgroup of* $\mathrm{GL}_n(A)$ *containing $E_n(A)$, and ψ an automorphism of the projective group* PH *which is identical on* $\mathrm{PE}_n(A)$. *Then ψ is identical on the whole group* PH.

Recently in [106] Golubchik has proven the following

Theorem 6.4. *Let A be a two-sided order in von Neumann regular ring Q, $1 \in R$, $n \geq 3$, H a subgroup of* $\mathrm{GL}_n(A)$ *containing $E_n(A)$, ψ an automorphism of the projective group* PH *identical on* $\mathrm{PE}_n(A)$. *Then ψ is also identical on the whole group* PH.

Note that in [141] Khlebutin has proved that $E_n(A)$ is a normal subgroup of $\mathrm{GL}_n(A)$ for $n \geq 2$ over any von Neumann regular ring A. As for automorphisms of unitary groups over associative rings with involution see [109, 111, 123].

7. On Malcev type theorems on elementary equivalences of linear groups and rings of matrices

In [175] Malcev proved the following remarkable theorem:

Theorem 7.1. *Two linear groups* $\mathrm{GL}_m(A)$ *and* $\mathrm{GL}_n(B)$ *over fields A and B, where $m, n \geq 3$, are elementary equivalent if and only if $m = n$ and the fields A and B are elementary equivalent.*

Taking into account recent progress in model theory, ring theory and the theory of linear groups over rings (see [109, 202, 105]), Beidar and Mikhalev [21, 22] have formulated a general approach to problems of this type and as a corollary obtained easy proofs of Malcev type theorems in rather general situations. We give here only two corollaries from this paper for linear groups over prime rings and skew fields.

Theorem 7.2. *Let A and B be prime associative rings with* 1. *Suppose that either $\frac{1}{2} \in A$, $\frac{1}{2} \in B$, and $m, n \geq 3$, or $m, n \geq 4$. Then the linear groups* $\mathrm{GL}_m(A)$ *and* $\mathrm{GL}_n(B)$ *are elementary equivalent if and only if the rings of matrices A_m and B_n are elementary equivalent, or the rings of matrices A_m and $(B^{\mathrm{op}})_n$ are elementary equivalent (here B^{op} is the opposite ring to B).*

Theorem 7.3. *Let A and B be skew fields and $m, n \geq 3$. Then the linear groups* $\mathrm{GL}_m(A)$ *and* $\mathrm{GL}_n(B)$ *are elementary equivalent if and only if $m = n$ and either skew fields A and B are elementary equivalent or the skew fields A and B^{op} are elementary equivalent.*

8. Related topics

8.1. Representation of rings as full matrix rings (hidden matrices)

Sometimes it is not very easy to determine whether or not certain rings are in fact full matrix rings.

Let $H_0 = \mathbb{Z}[i, j]$, the ring of Lipschitz quaternions. Chatters [33] asked whether the tiled matrix ring

$$\begin{pmatrix} H_0 & 3H_0 \\ H_0 & H_0 \end{pmatrix}$$

is isomorphic to a complete (2×2)-matrix A_2, and if so, what is A. This question has led to quite a bit of activity ([34], [233], [35], [159], [155], [37], [40]). Chatters [34], Levy, Robson and Stafford [159], and Robson [233] answer the question in the affirmative, and show, in particular, more generally that

$$\begin{pmatrix} H_0 & nH_0 \\ H_0 & H_0 \end{pmatrix}$$

is a complete (2×2)-matrix ring if and only if n is odd.

Note that in connection with the Chatter's problem Robson in [233] shows that a ring R is a complete $(n \times n)$-matrix ring A_n if and only if it contains elements $a_1, \dots, a_n, f$ such that $f^n = 0$ and

$$1 = a_1 f^{n-1} + f a_2 f^{n-2} + f^2 a_3 f^{n-3} + \cdots + f^{n-1} a_n.$$

Chatters in [37] considers the question how to find "hidden" matrix units in these rings. P. Fuchs [81] shows that $R \cong A_n$ for some ring A and $n \geq 2$ if and only if their exist $x, y \in R$ such that $x^{n-1} \neq 0$, $x^n = y^2 = 0$, $x + y$ is invertible and $\operatorname{Ann}(x^{n-1}) \cap Ry = 0$.

In [40] the above and related results carry over to generalized quaternion rings $H = \mathbb{Z}[i, j]$, where $i^2 = a$, $j^2 = b$ $(a, b \in \mathbb{Z})$ and $ij + ji = 0$. In particular, Chatters and Parmenter give necessary and sufficient conditions for H/nH to be a complete (2×2)-matrix ring. It gives a possibility to construct many examples of stably free non-free right ideals of H and also to obtain an example of a stably free non-free two-sided ideal in a maximal $\mathbb{Z}$-order. The question of the existence of such ideals (even in orders which are not necessarily maximal) was posed by Gustafson and Roggenkamp [117]; the examples were given by Guralnick and S. Montgomery in [116], where they compare the automorphism group $\operatorname{Aut}(A_n)$ of the ring of $(n \times n)$-matrices over A, and the automorphism group $\operatorname{Aut}(A)$ of A. Chatters in [38, Theorem 4.4], continuing the series of examples of non-isomorphic orders A and B in a finite dimensional central simple algebra such that $A_2 \cong B_2$ [34, 35, 267], has proved that for any positive integer n there exist a division algebra D of generalized rational quaternions and n pairwise matrix isomorphic non-isomorphic $\mathbb{Z}$-orders.

8.2. Isomorphisms and automorphisms of triangular and structural matrix rings

In [278] Wolfson proves that if A and B are skew fields, then any ring isomorphism of the rings of triangular matrices $T_n(A)$ and $T_m(B)$ implies $m = n$ and A and B isomorphic, provided m or n at least three (note that in [31] and [57] it was shown that any automorphism of the ring $T_n(A)$ fixing scalar matrices is inner). In [1] automorphisms of the ring of triangular endomorphisms of countable dimensional linear spaces were described.

Jondrup [133] proved for a simple Artinian ring A, finite dimensional over its center F, that every F-automorphism of $T_n(A)$, the ring of triangular $(n \times n)$-matrices over A, is inner.

As a result of the series of articles by Barker and Kezlan ([16], [14], [17], [140]) it is shown that if A is a commutative ring, then every A-algebra automorphism of the algebra $T_n(A)$ of upper triangular $(n \times n)$-matrices over A is inner. In [15] Barker considers automorphisms of triangular matrices over graphs.

Coelho in [41], [42] studied the automorphisms of some structural matrix rings (the base ring is taken to be a field, or an indecomposable semiprime ring with some condition on its center, or a simple algebra, finite dimensional over its center).

8.3. Isomorphisms and automorphisms of incidence rings and algebras

Let A be a ring with 1 and P a set with a given reflexive binary relation $\leq$ (in particular, a poset). For $x, y \in P$ with $x \leq y$ we denote by $[x, y]$ the interval $\{z \in P \mid x \leq z \leq y\}$. We suppose that $(P, \leq)$ is locally finite, i.e., every interval is finite. Consider the function

$$\xi(x, y) = \begin{cases} 1, & \text{if } x \leq y \\ 0, & \text{otherwise.} \end{cases}$$

Let

$$I(P, A) = \{f \colon P \times P \to A \mid f(x, y) = f(x, y)\xi(x, y)\}$$

with the operations

$$\begin{aligned} (f + g)(x, y) &= f(x, y) + g(x, y), \\ (fg)(x, y) &= \sum_{x \leq z \leq y} f(x, z)g(z, y) \end{aligned}$$

for $f, g \in I(P, A)$, $x, y \in P$. Then $I(P, A)$ is a ring called *the incedence ring of P over A*. If A is commutative we define

$$(a\,f)(x, y) = a\,f(x, y)$$

for $a \in A$ and $f \in I(P, A)$, and then $I(P, A)$ is *the incidence A-algera of P over A* (see [235] where incidence algebras were introduced by Rota).

The incidence algebras of locally finite posets over fields belong to fundamental structures of combinatorial theory. At the same time, as a nanural generalization of rings of matrices, they play an essential role in ring theory.

The isomorphism problem for incidence rings goes back to Stanley ([261], see also [56]): over a field A if the incidence A-algebras $I(P_1, A)$ and $I(P_2, A)$ of finite posets $(P_1, \leq)$ and $(P_2, \leq)$ are isomorphic, then posets $(P_1, \leq)$ and $(P_2, \leq)$ are isomorphic. In [26] this theorem was generalized on preorderings, and later was investigated for partial orders and preoders in [214], [269], [230], [118], [76], [158], [221]. A review of results of definability of graphs by semigroups of mappings can be found in [209].

Some articles are devoted to structure problems of the automorphism group $\mathrm{Aut}(I(P, A))$. In fact, in [261] Stanley has proved that if $|P| < \infty$ and A is a field, then any automorphism of the incidence algebra $I(P, A)$ is a standard one. Baclawski in [11] showed that Stanley's theorem still holds over fields if $|X|$ is not finite. Haack in [118] considered automorphisms of $I(P, A)$ for some noncommutative rings A. But if A is commutative and indecomposable and $|X| < \infty$ then every automorphism of $I(P, A)$ is again a standard one. Spiegel in [259] showed that not every automorphism of $I(P, A)$ for A a commutative ring is a standard one, but if $|P| < \infty$, all elements of $\mathrm{Aut}(I(P, A))$ are the products of standard automorphisms (if the cardinality of P is infinite, then even this statement is not, in general, true).

In [251] Shmatkov proves a Stanley type theorem for reflexive binary relations (moreover, he considers multiplicative isomorphisms of incidence rings and also shows that the theorem takes place for lattices of certain kind. In [252] he considers incidence algebras over lattices. In [253] he investigates algebras of binary functions and their isomorphisms. Marenich in [179] also considers isomorphisms of algebras of binary functions. Abrams in [5] investigates the structure of incidence rings in situations where the underlying set is not necessary preordered. Modifying the techniques used by Haack [118] in his study of incidence ring of a finite poset over semiperfect rings (in fact, Haack shows that for any semiperfect rings A_1 and A_2 and any finite poset P, if the incidence rings $I(P, A_1)$ and $I(P, A_2)$ are isomorphic, then A_1 and A_2 are necessarily isomorphic), he obtains a recovery result in this more general setting. Finally he uses these ideas to investigate a recovery question for group-graded rings.

Barker [15] investigates automorphisms of triangular matrices over graphs.

8.4. Isomorphisms and anti-isomorphisms of operator rings and semigroups of Banach and topological linear spaces

In 1940 Eidelheit [63, Theorem 2] proved that any real Banach space is completely determined by its ring of continuous operators, and furthermore any isomorphism between two operator rings of real Banach spaces is induced by an isomorphism of Banach spaces. In fact, Eidelheit in [63, Theorems 2 and 3] started to consider isomorphisms of multiplicative endomorphism semigroups. Another proof of Eidelheit's theorem for real normed linear spaces was suggested by Mackey [172]. Arnold in [9] considered the problem for complex Banach spaces, and Mackey in [173, Theorem III-4] generalized it to the endomorphism rings of linear systems, i.e., linear spaces with distinguished subspaces of their conjugate spaces (see also the paper of Rickart [231] for isomorphisms of linear groups of linear systems). In [207] Mikhaleva has received a p-adic analog of Eidelheit's theorem.

In [213] Mustafaev gave a positive solution for Eidelheit's problem on multiplicative isomorphisms, he showed that a real or complex Banach space is determined by its multiplicative semigroup of all continuous operators. A p-adic version of this result was proved by Mikhaleva in [207]. For real or complex Banach spaces Wolfson [277] under suitable dimensional restrictions identified the anti-isomorphism between the rings of all continuous linear transformations as the mapping of an operator upon its adjoint (each of Banach spaces is reflexive and they are mutually conjugate). In [145], [146] Knauer, Mikhalev and Mikhaleva have described anti-isomorphisms of operator semigroups of p-adic Banach spaces (in particular, p-adic Banach spaces with anti-isomorphic semigroups of continuous operators are finite dimensional).

As for (topological) isomorphisms of continuous endomorphism rings of locally convex and topological linear spaces see for example [139], [174], [8], [198]. A nonadditive version for topological acts over monoids can be found in the paper [143] by Knauer and Mikhalev.

8.5. Automorphisms and anti-automorphisms of endomorphism semigroups of partially ordered linear spaces

In [94], [95] Gluskin considered automorphisms of semigroups of matrices with nonnegative elements and semigroups of invertible matrices with nonnegative elements over an ordered field.

Mikhalev in [200] considered isomorphisms of positive endomorphisms of partially ordered modules over ordered rings. Fajans in [67] found a necessary and sufficient condition of definability of a finite dimensional real ordered linear space by its semigroup of positive endomorphisms of rank ≤ 1. In [68] he described isomorphisms of invertible linear transformations leaving invariant a cone of a Euclidean space (it was shown in [69] that the affine endomorphism semigroup in some cases also defines the affine structure). In [70] it is possible to find matrix presentation and description of automorphisms of endomorphism semigroups of ordered linear spaces over ordered skew fields.

Mikhalev and Shatalova in [205] described all automorphisms and anti-automorphisms of the semigroup $G_n(A)$ consisting of all invertible matrices from $\mathrm{GL}_n(A)$ with nonnegative elements over an ordered skew field A, when $n \geq 2$ (note that Cohn in [43, §8] developed a very useful technique to work with $\mathrm{GL}_2(A)$ over a discretely ordered ring A). In [71] Fajans clarified structure of quotient group for $G_2(A)$ over an ordered skew field (it is not coincide with $\mathrm{GL}_2(A)$) and found its automorphisms.

8.6. Frobenius type results

In 1897 Frobenius [75] proved that if the $\mathbb{C}$-linear mapping $T: \mathbb{C}_n \to \mathbb{C}_n$ preserved determinants (i.e., $\det(T(X)) = \det(X)$) then either $T(X) = PXQ$ for all $X \in \mathbb{C}_n$ or that $T(X) = PX^tQ$ for all $X \in \mathbb{C}_n$, where $P, Q \in \mathrm{GL}_n(\mathbb{C})$ and $\det(PQ) = 1$.

Many of the problems on R-linear mappings $T: R_n \to R_n$ preserving some invariants can be reduced to the problem of determining the set of linear mappings which carry the rank one matrices into themselves (see the survey papers [176], [177] by Marcus). Marcus and Moyls [178] found the linear mappings on the space of all $(n \times n)$-matrices A_n over a field A, which transform rank 1 matrices into rank 1 matrices. James [129] solved the problem for A an integral domain. Their result was extended to matrices over any commutative ring A, by Waterhouse [272] and McDonald [195]. Wong [288, 290, 291] started on the noncommutative case (skew fields, noncommutative local rings, semisimple algebras).

In [289] Wong considers mappings on simple algebras preserving zero products and associated Lie algebras; in [292] he determines the additive surjective mappings on the unitary Lie ring of skew-Hermitian transformations relative to a nondegenerate skew-Hermitian form on a finite dimensional vector space over a skew field, which preserve the set of rank 1 elements.

References

[1] J. Abdeljaouad and M. Abdeljaouad, Automorphismes de l'anneau des endomorphismes triangulaires d'un espace vectoriel de dimension dénombrable, Bull. Soc. Math. Belgique 28 (1976), 297–311.

[2] G. D. Abrams, Infinite matrix types which determine Morita equivalence, Arch. Math. 46 (1986), 33–37.

[3] G. D. Abrams, On the existence of rings R with R isomorphic to $\mathrm{RFM}(R)$, J. Austral. Math. Soc. (Ser. A) 42 (1987), 129–131.

[4] G. D. Abrams, On dence subrings of $\mathrm{RFM}(R)$, J. Algebra 110 (1987), 243–248.

[5] G. D. Abrams, Group gradings and recovery results for generalized incidence rings, J. Algebra 164 (1994), 859–876.

[6] S. A. Amitsur, Rings of quotients and Morita contexts, J. Algebra 17 (1971), 273–298.

[7] F. W. Anderson and K. R. Fuller, Rings and Categories of Modules, Springer-Verlag, 1973.

[8] T. Ando, Note on linear topological spaces, Proc. Japan Acad. 30 (1954), no. 6, 435–436.

[9] B. H. Arnold, Rings of operators on vector spaces, Ann. Math. 45 (1944), no. 1, 24–49.

[10] E. Artin, Geometric Algebra, Wiley Interscience, 1957.

[11] K. Baclawski, Automorphisms and derivations of incidence algebras, Proc. Amer. Math. Soc. 36 (1972), 351–356.

[12] R. Baer, Automorphism rings of primary Abelian operator groups, Ann. Math. 44 (1943), 192–227.

[13] R. Baer, Linear Algebra and Projective Geometry, Academic Press, 1952.

[14] G. P. Barker, Automorphism groups of algebras of triangular matrices, Linear Algebra Appl. 121 (1989), 207–215.

[15] G. P. Barker, Automorphisms of triangular matrices over graphs, Linear Algebra Appl. 160 (1992), 155–192.

[16] G. P. Barker and T. P. Kezlan, The automorphism group of a matrix algebra, in: Current Trends in Matrix Theory, North Holland, 1987, 33–39.

[17] G. P. Barker and T. P. Kezlan, Automorphisms of algebras of upper triangular matrices, Arch Math. 55 (1990), 38–43.

[18] H. Bass, Algebraic K-theory, Addison-Wesley, Reading 1968.

[19] S. Bazzoni and C. Metelli, On Abelian torsion-free separable groups and their endomorphism rings, Symp. Math. Ist Naz. Alta mat. Francesco Severi Roma 23 (1979), 259–285.

[20] K. I. Beidar, U. Knauer, and A. V. Mikhalev, An example of two von Neumann regular rings with nonisomorphic Morita equivalent multiplicative semigroups, Semigroup Forum 48 (1994), 381–383.

[21] K. I. Beidar and A. V. Mikhalev, On Malcev's theorem on elementary equivalence of linear groups, Int. Conf. in Algebra. Rings, algebras, modules, Novosibirsk, 1989, p. 19.

[22] K. I. Beidar and A. V. Mikhalev, On Malcev's theorem on elementary equivalence of linear groups, Contemp. Math. 131 (1992), Part 1, 29–35.

[23] K. I. Beidar and A. V. Mikhalev, Anti-isomorphisms of endomorphism rings of modules closed to free ones induced by Morita anti-equivalences, Trudy Sem. Petrovsk. 18 (1994), 1–7; English translation in J. Math. Sci.

[24] K. I. Beidar and A. V. Mikhalev, Anti-isomorphisms of endomorphism rings of modules and Morita anti-equivalences, Uspekhi Mat. Nauk 50 (1995), no. 1, 187–188; English translation in Russian Math. Surveys.

[25] I. H. Bekker and S. F. Kozuhov, Automorphisms of Torsion-free Abelian Groups, Tomsk Univ., 1988.

[26] W. Belding, Incidence rings of pre-ordered sets, Notre Dame J. Formal Logic 14 (1973), 481–509.

[27] M. L. Bolla, Isomorphisms between endomorphism rings of progenerators, J. Algebra 87 (1984), 261–281.

[28] M. L. Bolla, Isomorphisms between infinite matrix rings, Linear Alg. and Appl. 69 (1985), 239–247.

[29] M. L. Bolla, Isomorphisms of general linear groups over rings, J. Algebra 96 (1985), 592–602.

[30] L. Bonami, Morita-equivalent rings are isomorphic, Arch. Math. 34 (1980), 108–110.

[31] E. E. Bray, K. A. Byrd, and R. L. Bernhardt, The injective envelope of the upper triangular matrix ring, Amer. Math. Monthly 78 (1971), 883–886.

[32] V. Camillo, Morita equivalence and infinite matrix rings, Proc. Amer. Math. Soc. (2) 90 (1984), 186–188.

[33] A. W. Chatters, Representation of tiled matrix rings as full matrix rings, Proc. Cambridge Philos. Soc. 105 (1989), 67–72.

[34] A. W. Chatters, Matrices, idealizers, and integer quaternions, J. Algebra 150 (1992), 45–56.

[35] A. W. Chatters, Nonisomorphic rings with isomorphic matrix rings, Proc. Edinburgh Math. Soc. 36 (1993), 339–348.

[36] A. W. Chatters, Isomorphic subrings of matrix rings over the integer quaternions, Commun. Algebra 23(2) (1995), 783–802.

[37] A. W. Chatters, How to find hidden matrix units, Univ. of Bristol, preprint.

[38] A. W. Chatters, Matrix-isomorphic maximal $\mathbb{Z}$-orders, Math. Research Rep. no. PM 95-01, Univ. of Bristol, 1995, 10 pp.

[39] A. W. Chatters and C. R. Hajarnavis, Rings with Chain Conditions, Pitman, 1980.

[40] A. W. Chatters and M. M. Parmenter, Stably free modules over rings of generalized integer quaternions, preprint, Univ. of Bristol, 6 pp.

[41] S. P. Coelho, The automorphism group of a structural matrix algebra, Linear Algebra Appl. 195 (1993), 35–58.

[42] S. P. Coelho, Automorphism groups of certain structural matrix rings, Commun. Algebra 22(14) (1994), 5567–5586.

[43] P. M. Cohn, On the structure of the GL_2 of a ring, Publ. IHES 30 (1966), 365–413.

[44] P. M. Cohn, Some remarks on the invariant basis property, Topology 5 (1966), 215–228.

[45] P. M. Cohn, Free Rings and Their Relations. 2nd Ed, Academic Press, 1985.

[46] A. L. S. Corner, Every countable reduced torsion free ring is an endomorphism ring, London Math. Soc. (3) 13 (1963), 687–710.

[47] A. L. S. Corner, On a conjecture of Pierce concerning direct decompositions of Abelian groups, Proc. Colloq. Abelian Groups, Budapest, 1964, 43–48.

[48] A. L. S. Corner, Endomorphism rings of torsion-free Abelian groups, Proc. Intern. Conf. Theory of Groups, Gordon and Breach, 1967, 59–69.

[49] A. L. S. Corner, On endomorphism rings of primary Abelian groups, Quart. J. Math. Oxford (2) 20 (1969), 277–296.

[50] A. L. S. Corner, Additive categories and a theorem of W. G. Leavitt, Bull. Amer. Math. Soc. 75 (1969), no. 1, 78–82.

[51] A. L. S. Corner, Endomorphism algebras of large modules with distinguished submodules, J. Algebra 11 (1969), no. 2, 155–185.

[52] A. L. S. Corner and R. Göbel, Prescribing endomorphism algebras; a unified treatment, Proc. London Math. Soc. (3) 50 (1985), 447–479.

[53] J. Dieudonné, On the automorphisms of the classical groups, Mem. Amer. Math. Soc. 2 (1951), 1–95.

[54] J. Dieudonné, La Geometrie des Groupes Classiques, Springer-Verlag, 1955.

[55] M. R. Dixon and M. J. Evans, Divisible automorphism groups, Quart. J. Math. Oxford 41(162) (1990), 179–188.

[56] P. Doubilet, G.-C. Rota, and R. P. Stanley, On the foundations of combinatorial theory IV: the idea of generating functions, Univ. California Press, 1972, 267–318.

[57] R. Dubish and S. Perlis, On total nilpotent algebras, Amer J. Math. 73 (1951), 439–452.

[58] M. Dugas and R. Göbel, Every cotorsion-free algebra is an endomorphism algebra, Math. Z. 181 (1982), 451–470.

[59] M. Dugas and R. Göbel, Every cotorsion-free ring is an endomorphism ring, Proc. London Math. Soc. (3) 45 (1982), 319–336.

[60] M. Dugas and R. Göbel, On endomorphism rings of primary Abelian groups, Math. Ann. 261 (1982), 359–385.

[61] M. Dugas and R. Göbel, Endomorphism rings of separable torsion-free Abelian groups, Houston J. Math. 11 (1985), no. 4, 471–483.

[62] M. Dugas, J. Irwin, and S. Khabraz, Coutable rings as endomorphism rings, Quart J. Math. Oxford (2) 39 (1988), 201–211.

[63] M. Eidelheit, On isomorphisms of rings of linear operators, Stud. Math. (PRL) 9 (1940), 97–105.

[64] C. Faith. Algebra: Rings, Modules and Categories, I, Springer-Verlag, 1973, 565 pp.

[65] C. Faith. Algebra II, Ring Theory, Springer-Verlag, 1976, 302 pp.

[66] C. Faith and St. Page, FPF Ring Theory, London Math. Soc. Lecture Notes Ser. 88, Cambridge Univ. Press, Cambridge 1984, 1–6.

[67] V. G. Fajans, On endomorphism semigroups of partially ordered linear spaces, Uch. Zap. Leningrad Gos. Ped. Inst. 302 (1967), 124–141.

[68] V. G. Fajans, Isomorphisms of semigroups of invertible linear transformations with an invariant cone, Izv. Vysh. Uch. Zav. 12(91) (1969), 93–98.

[69] V. G. Fajans, Isomorphisms of semigroups of affine transformations, Sib. Mat. Zh. 11 (1970), no. 1, 193–198.

[70] V. G. Fajans, On endomorphisms of partially ordered linear spaces, Uch. Zap. Leningrad Gos. Ped. Inst. 404 (1971), 498–506.

[71] V. G. Fajans, The group of quotients of the semigroup of nonsingular matrices with non-negative elements, Uspekhi Mat. Nauk 28 (1973), 221–222.

[72] R. B. Feinberg, Characterization of incidence algebras, Discrete Math. 17 (1977), 47–70.

[73] V. Fleischer, Definability of a free act by its endomorphism semigroup, Uch. Zap. Tartu Univ. 366 (1975), 27–40.

[74] W. N. Fransen and P. Schultz, The endomorphism ring of a locally free module, J. Austral. Math. Soc. 35 (1983), 308–326.

[75] G. Frobenius, Über die Darstellung der endlichen Gruppen durch lineare Substitutionen, S.-B. Preuss. Akad. Will. Berlin (1897), 994–1015.

[76] J. Froelich, The isomorphism problem for incidence rings, Ill. J. Math. 29 (1985), 142–152.

[77] L. Fuchs, Abelian Groups, Akad. Kiado, 1958; Pergamon Press, 1967.

[78] L. Fuchs, Infinite Abelian Groups, Vol. 1, Academic Press, 1970.

[79] L. Fuchs, Infinite Abelian Groups, Vol. 2, Academic Press, 1973.

[80] L. Fuchs and P. Schultz, Endomorphism rings of valued vector spaces, Rend. Sem. Mat. Univ. Padova 65 (1981).

[81] P. R. Fuchs, A characterization result for matrix rings, Bull. Austral. Math. Soc. 43 (1991), 265–267.

[82] K. R. Fuller, Density and equivalence, J. Algebra 29 (1974), 528–550.

[83] J. L. Garsia, Morita-like equivalence of categories of modules, Proc. 3-ème Contact Franco-Belge on Algèbre, 1988, 147–156.

[84] J. L. Garsia, The infinite column matrix ring of a ring, Proc. 1st Belgian-Spanish Week on Algebra and Geom., Antwerpen, 1988, 64–79.

[85] J. L. Garsia, The characterization problem for endomorphism rings, J. Austral. Math. Soc. (Ser. A) 50 (1991), 116–137.

[86] J. L. Garsia and M. Saorin, Endomorphism rings and category equivalences, J. Algebra 127(1) (1989), 182–205.

[87] J. L. Garsia and J. J. Simon, Isomorphisms between endomorphism rings of projective modules, Glasgow Math. J. 35 (1993), 353–355.

[88] J. L. Garsia and J. J. Simon, Characterization and isomorphisms of endomorphism rings of locally free modules, J. Algebra, to appear.

[89] J. L. Garsia and J. J. Simon, Automorphisms of subrings of RFM(R), Arch. Math., to appear.

[90] E. R. Gentile, A uniqueness theorem on rings of matrices, J. Algebra 6 (1967), 131–134.

[91] V. N. Gerasimov, The group of units of a free product of rings, Mat. Sb. 134 (1987), no. 1, 42–65; English translation: Math. USSR-Sb. 62 (1989), 41–62.

[92] L. Gewirtzman, Anti-isomorphisms of the endomorphism rings of a class of free modules, Math. Ann. 159 (1965), 278–284.

[93] L. Gewirtzman, Anti-isomorphisms of endomorphism rings of torsion-free modules, Math. Z. 98 (1967), 391–400.

[94] L. M. Gluskin, Semigroup of matrices with non-negative coefficients, Uch. Zap. Kharkov Univ. 25 (1957), 167–173.

[95] L. M. Gluskin, Semigroups of nonsingular matrices with non-negative elements, Uch. Zap. Kharkov Pedagog. Inst. 21 (1957), 81–98.

[96] L. M. Gluskin, Semigroups and rings of endomorphisms of linear spaces, I, II, Izv. Akad. Nauk SSSR, Ser. Mat. 23 (1959), 841–870; 25 (1961), 809–814.

[97] L. M. Gluskin, On endomorphisms of modules, in: Algebra and Math. Logic, Kiev, 1966, 3–20.

[98] R. Göbel, Endomorphism rings of Abelian groups, Lecture Notes in Math. 1006 (1983), 340–353.

[99] R. Göbel and W. May, The construction of mixed modules from torsion-free modules, Arch. Math. 48 (1987), 476–490.

[100] R. Göbel and B. Wald, Separable torsion-free modules of small type, Houston J. Math. 16 (1990), no. 2, 271–287.

[101] A. W. Goldie, Non-commutative principal ideal rings, Arch. Math. 13 (1962), 214–221.

[102] A. W. Goldie, Torsion-free modules and rings, J. Algebra 1 (1964), 268–287.

[103] I. Z. Golubchik, Isomorphisms of the group $GL_n(R)$, $n \geq 4$, over an associative ring R, Int. Conf. in Algebra, Novosibirsk, 1989, Group theory, p. 33.

[104] I. Z. Golubchik, Isomorphisms of projective groups over associative rings, Int. Conf. in Algebra, Barnaul, 1991, p. 24.

[105] I. Z. Golubchik, Isomorphisms of the general linear group $GL_n(R)$, $n \geq 4$, over an associative ring, Contemp. Math. 131 (1992), Part 1, 123–136.

[106] I. Z. Golubchik, Isomorphisms of projective groups over associative rings, Fundamentalnaya i Prikladnaya Matematika 1 (1995), no. 1, 311–314.

[107] I. Z. Golubchik and A. V. Mikhalev, Epimorphisms of projective groups over associative rings, in: Algebra, Moscow State University, 1982, 34–45.

[108] I. Z. Golubchik and A. V. Mikhalev, Isomorphisms of the general linear group over an associative ring, Vestnik Mosk. Univ. Ser. I. Mat. Mekh. 1983, no. 3, 61–72; English translation: Moscow Univ. Math. Bull. 38 (1983), no. 3, 73–85.

[109] I. Z. Golubchik and A. V. Mikhalev, Isomorphisms of unitary groups over associative rings, Zap. Nauch. Sem. LOMI AN SSSR 132 (1983), 97–110; English translation: J. Soviet Math. 30 (1985), 1863–1871.

[110] I. Z. Golubchik and A. V. Mikhalev, On the group of elementary matrices over PI-rings, Research in Algebra, Tbilici, 1985, 20–24.

[111] I. Z. Golubchik and A. V. Mikhalev, The elementary subgroup of a unitary group over a PI-ring, Vestnik Mosk. Univ. Ser. I. Mat. Mekh. 1985, no. 1, 30–36; English translation: Moscow Univ. Math. Bull. 40 (1985), no. 1, 44–54.

[112] K. R. Goodearl, Simple Noetherian rings – the Zalesskii-Neroslavskii examples, Lecture Notes in Math. 734 (1979), 118–130.

[113] K. R. Goodearl, Simple Noetherian rings not isomorphic to matrix rings over domains, Commun. Algebra 12(12) (1984), 1421–1434.

[114] K. R. Goodearl, Von Neumann Regular Rings, Pitman, 1979; 2nd Ed., Kreger Publ. Co., 1991.

[115] S. Ya. Greenspon, Primary Abelian groups with isomorphic endomorphism groups, Mat. Zametki 14 (1973), no. 5, 733–741.

[116] R. M. Guralnick and S. Montgomery, On invertible bimodules and automorphisms of noncommutative rings, Trans. Amer. Math. Soc. 341 (1994), no. 2, 917–937.

[117] W. H. Gustafson and K. Roggenkamp, A Mayer–Vietoris sequence for Picard groups, with applications to integral group rings of dihedral and quaternion groups, Ill. J. Math. 32 (1988), 375–406.

[118] J. K. Haack, Isomorphisms of incidence rings, Ill. J. Math. 28 (1984), 676–683.

[119] M. Hacque, Isomorphismes et automorphismes des anneaux presque-simples, Commun. Algebra 12(16)1984, 1989–2036.

[120] A. J. Hahn, Category equivalences and linear groups over rings, J. Algebra 77 (1982), 505–543.

[121] A. J. Hahn, Algebraic K-theory, Morita theory and the classical groups, Lecture Notes in Math. 1185 (1986), 88–117.

[122] A. J. Hahn, D. G. James, and B. Weisfeiler, Homomorphisms of algebraic and classical groups: a survey, in: Canad. Math. Soc. Conf. Proc., 4, Amer. Math. Soc., 1984, 249–296.

[123] A. J. Hahn and O. T. O'Meara, The Classical Groups and K-theory, Springer-Verlag, 1989.

[124] C. J. Hauptfleisch, Torsion-free Abelian groups with isomorphic endomorphism rings, Arch. Math. (Basel) 24 (1973), 269–273.

[125] I. N. Herstein, Topics in Ring Theory, Univ. of Chicago Press, 1969.

[126] I. M. Isaacs, Automorphisms of matrix algebras over commutative rings, Linear Algebra Appl. 31 (1980), 215–231.

[127] N. Jacobson, On the theory of primitive rings, Ann. Math. 48 (1947), 8–21.

[128] N. Jacobson, Structure of Rings, Ann. Math. Soc. Colloq. Publ. 37, 1956.

[129] D. James, On the automorphisms of $\det(X_{ij})$, Math. Chronicle 9 (1980), 35–40.

[130] A. V. Jategaonkar, Left Principal Ideal Rings, Lecture Notes in Math. 123, 1970.

[131] A. V. Jategaonkar, Non-isomorphic right Noetherian domains with isomorphic matrix rings, preprint, Cornell Univ., 1970.

[132] A. V. Jategaonkar, Endomorphism rings of torsionless modules, Trans. Amer. Math. Soc. 161 (1971), 457–466.

[133] S. Jondrup, Automorphisms of upper triangular matrix rings, Arch. Math. 49 (1987), 497–502.

[134] I. Kaplansky, Some results on Abelian groups, Proc. Nat. Acad. Sci. USA 38 (1952), 538–540.

[135] I. Kaplansky, Infinite Abelian Groups, Univ. of Michigan Press, 1954; rev. ed. 1969.

[136] I. Kaplansky, Rings of Operators, The Univ. of Chicago, 1955.

[137] S. Kasahara, Sur l'espace des endomorphismes continus de l'espace vectoriel localement convexe, Math. Japonicae 3 (1955), 111–116.

[138] S. Kasahara, Representation of some topological algebras, II, Proc. Japan Acad. 35 (1959), 89–94.

[139] Y. Kawada, Über den Operatorenring Banachscher Räume, Proc. Imp. Acad. Tokyo 19 (1943), 616–621.

[140] T. P. Kezlan, A note on algebra automorphisms of triangular matrices over commutative rings, Linear Algebra Appl. 135 (1990), 181–184.

[141] S. G. Khlebutin, Some properties of the elementary subgroup, in: Algebra, Logic and Number Theory, Moscow State University, 1986, 86–90.

[142] L. Klinger and L. S. Levy, Wild torsion modules over Weyl algebras, and general torsion modules over HNPs, preprint, Univ. Wisconsin, 1993, 22 pp.

[143] U. Knauer and A. V. Mikhalev, Endomorphism monoids of acts over monoids, Semigroup Forum 6 (1973), no. 1, 50–58.

[144] U. Knauer and A. V. Mikhalev, Endomorphism monoids of generators in the category of S-acts, Proc. of the Conf. on Semigroups with Applications, World Scientific, Singapore 1992, 132–151.

[145] U. Knauer, A. V. Mikhalev, and L. I. Mikhaleva, Anti-isomorphisms of operator semigroups of p-adic Banach spaces, 3rd Int. Conf. in Algebra, Krasnoyarsk, 1993, 407–408.

[146] U. Knauer, A. V. Mikhalev, and L. I. Mikhaleva, Anti-isomorphisms of operator semigroups of p-adic Banach spaces, Proc. of the 2nd Math. Conf. of Moscow State Social Univ., 1994, 69–71.

[147] M.-A. Knus, Algèbres d'Azumaya et modules projectifs, Comment. Math. Helv. 45 (1970), 372–383.

[148] P. A. Krylov, Irreducible torsion-free Abelian groups and their endomorphism rings, Abelian Groups and Modules, Tomsk, 1986, 73–100.

[149] P. A. Krylov and A. V. Mikhalev, Endomorphism Rings of Abelian Groups, in preparation.

[150] T. Y. Lam, Series summation of stably free modules, Quart. J. Math. Oxford (2) 27 (1976), 37–46.

[151] T. Y. Lam, The Algebraic Theory of Quadratic Forms, Addison-Wesley, Reading 1980.

[152] T. Y. Lam, A First Course in Noncommutative Rings, Springer-Verlag, 1991.

[153] T. Y. Lam, A Second Course in Noncommutative Rings, to appear.

[154] T. Y. Lam, Modules with isomorphic multiples and rings with isomorphic matrix rings, a survey, to appear.

[155] T. Y. Lam, A lifting theorem, and rings with isomorphic matrix rings, preprint, Univ. of California, Berkeley, 1994, 12 pp.

[156] W. G. Leavitt, The module type of a ring, Trans Amer. Math. Soc. 103 (1962), 113–130.

[157] H. Leptin, Abelsche p-Gruppen und Ihre Automorphismengruppen, Math. Z. 73 (1960), 235–253.

[158] P. Leroux, The isomorphism problem for incidence algebras of Möbius categories, Ill. J. Math. 26 (1982), 52–61.

[159] L. S. Levy, J, C. Robson, and J. T. Stafford, Hidden matrices, Proc. London Math. Soc. 69 (1994), no. 2, 277–308.

[160] F. L. Li and Z. X. Li, Isomorphisms of GL_3 over commutative rings, Contemp. Math. 82 (1989), 47–52.

[161] M. S. Li and J. M. Zelmanowitz, Artinian rings with restricted primeness conditions, J. Algebra 124 (1989), 139–148.

[162] W. Liebert, Charakterisierung der Endomorphismenringe endlicher abelscher Gruppen, Arch. Math. 18 (1967), no. 2, 128–135.

[163] W. Liebert, Charakterisierung der Endomorphismenringe beschränkter abelscher Gruppen, Math. Ann. 174 (1967), 217–232.

[164] W. Liebert, Endomorphism rings of Abelian p-groups, in: Studies on Abelian Groups, Springer-Verlag, 1968, 239–258.

[165] W. Liebert, Characterization of the endomorphism rings of divisible torsion modules and reduced complete torsion-free modules over complete discrete valuation rings, Pacif. J. Math. 37 (1971), 141–170.

[166] W. Liebert, Endomorphism rings of reduced torsion-free modules over complete discrete valuation rings, Trans. Amer. Math. Soc. 169 (1972), 347–363.

[167] W. Liebert, Endomorphism rings of reduced complete torsion-free modules over complete discrete valuation rings, Proc. Amer. Math. Soc. 36 (1972), no. 2, 375–378.

[168] W. Liebert, Endomorphism rings of free modules over principal ideal domains, Duke Math. J. 41 (1974), 323–328.

[169] W. Liebert, Isomorphic automorphism groups of primary Abelian groups, in: Abelian Group Theory, Proc. 3rd Conf. Oberwolfach 1985, N. Y., 1987, 9–31.

[170] W. Liebert, Isomorphic automorphism groups of primary Abelian groups, II, Contemp. Math. 87 (1989), 51–59.

[171] T.-M. Lok, Morita equivalence and isomorphisms between general linear groups, preprint, Chin. Univ. Hong Kong, 1994, 75 pp.

[172] G. Mackey, Isomorphism of normed linear spaces, Ann. Math. 43 (1942), no. 2, 244–260.

[173] G. Mackey, On infinite dimensional linear spaces, Trans. Amer. Math. Soc. 57 (1945), 155–207.

[174] G. Mackey, Convex topological spaces, Trans. Amer. Math. Soc. 60 (1946), 519–537.

[175] A. I. Malcev, On elementary properties of linear groups, in: Some Problems in Math. and Mekh., Novosibirsk, 1981, 110–132.

[176] M. Marcus, Linear operations on matrices, Amer. Math. Monthly 69 (1962), 837–847.

[177] M. Marcus, Linear transformations on matrices, J. Nat. Bureau Standards 758 (1971), 107–113.

[178] M. Marcus and B. N. Moyls, Transformations on tensor product spaces, Pacif. J. Math. 9 (1959), 1215–1221.

[179] E. E. Marenich, K-automorphisms of algebras of binary functions, Mat. Zametki 55(6) (1994), 90–102.

[180] V. T. Markov, A. V. Mikhalev, L. A. Skornyakov, and A. A. Tuganbaev, Endomorphism rings of modules and lattices of submodules, Itogi Nauki i Tehniki, VINITI, 21 (1983), 183–254.

[181] W. May, Endomorphism rings of Abelian groups with ample divisible subgroups, Bull. London Math. Soc. 10 (1978), no. 3, 270–272.

[182] W. May, Endomorphism rings of mixed Abelian groups, Contemp. Math. 87 (1989), 61–74.

[183] W. May, Isomorphism of endomorphism algebras over complete discrete valuation rings, Math. Z. 204 (1990), no. 4, 485–499.

[184] W. May, Endomorphism algebras of not necessarily cotorsion-free modules, Contemp. Math. 130 (1992), 257–264.

[185] W. May and E. Toubassi, Endomorphisms of Abelian groups and the theorem of Baer and Kaplansky, J. Algebra 43 (1976), no. 1, 1–13.

[186] W. May and E. Toubassi, A result on problem 87 of L. Fuchs, Lecture Notes in Math. 616 (1977), 354–367.

[187] W. May and E. Toubassi, Isomorphism of endomorphism rings of rank one mixed groups, J. Algebra 71 (1981), no. 2, 508–514.

[188] W. May and E. Toubassi, Endomorphisms of rank one mixed modules over discrete valuation rings, Pacif. J. Math. 108 (1983), no. 1, 155–163.

[189] W. May and E. Toubassi, Classifying endomorphism rings of rank one mixed groups, in: Abelian Groups and Modules, New York, 1984, 252–263.

[190] J. C. McConnell and J. C. Robson. Noncommutative Noetherian Rings, Wiley Interscience, 1987.

[191] B. R. McDonald, Geometric Algebra over Local Rings, Marcel Dekker, New York, 1976.

[192] B. R. McDonald, Automorphisms of $GL_n(R)$, Trans. Amer. Math. Soc. 215 (1976), 145–159.

[193] B. R. McDonald, Endomorphism rings of infinitely generated projective modules, J. Algebra 45 (1977), no. 1, 69–82.

[194] B. R. McDonald, Automorphisms of $GL_n(R)$, Trans. Amer. Math. Soc. 246 (1978), 155–171.

[195] B. R. McDonald, R-linear Endomorphisms of $(R)_n$ Preserving Invariants, Mem. Amer. Math. Soc. 287, 1983.

[196] B. R. McDonald, Linear Algebra over Commutative Rings, Marcel Dekker, 1984.

[197] C. Metelli and L. Salce, The endomorphism ring of an Abelian torsion-free homogeneous separable group, Arch. Math. (Basel) 26 (1975), 480–485.

[198] A. V. Mikhalev, On isomorphism of continuous endomorphism rings, Sib. Mat. Zh. 4 (1963), no. 1, 177–186.

[199] A. V. Mikhalev, Isomorphisms of endomorphism semigroups of modules, I, Algebra i Logika 5 (1966), no. 5, 59–67.

[200] A. V. Mikhalev, Isomorphisms of endomorphism semigroups of modules, II, Algebra i Logika 6 (1967), no. 6, 38–47.

[201] A. V. Mikhalev, Endomorphism rings of modules and lattices of submodules, Algebra, Topology, Geometry. Itogi Nauki i Tehniki, VINITI, 12 (1974), 52–76.

[202] A. V. Mikhalev, Multiplicative classification of associative rings, Matem Sb. 135(77) (1988), no. 2, 210–224; English translation: Math. USSR Sbornik 63 (1989), no. 1, 205–218.

[203] A. V. Mikhalev, Isomorphisms of endomorphism rings of modules closed to free ones, Vestnik Moskov. Univ. Ser. 1. Mat. Mekh. 1989, no. 2, 20–27; English translation: in Moscow Univ. Math. Bull.

[204] A. V. Mikhalev and A. P. Mishina, Infinite Abelian groups: methods and results, Fundamental'naya i Prikladnaya Matematika 1 (1995), no. 2, 319–376.

[205] A.. V. Mikhalev and M.. A.. Shatalova, Automorphisms and anti-automorphisms of semigroups with non-negative elements, Mat. Sb. 81(123) (1970), no. 4, 600–609.

[206] L. I. Mikhaleva, Bounded subsets of non-archimedean Banach spaces and their rings of continuous operators, Vestnik Mosk. Univ. Ser. I. Mat. Mekh. 1982, no. 5, 39–43.

[207] L. I. Mikhaleva, Isomorphisms of rings and semigroups of continuous operators of non-archimedean Banach spaces, Vestnik Mosk. Univ. Ser. I. Mat. Mekh. 1985, no. 6, 7–10.

[208] B. Mitchell, Theory of Categories, Academic Press, 1965.

[209] V. A. Molchanov, Semigroups of mappings of graphs, Semigroup Forum 27 (1983), 155–192.

[210] K. Morita, Duality for modules and its application to the theory of rings with minimum condition, Sci. Rep. Tokyo Kyoiku Daigaku, Ser. A, 6 (1958), 83–142.

[211] K. Morita, Category-isomorphisms and endomorphism rings of modules, Trans. Amer. Math. Soc. 103 (1962), 451–469.

[212] I. Murase, On the structure of generalized uniserial rings, I, Sci. Papers College Gen., Ed. Univ. Tokyo, 13 (1963), 1–22.

[213] L. G. Mustafaev, Semigroups of linear continuous transformations, I, II, in: Special Problems of Algebra and Topology, Baku, 1950, 56–60; 61–65.

[214] N. A. Nachev, Incidence rings, Vestnik Moskov. Univ. Ser. 1. Mat. Mekh. 1977, 36–42.

[215] H. Nagao and T. Nakayama, On the structure of M_0 and M_u-modules, Math. Z. 59 (1953), 164–170.

[216] O. M. Neroslavskii and A. E. Zalesskii, On simple Noetherian rings, Vestsi Akad. Nauk BSSR 5 (1975), 38–42.

[217] V. A. Nikiforov, Torsion-free Abelian groups with finite automorphism groups, Mat. Zametki 39 (1986), no. 5, 641–646.

[218] O. T. O'Meara, Lectures on Linear Groups, Reg. Conf. Ser. Math. Amer. Math. Soc., 1974.

[219] O. T. O'Meara, A general isomorphism theory for linear groups, J. Algebra 49 (1977), 93–142.

[220] O. T. O'Meara, A survey of the isomorphism theory of the classical groups, in: Ring Theory and Algebra III, Marcel Dekker, 1980, 225–241.

[221] M. Parmenter, J. Schmerl, and E. Spiegel. Isomorphic incidence algebras, Adv. Math. 84 (1990), 226–236.

[222] V. M. Petechuk, Automorphisms of $SL_3(K)$, $GL_3(K)$, Metem. Zametki 31 (1982), no. 5, 657–667; English translation: Math. Notes 31 (1982), 335–340.

[223] V. M. Petechuk, Automorphisms of matrix groups over commutative rings, Mat. Sb. 117(159) (1982), 534–547; English translation: Math. USSR-Sb. 45 (1983), 527–542.

[224] V. M. Petechuk, Isomorphisms of groups rich of transvections, Math. Notes 39 (1986), no. 1-2, 103–107.

[225] M. Petrich, Rings and Semigroups, Lecture Notes in Math. 380, 1974.

[226] M. Petrich, Categories of algebraic systems, Lecture Notes in Math. 553, 1976.

[227] B. I. Plotkin, Groups of Automorphisms of algebraic systems, Moscow, Nauka, 1966; English translation: Wolters-Noordhoff Publ., 1972.

[228] P. Puusemp, On definability of a torsion Abelian group by its endomorphism semigroup in the class of all torsion Abelian groups, Izv. AN Est. SSR 29 (1980), no. 3, 246–253.

[229] P. Puusemp, On one theorem of May, Izv. AN Est. SSR 38 (1989), no. 2, 139–145.

[230] P. Ribenboim, Algebraic structures on graphs, Algebra Univers. 16 (1983), 105–123.

[231] C. E. Rickart, Isomorphic groups of linear transformations, I, II, Amer. J. Math. 72 (1950), 451–464; 73 (1951), 697–716.

[232] J. C. Robson, Rings in which finitely generated right ideals are principal, Proc. London Math. Soc. 17 (1967), 617–628.

[233] J. C. Robson, Recognition of matrix rings, Commun. Algebra 19(7) (1991), 2113–2124.

[234] A. Rosenberg and D. Zelinsky, Automorphisms of separable algebras, Pacif. J. Math. 11 (1961), 1109–1117.

[235] G. Rota, On the foundations of combinatorial theory, Z. Wahrsch. Verw. Gebiete 2 (1964), 340–368.

[236] L. H. Rowen, Ring Theory, I, II, Academic Press, 1988.

[237] O. Schreier and Van der Waerden, Die Automorphismen der projektiven Gruppen, Abh. Math. Sem. Univ. Hamburg 6 (1928), 303–322.

[238] A. M. Sebeldin, Isomorphism conditions for completely decomposable torsion-free Abelian groups with isomorphic endomorphism rings, Mat. Zametki 11 (1972), no. 4, 403–408.

[239] A. M. Sebeldin, Complete direct sums of torsion-free rank 1 Abelian groups with isomorphic endomorphism groups or rings, 1, Mat. Zametki 14 (1973), no. 6, 867–878.

[240] A. M. Sebeldin, On definability of torsion-free Abelian groups by their endomorphism rings and groups, Groups and Modules, Tomsk, 1976, 70–77.

[241] A. M. Sebeldin, Complete direct sums of torsion-free rank 1 Abelian groups with isomorphic endomorphism groups or rings, 2, Abelian Groups and Modules, Tomsk, 1979, 159–164.

[242] A. M. Sebeldin, Torsion-free Abelian groups with isomorphic endomorphism rings, Abelian Groups and Modules, Tomsk, 1979, 165–170.

[243] A. M. Sebeldin, Definability of a torsion-free nonreduced Abelian group by its endomorphism group, Abelian Groups and Modules, Tomsk, 1980, 102–108.

[244] A. M. Sebeldin, Definability of Abelian groups by their endomorphism rings, Abelian Groups and Modules, Tomsk, 1989, 113–123.

[245] A. M. Sebeldin, On determining Abelian groups by their semigroups of endomorphisms, Abelian Groups and Modules, Tomsk, 1991, 125–133.

[246] A. M. Sebeldin, Definability of vector groups by their endomorphism semigroups, Algebra i Logika 33 (1994), no. 4, 422–428.

[247] A. M. Sebeldin, Abelian groups with isomorphic endomorphism semigroups, Uspekhi Mat. Nauk 49 (1994), 211–212.

[248] A. M. Sebeldin, Abelian groups of some classes with isomorphic endomorphism rings, Uspekhi Mat. Nauk 50 (1995), no. 1, 207–208.

[249] A. M. Sebeldin, Definability of Abelian groups by endomorphism rings, Fundamentalnaya i Prikladnaya Matematika, in press.

[250] S. Shelah, A combinatorial theorem and endomorphism rings of Abelian groups, II, Abelian Groups and Modules, 1984, 37–86.

[251] V. D. Shmatkov, On isomorphisms of incidence algebras, Diskret. Mat. 3 (1991), no. 1, 133–144; English translation: Discrete Math. Appl. 2 (1992), no. 4, 395–406.

[252] V. D. Shmatkov, Incidence algebras over lattices, Uspekhi Mat. Nauk 47 (1992), no. 4, 217–218.

[253] V. D. Shmatkov, Properties of algebras of binary functions, Mat. Zametki 55 (1994), no. 1, 130–140.

[254] V. D. Shmatkov, Isomorphisms of algebras of binary functions, Mat. Zametki 55 (1994), no. 4, 120–127.

[255] J. J. Simon, Subrings of RFM(R) and equivalences, Commun. Algebra 21(9) (1993), 3183–3195.

[256] K. C. Smith and L. van Wyk, An internal characterization of structural matrix rings, Commun. Algebra 22(14) (1994), 5599–5622.

[257] S. P. Smith, An example of a ring Morita equivalent to the Weyl algebra A_1, J. Algebra 73 (1981), 552–555.

[258] A. G. Solonina, On definability of p-adic modules by their automorphism groups, preprint, Mosk. Gos. Ped. Inst., Moscow, 1975, Dep., 15 pp.

[259] E. Spiegel, Automorphisms of incidence algebras, Commun. Algebra 21 (1993), 2973–2981.

[260] J. T. Stafford, Endomorphisms of right ideals of the Weyl algebra, Trans. Amer. Math. Soc. 299 (1987), 623–639.

[261] R. P. Stanley, Structure of incidence algebras and their automorphism groups, Bull. Amer. Math. Soc. 76 (1970), 1236–1239.

[262] W. Stephenson, Characterization of rings and modules by means of lattices, Ph. D. Thesis, Bedford College, Univ. of London, 1965.

[263] W. Stephenson, Unique addition rings, Canad. J. Math. 21 (1969), 1455–1461.

[264] W. Stephenson, Lattice isomorphisms between modules, (1) Endomorphism rings, J. London Math. Soc. (2) 1 (1969), 177–183.

[265] D. A. Suprunenko, Matrix Groups, Transl. Math. Monographs 45, Amer. Math. Soc., 1976.

[266] A. A. Suslin, On the structure of the special linear group over polynomial rings, Izv. AN SSSR 41 (1977), no. 2, 235–252; English translation: Math. USSR Izv. 11 (1977), no. 2, 221–238.

[267] R. G. Swan, Projective modules over group rings and maximal orders, Ann. Math. 76 (1962), 55–61.

[268] S. N. Tronin, The Morita-contexts arising from isomorphisms between rings of endomorphisms, Int. Conf. Algebra and Analysis, Kazan, 1994, p. 151.

[269] E. R. Voss, On the isomorphism problem for incidence rings, Ill. J. Math. 24 (1980), no. 4, 624–638.

[270] W. C. Waterhouse, Automorphisms of $\mathrm{GL}_n(R)$, Proc. Amer. Math. Soc. 79 (1980), no. 3, 347–351.

[271] W. C. Waterhouse, Linear transformations on self-adjoint matrices: The preservation of rank-one-plus-scalar, Linear Algebra Appl. 74 (1986), 73–85.

[272] W. C. Waterhouse, Automorphisms of $\det(X_{ij})$: The group scheme approach, Adv. Math. 65 (1987), 171–203.

[273] D. B. Webber, Ideals and modules of simple hereditary rings, J. Algebra 16 (1970), 239–242.

[274] D. A. F. Wehrfritz, Infinite Linear Groups, Ergeb. Math. Grenzgeb. 76, Springer-Verlag, 1973.

[275] R. Wisbauer. Foundations of Module and Ring Theory. A Handbook for Study and Research, Gordon and Breach Sci. Publ., 1991, 606 pp.

[276] K. G. Wolfson, An ideal-theoretic characterization of the ring of all linear transformations, Amer. J. Math. 75 (1953), 358–386.

[277] K. G. Wolfson, Anti-isomorphisms of the ring and lattice of a normed linear space, Port. Math. 14 (1955), no. 1, 1–7.

[278] K. G. Wolfson, Baer subrings of the ring of linear transformations, Math. Z. 75 (1961), 328–332.

[279] K. G. Wolfson, Isomorphisms of the endomorphism rings of torsion-free modules, Proc. Amer. Math. Soc. 13 (1962), 712–714.

[280] K. G. Wolfson, Isomorphisms of the endomorphism ring of a free module over a principal left ideal domain, Michigan Math. J. 9 (1962), 69–75.

[281] K. G. Wolfson, Isomorphisms of the endomorphism rings of a class of torsion free modules, Proc. Amer. Math. Soc. 14 (1963), 589–594.

[282] K. G. Wolfson, Isomorphisms of prime Goldie semi-principal left ideal rings, Proc. Amer. Math. Soc. 104 (1988), 25–29.

[283] K. G. Wolfson, Isomorphisms of prime Goldie semi-principal left ideal rings, II, Canad. Math. Bull. 31(3) (1988), 374–379.

[284] K. G. Wolfson, Anti-isomorphisms of endomorphism rings of locally free modules, Math. Z. 202 (1989), 151–159.

[285] K. G. Wolfson, The endomorphism ring of a free module over a principal left ideal domain, Commun. Algebra 18(2) (1990), 357–370.

[286] K. G. Wolfson, Locally free modules, Port. Math. 47 (1990), Fasc. 1, 1–11.

[287] K. G. Wolfson, Endomorphism rings of locally free modules, Commun. Algebra 19(12) (1991), 3465–3486.

[288] W. J. Wong, Maps on simple algebras preserving zero products, I: the associative case, Pacif. J. Math. 89 (1980), 229–247.

[289] W. J. Wong, Maps on simple algebras preserving zero products, II: Lie algebras of linear type, Pacif. J. Math. 92 (1981), 469–488.

[290] W. J. Wong, Rank 1 preserving maps of linear transformations over noncommutative local rings, J. Algebra 113 (1988), 263–293.

[291] W. J. Wong, Maps on spaces of linear transformations over semisimple algebras, J. Algebra 115 (1988), 386–400.

[292] W. J. Wong, Rank 1 preservers on the unitary Lie ring, J. Austral Math. Soc. (Ser. A) 49 (1990), 399–417.

[293] Y. Xu, An equivalence between ring F and infinite matrix, Chin. Ann. Math. 11B:1 (1990), 66–69.

[294] Y. Xu, Isomorphisms between endomorphism rings of free modules, Sci. China (Ser. A) 35 (1992), 769–779.

[295] Y. Xu, K. P. Shum, and Y. Fong, Infinite matrix rings and Morita-like equivalence, preprint, Chin. Univ. Hong Kong.

[296] Y. Xu, K. P. Shum, and R. F. Turner-Smith, Morita-like equivalence of infinite matrix subrings, J. Algebra 159 (1993), 425–435.

[297] A. E. Zaleskii, Linear groups, J. Soviet Math. 31:3 (1985), 2974–3004.

[298] E. I. Zelmanov, Isomorphisms of a complete linear group over an associative ring, Sibirsk. Mat. Zh. 26 (1985), no. 4, 49–67; English translation: Siberian Math. J. 26 (1985), no. 4, 515–530.

Department of Mechanics and Mathematics
Chair of Higher Algebra
Moscow State University
Moscow 119899, Russia

Algorithmic solution of the quantum Serre conjecture

*V. A. Artamonov**

Serre's conjecture states that finitely generated projective modules over polynomial rings with coefficients in a field k are free [L], [S1]. The quantum version of this conjecture states that finitely generated projective modules over Q-commutative polynomials are free, where $Q = (q_{ij}) \in \mathrm{Mat}(n, k)$ is a matrix of multiparameters. This problem has a positive answer in the following cases:

(i) each entry of Q is a root of unity [A2];

(ii) elements q_{ij}, $1 \leq i < j \leq n$, are independent in the multiplicative abelian group k^* [A3].

The proof in both cases is based on a study of actions of groups of elementary matrices on unimodular rows. Thus given a unimodular row, it is necessary to find a sequence of elementary transformations which reduce the row to the standard one, namely, to $(1, 0, \ldots, 0)$. The aim of this paper is to present an algorithm which enables us to find an explicit form of these transformations.

For all necessary definitions concerning projective modules and general linear groups consult [A1], [A2], [A4], [L], [S1].

§1. Preliminaries

The solution of the Serre conjecture and some related problems is based on the following two Theorems [A1], [A2], [A4], [S2].

Theorem 1 ([A1]). *Let B be a semilocal left Noetherian ring with an automorphism α and Jacobson radical J. Let $A = B[T^{\pm 1}, \alpha]$ be the Laurent skew polynomial ring.*

---

* Research partially supported by Grant 93-01-1544 for fundamental sciences of Russian Academy of Sciences, by INTAS Grant 93-2681, by International Science Foundation and Russian government Grant JBX100

If $n \geq 4$ then

$$E(n, A) \cap \mathrm{GL}(n, A, JA) \subseteq \\ E(n, B[T^{-1}, \alpha^{-1}], JB[T^{-1}, \alpha^{-1}])\, \mathrm{GL}(n, B, J) E(n, B[T, \alpha, JB[T, \alpha]).$$

Corollary 2 ([A1]). *Let B, α, J, A, n be from Theorem 1. Suppose that S is the set of all monic polynomials in $B[T, \alpha]$. Then S is the Ore set and*

$$\mathrm{Mat}(n, B[T, \alpha]) \cap E(n, S^{-1}A) \cap \mathrm{GL}(n, S^{-1}A, JS^{-1}A) \\ \subseteq E(n, B[T, \alpha], JB[T, \alpha])\, \mathrm{GL}(n, B, J).$$

Theorem 3 ([A1], [A2], [A4]). *Let*

$$A = \oplus_{n \geq 0} A(n)$$

be a graded ring with central homogeneous subalgebra Z which has no zero divisors in A. Let N be a finitely presented left $A(0)$-module and

$$f : A \otimes_{A(0)} N \to A^r, \quad r \geq 3$$

be a homomorphism of left $A(0)$-modules. For any integer $d \geq 0$ denote by $V(d)$ a subset of $Z(d)$ consisting of zero and all nonzero elements $b \in Z(d)$ such that $gf(N) \subseteq A(0)^r$ for some $g \in E(r, A_b^+)$. Then

$$V = \oplus_{n \geq 0} V(d)$$

is a homogeneous semiprime ideal in Z. Here A_b^+ is the non-negative part of the $\mathbb{Z}$-graded localized ring A_b.

Note that Theorem 3 reduces the proof of the quantum Serre conjecture in the case (i) to the situation when we take an algebra $B[T, \alpha]$ as in Theorem 1 instead of a polynomial algebra. Then we apply Corollary 2. See for details [A1], [A2], [A4].

The proofs of Theorems 1 and 3 in [A1] are not constructive. In the present paper we can offer the constructive versions of these theorems. Note that similar problems have been considered in [LS]

§2. Matrices over skew polynomials

Let B be a left Noetherian ring with an automorphism α. Let $A = B[T, \alpha]$ be the ring of skew polynomials, where $Tb = \alpha(b)T$ for all $b \in B$. Denote by S the set of all polynomials from A whose leading coefficients are invertible in B. An easy exercise shows that S satisfies the left Ore condition. Similarly the multiplicative monoid generated by T satisfies the Ore condition. Denote by A_T the corresponding localization. Note that A_T is the Laurent skew polynomial algebra $B[T^{\pm 1}, \alpha]$.

Let D (respectively, I) be the subset of $S^{-1}A$ consisiting of all fractions $f^{-1}g$, where $f \in S$, $g \in A$ and $\deg f \geq \deg g$ (respectively, $\deg f > \deg g$). Clearly D is a subalgebra of $S^{-1}A$, $I = T^{-1}D$ is an ideal of D,

$$S^{-1}A = A + I, \quad I \cap A = 0, A \cap D = B. \tag{1}$$

and I is contained in the Jacobson radical of D.

For any ring R denote by s.r. R the stable rank of R.

Proposition 4. *Let $n \geq 1 +$ s.r. B and let B be left regular in the sense of* [S1]. *Then*

$$E(n, D, T^{-1}D) \cap E(n, A_T) = E(n, B[T^{-1}, \alpha^{-1}], T^{-1}B[T^{-1}, \alpha^{-1}]),$$
$$E(n, D, T^{-1}D) = E(n, D) \cap \mathrm{GL}(n, D, T^{-1}D).$$

Proof. Let $M \in E(n, D, I) \cap E(n, A_T)$. Then

$$M \in \mathrm{GL}(n, B[T^{-1}, \alpha^{-1}], T^{-1}B[T^{-1}, \alpha^{-1}]).$$

The image of M in $K_1(B[T^{-1}, \alpha^{-1}])$ is trivial, since by [Q, p. 114]

$$K_1(B[T^{-1}, \alpha^{-1}]) = K_1(B),$$

because by the assumption the ring B is left regular. Since $n \geq 1 +$ s.r. B by [S1, ch.1, §4] the group $E(n, B[T^{-1}, \alpha^{-1}])$ is normal in $\mathrm{GL}(n, B[T^{-1}, \alpha^{-1}])$ and

$$\mathrm{GL}(n, B[T^{-1}, \alpha^{-1}])/E(n, B[T^{-1}, \alpha^{-1}]) \simeq K_1(B).$$

It follows that $M \in E(n, B[T^{-1}, \alpha^{-1}])$. Since $M \equiv 1 \bmod T^{-1}B[T^{-1}, \alpha^{-1}]$ we have

$$M \in E(n, B[T^{-1}, \alpha^{-1}], T^{-1}B[T^{-1}, \alpha^{-1}]).$$

The result follows. □

Theorem 5 ([A1, Corollary 1 from Theorem 1]). *Let J be an ideal in B invariant under α and $n \geq \max(3, 1 + \text{s.r.}(B/J)$. Then*

$$E(n, S^{-1}A, JS^{-1}A) = E(n, A_T, JA_T)[\mathrm{GL}(n, D, JT^{-2}A) \cap E(n, S^{-1}A, JS^{-1}A)];$$

and

$$E(n, S^{-1}A) \cap \mathrm{GL}(n, S^{-1}A, JS^{-1}A)$$
$$= [E(n, A_T) \cap \mathrm{GL}(n, A_T, JA_T)][\mathrm{GL}(n, D, T^{-1}JD) \cap E(n, S^{-1}A)].$$

Proof. Let R be any ring. Then standard generators $1 + ae_{ij}, a \in R, 1 \leq i \neq j \leq n$, of the group $E(n, R)$ satisfy the Steinberg relations

$$[1 + ae_{ij}, 1 + be_{rs}] = \begin{cases} 1, & \text{if } i \neq s, j \neq r; \\ 1 + abe_{is}, & \text{if } i \neq s, j = r; \\ 1 - bae_{rj}, & \text{if } i = s, j \neq r. \end{cases} \tag{2}$$

In particular, if the ring R is generated by a subset R_0, then the group $E(n, R)$, $n \geq 3$, is generated by all matrices $1 + re_{ij}, 1 \leq i \neq j \leq n, r \in R_0$.

The ring $S^{-1}A$ is generated by S^{-1}, T and by elements from B. The ring D is generated by T^{-1}, by elements from B and by all elements

$$f^{-1}g, \quad f \in S, \ g \in A, \ \deg f - \deg g \geq 2.$$

Thus the group $E(n, S^{-1}A, JS^{-1}A)$ is generated by all elements

$$U(1 + be_{ij})U^{-1}, \quad U \in E(n, S^{-1}A), \ b \in J. \qquad \square$$

Lemma 1. *Let $M \in \mathrm{GL}(n, D, JT^{-1}D)$. Then $M = M_0M_1$ where*

$$M_0 \in E(n, B[T^{-1}, \alpha^{-1}], JT^{-1}B[T^{-1}, \alpha^{-1}]), \quad M_1 \in \mathrm{GL}(n, D, T^{-2}D).$$

The proof is an easy exercise.

Lemma 2. *Let $M \in \mathrm{GL}(n, D, T^{-2}D)$. Then*

$$(1 - Te_{ij})M(1 + Te_{ij}) \in \mathrm{GL}(n, D, JT^{-1}D).$$

Proof. By the assumption $M = 1 + T^{-2}N$, where $N \in \mathrm{Mat}(n, JD)$. Then

$$\begin{aligned} &(1 - Te_{ij})(1 + T^{-2}N)(1 + Te_{ij}) \\ &\qquad = 1 + (1 - Te_{ij})T^{-2}N(1 + Te_{ij}) \in \mathrm{GL}(n, D, JT^{-1}D) \end{aligned}$$

$\square$

To end the proof of Theorem 5 take $U \in E(n, D, T^{-1}D)$. By Lemma 1, $U = U_0U_1$, where

$$U_0 \in E(n, B[T^{-1}, \alpha^{-1}], JT^{-1}B[T^{-1}, \alpha^{-1}]),$$

and

$$U_1 \in E(n, B[T^{-1}, \alpha^{-1}], JT^{-2}B[T^{-1}, \alpha^{-1}]).$$

In particular for $b \in J$ we have

$$\begin{aligned} &(1 - Te_{ij})U_0U_1(1 + be_{ij})U_1^{-1}U_0^{-1}(1 + Te_{ij}) \\ &\quad = (1 - Te_{ij})U_0(1 + be_{ij})U_0^{-1}(1 + Te_{ij})(1 - Te_{ij})U_0(1 - be_{ij}) \\ &\qquad\qquad \times U_1(1 + be_{ij})U_1^{-1}U_0^{-1}(1 + Te_{ij}), \end{aligned}$$

where

$$(1 - Te_{ij})U_0(1 + be_{ij})T_{)}-1(1 + Te_{ij}) \in E(n, A_T, JA_T).$$

Moreover, modulo T^{-2} we have

$$U_0(1 - be_{ij})U_1(1 + be_{ij})U_1^{-1}U_0^{-1} \equiv U_0(1 - be_{ij})(1 + be_{ij})U_0^{-1} \equiv 1$$

The first assertion now follows from Lemma 2.

In particular, if $J = B$ then

$$E(n, S^{-1}A) = E(n, A_T)[\mathrm{GL}(n, B) \cap E(n, S^{-1}A)]E(n, D, T^{-1}D).$$

Assume now that J is an arbitrary α-stable ideal of B and

$$M \in E(n, S^{-1}A) \cap \mathrm{GL}(n, S^{-1}A, JS^{-1}A).$$

Let "$\overline{}$" denote the factorization modulo J. It is already proved that $M = M_1M_2M_3$, where $M_1 \in E(n, A_T)$, $M_2 \in \mathrm{GL}(n, B)$ and $M_3 \in E(n, D, T^{-1})$. It follows that $1 = \overline{M_1M_2M_3}$ and

$$\overline{M_3} = \overline{(M_1M_2)}^{-1} \in E(n, \overline{D}, \overline{T^{-1}D}) \cap E(n, \overline{A_T}).$$

By Proposition 1 there exists

$$U \in E(n, B[T^{-1}, \alpha^{-1}], T^{-1}B[T^{-1}, \alpha^{-1}])$$

such that $U^{-1}M \equiv 1 \bmod J$. Replacing M_3 by $U^{-1}M_3$, we finish the proof. $\sqcup$

We are now going to establish some properties of the group $E(n, A_T, JA_T)$, $n \geq 3$, where J is an α-stable ideal of B. First we need some auxilary statements. If $1 \leq i \leq n$, $a \in A_T$ denote by $d_i(a)$ the diagonal matrix in which the ith entry of the principal diagonal is equal to a, the other diagonal entries are equal to 1. This means that $d_i(a) = 1 + (a - 1)e_{ii}$. If $1 \leq i, j \leq n$ and $a, b \in A_T$, then put $d_{ij}(a, b) = d_i(a)d_j(b)$, $d_i = d_i(T)$.

Proposition 6. *Let* $M \in \mathrm{GL}(n, A, TJA)$. *Then*

$$d_iMd_i^{-1}, d_i^{-1}Md_i \in \mathrm{GL}(n, A, TJA)E(n, B, J) \cap E(n, B, J)\,\mathrm{GL}(n, A, TA).$$

Proof. By the assumption $M = 1 + UT + VT^2$, where

$$U = (u_{ij}) \in \mathrm{Mat}(n, J), \quad V = (v_{ij}) \in \mathrm{Mat}(n, JA).$$

Suppose for simplicity that $i = 1$. Then $d_1Md_1^{-1}$ is equal to

$$\begin{pmatrix} 1 + Tu_{11} + Tv_{11}T & Tu_{12}T + Tv_{12}T^2 & \dots & Tu_{1n} + Tv_{1n}T^2 \\ u_{21} + v_{21}T & 1 + u_{22}T + v_{22}T^2 & \dots & u_{2n}T + v_{2n}T^2 \\ \vdots & \vdots & \ddots & \vdots \\ u_{n1} + v_{n1}T & u_{n2}T + v_{2n}T^2 & \dots & 1 + u_{nn}T + v_{nn}T^2 \end{pmatrix}$$

and belongs to $\mathrm{Mat}(n, A)$ Similarly

$$(d_1 M d_1^{-1})^{-1} = d_1 M^{-1} d_1^{-1} \in \mathrm{Mat}(n, A),$$

and therefore

$$d_1 M d_1^{-1} \in \mathrm{GL}(n, A, JA).$$

Moreover

$$d_1 M d_1^{-1} = M_1(1 + u_{21}e_{21}) \dots (1 + u_{n1}e_{n1}) = (1 + u_{21}e_{21}) \dots (1 + u_{n1}e_{n1})M_2,$$

where $M_1, M_2 \in \mathrm{GL}(n, A, TJA)$. The case of $d_1^{-1} M d_1$ is similar. □

Proposition 7. *Let*

$$M \in E(n, A), \quad n \geq 3, M \equiv 1 \bmod TA.$$

Then

$$d_j M d_j^{-1}, \ d_j^{-1} M d_j$$

are products of matrices of the form

$$1 + \begin{pmatrix} T^{-1}u_1 T \\ u_2 T \\ \vdots \\ u_n T \end{pmatrix} (vw_1 T, \quad vw_2, \quad \dots, \quad vw_n) \tag{3}$$

where $U_i, w_i \in B$, $v \in A$, the column and the row

$$\begin{pmatrix} u_1 \\ u_2 \\ \vdots \\ u_n \end{pmatrix}, \quad (w_1, \quad w_2, \quad \dots, \quad w_n)$$

are unimodular and $\sum_{i=1}^{n} w_i u_i = 0$. If $n \geq 2 + \mathrm{s.r.}\, B$, then the matrix (3) *belongs to $E(n, A)$.*

Proof. Suppose that $j = 1$ and $M = \prod_i (1 + a_i e_{r_i s_i})$, $a_i = b_i + T v_i$, $b_i \in B$, $v_i \in A$. By the assumption

$$\prod_i (1 + b_i e_{r_i s_i}) = 1.$$

Note that

$$1 + a_i e_{r_i s_i} = (1 + b_i e_{r_i s_i})(1 + T v_i e_{r_i s_i})$$

for all i. For $q \geq 1$ put

$$h_q = \prod_{i=1}^{q-1} (1 + b_i e_{r_i s_i}).$$

Then

$$M = \prod_i [1 + h_i T v_i e_{r_i s_i} h_i^{-1}].$$

It follows that without loss of generality we can assume that

$$M = 1 + hTve_{r_i s_i}h^{-1}, \quad h \in E(n, B).$$

Let

$$\begin{pmatrix} u_1 \\ u_2 \\ \vdots \\ u_n \end{pmatrix}, \quad (\, w_1, \quad w_2, \quad \dots, \quad w_n \,)$$

by the ith column and the jth row of h. Since $i \neq j$ we have $w_1 u_1 + \cdots + w_m v_n = 0$. These column and row are unimodular,

$$M = 1 + \begin{pmatrix} u_1 \\ u_2 \\ \vdots \\ u_n \end{pmatrix} Tv\,(\, w_1, \quad w_2, \quad \dots, \quad w_n \,).$$

In particular,

$$d_1^{-1} M d_1 = 1 + \begin{pmatrix} T^{-1} u_1 T \\ u_2 T \\ \vdots \\ u_n T \end{pmatrix} (\, vw_1 T, \quad vw_2, \quad \dots, \quad vw_n \,).$$

Suppose now that $n \geq 2 + \text{s.r. } B$. There exist elements $q_2, \dots, q_n \in B$ such that

$$B = B(u_2 + q_2 u_1) + \cdots + B(u_n + q_n u_1).$$

If we put

$$N = (1 + q_2 T e_{21}) \dots (1 + q_n T e_{n1})$$

then $N(d_1^{-1} M d_1) N^{-1}$

$$1 + \begin{pmatrix} T^{-1} u_1 T \\ (u_2 + q_2 u_1) T \\ \vdots \\ (u_n + q_n u_1) T \end{pmatrix} (\, vw_1 T, \quad vw_2, \quad \dots, \quad vw_n \,) N^{-1}.$$

Note that $n - 1 \geq 1 + \text{s.r. } B$ and the column

$$\begin{pmatrix} (u_2 + q_2 u_1) \\ \vdots \\ (u_n + q_n u_1) \end{pmatrix} \tag{4}$$

is unimodular over B. Applying elementary transformations and conjugations by elementary matrices as above we can assume that the column (4) is equal to

$$\begin{pmatrix} 1 \\ 0 \\ \vdots \\ 0 \end{pmatrix}.$$

It follows that

$$d_1^{-1} M d_1 = 1 + \begin{pmatrix} T^{-1}u_1 T \\ T \\ 0 \\ \vdots \\ 0 \end{pmatrix} (l_1, \quad \dots, \quad l_n), \tag{5}$$

where $l_i \in A$. Since $n \geq 3$ by [A1,S1] we have $d_1^{-1} M d_1 \in E(n, A)$. □

Proposition 8. *Let* $M = (m_{rs}) \in \mathrm{GL}(n, A, JTA)$ *and* $1 \leq i \neq j \leq n$. *Suppose that*

$$m_{ij} \equiv t_{ij} T \pmod{T^2 A}, \quad t_{ij} \in B.$$

Then

$$(1 + T^{-1} e_{ij}) M = [d_i^{-1} M' d_i] M'' (1 + T^{-1} e_{ij})(1 + t_{ij} T e_{ji}),$$

for some $M' \in ME(n, J)$, $M'' \in E(n, J + JT)$.

Proof. Without loss of generality as in [A1,S2] we can assume that $i = 1, j = n$. By Proposition 4 and (5) we have

$$\begin{aligned} w = (1 + T^{-1} e_{1n}) M &= d_1^{-1}(1 + e_{1n}) d_1 M d_1^{-1} d_1 \\ &= d_1^{-1}(1 + e_{1n}) M' (1 + t_{21} e_{21}) \dots (1 + t_{n1} e_{n1}) d_1 \\ &= d_1^{-1}(1 + e_{1n}) M' (1 - e_{1n}) d_1 d_1^{-1} (1 + e_{1n})(1 + t_{21} e_{21}) \dots (1 + t_{n1} e_{n1}) d_1, \end{aligned} \tag{6}$$

where $M' = (1 + e_{1n}) M_1 (1 - e_{1n}) \in \mathrm{GL}(n, A, TJA)$. Moreover, if $1 < s < n$ then

$$(1 + e_{1n})(1 + t_{s1} e_{s1}) = (1 - t_{sn} e_{sn})(1 + t_{s1} e_{s1})(1 + e_{1n}).$$

Thus in (6) we have

$$\begin{aligned} w &= [d_1^{-1} M' d_1] d_1^{-1} [\prod_{s=2}^{n-1} ((1 - t_{sn})(1 + t_{s1} e_{s1}))] (1 + e_{1n})(1 + t_{n1} e_{n1}) d_1 \\ &= [d_1^{-1} M' d_1] [\prod_{s=2}^{n-1} ((1 - t_{sn})(1 + t_{s1} e_{s1}))] (1 + T^{-1} e_{1n})(1 + t_{n1} T e_{n1}). \end{aligned}$$

The assertion follows. □

Proposition 9. *Let* $1 \le i \ne j \le n$ *and* $t \in B$. *Then*

$$(1 + T^{-1}e_{ij})(1 + tTe_{ji})$$
$$= [1 + t(1+t)^{-1}Te_{ji}]d_{ji}((1+t)^{-1}, 1 + T^{-1}tT)[1 + T^{-1}(1+t)^{-1}e_{ij}].$$

Proof. Assume that $i = 2, j = 1$. Without loss of generality we can suppose that $n = 2$. A direct calculation ends the proof. □

Propositions 6–9 imply

Proposition 10. *Let* $U \in \mathrm{GL}(n, A, JA)$ *and* $U = MV$, *where*

$$M = (m_{ij}) \in \mathrm{GL}(n, A, TJA), \quad V \in \mathrm{GL}(n, B, J).$$

If $1 \le i \ne j \le n$ *and* $M_{ij} \equiv t_{ij}T \pmod{T^2A}$, $t_{ij} \in B$ *then*

$$(1 + T^{-1}e_{ij})U = [d_i^{-1}M'd_i]M''[1 + t_{ji}(1+t_{ji})^{-1}e_{ji}]$$
$$\times d_{ji}((1+t_{ji})^{-1}, 1 + T^{-1}t_{ji}T)[1 + T^{-1}(1+t_{ji})^{-1}e_{ij}]V,$$

where $M' \in ME(n, J)$, $M'' \in E(n, J + JT)$. *If* $U \in E(n, A)$ *and* $n \ge 2 + \mathrm{s.r.}\, B$ *then*

$$d_i^{-1}M'd_i \in E(n, A) \cap \mathrm{GL}(n, A, JA).$$

Corollary 11. *Let* $n \ge 2 + \max(1, \mathrm{s.r.}(B/J))$ *and* $1 + J$ *satisfy the left Ore condition in* B. *Then* $E(n, A_T, JA_T) \subseteq E(n, (1+J)^{-1}A, (1+J)^{-1}JA)$ *is included in*

$$[\mathrm{GL}(n, (1+J)^{-1}B, (1+J)^{-1}J) \cap E(n, (1+J)^{-1}A_T)]$$
$$\times E(n, (1+J)^{-1}B[T^{-1}, \alpha^{-1}], (1+J)^{-1}JB[T^{-1}, \alpha^{-1}]).$$

Remark. If $1 + J$ satisfies the left Ore condition in B then $1 + J$ satisfies the left Ore condition in $A, A_T, S^{-1}A$.

The next proposition has been proved in [A1].

Proposition 12. *Let*

$$n \ge 2 + \max(1, \mathrm{s.r.}(B/J), \mathrm{s.r.}(A/JA)_T).$$

Then

$$E(n, A_T) \cap \mathrm{GL}(n, A_T, JA_T) = E(n, A_T, JA_T);$$
$$E(n, A) \cap \mathrm{GL}(n, A, JA) = E(n, A, JA).$$

A similar statement is valid for $(1+J)^{-1}B$, $(1+J)^{-1}A$ *and* $(1+J)^{-1}A_T$.

Proposition 12, Theorem 5 and Corollary 11 imply

Theorem 13. *Let* $1+J$ *satisfy the left Ore condition in* B *and*

$$n \geq 2 + \max(1, \text{s.r.}\, B, \text{s.r.}(B/J), \text{s.r.}(A/JA)_T).$$

Then

$$E(n, S^{-1}A) \cap \mathrm{GL}(n, S^{-1}A, JS^{-1}A) \subseteq E(n, (1+J)^{-1}A, TJ(1+J)^{-1}A)\,\mathrm{GL}(n, B, J)E(n, D, T^{-1}JD).$$

Proof. By Proposition 12 and Corollary 11 it suffices to notice that

$$E(n, (1+J)^{-1}A, J(1+J)^{-1}A) = E(n, (1+J)^{-1}A, TJ(1+J)^{-1}A)E(n, (1+J)^{-1}B, J(1+J)^{-1}B).$$

□

Apply the previous results to the study of epimorphisms of free left A-modules $f : A^n \to A$, $n \geq 3$. Here A^n and A are free arithmetic left A-modules of ranks n and 1, respectively. Denote by $p : A^n \to A$ the natural projection of A^n onto A as the first summand in A^n. In other words in e_i is the ith basic vector in A^n then

$$p(e_i) = \begin{cases} e_1, & \text{if } i = 1 \\ 0, & \text{if } i > 1. \end{cases}$$

Proposition 14. *Let* $f \equiv p \pmod{JTA}$ *in the assumption of Theorem* 12. *Suppose that there exist* $M \in E(n, S^{-1}A) \cap \mathrm{GL}(n, S^{-1}A, JS^{-1}A)$ *such that* $fM = p$. *Then there exists* $U \in E(n, (1+J)^{-1}A, TJ(1+J)^{-1}A)$ *such that* $fU = p$.

Proof. By Theorem 13, $M = UV$, where

$$U \in E(n, (1+J)^{-1}A, TJ(1+J)^{-1}A), \quad V \in \mathrm{GL}(n, (1+J)^{-1}D, J(1+J)^{-1}D).$$

Then $p = fM = fUV$ and therefore $pV^{-1} = fU$ is the matrix whose entries belong to $(1+J)^{-1}D \cap (1+J)^{-1}A = (1+J)^{-1}B$. But $M \equiv 1 \pmod{TA}$. Since $f \equiv p \pmod{TA}$ we have $p = fU$. □

Theorem 15. *Suppose that in the assumptions of Theorem* 13 *and Proposition* 14 *we have in addition* $n \geq 2 + \text{s.r.}\, S^{-1}(A/JA)$. *Assume that* $fH = p$ *for some* $H \in E(n, S^{-1}A)$. *Then there exists* $V \in E(n, (1+J)^{-1}A)$ *such that* $fV = p$.

Proof. Modulo $JS^{-1}A$ the image of H has the form

$$\overline{H} = \begin{pmatrix} 1 & 0 \\ \overline{N_1} & \overline{N} \end{pmatrix} \in E(n, \overline{S^{-1}A}),$$

where $\overline{N} \in \mathrm{GL}(n-1, \overline{S^{-1}A})$, and $\overline{N_1}$ is a column of height $n-1$. Since $n-1 >$ s.r. $\overline{S^{-1}A}$ by [S1,Theorem 4.4] we have $\overline{N} \in E(n-1, \overline{S^{-1}A})$. There exists a matrix

$L \in E(n-1, S^{-1}A) \supseteq E(n, S^{-1}A)$ whose image modulo J is equal to N and $pL = p$. It is a routine to show that changing H we can assume that $\overline{H} = 1$, i.e.

$$H \in E(n, S^{-1}A) \cap \mathrm{GL}(n, S^{-1}A, JS^{-1}A).$$

Apply now Proposition 14. □

Corollary 16 ([A1]). *Suppose that in the assumptions of Theorem* 15 *the ideal J is contained in the Jacobson radical of B and B/J is finite dimensional. Assume that $n \geq 4$ and $fH = p$ for some $H \in E(n, S^{-1}A)$. Then $fU = p$ for some $U \in E(n, A, TA)$.*

Proof. $1 + J$ is invertible in B since J is radical. By [B, Proposition 8.2] $\text{s.r.}\, B \leq 1$ and $\text{K.}\dim S^{-1}(A/JA) \leq \text{K}\dim(B/J) = 0$ [RSS]. Hence $\text{s.r.}\, S^{-1}(A/JA) \leq 1 + \text{K}\dim(B/J) = 1$ [S], and

$$\text{s.r.}(A/JA)_T \leq 1 + \text{K}\dim(A/JA)_T \leq 2.$$

Apply Theorem 14. □

Theorem 17. *Let J, n be from Theorem* 14 *and $f : A_T^n \to A_T$ be an epimorphism of free left A_T-modules, $f \equiv p \pmod{JA_T}$, where $p : A_T^n \to A_T$ is the natural projection as above. Suppose that there exists $M \in E(n, S^{-1}A)$ such that $fM = p$. Then there exists*

$$U \in \mathrm{GL}(n, (1+J)^{-1}A_T, J(1+J)^{-1}A_T)$$

such that $fU = p$. If $n \geq 2 + \text{s.r.}\, B$ then we can take

$$U \in E(n, (1+J)^{-1}A_T, J(1+J)^{-1}A_T).$$

The proof follows the lines of the proof of Theorem 15, see [A1].

§3. Patching theorem

In this section we expose a constructive proof of Theorem 3. In order to do that we shall prove some formulae for matrices over free associative rings whose coefficients are integer polynomials. Let

$$A = \oplus_{n \geq 0} A(n)$$

be a graded R-algebra over a homogeneous central subalgebra R such that each non-zero element of R is not a zero divisor in A. If b is a non-zero homogeneous element in R then A_b^+ will denote the non-negative part of the $\mathbb{Z}$-graded localized algebra A_b. If M is a finitely presented $A(0)$-module, then $\mathrm{End}(A \otimes_{A(0)} M)$ is a graded algebra whose n^{th} component is $\hom(M, A(n) \otimes_{A(0)} M)$. Let f be an endomorphism of $A \otimes_{A(0)} M$ and let $f = f_0 + f_1 + \dots$ be its decomposition induced by this grading. For $c \in R$ put

$$f(c) = f_0 + cf_1 + c^2 f_2 + \cdots \in \operatorname{End}(A \otimes_{A(0)} M).$$

Note that a map $f \rightsquigarrow f(c)$ for a given c is an endomorphism of the algebra $\operatorname{End}(A \otimes_{A(0)} M)$. In particular, if $U \in \operatorname{Mat}(n, A)$ and $c \in R$ we can define $U(c) \in \operatorname{Mat}(n, A)$.

From now on we fix in this section the following notations. Let $B = \mathbb{Z}\langle T_1, T_2, \dots \rangle$ be a free associative ring and $A = B[X, Y, Z]$ be the polynomial ring.

Proposition 18. *Let* $1 \le s \ne r, i \ne j \le n$ *and*

$$w = (1 + XT_1 e_{rs})(1 + YZT_2 e_{ij})(1 - XT_1 e_{rs}).$$

If $s \ne i \ne j \ne r$, *then*

$$w = 1 + YZT_2 e_{ij}.$$

If $s \ne i \ne j = r$, *then*

$$w = (1 + YZXT_2 T_1 e_{is})(1 + YZT_2 e_{ij}).$$

If $s = i \ne j \ne r$, *then*

$$w = (1 - YZXT_1 T_2 e_{rj})(1 + YZT_2 e_{ij}).$$

If $s = i \ne j = r$ *and an index* q *is different from* i, j *then*

$$w = [(1 + Ye_{ij})(1 - YXT_1 e_{jq}), (1 + ZT_2 e_{qi})(1 + ZXT_2 T_1 e_{qi})].$$

The proof follows from the Steinberg relations (2) by direct calculations.

According to [A1,Proposition 9] for any positive integer m there exists an integer homogeneous polynomial $h \in \mathbb{Z}[X, Y]$ of total degree m such that for all positive integers L

$$(X + Y)^{2mL} = (X^m h)^L + y^m g \tag{7}$$

where $g = g_L \in \mathbb{Z}[X, Y]$ is a homogeneous polynomial of total degree $2mL - m$.

Consider now in A a grading such that T_i is a homogeneous element of degree $d_i \ge 0, i = 1, 2, \dots$, and X, Y, Z are homogeneous elements of degree d.

Theorem 19 ([A1, Theorem 4]). *Let* $n \ge 3$ *and*

$$G = \prod_{r=1}^{l} (1 + T_r e_{i_r, j_r}) \in E(n, B),$$

where $1 \le i_r \ne j_r \le n$. *If* $L \ge l, m = 2^l$ *and* $h = h(X, Y)$ *from* (7) *then*

$$G((X + Y)^{2^{L+1}}) G(X^{2^L} h)^{-1}$$

is a product of elementary matrices of the form

$$1 + Y^{2^{L-l}} v_j u_j e_{s_j r_j}.$$

Here v_j is a monomial in X, Y and u_j is a monomial in $T_1, T_2, \ldots$.

Proof. Suppose first that $l = 1$. By (7) with $m = 2^L$ we have

$$\begin{aligned} G((X+Y)^{2^{L+1}})G(X^{2^L}h)^{-1} &= [1 + (X+Y)^{2^{L+1}d_1} T_1 e_{i_1 j_1}][1 - (X^{2^L}h)^{d_1} T_1 e_{i_1 j_1}] \\ &= 1 + T_1(-(X^{2^L}h)^{d_1} + (X+Y)^{2^{L+1}d_1}) e_{i_1 j_1} \\ &= 1 + T_1 Y^{2^L} g e_{i_1 j_1}, \end{aligned}$$

where $g(X, Y) \in \mathbb{Z}[X, Y]$ is a homogeneous polynomial of degree $d_1 - 1$ if $d_1 \geq 1$ and $g(X, Y) = 0$ if $d_1 = 0$. Thus the case $l = 1$ is proved.

Suppose that for $l - 1$ the statement is proved. Then

$$G = (1 + T_1 e_{i_1 j_1}) G_1,$$

where G_1 is a product of the last $l - 1$ factors of G. By induction on l we have

$$\begin{aligned} &G((X+Y)^{2^{L+1}})G(X^{2^L}h)^{-1} \\ &= [1 + (X+Y)^{2^L d_1} T_1 e_{i_1 j_1}] G_1((X+Y)^{2^{L+1}}) G_1(X^{2^L}h)^{-1} [1 - (X^{2^L}h)^{d_1} T_1 e_{i_1 j_1}] \\ &= [1 + (X+Y)^{2^L d_1} T_1 e_{i_1 j_1}] \Big\{ \prod [1 + Y^{2^{L-l+1}} v_j u_j e_{s_j r_j}] \Big\} [1 - (X^{2^L}h)^{d_1} T_1 e_{i_1 j_1}] \\ &= [1 + (X+Y)^{2^L d_1} T_1 e_{i_1 j_1}][1 - (X^{2^L}h)^{d_1} T_1 e_{i_1 j_1}] \\ &\quad \times \prod \Big\{ (1 + (X^{2^L}h)^{d_1} T_1 e_{i_1 j_1})(1 + Y^{2^{L-l+1}} v_j u_j e_{s_j r_j})(1 - (X^{2^L}h)^{d_1} T_1 e_{i_1 j_1}) \Big\}, \end{aligned}$$

where u_i, v_i are as above. By Proposition 18 each bracket $\{\ldots\}$ is a product of elementary matrices

$$1 \pm Y^{2^{L-l}} p f e_{\alpha,\beta},$$

where p is a monomial in X, Y and f is a monomial in $T_1, T_2, \ldots$. □

Corollary 20. *Let G, L, l, h be from the assumption of Theorem* 19. *Then*

$$G = UG\Big(\big(\frac{X}{X+Y}\big)^{2^L} h\big(\frac{X}{X+Y}, \frac{Y}{X+Y}\big)\Big)$$

where

$$U = GG\Big(\big(\frac{X}{X+Y}\big)^{2^L} h\big(\frac{X}{X+Y}, \frac{Y}{X+Y}\big)\Big)^{-1}$$

is a product of elementary matrices of the form

$$1 + \big(\frac{Y}{X+Y}\big)^{2^{L-l}} v_j\big(\frac{X}{X+Y}, \frac{Y}{X+Y}\big) u_j e_{s_j r_j},$$

where v_j, u_j are from the assumption of Theorem 19.

Proof. Apply in Theorem 20 the substitution

$$X \to \frac{X}{X+Y}, \quad Y \to \frac{Y}{X+Y}.$$

Then $G((X+Y)^{2^L}) \to G$. □

Theorem 21 (see [A1, Theorem 5]). *Let M be a finitely generated left module over a ring $A(0)$ and $n \geq 3$. Suppose that*

$$f : A \otimes_{A(0)} M \to A^n$$

is a homomorphism of left A-modules and $f = Sf(0)^{\cdot} = Gf(0)$, where $S, G \in E(n, A)$. If L, l, h, U are from Theorem 19 and Corollary 20 then

$$f = US\Big(\big(\frac{X}{X+Y}\big)^{2^L} h\big(\frac{X}{X+Y}, \frac{Y}{X+Y}\big)\Big) f(0).$$

Proof. By the assumption $S^{-1}Gf(0) = f(0)$. Hence by Theorem 19 and Corollary 20 we have

$$S\Big(\big(\frac{X}{X+Y}\big)^{2^L} h\big(\frac{X}{X+Y}, \frac{Y}{X+Y}\big)\Big)^{-1} G\Big(\big(\frac{X}{X+Y}\big)^{2^L} h\big(\frac{X}{X+Y}, \frac{Y}{X+Y}\big)\Big) f(0) = f(0)$$

and therefore

$$\begin{aligned}
&US\Big(\big(\frac{X}{X+Y}\big)^{2^L} h\big(\frac{X}{X+Y}, \frac{Y}{X+Y}\big)\Big) f(0) \\
&= GG\Big(\big(\frac{X}{X+Y}\big)^{2^L} h\big(\frac{X}{X+Y}, \frac{Y}{X+Y}\big)\Big)^{-1} S\Big(\big(\frac{X}{X+Y}\big)^{2^L} h\big(\frac{X}{X+Y}, \frac{Y}{X+Y}\big)\Big) f(0) \\
&= GG\Big(\big(\frac{X}{X+Y}\big)^{2^L} h\big(\frac{X}{X+Y}, \frac{Y}{X+Y}\big)\Big)^{-1} S\Big(\big(\frac{X}{X+Y}\big)^{2^L} h\big(\frac{X}{X+Y}, \frac{Y}{X+Y}\big)\Big) \\
&\quad \times S\Big(\big(\frac{X}{X+Y}\big)^{2^L} h\big(\frac{X}{X+Y}, \frac{Y}{X+Y}\big)\Big)^{-1} G\Big(\big(\frac{X}{X+Y}\big)^{2^L} h\big(\frac{X}{X+Y}, \frac{Y}{X+Y}\big)\Big) f(0) \\
&= Gf(0) = f.
\end{aligned}$$

□

The proof of Theorem 3 follows from Theorem 21, see [A1].

References

[A1] Artamonov, V. A., Projective modules and elementary matrices over rings of skew polynomials, in: Sbornik rabot po algebre, Moscow, Moscow State University, 1989, 6–49.

[A2] Artamonov, V. A., Projective modules over crossed products, J. Algebra 173 (1995), 696–714.

[A3] Artamonov, V. A., Projective modules over quantum polynomials, Mat. Sb. 185 (1994), no. 7, 3–12.

[A4] Artamonov, V. A., Automorphisms of decomposable projective modules, Fundamentalnaya i Prikladnaya Matematika 1 (1995), no. 1, 63–69.

[B] Bass, H., Algebraic K-theory, Benjamin, New York 1968.

[L] Lam, T. Y., Serre's conjecture, Lecture Notes in Math. 635, Springer-Verlag, 1978.

[LS] Logar, A., Sturmfels, B., Algorithms for the Quillen–Suslin theorem, J. Algebra 145 (1992), 231–239.

[Q] Quillen, D., Higher algebraic K-theory, Lecture Notes in Math. 341, Springer-Verlag, 85–147.

[RSS] Resco, R., Small, S., Stafford, J. T., Krull and global dimensions of semiprime Noetherian PI-rings, Trans. Amer. Math. Soc 274 (1982), 285–295.

[S] Stafford, J. T., On the stable range of right Noetherian rings, Bull. London Math. Soc. 13 (1981), 39–41.

[S1] Suslin, A. A., Algebraic K-theory, in: Algebra. Topology. Geometry, Moscow VINITI 20, 1982, 71–152.

[S2] Suslin, A. A., On the structure of special linear group over polynomial ring, Izv. Akad. Nauk SSSR Ser. Mat. 41 (1977), no. 2, 235–252.

Department of Mechanics and Mathematics
Chair of Higher Algebra
Moscow State University
Moscow 119899, Russia
e-mail: artamon@nw.math.msu.su

On the simplicity of centralizer nearrings

*K. I. Beidar**, *Y. Fong and W.-F. Ke*

§1. Introduction and background

Given a set X, we denote by $|X|$ the cardinality of X. Let G be a not necessarily abelian group written additively and let $\mathcal{A}$ be a subgroup of the group $\operatorname{Aut} G$ of all automorphisms of G. Also, set $G^* = G \setminus \{0\}$ and let $\mathcal{C}(\mathcal{A}) = \mathcal{C}_G(\mathcal{A}) = \{f : G \to G \mid f(0) = 0 \text{ and } fa = af \text{ for all } a \in \mathcal{A}\}$. The set $\mathcal{C}(\mathcal{A})$ with pointwise addition and composition is a right zero-symmetric nearring, and is usually referred to as the centralizer nearring on G determined by $\mathcal{A}$ (see [2, 9, 12]). For any $g \in G$, let $\mathrm{St}(g) = \{a \in \mathcal{A} \mid ag = g\}$ be the $\mathcal{A}$-stabilizer of g. Clearly, $\mathrm{St}(g)$ is a subgroup of $\mathcal{A}$. We call a stabilizer $\mathrm{St}(g)$ maximal if it is not a proper subgroup of any other stabilizer.

Recall that a right nearring N is said to be regular if for any element $n \in N$ there exists an element $x \in N$ such that $nxn = n$.

The ideal structure of rings of linear transformations of vector spaces over skew fields is well-known and is easily determined (see [6]). In contrast with this, the ideal structure of centralizer nearrings (a nonlinear analogy of linear transformation rings) is very complicated and, in fact, is not completely described up to now, although these nearrings play an important role in the structure theory of nearrings (see [2, 3, 9, 12]). In a number of papers the question when $\mathcal{C}(\mathcal{A})$ is simple was investigated [1, 5, 7, 8, 10, 11, 14]. Berman and Silverman [1] proved the simplicity of $\mathcal{C}(\mathcal{A})$ in the case $\mathcal{A} = \{\mathrm{id}_G\}$. The case with nontrivial automorphism groups $\mathcal{A}$ is much more complicated and there is no satisfactory description of it. But there are important partial results (see [12, p. 370], [5, 7, 8, 10, 11, 14]). We mention some of these results which are close to the study of this paper.

Lemma A (Betsch [4]). *Let $u, v \in G$. Then $St(u) \subseteq St(v)$ if and only if there exists an $f \in \mathcal{A}$ such that $f(u) = v$.* □

*Partially supported by the National Science Council, R. O. C., grant #NSC-84-2121-M-006-024.

Theorem B (Meldrum–Oswald [10, Theorem 1]). *$C(\mathcal{A})$ is regular if and only if all stabilizer subgroups are maximal.* □

Theorem C (Maxson–Smith [8, Theorems 1 and 4,]). *Let G be a finite group. Then the following are equivalent.*

(1) *$C(\mathcal{A})$ is simple.*

(2) *All stabilizers $St(g)$, $g \in G$, are maximal and conjugate.*

The following question was raised by Meldrum–Zeller [11] (also, see [12, Problem 22]).

Does $C(\mathcal{A})$ simple imply all stabilizers maximal and conjugate?

In this paper we shall present results which give this question both affirmative and negative answers, namely, we shall prove the following two theorems.

Theorem 1. *If $\mathcal{A}$ is a nilpotent group and $C(\mathcal{A})$ is simple, then* (1) *all stabilizers are maximal and conjugate, and* (2) *$C(\mathcal{A})$ is regular.*

Theorem 2. *Let F be a countable field and V a countable dimensional linear space over F. Then there exist a solvable subgroup $\mathcal{A}$ of the automorphism group $\mathrm{GL}_F(V)$ of class 2 such that:*

(1) *$C(\mathcal{A})$ is simple;*

(2) *$C(\mathcal{A})$ is not regular;*

(3) *there is no nonzero elements of V with maximal stabilizer;*

(4) *not all stabilizers are conjugate.*

Thus, Theorem 1 shows that nilpotent groups are examples to Meldrum–Zeller question on the affirmative side, while Theorem 2 says that a solvable subgroup of $\mathrm{GL}_F(V)$ of class 2, F a countable field, gives a counterexample to the question. Moreover, from Theorem 1, we understand that an example of Theorem 2 is in some sense a minimal one.

§2. The case of nilpotent groups

We are going to prove Theorem 1 in this section. Hence, if G is a nilpotent group, then the simplicity of $C(\mathcal{A})$ implies that all stabilizers are maximal and conjugate. The proof will follow from the next two results. The first proposition is inspired by the proof of Theorem 1 from [8].

(2.1) Proposition. *Suppose that $C(\mathcal{A})$ is simple. Then for all nonzero elements $u, v \in G$ there exists an element $a \in \mathcal{A}$ such that $a\,\mathrm{St}(u)a^{-1} \supseteq \mathrm{St}(v)$.*

Proof. Assume the contrary, and let $u, v \in G \setminus \{0\}$ such that $a\,\mathrm{St}(u)a^{-1} \not\supseteq \mathrm{St}(v)$ for all $a \in G$. Set

$$K = \{w \in G \mid \text{there exists an element } a \in \mathcal{A} \text{ such that } a\,\mathrm{St}(w)a^{-1} \supseteq St(v)\}$$

and let $I = \{f \in C(\mathcal{A}) \mid f(K) = 0\}$. Notice that $u \notin K$ and $v \in K$. It is easy to verify that I is a left ideal. We want to show a contradiction that I is a proper ideal of $C(\mathcal{A})$.

For $w \in K$, $\alpha, \beta \in \mathcal{A}$ and $\alpha St(w)\alpha^{-1} \supseteq St(v)$, we have

$$\begin{aligned}(\alpha\beta^{-1})St(\beta w)(\alpha\beta^{-1})^{-1} &= (\alpha\beta^{-1})(\beta St(w)\beta^{-1})(\alpha\beta^{-1})^{-1}\\ &= \alpha St(w)\alpha^{-1} \supseteq St(v).\end{aligned}$$

Hence $\beta w \in K$ and $\mathcal{A}K = K$, and so there exists an element $f \in C(\mathcal{A})$ such that $f(K) = 0$ and $f(x) = x$ for all $x \in G \setminus K$. Since $u \in G \setminus K$, we have $f \neq 0$. Therefore, $I \neq 0$. As $K \neq 0$, we conclude that $\mathrm{id}_G \notin I$; hence I is a proper left ideal of $C(\mathcal{A})$. Now let $g \in C(\mathcal{A})$, $i \in I$ and $k \in K$. Since $\mathrm{St}(g(k)) \supseteq \mathrm{St}(k)$, we infer that $g(k) \in K$. Therefore, $ig(k) = i(g(k)) = 0$ for all $k \in K$. This implies that $ig \in I$ and so I is a right ideal. Therefore I is a proper ideal of $C(\mathcal{A})$, a contradiction to our hypothesis that $C(\mathcal{A})$ is simple. Thus we conclude that there indeed exists an element $a \in G$ such that $a\,\mathrm{St}(u)a^{-1} \supseteq \mathrm{St}(v)$. □

(2.2) Lemma. *Let $\mathcal{A}$ be a nilpotent group, U and V subgroups of $\mathcal{A}$ and $\alpha \in \mathcal{A}$. Suppose that $\alpha U\alpha^{-1} \supseteq V \supseteq U$. Then $U = V$.*

Proof. Let Z be the center of $\mathcal{A}$. Then

$$\begin{aligned}U \cap Z = \alpha(U \cap Z)\alpha^{-1} &= (\alpha U\alpha^{-1}) \cap (\alpha Z\alpha^{-1})\\ &= (\alpha U\alpha^{-1}) \cap Z \supseteq V \cap Z \supseteq U \cap Z;\end{aligned}$$

hence $K = U \cap Z = V \cap Z$.

We proceed by induction on the class of nilpotency $\mathrm{cl}(\mathcal{A})$ of $\mathcal{A}$, i.e., the smallest integer n such that $Z_n = \mathcal{A}$, where $Z_0 = \{1\}$, and $Z_{t+1} = \{x \in \mathcal{A} \mid [x, \mathcal{A}] \subseteq Z_t\}$ for $t \geq 1$.

Clearly, if $\mathrm{cl}(a) = 1$, then $\mathcal{A}$ is abelian and $U = \alpha U\alpha \supseteq V \supseteq U$. Therefore, in this case, we have $U = V$.

Suppose now that the statement holds for all nilpotent groups of class n, where $n \geq 1$, and let $\mathrm{cl}(\mathcal{A}) = n + 1$. Note that $Z_1 = Z$. In the factor group $\mathcal{A}/Z$, we have

$$(\alpha Z)(UZ/Z)(\alpha Z)^{-1} \supseteq (VZ)/Z \supseteq (UZ)/Z.$$

Since $\mathrm{cl}(\mathcal{A}/Z) = n$, we have, by the induction hypothesis, that $(UZ)/Z = (VZ)/Z$. Hence $UZ = VZ$, and for every $v \in V$ there exist $u \in U$ and $z \in Z$ such that $v = uz$.

Therefore, $z = u^{-1}v \in (V \cap Z) = K \subseteq U$ and $v = uz \in U$. Thus, $V = U$, and the proof is completed. □

We can now prove Theorem 1.

Proof of Theorem 1. Suppose that $C(\mathcal{A})$ is simple, $\mathcal{A}$ is nilpotent, and let $u, v \in G \setminus \{0\}$. By (2.1), there exist elements $\alpha, \beta \in \mathcal{A}$ such that $\alpha\,\mathrm{St}(u)\alpha^{-1} \supseteq \mathrm{St}(v)$, and $\beta\,\mathrm{St}(v)\beta^{-1} \supseteq \mathrm{St}(u)$. Hence

$$(\alpha\beta)(\beta^{-1}\,\mathrm{St}(u)\beta)(\alpha\beta)^{-1} \supseteq \mathrm{St}(v) \supseteq \beta^{-1}\,\mathrm{St}(u)\beta.$$

From (2.2), it follows that $\mathrm{St}(v) = \beta^{-1}\,\mathrm{St}(u)\beta$. This shows that all stabilizers are conjugate.

Assume now that $\mathrm{St}(v) \supseteq \mathrm{St}(u)$. By (2.1), we have that $\alpha\,\mathrm{St}(u)\alpha^{-1} \supseteq \mathrm{St}(v) \supseteq \mathrm{St}(u)$ for some $\alpha \in \mathcal{A}$. From (2.2), we now deduce that $\mathrm{St}(u) = \mathrm{St}(v)$. Hence all stabilizers are maximal, and the proof is complete. □

§3. A counterexample

Let $\mathbb{Z}$ be the ring of integers. Let $A_i = \langle a_i \rangle$ be infinite cyclic groups with generators a_i, $i \in \mathbb{Z}$, and set $D = \biguplus_{i\in\mathbb{Z}} A_i$. Note that every element $d \in D$ can be uniquely represented in the form

$$d = a_{i_0}^{\alpha_0} a_{i_1}^{\alpha_1} \dots a_{i_n}^{\alpha_n},$$

where $n, \alpha_i \in \mathbb{Z}$, $\alpha_t \neq 0$ for all t, and $i_0 < i_1 < \cdots < i_n$. The group D has an automorphism $\sigma : D \to D$ given by

$$\sigma(a_i) = a_{i+2} \quad \text{for all } i \in \mathbb{Z}.$$

Let $C = \langle c \rangle$ be an infinite cyclic group with a generator c. Now consider the semidirect product $\mathcal{A} = D \rtimes_\sigma C$, where

$$(u, c^i)(v, c^j) = (u(\sigma^i v), c^{i+j}) \quad \text{for all } (u, c^i), (v, c^j) \in \mathcal{A}.$$

We'll identify the groups D and C with the subgroups $(D, 1)$ and $(1, C)$ of $\mathcal{A}$, respectively. Then

$$cdc^{-1} = (1, c)(d, 1)(1, c)^{-1} = \sigma d \quad \text{for all } d \in D$$

and

$$dc^i = c^i\sigma^{-i}d \quad \text{for all } i \in z. \tag{3:1}$$

Furthermore, any element $a \in \mathcal{A}$ can be expressed as $c^i d$ for some $d \in D$ and $i \in \mathbb{Z}$. Finally, for any integer $m \in \mathbb{Z}$, set

$$D_m = \oplus_{i\leq m} A_i.$$

(3.1) Lemma. *Let* $m, l \in \mathbb{Z}$. *Then:*

(i) D_m *and* D_l *are conjugate if and only if* $m \equiv l \pmod 2$.

(ii) $D_m \cap D_l = D_{\min\{m,l\}}$.

Proof. (i) Assume that D_m and D_l are conjugate. Then $D_m = aD_la^{-1}$ for some $a \in \mathcal{A}$. Let $a = c^{\mathbf{Z}}$ and $d \in D$. Since D is abelian, it follows from (3:1) that

$$aD_la^{-1} = c^iD_lc^{-i} = \sigma^iD_l = D_{l+2i}.$$

Hence $D_m = D_{l+2i}$, and so $m = l + 2i$, i.e., $m \equiv l \pmod 2$.

Conversely, suppose $m \equiv l \pmod 2$. Then $m = l + 2i$ for some $i \in z$. Therefore, $c^iD_lc^{-i} = \sigma^iD_l = D_{l+2i} = D_m$.

(ii) Obvious. □

Let E be the disjoint union of the left cosets in $\mathcal{A}/D_0$ and $\mathcal{A}/D_1$. We shall denote by $[aD_i]$ ($i \in \{0, 1\}$) the left coset in $\mathcal{A}/D_i$ containing $a \in \mathcal{A}$.

Let F be a field with countably many elements. Consider the F-linear space G with E as its basis. Thus, every element $g \in G$ is a linear combination of elements of E:

$$g = \sum_i \lambda_i[c^{m_i}u_iD_0] + \sum_j \mu_j[c^{n_j}v_jD_1], \tag{3:2}$$

where $m_i, n_j \in \mathbb{Z}$, $\lambda_i, \mu_j \in F$ and $u_i, v_j \in D$. Define

$$\mathrm{tr}(g) = \sum_i \lambda_i + \sum_j \mu_j.$$

Clearly, $\mathrm{tr} : G \to F$ is a linear mapping.

For $k \in \mathbb{Z}$ and $i \in \{0, 1\}$, let $G^{(k,i)} = \sum_{d \in D} F \cdot [c^kdD_i]$. It is clear that

$$G = \oplus_k\big(G^{(k,0)} + G^{(k,1)}\big). \tag{3:3}$$

Let $\pi_{k,i}$ be the canonical projection of G onto $G^{(k,i)}$.

Notice that any coset $[aD_i] \in E$ can be uniquely written as

$$[aD_i] = [c^ka_{i_0}^{\alpha_0}a_{i_1}^{\alpha_1}\dots a_{i_n}^{\alpha_n}D_i] \tag{3:4}$$

with $k \in \mathbb{Z}$, $i < i_0 < i_1 < \cdots < i_n$ and $\alpha_j \neq 0$ for all j, $j = 0, 1, \dots, n$. Define an action of $\mathcal{A}$ on G by setting $\beta[aD_i] = [\beta aD_i]$ for all $\beta, a \in \mathcal{A}$, $i = 1, 2$. It is easy to see that

$$\mathrm{tr}(h) = \mathrm{tr}(ah) \quad \text{for all } a \in \mathcal{A} \text{ and } h \in G. \tag{3:5}$$

(3.2) Lemma. *Let* $i \in \mathbb{Z}$, $d_s \in D$, $s = 1, 2, \dots, r$, *and let* g *be as given in* (3:2). *Then:*

(i) $\mathrm{St}(\sum_s \tau_s[c^id_sD_j]) = D_{j+2i}$ *where* t_s*'s are in* F;

(ii) $\mathrm{St}(g) = D_k$, *where* $k = \min_{i,j}\{2m_i, 1 + 2n_j\}$.

Proof. (i) Let $\alpha = c^t b \in \mathrm{St}(\sum_s \tau_s[c^i d_s D_j])$, where $t \in \mathbb{Z}$ and $b \in D$. Then by (3:1) we have

$$\sum_s \tau_s[c^i d_s D_j] = \alpha \sum_s \tau_s[c^i d_s D_j] = \sum_s \tau_s \alpha[c^i d_s D_j] = \sum_s \tau_s[c^{t+i} d_s(\sigma^{-i} b) D_j].$$

Hence $t = 0$. It also follows that in the factor group D/D_j, the multiplication by $x = (\sigma^{-i} b) D_j$ leaves the finite subset $\{d_s D_j\}$ fixed. Hence the multiplication by some power of x leaves all elements of this subset fixed. Since D/D_j is a free abelian group, we conclude that $(\sigma^{-i} b) D_j = 1$, and so $\alpha = b \in \sigma^i D_j = D_{j+2i}$. Therefore, $\mathrm{St}(\sum_s \tau_s[c^i d_s D_j]) \subseteq D_{j+2i}$. The converse inclusion is obvious; hence $\mathrm{St}(\sum_s [c^i d_s D_j]) = D_{j+2i}$ as required.

(ii) As above, one can show that $\mathrm{St}(g) \subseteq D$. Since $DG^{(k,j)} \subseteq G^{(k,j)}$ for all $k \in \mathbb{Z}$ and $j = 0, 1$, we see that $\mathrm{St}(g) \subseteq \mathrm{St}(\pi_{k,j}(g))$ for all k and j, and so $\mathrm{St}(g) = \bigcap_{k,j} \mathrm{St}(\pi_{k,j}(g))$. Now the statement follows from (i) and (3.1(ii)). □

We can view $\mathcal{A}$ as a group of automorphisms of the linear space G over F, i.e., $\mathcal{A} \leq \mathrm{GL}_F(G)$. Indeed, since $\bigcap_{i \in \mathbb{Z}} D_i = 1$, it follows from (3.1) and (3.2) that for any $\alpha \in \mathcal{A} \setminus \{1\}$, there exists an element $g \in G$ such that $\alpha g \neq g$. Thus the homomorphism from $\mathcal{A}$ into $\mathrm{GL}_F(G)$ has trivial kernel, and so $\mathcal{A}$ can be viewed as a subgroup of $\mathrm{GL}_F(G)$.

We proceed to show that the groups G and $\mathcal{A}$, and the nearring $N = C(\mathcal{A})$ fulfill the statements of Theorem 2.

Proof of Theorem 2. First of all, as $[\mathcal{A}, \mathcal{A}] \subseteq D$ and D is abelian, the group $\mathcal{A}$ is solvable of class 2. By (3.2), $\mathrm{St}([D_j]) = D_j$ for $j \in \{0, 1\}$. It follows from (3.1(i)) that $\mathrm{St}([D_0])$ and $\mathrm{St}([D_1])$ are not conjugate. Again, from (3.2), we conclude that there is no nonzero element in G with maximal stabilizer. Now from Theorem B, it follows that $C(\mathcal{A})$ is not regular.

It remains to show that $C(\mathcal{A})$ is simple. Let $K \subseteq N = C(\mathcal{A})$ be an nonzero ideal. We need to show that $K = N$.

Let $G = \bigcup_{i \in I} \mathcal{A} y_i$ be a disjoint union of the $\mathcal{A}$-orbits such that $y_0 = [D_0]$. Choose any $i \in I$. By (3.2), we have $\mathrm{St}(y_i) = D_{k_i}$ for some $k_i \in \mathbb{Z}$. Write $k_i = 2t_i + \epsilon_i$, where $\epsilon_i \in \{0, 1\}$. Then

$$\mathrm{St}(c^{-t_i} y_i) = c^{-t_i} D_{k_i} c^{t_i} = \sigma^{-t_i} D_{k_i} = D_{\epsilon_i}.$$

Replacing y_i by $c^{-t_i} y_i$, we can assume that $\mathrm{St}(y_i) = D_{\epsilon_i}$ for all $i \in I$. Now, set $x_p = [D_0] + [c^p D_1]$ and $z_p = [D_1] + [c^p D_0]$ for $p = 1, 2, \ldots$. By (3.2), $\mathrm{St}(x_p) = D_0$ and $\mathrm{St}(z_p) = D_1$ for all p. From the definition of the action of $\mathcal{A}$ on G it follows that

$$\mathcal{A} x_p \cap \mathcal{A} x_q = \mathcal{A} z_p \cap \mathcal{A} z_q = \mathcal{A} x_p \cap \mathcal{A} z_q = \emptyset \quad \text{for all } p \neq q.$$

Furthermore by (3.2) we have that $\mathrm{St}(x) = D_{2i}$ and $\mathrm{St}(z) = D_{2j+1}$ for all $x \in \mathcal{A}x_p$ and $z \in \mathcal{A}z_q$. Therefore

$$|\{i \in I \mid \mathrm{St}(y_i) = D_0\}| = \infty = |\{i \in I \mid \mathrm{St}(y_i) = D_1\}|.$$

Since F is countable, G and I are both countable also. By Lemma A, there exists an element $f \in N$ such that the induced mapping $f : \{y_i \mid i \in I\} \to \{x_p \mid p = 1, 2, \dots\} \cup \{z_p \mid p = 1, 2, \dots\}$ is a bijection. We set $U = \cup_p(\mathcal{A}x_p \cup \mathcal{A}z_p)$ and define $h : U \to \cup_i \mathcal{A}y_i$ via

$$h(n) = \begin{cases} f^{-1}(n), & \text{if } n \in U \\ h(n) = 0, & \text{otherwise.} \end{cases}$$

Clearly, $h \in N$.

Since $\mathrm{St}(y_i) = D_{\epsilon_i} \subseteq D_2 = \mathrm{St}([cD_0])$, there exists an element $v \in N$ such that $v(y_0) = y_0$ and $v(y_i) = [cD_0]$ for all $i \neq 0$. Therefore, $v(G^*) \subseteq \mathcal{A}[D_0]$.

As $K \neq 0$, we have $k(g) \neq 0$ for some $k \in K$ and $g \in G$. From (3.2), we conclude that $\mathrm{St}(g) = D_r$ for some $r \in \mathbb{Z}$. Let $r = 2t + i$, where $i \in \{0, 1\}$ and $t \in \mathbb{Z}$. Since $\mathrm{St}([c^t D_0]) = D_{2t} \subseteq D_{2t+i} = D_r$, by Lemma A, there exists an element $u \in N$ such that $u([c^t D_0]) = g$. It is clear that $k \cdot u \in K$ and

$$k \cdot u([c^t D_0]) = k(u([c^t D_0])) = k(g) \neq 0.$$

Hence, without loss of generality, we can assume that

$$k([c^t D_0]) \neq 0 \quad \text{for some } t \in \mathbb{Z}.$$

Note that $c^t k([D_0]) = k([c^t D_0]) \neq 0$ and so $k([D_0]) \neq 0$. From $v(G^*) \subseteq \mathcal{A}[D_0]$ and $k([D_0]) \neq 0$, it follows that $v \cdot k \cdot v(G^*) \subseteq \mathcal{A}[D_0]$. By replacing k with $v \cdot k \cdot v$, we may assume that

$$k(G^*) \subseteq \mathcal{A}[D_0].$$

Denote define a mapping $w : G \to G$ by the rule

$$w(g) = \begin{cases} g, & \text{if } x \in U; \\ 0, & \text{if } g \in G \setminus U. \end{cases}$$

It is easy to see that $w \in N = C(\mathcal{A})$, $k \cdot w \in K$ and $w' = w \cdot w - w \cdot (w + k \cdot w) \in K$. If $x \notin U$, then $w'(x) = 0$ since $w(x) = 0$.

Now take an $x \in \{x_p \mid p = 1, 2, \dots\} \cup \{z_p \mid p = 1, 2, \dots\} \subseteq U$. Then $w(x) = x$ and $k(x) \in \mathcal{A}[D_0]$; hence $\mathrm{tr}(k(x)) = 1$ and $\mathrm{tr}(x) = 2 \neq 3 = \mathrm{tr}(x + k(x))$. From (3:5) we conclude that $x + k(x) \notin U = \cup_p(\mathcal{A}x_p \cup \mathcal{A}z_p)$. Therefore, $w(x + k(x)) = 0$, and so $w'(x) = x$ for all $x \in U$. Recall that $f(y_i) \in U$ and $h \cdot f(y_i) = y_i$ for all $i \in I$. Hence, $h \cdot w' \cdot f(y_i) = y_i$ for all $i \in I$, and so $\mathrm{id}_G = h \cdot w' \cdot f \in K$. This implies that $K = N$ and the proof of Theorem 2 is completed. □

References

[1] G. Berman and R. J. Silverman, Simplicity of near-rings of transformations, Proc. Amer. Math. Soc. 10 (1959), 456–459.

[2] G. Betsch, Primitive near-rings, Math. Z. 78 (1973), 351–361.

[3] G. Betsch, Some structure theorems on 2-primitive near-rings, in: A. Kertesz, editor, Rings, Modules and Radicals, Coll. Math. Soc. Janus Bolyai, No. 6. North Holland Pub. Co., Amsterdam 1973.

[4] G. Betsch, Near-rings of group mappings, Oberwolfach, 1976.

[5] P. Fuchs, C. J. Maxson, M. R. Pettet, and K. C. Smith, Centralizer near-rings determined by fixed point free automorphism groups, Proc. Royal Soc. Edinburgh 107 (1987), 327–337.

[6] N. Jacobson, Structure of Rings, Amer. Math. Soc., Providence, RI, 1956.

[7] C. J. Maxson and K. C. Smith, The centralizer of a group automorphism, J. Algebra 54 (1978), 27–41.

[8] C. J. Maxson and K. C. Smith, The centralizer of a set of group of automorphisms. Comm. Algebra 8 (1980), 211–230.

[9] J. D. P. Meldrum, Near-rings and Their Links with Groups, Research Note Series 134, Pitman Publ. Co., 1985.

[10] J. D. P. Meldrum and A. Oswald, Near-rings of mappings, Proc. Royal Soc. Endinburgh 83 (1979), 213–223.

[11] J. D. P. Meldrum and M. Zeller, The simplicity of near-rings of mappings. Proc. Royal Soc. Endinburgh 90 (1981), 185–193.

[12] G. Pilz, Near-rings, North-Holland/American Elsevier, second, revised edition, 1983.

[13] W. R. Scott, Group Theory, Dover publications, New York, Dover edition, 1987.

[14] M. Zeller, Centralizer Near-rings of Infinite Groups, Ph. D. Thesis, Texas A&M University, 1980.

Department of Mathematics
National Cheng Kung University
Tainan
Taiwan 701, ROC

Semigroups with generalized polynomial identities

K. I. Beidar, Y. Fong, U. Knauer, A. V. Mikhalev and K. P. Shum

Abstract. The structure of semigroups with a multilinear generalized polynomial identity and some primeness condition is described thereby providing an analog of corresponding results on prime rings with generalized polynomial identity.

1. Introduction

A well-known theorem by Posner states that a prime ring with polynomial identity is a two sided Goldie ring whose classical ring of quotients is a PI-ring [11]. Analogous results are proved for semiprime rings with polynomial identities [4], [9], [1] and [7]. But there are no parallel results for semigroups with (general) polynomial identities. The reason of their absence is probably the following: every ring with polynomial identity has a multilinear polynomial identity. The last property is no longer true for semigroups. Therefore, when studying structure of semigroups with polynomial identities, it is natural to restrict the consideration to multilinear polynomial identities. As it was shown by R. Wiegandt [14], a semigroup with multilinear identity and some primeness condition is commutative (see also [12], [10] and [15]). It can be shown that in this case a semigroup has a classical group of quotients and so this result is a version of Posner's theorem for semigroups. On the other hand the structure of prime rings with generalized polynomial identities was described by W. S. Martindale [8]. The goal of the present paper is to prove an analog of Martindale's theorem for semigroups with multilinear generalized polynomial identities.

2. Definitions and preliminary results

For concepts in semigroup theory which are not defined here compare for example [3]. We first introduce the concept of a left (right) equiprime semigroup which simulates the ring theoretic definition of a prime ring R.

A semigroup S with 0 is called *left equiprime* (*right equiprime*) if for all $a \neq b \in S$ and $0 \neq c \in S$ there exists an element $x \in S$ such that $axc \neq bxc$ (respectively,

$cxa \neq cxb$). Further a semigroup S with 0 is said to be *equiprime* if it is left and right equiprime. We note that for near-rings the corresponding concept was introduced in [2]. Further, if S is a multiplicative semigroup of a ring, then S is equiprime if and only if the corresponding ring is prime. It is clear that, for example, right cancellative semigroups with zero and left zero semigroups are left equiprime. Moreover, if a semigroup with zero is left equiprime, then it is right reductive, i.e., $as = bs$ for all $s \in S$ implies $a = b$, for $a, b \in S$, in other words S_S as a right S-act is dense in S_S, cf. [13] or, S_S is a faithful right S-act, cf. [3]. The definitions of right or left S-acts can, for example, be found in [6] or in [13].

In what follows S will be a left equiprime semigroup with 0. We proceed now to define the right Martindale semigroup of quotients $Q_r(S)$ of S and the extended (Martindale) centroid of S. To this end we denote the set of all nonzero two-sided ideals of S by $\mathcal{I}(S)$. Given an ideal I of S we denote by $\mathrm{Hom}(I_S, S_S)$ the set of all mappings $f : I \to S$ such that $f(xs) = f(x)s$ for all $x \in I$, $s \in S$, i.e., the set of all right S-act homomorphisms from $I_S \to S_S$, cf. [6]. Further, given an element $a \in S$ we denote by l_a the left multiplication by a (i.e., $l_a : S \to S$, $l_a(x) = ax$ for all $x \in S$) and note that $l_a \in \mathrm{Hom}(S_S, S_S)$. We set

$$H = \{(I; f) \mid I \in \mathcal{I}(S),\ f \in \mathrm{Hom}(I_S, S_S)\}$$

and define the binary relation $\sim$ on H as follows:

$$(I; f) \sim (J; g) \iff \text{there exists } K \in \mathcal{I}(S) \text{ such that } K \subseteq I \cap J \text{ and } f|_K = g|_K.$$

One can easily check that $\sim$ is an equivalence relation on H. Denote by $[I; f]$ the equivalence class of $(I; f) \in H$. Then clearly $[S; f] = [I; f]$ if $(S; f), (I; f) \in H$. Set $Q_r(S) = H/{\sim}$. Define a multiplication on $Q_r(S)$ as follows:

$$[I; f][J; g] = [JI; fg] \quad \text{for all} \quad [I; f], [J; g] \in Q_r(S).$$

It is easy to see that this operation is well-defined and associative taking into account that $fg(ji) = f(g(j)i)$ and $g(j)i \in I$ for all $i \in I$ and $j \in J$. We call $Q_r(S)$ with this multiplication *the right Martindale semigroup of quotients of S*.

Consider the mapping $F : S \to Q_r(S)$ given by $F(x) = [S; l_x]$ for all $x \in S$. Clearly F is a homomorphism of semigroups.

Lemma 2.1. *The homomorphism F is injective.*

Proof. Suppose $F(a) = F(b)$ for some $a \neq b \in S$. Then $(S; l_a) \sim (S; l_b)$ and so there exists $K \in \mathcal{I}(S)$ such that $l_a|_K = l_b|_K$. This means that $as = bs$ for all $s \in K$. Choose any $0 \neq c \in K$. By hypothesis S is left equiprime, i.e. there exists $x \in S$ such that $axc \neq bxc$. But $xc \in K$, and thus we get a contradiction. □

We identify S and $F(S) \subseteq Q_r(S)$ via F. That is, we will always omit F and write $[I; f]a = [I; f][S; l_a]$ for all $[I; f] \in Q_r(S)$ and $a \in S$.

Lemma 2.2. *Take $p = [I; f], q = [J; g] \in Q_r(S)$ and $K \in \mathcal{I}(S)$. Then:*

(i) $[I; f]a = f(a) \in S$ *for all* $a \in I$, *and thus* $[I; f]I = f(I) \subseteq S$;

(ii) $qx = px$ *for all* $x \in K$ *if and only if* $p = q$;

(iii) $Q_r(S)$ *is left equiprime. Furthermore, given* $p \neq q \in Q_r(S)$, $L \in \mathcal{I}(S)$ *and* $0 \neq r \in Q_r(S)$, *there exists* $u \in L$ *such that* $pur \neq qur$ *(i.e.,* $p[S; l_u]r \neq q[S; l_u]r$*).*

Proof. (i) Let $a \in I$. Then

$$[I; f]a = [I; f][S; l_a] = [SI; fl_a].$$

Further $(fl_a)(x) = f(ax) = f(a)x = l_{f(a)}(x)$ for $x \in S$ and so $[SI; fl_a] = [SI; l_{f(a)}] = f(a)$.

(ii) Let $px = qx$ for all $x \in K$. Setting $L = K \cap I \cap J$, we note that $L \in \mathcal{I}(S)$ and $f(x) = px = qx = g(x)$ for all $x \in L$. Therefore $f|_L = g|_L$ and so $p = [I; f] = [J; g] = q$. The converse is obvious.

(iii) Suppose that $p \neq q$ and let $0 \neq r = [K; h] \in Q_r(S)$. Then for all $L \in \mathcal{I}(S)$ there exist $x \in I \cap J \cap L$ such that $f(x) \neq g(x)$. Choose $y \in K$ such that $h(y) \neq 0$. Since S is left equiprime, $f(x)zh(y) \neq g(x)zh(y)$ for some $z \in S$. We compute $pxzry$ and $qxzry$

$$[I; f][S; l_x][S; l_z][K; h][S; l_y] = [S; l_{f(x)}][S; l_z][S; l_{h(y)}] = [S; l_{f(x)}l_zl_{h(y)}],$$
$$[J; g][S; l_x][S; l_z][K; h][S; l_y] = [S; l_{g(x)}][S; l_z][S; l_{h(y)}] = [S; l_{g(x)}l_zl_{h(y)}].$$

Assume that $[S; l_{f(x)}l_zl_{h(y)}] = [S; l_{g(x)}l_zl_{h(y)}]$. Then there exists $P \in \mathcal{I}(S)$ such that $(l_{f(x)}l_zl_{h(y)})|_P = l_{g(x)}l_zl_{h(y)}|_P$, that is $f(x)zh(y)u = g(x)zh(y)u$ for all $u \in P$, a contradiction to (ii). Therefore $p(xz)r \neq q(xz)r$. Thus $Q_r(S)$ is left equiprime. Noting that $xz \in L$, we complete the proof. □

Given a semigroup S, we define *the center of* S as

$$Z(S) = \{c \in S \mid cs = sc \text{ for all } s \in S\}.$$

We are now in a position to prove the following result (for the definition of free right S-acts compare, for example, [6]).

Proposition 2.3. *Let S be a left equiprime semigroup with* 0 *and let* $K \in \mathcal{I}(S)$. *Then:*

(i) $C = Z(Q_r(S))$ *is an abelian group with* 0;

(ii) $C = \{q \in Q_r(S) \mid qx = xq \text{ for all } x \in K\}$;

(iii) *Given any* $1 \neq c \in C$ *and* $0 \neq q \in Q_r(S)$ *one has* $cq \neq q$, *i. e.* $Q_r(S)$ *is a free left (and right) C-act with zero;*

(iv) $T = SC$ *is a left equiprime subsemigroup of* $Q_r(S)$.

Proof. (i) Clearly $Q_r(S)$ has the identity element $1 = [S; \mathrm{id}_S] \in C$. For $0 \neq q = [I; f] \in C$ the mapping $l_q : I \to S$ defined by $l_q(a) = qa = [I; f](a) = f(a)$

is injective (see Lemma 2.2 (i)). Indeed, let $a \neq b \in I$ and choose $c \in I$ such that $qc = f(c) \neq 0$. Then $ax(qc) \neq bx(qc)$ for some $x \in S$ because S is left equiprime. Using that $q \in C = Z(Q_r(S))$, we get that

$$(qa)xc = ax(qc) \neq bx(qc) = (qb)xc$$

and so $qa \neq qb$. Therefore $l_q = f$ is injective. Since $0 \neq q \in C$, we conclude that qI is a nonzero ideal of S. Define the mapping $h : qI \to S$ by the rule $h(qx) = x$ for all $x \in I$. Setting $p = [qI; h] \in Q_r(S)$, we get $pq = 1 = qp$. Therefore C is a group with 0.

(ii) Take $q \in Q_r(S)$ and suppose that $qx = xq$ for all $x \in K$. For $r = [J; g] \in Q_r(S)$ and $y \in J, x \in K$, we have $ry \in S$ and $(ry)x \in K$. It follows that

$$(qr)(yx) = q[(ry)x] = [(ry)x]q = r[(yx)q] = r[q(yx)] = (rq)(yx)$$

and so $qr = rq$ for all $r \in Q_r(S)$. Consequently $q \in C$. The converse implication is clear.

(iii) Let $1 \neq c = [K; h] \in C$ and $0 \neq q = [I; f] \in Q_r(S)$. If $cq = q$, then there exists $J \in \mathcal{I}(S)$ such that $J \subseteq I$ and $c(qx) = qx$ for all $x \in J$. Choose $x \in J$ such that $qx \neq 0$. By Lemma 2.2 (i) we have that $0 \neq qx = f(x) \in S$. It follows that $cy = y$ for all $y \in S(qx)S \cap K$. But then $h(y) = y$ for all $S(qx)S \cap K$ which implies that $c = [K; h] = 1$, a contradiction.

(iv) follows directly from Lemma 2.2 (iii). □

For a left equiprime semigroup S we call the abelian group with zero $C = C(S) = Z(Q_r(S)))$ the *extended (Martindale) centroid of* S. Further, the subsemigroup

$$SC = \{sc \mid s \in S,\ c \in C\} \subseteq Q_r(S)$$

is called the (*Martindale*) *central closure of* S. It follows from Lemma 2.2 (iii) that SC is left equiprime. The semigroup S is said to be *centrally closed* if $SC = S$.

3. Main results

Lemma 3.1, Theorem 3.3, and Theorem 3.4 (i) are a semigroup-theoretic analogs of [8, Theorem 1, Theorem 2 and Theorem 3] while Theorem 3.5 is an analog of [5, Theorem 3.2]. Without explicitly mentioning we will always use that the extended (Martindale) centroid is an abelian group with zero (see Proposition 2.3 (i)).

Lemma 3.1. *Let S be a left equiprime semigroup with extended centroid C, let $K \in \mathcal{I}(S)$ and let $a, b, c, d \in Q_r(S)$. If $d \neq 0$ and $axb = cxd$ for all $x \in K$, then there exists $\lambda \in C$ such that $c = \lambda a$.*

Proof. Note that the intersection of a finite number of nonzero ideals of a left equiprime semigroup contains their product, which is a nonzero ideal since S is left equiprime. Now it follows from Lemma 2.2 (i) that there exists a nonzero ideal J of S such that $aJ, bJ, cJ, dJ \subseteq S$. We set $I = SaJK \cup aJK$ and define the mapping $\alpha : I \to S$ as follows: $\alpha(sajk) = scjk$ for all $s \in S \cup \{1\}$, $j \in J$, $k \in K$. We claim that α is well-defined. Indeed, let $xay = uav$ for some $x, u \in S \cup \{1\}$ and $y, v \in JK \subseteq K$. Assume that $xcy \neq ucv$. Then $xcyzd \neq ucvzd$ for some $z \in S$ since S is left equiprime. On the other hand $xayzb = uavzb$. Since $ayzb = cyzd$ and $avzb = cvzd$ by hypothesis, we conclude that $xcyzd = ucvzd$, a contradiction. Therefore α is well-defined. Obviously $\alpha \in \text{Hom}(I_S, S_S)$. Set $\lambda = [I; \alpha] \in Q_r(S)$. Next we claim that $\lambda s = s\lambda$ for all $s \in S$. Indeed, if $\lambda s \neq s\lambda$ for some $s \in S$, then by Lemma 2.2 (iii) we get $(\lambda s)u \cdot 1 \neq (s\lambda)u \cdot 1$ for some $u \in I$, where $1 \in Q_r(S)$ is the identity. Write $u = xay$ where $x \in S \cup \{1\}$, $y \in JK$. Then $\lambda(su) = \alpha(su) = sxcy$ and $(s\lambda)u = s\alpha(u) = sxcy$ by Lemma 2.2 (i). Therefore $(\lambda s)u \cdot 1 = sxcy = (s\lambda)u \cdot 1$, a contradiction. Hence our claim is proved. It follows from Proposition 2.3 that $\lambda \in C$. Finally we claim that $c = \lambda a$. Indeed, if not, then again $cu \cdot 1 \neq (\lambda a)u \cdot 1$ for some $u \in JK$ by Lemma 2.2 (iii). But $(\lambda a)u = \lambda(au) = \alpha(au) = cu$, a contradiction. Thus $\lambda a = c$. The proof is complete. □

Given a left equiprime semigroup S, an idempotent $e \in SC$ is said to be *of C-rank* 1 if $e \neq 0$ and $eSCe = Ce$.

Lemma 3.2. *Let S be a left equiprime semigroup with* 0 *and extended centroid C. If there exist nonzero elements $a, b, d \in Q_r(S)$ such that $aSb \subseteq Cd$ and $d^2 \neq 0$, then there exists a nonzero idempotent $e \in SC$ such that $eSe \subseteq Ce = Cd$, that is e is of C-rank* 1.

Proof. First of all we note that $aSb \neq 0$ because $a \neq 0$ and so $axb \neq 0 \cdot xb = 0$ for some $x \in S$ (see Lemma 2.2 (iii)). Next we remark that Cd does not contain nonzero elements whose squares are equal to zero because $(\lambda d)^2 = \lambda^2 d^2 \neq 0$ for all $0 \neq \lambda \in C$ since $\lambda^2 d^2 = 0$ would imply $d^2 = 0$ contradicting the hypothesis. From the hypothesis it follows that for each $x \in S$ there exists $\lambda(x) \in C$ such that $axb = \lambda(x)d$. If $x \in S$ is such that $axb \neq 0$, we have that $\lambda(x) \neq 0$. Now

$$\lambda(x)^2 d^2 = (axb)^2 = a(xbax)b = \lambda(xbax)d$$

and so $d^2 = \lambda d$ with $0 \neq \lambda = \lambda(x)^{-2}\lambda(xbax) \in C$. Setting $e = \lambda^{-1}d$, we note that $e^2 = e$, because $\lambda \in C = Z(Q_r(S)$ and $d^2 = \lambda d$. Since $e \in Cd$, $Cd = Ce$. Further, $a(SC)b = CaSb = Cd = Ce$ by hypothesis and hence $axb = e$ for some $x \in SC$. Therefore

$$eSe \subseteq eSCe = a(xbSCax)b \subseteq aSCb = Ce = Cd.$$

The proof is complete. □

For an infinite set X we denote by $\langle X\rangle$ the free semigroup with 0 and 1 on X. Further we denote by $Q_r(S)\langle X\rangle$ the free product (i. e. the coproduct in the category of semigroups with zero) of the semigroups $Q_r(S)$ and $\langle X\rangle$. Clearly for every subset $\{s_x \mid x \in X\} \subseteq Q_r(S)$ there exists a unique homomorphism $\phi : Q_r(S)\langle X\rangle \to Q_r(S)$ such that $\phi(x) = s_x$ and $\phi(a) = a$ for all $x \in X$ and $a \in Q_r(S)$. Such homomorphism we will call *substitutions*. Let σ be a permutation of $\{1, 2, \dots, n\}$, $K \in \mathcal{I}(S)$ and

$$u = a_1x_1a_2x_2 \dots a_nx_na_{n+1},\ v = b_1x_{\sigma(1)}b_2x_{\sigma(2)} \dots b_nx_{\sigma(n)}b_{n+1} \in Q_r(S)\langle X\rangle.$$

The expression

$$a_1x_1a_2x_2 \dots a_nx_na_{n+1} = b_1x_{\sigma(1)}b_2x_{\sigma(2)} \dots b_nx_{\sigma(n)}b_{n+1}$$

is called *a multilinear generalized polynomial identity* (MGPI) on K with coefficients in $Q_r(S)$ if $\phi(u) = \phi(v)$ for every substitution ϕ with $\phi(X) \subseteq K$, that is

$$a_1s_1a_2s_2 \dots a_ns_na_{n+1} = b_1s_{\sigma(1)}b_2s_{\sigma(2)} \dots b_ns_{\sigma(n)}b_{n+1}$$

for all s_i's in K. An MGPI is said to be *a nontrivial* MGPI if $\sigma \neq \mathrm{id}$ and all elements a_i are nonzero.

The next result concerns trivial MGPI (with $\sigma = \mathrm{id}$) on equiprime semigroups showing that they do not contain any important information about the semigroup.

Theorem 3.3. *Let S be an equiprime semigroup with extended centroid C and take $K \in \mathcal{I}(S)$. If*

$$a_1x_1a_2x_2 \dots a_nx_na_{n+1} = b_1x_1b_2x_2 \dots b_nx_nb_{n+1}$$

is a nonzero MGPI *on K with coefficients in $Q_r(S)$, then there exist $\lambda_i \in C$, $i = 1, 2, \dots, n+1$, such that $\lambda_1\lambda_2 \dots \lambda_{n+1} = 1$ and $a_i = \lambda b_i$ for all $i = 1, 2, \dots, n+1$.*

Proof. We note that all the elements a_i and b_i are nonzero (see Lemma 2.2 (iii)). We proceed by induction on n. First take $n = 1$. Then $a_1xa_2 = b_1xb_2$ is an MGPI on K. By Lemma 3.1 we have that $a_1 = \lambda b_1$ and so $b_1x(\lambda a_2) = b_1xb_2$ is an MGPI on K. If $\lambda a_2 \neq b_2$, then we can choose $s \in S$ such that $u = (\lambda a_2)s$ and $v = b_2s$ are distinct elements of S (see Lemma 2.2 (ii)). Clearly $b_1xu = b_1xv$ for all $x \in K$. Since $u \neq v$ and S is right equiprime, there exists $z \in S$ such that $(b_1x)zu \neq (b_1x)zv$. But $xz \in K$ and so $b_1(zx)u = b_1(zx)v$, a contradiction. Setting $\lambda_1 = \lambda$ and $\lambda_2 = \lambda^{-1}$, we see that $a_1 = \lambda_1b_1$, $a_2 = \lambda_2b_2$ and $\lambda_1\lambda_2 = 1$.

Next consider the general case. Let $s_2, s_3, \dots, s_n \in K$. Suppose that

$$b_2s_2b_3 \dots b_ns_nb_{n+1} \neq 0.$$

Then by hypothesis,

$$a_1x_1a_2s_2 \dots a_ns_na_{n+1} = b_1x_1b_2s_2 \dots b_ns_nb_{n+1}$$

for all $x_1 \in K$. Again by Lemma 3.1 we get that $b_1 = \lambda a_1$ for some $\lambda \in C$. Since $b_1 \neq 0$, we conclude that $\lambda \neq 0$. We set $\lambda_1 = \lambda^{-1}$ and note that λ_1 is uniquely determined by the equation $a_1 = \lambda_1 b_1$ (see Proposition 2.3 (iii)). Then

$$b_1 x_1 (\lambda_1 a_2 s_2 \dots a_n s_n a_{n+1}) = b_1 x_1 b_2 s_2 \dots b_n s_n b_{n+1}$$

and so

$$\lambda_1 a_2 s_2 \dots a_n s_n a_{n+1} = b_2 s_2 \dots b_n s_n b_{n+1}$$

by the first case. Since λ_1 is independent of $s_2, s_3, \dots, s_n$, we conclude that the above equation holds for all $s_2, s_3, \dots, s_n \in K$ such that $b_2 s_2 b_3 \dots b_n s_n b_{n+1} \neq 0$. If $b_2 s_2 b_3 \dots b_n s_n b_{n+1} = 0$, then $a_1 x_1 (\lambda_1 a_2 s_2 \dots a_n s_n a_{n+1}) = 0$ for all $x_1 \in K$. Assume that $a = \lambda_1 a_2 s_2 \dots a_n s_n a_{n+1} \neq 0$. Then by Lemma 2.2 (iii) we get $a_1 x a \neq 0 \cdot xa = 0$ for some $x \in K$, a contradiction. Therefore $\lambda_1 a_2 s_2 \dots a_n s_n a_{n+1} = 0$ and so

$$\lambda_1 a_2 x_2 a_3 x_3 \dots a_n x_n a_{n+1} = b_2 x_2 b_3 x_3 \dots b_n x_n b_{n+1}$$

is a nonzero MGPI on K. Applying the induction hypothesis completes the proof. □

As the main result we now prove a structure theorem whose first statement is analogous to Martindale's results on prime rings with generalized polynomial identity [8].

Theorem 3.4. *Let S be a left equiprime semigroup with extended centroid C, and Martindale central closure $T = SC$ and let $K \in \mathcal{I}(S)$. If*

$$a_1 x_1 a_2 x_2 \dots a_n x_n a_{n+1} = b_1 x_{\sigma(1)} b_2 x_{\sigma(2)} \dots b_n x_{\sigma(n)} b_{n+1}$$

is a nontrivial MGPI *on K with coefficients in $Q_r(S)$, then*

(i) *every nonzero left (right) ideal of T contains an idempotent of C-rank* 1;

(ii) $a_1 = 1$ *and* $\sigma(1) \neq 1$ *imply* $T = C$;

(iii) *for any idempotents u, v in SC of C-rank* 1 *one has*

  (a) $TuT = TvT$;

  (b) $Tu \subseteq Tv$ $(uT \subseteq vT)$, *implies* $Tu = Tv$ *(respectively,* $uT = vT$*)*;

  (c) *every nonzero principal left (right) ideal of S contained in TuT is generated by an idempotent of C-rank* 1;

  (d) TuT *is a completely* 0*-simple semigroup;*

  (e) TuT *is contained in any nonzero ideal of* T;

  (f) Tu *is a free right C-act and* $Q_r(T) \cong \mathrm{End}_C(Tu)$.

Proof. (i) We choose $1 \leq k \leq n$ such that $\sigma(i) = i$ for all $i < k$ but $\sigma(k) \neq k$. Clearly $k = \sigma(m)$ for some $m > k$. It follows from Lemma 2.2 (iii) that $b_1 d_{\sigma(1)} b_2 \dots d_{\sigma(n)} b_{n+1} \neq 0$ for some $d_1, d_2, \dots, d_n \in K$. Then $a_1 d_1 a_2 \dots d_n a_{n+1} \neq 0$ as well. Making use of the substitution $x_i \mapsto d_i$, $i \neq k$, we get that

$$a_1 d_1 \dots a_k x_k a_{k+1} d_{k+1} \dots a_n d_n a_{n+1} =$$
$$b_1 d_1 \dots b_{k-1} d_{k-1} b_k d_{\sigma(k)} \dots d_{\sigma(m-1)} b_m x_k b_{m+1} d_{\sigma(m+1)} \dots b_n d_{\sigma(n)} b_{n+1}$$

for all $x_k \in K$. We set $a = a_1 d_1 a_2 \dots d_{k-1} a_k$ and $b = b_1 d_1 b_2 \dots d_{k-1} b_k$ where $a \neq 0 \neq b$. Now

$$ax_k a_{k+1} d_{k+1} \dots a_n d_n a_{n+1} =$$
$$bd_{\sigma(k)} b_{k+1} \dots d_{\sigma(m-1)} b_m x_k b_{m+1} d_{\sigma(m+1)} \dots b_n d_{\sigma(n)} b_{n+1}$$

for all $x_k \in K$. It follows from Theorem 3.3 that $bd_{\sigma(k)} \dots d_{\sigma(m-1)} b_m \in Ca$ for all $d_{\sigma(k)}, \dots, d_{\sigma(m-1)} \in K$ such that $bd_{\sigma(k)} \dots d_{\sigma(m-1)} b_m \neq 0$. Therefore $bx_{\sigma(k)} \dots x_{\sigma(m-1)} b_m \in Ca$ for all $x_{\sigma(k)}, \dots, x_{\sigma(m-1)} \in K$. Setting

$$c = b_{k+1} d_{\sigma(k+1)} \dots d_{\sigma(m-1)} b_m,$$

we see that

$$b, c \neq 0 \text{ and } bxc \in Ca \text{ for all } x \in K \tag{1}$$

and $bxc \in Ca$ for all $x \in K$. Let L be a nonzero left ideal of SC. We choose $I \in \mathcal{I}(S)$ such that $aI \subseteq S$. From Lemma 2.2 (iii) we infer that $czb \neq 0$ for some $z \in I(L \cap S)S$, that is $z = uv$ for some $u \in I(L \cap S)$ and $v \in S$. Note that $cu \neq 0 \neq vb$ because $cuvb \neq 0$. Further, $vbxcu \in Cvau$ for all $x \in K$. If $(vau)^2 = 0$, then $(Cvau)^2 = 0$ and so $vbxcuvbycu = 0$ for all $x, y \in K$. On the other hand, by Lemma 2.2 (iii) we have that $vbxcuvb \neq 0$ for some $x \in K$ and so $(vbxcuvb)y(cu) \neq 0$ for some $y \in K$, a contradiction. Therefore $(vau)^2 \neq 0$. By Theorem 3.3 there exists an idempotent e of C-rank 1 such that $Ce = Cvau$. Since $u \in I(L \cap S)$, we see that $u = il$ for some $i \in I$ and $l \in L$. Therefore $vau = vail \in SL \subseteq L$. It follows that $e \in L$.

Now let R be a nonzero right ideal of $T = SC$. Choose $0 \neq r \in R$. By the above result SCr contains an idempotent t of C-rank 1. Clearly $t = sr$ for some $s \in SC$. Setting $e = rts$, we see that $e \in T$ and $e^2 = rt(sr)ts = rt^3 s = rts = e$ and $eSCe \subseteq r(tSCt)s = r(Ct)s = C(rts) = Ce$. Finally, since $0 \neq t = t^3 = ser$, we conclude that $e \neq 0$.

(ii) By the hypothesis of (ii) it follows that $k = 1$ and $a = 1$ and so (1) turns into

$$b, c \neq 0 \text{ and } bxc \in C \text{ for all } x \in K.$$

Since $1^2 = 1 \neq 0$, we infer from Theorem 3.3 that there exists an idempotent $e \in SC$ of C-rank 1 such that $Ce = C$. It follows that $e = 1$ and so $C = eSCe = SC$.

(iii) (a) Clearly $L = TuT \cap S \in \mathcal{I}(S)$. By Lemma 2.2 (iii) there exists $x \in L$ such that $vxv \neq 0$. Recalling that $vxv = \lambda v$ for some $\lambda \in C$, we conclude that $\lambda \neq 0$ and so $v = \lambda^{-1} vxv \in TuT$. It follows that $TvT \subseteq TuT$. Analogously one can show that $TuT \subseteq TvT$ and hence $TuT = TvT$.

(b) Suppose that $Tu \subseteq Tv$. Then $u = xv$ for some $x \in T$. By Lemma 2.2 (iii) there exists $z \in S$ such that $vzxv \neq 0$. Since $vzxv = \lambda v$ for some $\lambda \in C$, we conclude that $\lambda \neq 0$ and so $v = \lambda^{-1} vzxv = vzu \in Tu$. Hence $Tv \subseteq Tu$. The case $Tu \subseteq Tv$ is considered analogously.

(c) We set $T^\# = T \cup \{1\}$. Let $0 \neq a \in T$ be such that $aT^\# \subseteq TuT$. Then $a \in TuT$ and so $a = xuy$ for some $x, y \in T$. By (i) there exists an idempotent w of C-rank 1 belonging to $aT^{\#}$. We claim that $aT^{\#} = wT$. Clearly $w = xuyz$ for some $z \in T^{\#}$. By Lemma 2.2 (iii) we get $uyzru \neq 0$ for some $r \in S$. Therefore $uyzru = \lambda u$ for some $0 \neq \lambda \in C$ and hence $xu\lambda = xuyzru = wru$. Consequently $xu = \lambda^{-1}wru \in wT$ and so $aT^{\#} \subseteq xuT^{\#} \subseteq wT$. Thus $aT^{\#} = wT$. The case $T^{\#}a \subseteq TuT$ is considered analogously.

(d) Let I be a nonzero ideal of the semigroup $H = TuT$. It follows directly from Proposition 2.3 (iv) that $IH \neq 0 \neq HIH$. Clearly HIH is a nonzero ideal of S. By (i) now HIH contains an idempotent w of C-rank 1. According to (a) we have that $TwT = H$ and so $H \supseteq I \supseteq HIH \supseteq TwT = H$. Since H has nonzero idempotents, $H^2 \neq 0$. Therefore H is a 0-simple semigroup. It follows directly from (b) and (c) that H satisfies the minimum conditions on principal left ideals and on principal right ideals. Thus H is a completely 0-simple semigroup.

(e) Let I be a nonzero ideal of T. By (i) the ideal I contains an idempotent w of C-rank 1. According to (a) we get $TuT = TwT \subseteq I$.

(f) It follows from Proposition 2.3 (iii) that Tu is a free C-act. Take $q = [I; f] \in Q_r(T)$. We know that $qI \subseteq T$ and $TuT \subseteq I$. Therefore $qTu = qTu^2 = (qTu)u \subseteq Tu$. Hence the mapping $\alpha : Q_r(T) \to \mathrm{End}_C(Tu)$ for $q \mapsto l_q|_{Tu}$, $q \in Q_r(T)$, is a homomorphism of semigroups. Let $p \neq q \in Q_r(T)$. By Lemma 2.2 (iii) we get $qz \neq pz$ for some $z \in TuT$. Clearly $z = xuy$ for some $x, y \in T$. It follows that $qxu \neq pxu$ and so $l_q|_{Tu} \neq l_p|_{Tu}$. Hence α is injective. We now show that α is surjective. Indeed, let $g \in \mathrm{End}_C(Tu)$. We define the mapping $h : TuT \to S$ by the rule $h(xuy) = g(xu)y$ and claim that h is well-defined. If not, then $xuy = aub$ for some $x, y, a, b \in T$ but $g(xu)y \neq g(au)b$. Since T is left equiprime by Proposition 2.3 (iv), we conclude that $g(xu)yzu \neq g(au)bzu$ for some $z \in T$. As $g(xu), g(au) \in Tu$, we infer that $g(xu) = g(xu)u$ and $g(au) = g(au)u$. Recalling that $uTu = Cu$, we get that $uyzu = \lambda u$ and $ubzu = \mu u$ for some $\lambda, \mu \in C$. We now have that

$$\begin{aligned} g(xu)yzu &= g(xu)uyzu = g(xu)\lambda u = g(xu\lambda u) = g(xuyzu) \\ &= g(aubzu) = g(au\mu u) = g(au)\mu u = g(au)ubzu \\ &= g(au)bzu, \end{aligned}$$

a contradiction to $g(xu)yzu \neq g(au)bzu$. Therefore h is well-defined. Setting $q = [TuT, h]$, we get that $\alpha(q) = g$. Thus α is surjective and the proof is complete. □

Theorem 3.5. *Let S be a left equiprime semigroup and take $K \in \mathcal{I}(S)$. If S is cancellative and*

$$a_1x_1a_2x_2\dots a_nx_na_{n+1} = b_1x_{\sigma(1)}b_2x_{\sigma(2)}\dots b_nx_{\sigma(n)}b_{n+1}$$

is a nontrivial MGPI *on K then S is commutative.*

Proof. By Theorem 3.4 (i) we get that $T = SC$ contains an idempotent $e \in T$ of C-rank 1, i. e. $eSCe = Ce$. Clearly $e = \lambda a$ for some $\lambda \in C$ and $a \in S$. We claim that $K = Sa$ is a commutative semigroup. Indeed, $eSe \subseteq eTe = Ce$. Let $x, y \in S$. Then

$$exeye = (exe)(eye) = (eye)(exe) = eyexe.$$

Multiplying first and last terms by λ^{-3} and using that $a = \lambda^{-1}e$, we get that $axaya = ayaxa$, that is $a(xaya) = a(yaxa)$. Since S is cancellative, we conclude that $(xa)(ya) = (ya)(xa)$ for all $x, y \in S$ which proves our claim. Now let $s, t \in S$ and $x, y \in K$. Then $sx, ty \in K$ and so

$$\begin{aligned}(st)(yx) &= s(ty)x = (sx)(ty) = (ty)(sx)\\ &= t[y(sx)] = t[(sx)y] = (ts)(xy) = (ts)(yx)\end{aligned}$$

and so $st = ts$ since S is cancellative. Thus S is commutative.

References

[1] E. P. Armendariz and S. A. Steinberg, Regular Selfinjective Rings with Polynomial identity, Trans. Amer. Math. Soc. 190 (1974), 417–425.

[2] G. L. Booth, N. J. Groenewald and S. Veldsman, A Kurosh–Amitsur Prime Radical for Nearrings, Commun. Algebra 18 (1990), 3111–3122.

[3] A. H. Clifford and G. B. Preston, The algebraic Theory of Semigroups I and II, Amer. Math. Soc., Providence 1961, 1967.

[4] J. W. Fisher, Structure of Semiprime P.I. Rings, Proc. Amer. Math. Soc. 39 (1973), 465–467.

[5] S. Jain, Prime Rings Having One-sided Ideal with Polynomial Identity Coincide with Special Johnson rings, J. Algebra 19 (1971), 125–130.

[6] U. Knauer, Projectivity of acts and Morita Equivalence of Monoids, Semigroup Forum 3 (1972), 359–370.

[7] V. T. Markov, Rings of Quotients of Semiprime P.I. Rings and Irreducible Subdirect Products, Russian Math. Surveys 30(4) (1975), 253–254 (Russian).

[8] W. S. Martindale 3rd, Prime Rings Satisfying a Generalized Polynomial Identity, J. Algebra 12 (1969), 576–584.

[9] W. S. Martindale 3rd, On Semiprime P.I. Rings, Proc. Amer. Math. Soc. 40 (1973), 365–369.

[10] M. M. Parmenter, P. N. Stewart and R. Wiegandt, On the Groenewald-Heyman Strongly Prime Radical, Quaestiones Math. 7 (1984), 225–240.

[11] E. C. Posner, Prime Rings Satisfying a Polynomial Identity, Proc. Amer. Math. Soc. 11 (1960), 180–184.

[12] M. S. Putcha and A. Yaqub, Semigroups Satisfying Permutation Identities, Semigroup Forum 3(1971), 68–73.

[13] H. J. Weinert, S-sets and semigroups of quotients, Semigroup Forum, 19 (1980), 1–78. Theory of Semigroups, Proc. Conf. at Greifswald, GDR, 1984, 125–127.

[14] R. Wiegandt, On the Commutativity of Certain Semigroups, Theory of Semigroups, Proc. Conf. at Greifswald, GDR, 1984, 125–127.

[15] R. Wiegandt, On Subdirectly Irreducible Near-rings which are Fields, in: Near-rings and Near-fields, G. Betsch (editor), Elsevier Science Publisher B.V. (North-Holland), 1987, 295–298.

K. I. Beidar, Y. Fong
Department of Mathematics
National Cheng-Kung University
Tainan
Taiwan 701, ROC

U. Knauer
Universität Oldenburg
Fachbereich Mathematik
Carl-von-Ossietzky-Straße
D-26129 Oldenburg, Germany

A. V. Mikhalev
Chair of Algebra
Department of Mechanics and Mathematics
Moscow State University
Moscow 11989, Russia

K. P. Shum
Department of Mathematics
The Chinese University of Hong Kong
Shatin, N.T., Hong Kong

Radicals of monomial algebras

A. Belov and T. Gateva-Ivanova**

Abstract. It is proved that the Jacobson radical of a monomial algebra over a field is nil and a description of elements of the Jacobson radical is obtained.

In what follows $\|D\|$ will denote the cardinality of a set D, $\mathbb{Z}$ the ring of integers, F a field, $Z = \{z_1, z_2, \dots, z_n\}$ a finite set, $\langle Z\rangle$ the free semigroup on Z with unit element, and $F\langle Z\rangle$ the free F-algebra on Z. Given any subset $Y \subseteq \langle Z\rangle$, the factor algebra $A = F\langle Z\rangle/(Y)$ is called *the monomial algebra defined by the set Y of words of $\langle Z\rangle$*, where (Y) is the ideal of $F\langle Z\rangle$ generated by Y. Note that

$$\mathcal{W} = \{w + (Y) \mid w \in \langle Z\rangle,\ w \notin (Y)\} \text{ is a basis of } A. \tag{1}$$

We denote by $\mathrm{Nil}(A)$ and $J(A)$ the nil-radical and the Jacobson radical of A respectively. Next, we set $X = \{x_i = z_i + (Y) \mid i = 1, 2, \dots, n\}$. Let $x_{i_1}, x_{i_2}, \dots, x_{i_m} \in X$. The product $u = x_{i_1}x_{i_2}\dots x_{i_m} \in A$ is called a *word in A*, and we set

$$|u| = \begin{cases} m, & \text{if } u \neq 0;\\ 0, & \text{if } u = 1 \ \ (\text{i.e., } m = 0);\\ -1, & \text{if } u = 0. \end{cases}$$

In view of (1) the number $|u|$ is well-defined. The number $|u|$ is called *the length of* u. Let $1 \le k \le l \le m$ and $w = x_{i_k}x_{i_{k+1}}\dots x_{i_l}$. We shall call w *a subword of u* and denote this fact by $w \ll u$. An infinite sequence $W = \{w_j \mid w \in \mathcal{W},\ j \in \mathbb{Z}\}$ is said to be *an infinite word in A*. A subword w of a product of the form $w_j w_{j+1}\dots w_{j+k}$ is called *a finite subword* of W (and this fact is denoted by $w \ll W$). An infinite word W is said to be *nonzero* if all of its finite subwords are nonzero.

Definition 1. A nonzero infinite word W is called *uniformly recurrent* if for any integer k there exists an integer $L = L(k)$ such that for any two finite subwords u and v of W with $|u| = k$ and $|v| = L$ we have that $u \ll v$.

The following theorem is the main result of the present paper.

Theorem 1. *Let A be a finitely generated monomial algebra. Then $J(A) = \mathrm{Nil}(A)$. Furthermore, as a linear space, $J(A)$ is spanned by all words which are not subwords of any uniformly recurrent word in A.*

*Partially supported by the Russian Foundation for Fundamental Research.

This theorem was proved by T. Gateva-Ivanova [1] and [2] in the finitely presented case (see also [4, Proposition 15, Chapter 24]). The reduction from the general case to that case belongs to A. J. Belov.

Definition 2. An infinite word $\mathcal{W}$ is called *periodic* if there exists an element $u = u(W) \in \mathcal{W}$ such that for any subword $w \ll W$ there exists a natural number k such that $w \ll u^k$.

Lemma 2. *Let W be a nonzero infinite periodic word and $u = u(W)$. Then $u^k \ll W$ for any natural number k.*

Proof. Let k be a natural number and w any subword of W with $|w| = (k+2)|u|$. It follows that $w \ll u^m$ for some $m \geq k+2$. Therefore $w = au^kb$ for some $a, b \in \mathcal{W}$ with $|a| < |u|$ and $|b| < |u|$. In particular $u^k \ll w$ and so $u^k \ll W$. □

Corollary 3. *Let W be a nonzero uniformly recurrent word. Then W is periodic if and only if there exists $1 \neq u \in \mathcal{W}$ such that $u^k \ll W$ for all $k = 1, 2, \ldots$.*

Proof. In view of Lemma 2 it is enough to show that W is periodic if there exists $1 \neq u \in \mathcal{W}$ such that $u^k \ll W$ for all $k = 1, 2, \ldots$. Indeed, let u satisfies the above property. Further, let $w \ll W$ be any subword and let $L = L(|w|)$ be as in Definition 1. Choose k such that $k|u| \geq L$. Since $u^k \ll W$ and W is a nonzero infinite word, $u^k \neq 0$ and so $|u^k| = k|u| \geq L$. It follows that $w \ll u^k$ and thus W is periodic. □

We now introduce the *pure lexicographic order* $<^*$ on the set $\mathcal{W}$ of all nonzero words in A. Given any $u, v \in \mathcal{W}$, we write $u <^* v$ if there exist $s_1, s_2, t_1, t_2 \in \mathcal{W}$ such that $u = s_1t_1$, $v = s_2t_2$, $|s_1| = |s_2|$ and s_1 is greater than s_2 with respect to the lexicographic order (which is induced on $\mathcal{W}$ by the lexicographic order on $\langle Z \rangle$ in view of (1)).

Lemma 4. *Let W be a nonzero nonperiodic uniformly recurrent word, $u, v \ll W$ and $u \neq v$. Then there exist $s, t \in \mathcal{W}$ such that $sut \ll W$ and $svt \not\ll W$.*

Proof. First of all, we note that we may assume without loss of generality that $A = F\langle Z \rangle$ (i.e., $\mathcal{W} = \langle Z \rangle$). Next, suppose that the lemma is false. Then

$$\text{for any } s, t \in \mathcal{W} \text{ the inclusion } sut \ll W \text{ implies } svt \ll W. \tag{2}$$

The following three cases are the only possibilities.

Case 1. $v \notin u\mathcal{W}$ and $u \notin v\mathcal{W}$. Renumbering the generators $z_1, z_2, \ldots, z_n$, if necessary, we may assume without loss of generality that that $v <^* u$. Since W is uniformly recurrent,

$$\text{there exists } L = L(|u|) \text{ such that } u \ll w \text{ for any } w \ll W \text{ with } |w| \geq L. \tag{3}$$

If $w \ll W$ with $|w| \geq L$, then w can be uniquely written in the form $w = w(L)w'$ where $w(L), w' \ll W$ and $|w(L)| = L$. Now choose any finite subword w of W with $|w| = L^{n+1}$. Then by (2), $u \ll w(L)$ and so $w = s_1ut_1$ for some $s_1, t_1 \ll W$ with $|s_1u| \leq L$. It follows from (2) that $w_1 = s_1vt_1 \ll W$. Clearly $|w_1| > |w| - |u| > L^{n+1} - L$. The inequality $v <^* u$ implies $s_1v <^* s_1u$. Since $|s_1u| \leq L$, we conclude that $w_1(L) <^* w(L)$. Therefore we have that

$$w_1 \ll W, |w_1| > L^{n+1} - L, \text{and } w_1(L) <^* w(L).$$

Again, as $|w_1| > L$ we have that $u \ll w_1(L)$ and $w_1 = s_2ut_2$ where $|s_2u| \leq L$. Setting $w_2 = s_2vt_2$, as above we obtain that

$$w_2 \ll W, |w_2| > L^{n+1} - 2L, \text{and } w_2(L) <^* w_1(L).$$

Assume that we have found $w_1, w_2, \ldots, w_k$ such that $w_i(L) > L^{n+1} - iL$ for all $i = 1, 2, \ldots, k$ and

$$w_1, w_2, \ldots, w_k \ll W, \ w_k(L) <^* w_{k-1}(L) <^* \cdots <^* w_1(L).$$

If $k < L^n$, then $|w_k| > L$ and so $u \ll w_k(L)$ by (3). It follows that $w_k = s_{k+1}ut_{k+1}$ for some $s_{k+1}, t_{k+1} \ll W$ with $|s_{k+1}u| \leq L$. Setting $w_{k+1} = s_{k+1}vt_{k+1}$, we see that

$$|w_{k+1}| > |w_k| - |u| > L^{n+1} - (k+1)L.$$

Next, the inequality $v <^* u$ implies $s_{k+1}v <^* s_{k+1}u$. As $|s_{k+1}u| \leq L$, we conclude that $w_{k+1}(L) <^* w_k(L)$. It follows that there exists a sequence $w_1, w_2, \ldots, w_{L^n}$ such that $w_i(L) > L^{n+1} - iL$ for all $i = 1, 2, \ldots, L^n$ and

$$w_{L^n}(L) <^* w_{L^n-1}(L) <^* \cdots <^* w_1(L) <^* w(L).$$

In particular $L^n + 1$ words $w(L), w_1(L), \ldots, w_{L^n}(L)$ are distinct, contradicting the fact that there exist at most L^n distinct nonzero words in $\{z_1, z_2, \ldots, z_n\}$ of length L. Therefore there exist $s, t \in \mathcal{W}$ such that $sut \ll W$ but $svt \not\ll W$.

Case 2. $v \in u\mathcal{W}$. Then $v = us$ for some $1 \neq s \in \mathcal{W}$. Since $us = v \ll W$, it follows from (2) that $us^2 = vs \ll W$. Hence again by (2) we have that $us^3 = vs^2 \ll W$. Continuing in this fashion we obtain that $us^{k+1} = vs^k \ll W$. In particular $s^k \ll W$ for all $k = 1, 2, \ldots$. Applying Corollary 3 we conclude that W is periodic, contradicting our assumptions. Hence $sut \ll W$ but $svt \not\ll W$ for some $s, t \in \mathcal{W}$.

Case 3. $u \in v\mathcal{W}$. Then $u = vs$ for some $1 \neq s \in \mathcal{W}$. We claim that there exists $t \in \mathcal{W}$ such that $ut \ll W$ and $vt \notin ut\mathcal{W}$. Assume that our claim is false. Then for each $t \in \mathcal{W}$

$$\text{the relation } ut \ll W \text{ implies } vt \in ut\mathcal{W} \tag{4}$$

Fix any natural number k. Since $u \ll W$, $ut \ll W$ for some t such that $|t| = |s^k|$. As $u = vs$, it follows from (4) that $vt \in ut\mathcal{W} = vst\mathcal{W}$ and hence $vt \ll vst$. Therefore $t \ll st$, i.e., $st = ta$ for some $a \in \mathcal{W}$. It follows that

$$s^kt = s^{k-1}(st) = s^{k-1}ta = s^{k-2}(st)a = s^{k-2}ta^2 = \cdots = ta^k.$$

Since $|t| = |s^k|$, we conclude that $s^k = t \ll ut \ll W$ and so $s^k \ll W$. According to Corollary 3, W is periodic, contradicting our assumptions. Therefore $ut \ll W$ and $vt \notin ut\mathcal{W}$ for some $t \in \mathcal{W}$. If $vt \not\ll W$, then there is nothing to prove. Suppose that $vt \ll W$. Setting $u' = ut$ and $v' = vt$, we see that $u', v' \ll W$ and $v' \notin u'\mathcal{W}$. If $u' \notin v'\mathcal{W}$, then by Case 1 there exist $s', t' \in \mathcal{W}$ such that $s'u't' \ll W$ but $s'v't' \not\ll W$. If $u' \in v'\mathcal{W}$, then by Case 2, $s'u't' \ll W$ but $s'v't' \not\ll W$ for some $s', t' \in \mathcal{W}$. The proof is complete. □

Definition 3. An element $u \in \mathcal{W}$ in A is called *dense in* A, if there exists an integer $L' = L'(u)$, $L' > |u|$, such that for any $w \in \mathcal{W}$, $|w| = L'$ implies $u \ll w$.

Definition 4. An element $u \in \mathcal{W}$ is said to be *almost dense in* A if there exists an integer $L'' = L''(u)$ such that for any integer $s \geq L''$ there exists a $w_s \in \mathcal{W}$ of length s with the property that every subword of w_s of length L'' contains u as a subword.

Lemma 5. *Let* $u \in \mathcal{W}$ *be a word such that* $u \in \mathrm{Nil}(A)$, L *a natural number and* $N_L = \{v \in \mathcal{W} \mid |v| \leq L\}$. *Then there exists a natural number k such that*

$$uv_1uv_2\dots uv_k = 0$$

for any sequence of k monomials $v_1, v_2, \dots, v_k \in N_L$.

Proof. Clearly N_L is a finite set. Let $f = \sum_{v \in N_L} v$. Then $uf \in \mathrm{Nil}(A)$, and so $(uf)^k = 0$ for some natural number k. Given any $v_i, v_i' \in N_L$, $i = 1, 2, \dots, m$, such that

$$w = uv_1uv_2\dots uv_m \neq 0 \neq uv_1'uv_2'\dots uv_m' = w',$$

we get from (1) that either $w \neq w'$ or $v_i = v_i'$ for all i. Now the equation $(uf)^k = 0$ implies that $uv_1uv_2\dots uv_k = 0$ for all $v_1, v_2, \dots, v_k \in N_L$. □

Proposition 6. *Let* $u \in \mathcal{W}$. *Then* $u \in \mathrm{Nil}(A)$ *if and only if u is almost dense.*

Proof. Suppose that $u \in \mathrm{Nil}(A)$, but u is almost dense. Let $L = L''(u)$ be as in Definition 4. According to Lemma 5 there exists a natural number k such that

$$uv_1uv_2\dots uv_k = 0 \text{ for all } v_1, v_2, \dots, v_k \in N_L$$

Set $s = Lk$. By Definition 4, there exists a $w_s \in \mathcal{W}$ with $|w_s| = s$ such that u is a subword of any subword v of w_s with $|v| = L$. It follows that w_s can be written in the form $w_s = auv_1uv_2\dots uv_kb$ where

$$|au| \leq L,\ |v_1u| \leq L,\ |v_2u| \leq L, \dots,\ |v_{k-1}u| \leq L \text{ and } |v_kb| \leq L.$$

In particular, $v_1, v_2, \dots, v_k \in N_L$. This implies that $w_s = 0$, contradicting the fact that $|w_s| = s > -1$. Therefore the inclusion $u \in \mathrm{Nil}(A)$ implies that u is not almost dense.

Now assume that u is not almost dense. We show that $u \in \mathrm{Nil}(A)$. Choose any $v_i, w_i \in \mathcal{W}$ and $\alpha_i \in F$, $i = 1, 2, \dots, m$. It is enough to show that $\sum_{i=1}^m \alpha_i v_i u w_i$ is nilpotent. Set $L = \max\{|v_i|\} + \max\{|w_i|\} + 2|u|$. Since u is not dense, there exists a natural number k such that each $w \in \mathcal{W}$ with $|w| \geq k$ contains a subword v of length L and $u \not\ll v$. Now consider

$$w = v_{i_1} u w_{i_1} v_{i_2} u w_{i_2} \dots v_{i_k} u w_{i_k},$$

where $1 \leq i_j \leq m$ for all $j = 1, 2, \dots, k$. If $w \neq 0$, then $|w| > k$ and so w contains a subword v of length L such that $u \not\ll v$. This cannot be true. Therefore $w = 0$ and so $(\sum_{i=1}^{m} \alpha_i v_i u w_i)^k = 0$. The proof is now complete. □

Lemma 7. *Let $\mathcal{U} = \{u_i \in \mathcal{W} \mid i = 1, 2, \dots, n, \dots\}$. Suppose that for every natural number k there exists a natural number i such that $|u_i| \geq k$. Then there exists a nonzero infinite word W such that each subword $w \ll W$ is a subword of some u_i.*

Proof. Given $w \in \mathcal{W}$, we set

$$\mathcal{U}_w = \{u_i \in \mathcal{U} \mid u_i \in w\mathcal{W}\}.$$

Clearly $\mathcal{U} = \mathcal{U}_1$ and

$$\| \mathcal{U} \setminus \cup_{i=1}^{n} \mathcal{U}_{x_i} \| < \infty$$

(in fact, this set is either empty or equal to $\{1\}$). Therefore $\| \mathcal{U}_{x_{i_1}} \| = \infty$ for some $1 \leq i_1 \leq n$. Analogously, one can find x_{i_2} such that $\| \mathcal{U}_{x_{i_1} x_{i_2}} \| = \infty$. Continuing in this fashion, we find x_{i_j}, $j = 1, 2, \dots$ such that $\| \mathcal{U}_{x_{i_1} x_{i_2} \dots x_{i_k}} \| = \infty$ for each $k = 1, 2, \dots$. Let

$$w_j = \begin{cases} x_{i_j}, & \text{if } j > 0; \\ 1, & \text{if } j \leq 0. \end{cases}$$

Then $W = \{w_j \mid j \in \mathbb{Z}\}$ is an infinite word such that each of its subword is a subword of some u_i. Since $u_i \in \mathcal{W}$ for $i = 1, 2, \dots$, we conclude that W is a nonzero infinite word. □

Lemma 8 ([5, P. 101]). *For any infinite word W there exists a uniformly recurrent word $\overline{W}$ such that $u \ll W$ for any $u \ll \overline{W}$.*

The following result follows directly from Lemma 7 and Lemma 8.

Corollary 9. *Let $\mathcal{U} = \{u_i \in \mathcal{W} \mid i = 1, 2, \dots, n, \dots\}$. Suppose that for every natural number k there exists a natural number i such that $|u_i| \geq k$. Then there exists a nonzero uniformly recurrent word W such that each subword $w \ll W$ is a subword of some u_i.*

Corollary 10. *Let A be a monomial algebra and $u \in \mathcal{W}$. Then u is almost dense in A if and only if there exists a uniformly recurrent word $W = W(u)$ such that $u \ll W$.*

Proof. The "only if" part follows directly from Definitions 1 and 4. For the "if" part, let u be almost dense and $L'' = L''(u)$. Then for any $s \geq L''$ there exists a $w_s \in \mathcal{W}$

such that any subword of w_s of length L'' contains u as a subword (see Definition 4). Set $w_{-s} = w_s$ and $W' = \{w_s \mid s \in \mathbb{Z}\}$. By Corollary 9 there exists a uniformly recurrent word W such that each $w \ll W$ is a subword of some w_s. In particular, if $|w| = L''$ and $w \ll W$, then $u \ll w$. Therefore $u \ll W$. □

Definition 5. An infinite dimensional monomial algebra A is called *monomially almost simple* (*almost simple*) if in any factor algebra of A by an ideal generated by a nonzero word (by any nonzero ideal, respectively) the image of X generates a nilpotent ideal.

Lemma 11. *A monomial algebra A is monomially almost simple if and only if any word $u \in \mathcal{W}$ is dense in A.*

Proof. Suppose that A is monomially almost simple and $u \in \mathcal{W}$. Then the image $\overline{X}$ of X in $A/(u)$ generates a nilpotent ideal $(\overline{X})$, where $(u) = AuA$. It follows that there exists a natural number k such that $(\overline{X})^k = 0$. That is to say $(X)^k \subseteq (u)$. Setting $L'(u) = k$, we see that $u \ll w$ for any $w \in \mathcal{W}$ with $|w| \geq L'(u)$ because $w \in (X)^k$. Therefore u is dense in A.

Now assume that every $u \in \mathcal{W}$ is dense in A. Let $u \in \mathcal{W}$ and set $k = L'(u)$. For any $w \in \mathcal{W} \cap (X)^k$ we have $|w| \geq k$, and so $u \ll w$. It follows that $(X)^k \subseteq (u)$. This completes the proof. □

Given a subset $S \subseteq A$, we denote by $\mathrm{Span}(S)$ the linear span of S.

Corollary 12. *If A is a monomially almost simple algebra, then* $\mathrm{Nil}(A) = 0$.

Proof. Suppose $\mathrm{Nil}(A) \neq 0$. From (1) we conclude that

$$\mathrm{Nil}(A) = \mathrm{Span}(\mathrm{Nil}(A) \cap \mathcal{W}).$$

Choose any $u \in \mathrm{Nil}(A) \cap \mathcal{W}$. According to Proposition 6, u is not almost dense in A. On the other hand, u is dense in A by Lemma 11, which is a contradiction. □

Let W be an nonzero infinite word in A. We set

$$I_A(W) = \mathrm{Span}(\{u \in \mathcal{W} \mid u \not\ll W\}).$$

If $A = F\langle Z\rangle$, we shall write $I(W)$ for $I_A(W)$.

Lemma 13. *Let W be an nonzero infinite word in A. Then $I_A(W)$ is an ideal of A. Furthermore, if W is periodic, then $I_A(W)$ is finitely generated.*

Proof. The first statement is obvious. Now suppose that W is periodic and set $u = u(W)$ (see Definition 2). Without loss of generality, we may assume that $u \neq v^k$ for all $v \in \mathcal{W}$ and $k > 1$. Let $w \in \mathcal{W} \cap I_A(W)$, $k > 1$ and $a, b \in \mathcal{W}$. Suppose that $aub \ll u^k$. We claim that there exist $a', b' \in \mathcal{W}$ such that $a'a = u^r$, $bb' = u^s$, $|a'| < |u|$ and $|b'| < |u|$. Indeed, since any element in $\mathcal{W}$ has a unique preimage in

$\langle Z \rangle$, without loss of generality we may assume that $\mathcal{W} = \langle Z \rangle$. Choose $x, y \in \langle Z \rangle$ such that $xauby = u^k$. Write $x = u^p a'$ and $y = b'u^q$ where $|a'| < |u|$ and $|b'| < |u|$. It follows that $a'aubb' = u^m$ where $m = k - p - q$. Write $a'a = u^\alpha v$ and $bb' = wu^\beta$ where $|v| < |u|$ and $|w| < |u|$. It is enough to show that $v = w = 1$. Suppose that $|v| + |w| \neq 0$ (i.e., $vu \neq 1$). We have that $vuw = u^t$ where $t = m - \alpha - \beta$. Since $|vuw| = |v| + |u| + |w| < 3|u|$ and $|v| + |w| \neq 0$, we conclude that $t = 2$ and so

$$|v| + |w| = |u| \tag{5}$$

As $vuw = u^2$, it follows that there exists $v', w' \in \langle Z \rangle$ such that $u = v'w'$, $vv' = u$, and $w'w = u$. From $vv' = u = w'w$ we conclude that $|v| + |v'| = |u| = |w'| + |w|$. Taking (5) into account, we obtain $|v'| = |w|$ and $|w'| = |v|$. Now $vv' = u = w'w$ imply that $v = w'$ and $v' = w$. Therefore $u = v'w' = wv$ and so

$$u^2 = vuw = v(wv)w = (vw)^2.$$

It follows that $u = vw = wv$. According to [3], there exists $c \in \langle Z \rangle$ such that $u = c^\gamma$, $v = c^\delta$ and $w = c^\epsilon$ for some natural number γ, δ, ϵ. It follows that $c \neq u$, contradicting the choice of u. Therefore $v = w = 1$ and our claim is proved.

Next we claim that for each $w \in \mathcal{W} \cap I_A(W)$ with $|w| > 2|u|$ there exists $w' \ll w$ with $|w'| \leq 2|u|$ and $w' \in I_A(W)$. We proceed by induction on $|w|$. If $|w| \leq 2|u|$, then there is nothing to prove. Suppose that our claim is proven for $v \in \mathcal{W} \cap I_A(W)$ with $|v| < k$ (where $k > 2|u|$) and $w \in \mathcal{W} \cap I_A(W)$ with $|w| = k$. It follows from the induction assumption that without loss of generality we may assume that $v \notin I_A(W)$ for each $v \ll w$ with $|v| < |w|$ and so

$$v \ll u^r \text{ for some natural number } k. \tag{6}$$

Write $w = vx$ where $x \in X$ and $v \in \mathcal{W}$. By (6), $v \ll u^r$ for some natural number r. Since $|v| = k - 1 \geq 2|u|$, we conclude that $v = aub$ for some $a, b \in \mathcal{W}$. By the above result there exist $a', b' \in \mathcal{W}$ such that $a'a = u^s$, $bb' = u^t$, $|a'| < |u|$, and $|b'| < |u|$. We show that $a \neq 1$. If $a = 1$, $v = ub$. Since $|v| \geq 2|u|$, $v = ub$, and $bb' = u^t$, we conclude that $|b| \geq |u|$ and $b = ub''$ for some $b'' \in \mathcal{W}$. Therefore $w = u^2 b'' x$. As $|ub''x| = |w| - |u|$, we obtain that $ub''x \ll u^\alpha$ for some natural number α (see (6)). By the above result there exist a $b''' \in \mathcal{W}$ and a natural number β such that $b''xb''' = u^\beta$. Then $wb''' = u^2(b''xb''') = u^{\beta+2}$ and hence $w \ll u^{\beta+2}$, contradicting $w \in I_A(W)$. Therefore $a \neq 1$. Then $|ubx| = |w| - |a| < |w|$. It follows from (6) that $ubx \ll u^m$ for some m. By the above result there exist $y \in \mathcal{W}$ and natural number γ such that $bxy = u^\gamma$. Then $a'wy = (a'a)u(bxy) = u^{\gamma+s+1}$ and again $w \ll u^{\gamma+s+1}$, a contradiction. Therefore there exists $w' \ll w$ with $w' \notin I_A(W)$ and $|w'| \leq 2|u|$. It follows that

$$I_A(W) = (\{w \in \mathcal{W} \cap I_A(W) \mid |w| \leq 2|u|\})$$

which proves our lemma. □

Lemma 14. *The following conditions are equivalent for a monomial algebra* $A = F\langle Z\rangle/(Y)$*:*

(i) *A is monomially almost simple.*

(ii) *There exists a uniformly recurrent word W in $\langle Z\rangle$ such that* $\mathcal{W} = \{u + (Y) \mid u \ll W\}$.

(ii) *There exists a uniformly recurrent word W in $\langle Z\rangle$ such that $I(W) = (Y)$ and so* $A = F\langle Z\rangle/I(W)$.

Proof. Clearly (ii)$\Longleftrightarrow$(iii). We show that (i)$\Longrightarrow$(ii). Since A is of infinite dimensional, $\|\langle Z\rangle \setminus (Y)\| = \infty$. Let

$$\langle Z\rangle \setminus (Y) = \{u_i \mid i = 0, 1, 2, \ldots, m, \ldots\}.$$

Since in $\langle Z\rangle$ there are only finitely many words of length $\leq k$, for any $k > 0$ there exists i such that $|u_i| \geq k$. By Corollary 9 there exists a uniformly recurrent word W such that each $w \ll W$ is a subword of some u_i and so $w \notin (Y)$ which implies that $w = u_j$ for some j. On the other hand, by Lemma 11, for each $u_i + (Y)$ there exists an integer $L' = L'(u_i + (Y))$ such that $u_i + (Y) \ll w + (Y)$ (and so $u_i \ll w$) for every $w \in \langle Z\rangle \setminus (Y)$ with $|w| \geq L'$. Choose any $w \ll W$ such that $|w| = L'$. Then $u_i \ll w$ and so $u_i \ll W$. It follows that

$$\mathcal{W} = \{u_i + (Y) \mid i = 0, 1, 2, \ldots, m, \ldots\} = \{w + (Y) \mid w \ll W\}$$

and so (i)$\Longrightarrow$(ii).

Finally, we show that (iii)$\Longrightarrow$(i). Let $u \in \mathcal{W}$ and let v be the preimage of u in $\langle Z\rangle$. As $u \neq 0$, $v \ll W$. Since W is uniformly recurrent, there exists an integer $L = L(|v|)$ such that $v \ll w$ for each $w \ll W$ with $|w| \geq L$. It follows that $(X)^{L'} \subseteq (u)$. The proof is complete. □

Theorem 15. *Let A be a monomially almost simple algebra. Then there exists a uniformly recurrent word W in $F\langle Z\rangle$ such that $A = F\langle Z\rangle/I(W)$ and $\mathcal{W} = \{w + (Y) \mid w \ll W\}$. Moreover, either A is almost simple or W is periodic.*

Proof. By Lemma 14 (ii) there exists a uniformly recurrent word $W = \{w_i \mid i \in \mathbb{Z}\}$ in $F\langle Z\rangle$ such that $A = F\langle Z\rangle/I(W)$. Clearly

$$\mathcal{W} = \{w + (Y) \mid w \ll W\}.$$

Next, suppose that W is not periodic. Let J be any nonzero ideal of A. It is enough to show that $J \cap \mathcal{W} \neq 0$. Choose any $0 \neq f = \sum_{i=1}^{m} \lambda_i u_i \in J$, $\lambda_i \in F$, $u_i \in \mathcal{W}$, $i = 1, 2, \ldots, m$, with minimal m. Clearly $\lambda_i \neq 0$ and $u_i \neq u_j$ for all $i \neq j$. If $m = 1$, then there is nothing to prove. Assume that $m \neq 1$. By Lemma 4 there exist $s, t \in \mathcal{W}$ such that $su_m t \not\ll W$ and $su_{m-1}t \ll W$. Hence $su_m t = 0$ and $su_{m-1}t \neq 0$. Clearly $su_{m-1}t \neq su_i t$ for all $i < m-1$ (see (1)). Therefore $0 \neq sft = \sum_{i=1}^{m-1} \lambda_i su_i t \in J$, contradicting the choice of f. The proof is complete. □

We note that in general it is not true that for any monomial algebra A there exists a nonzero infinite word W such that a word w is nonzero in A if and only if $w \ll W$. For example, if $A = F[x] \oplus F[y]$, the direct sum of two polynomial rings in x and y respectively, then there is no such W.

Given a nonempty subset $S \subseteq \mathcal{W}$, the ideal (S) of A is called a monomial ideal.

Definition 6. A monomial ideal I of A is called *almost maximal* if A/I is monomially almost simple.

Lemma 16. *Let A be a monomial algebra. A monomial ideal J of A is almost maximal if and only if there exists a uniformly recurrent word W in A such that* $J = I_A(W)$.

Proof. Let A/J be monomially almost simple. It follows from Lemma 14 (iii) that $A/J = F\langle Z\rangle/I(W')$ for some uniformly recurrent word $W' = \{u_i \mid i \in \mathbb{Z}\}$ in $F\langle Z\rangle$. Since $Y = 0$ in A/J, we conclude that $(Y) \subseteq I(W')$. It follows that $y \not\ll W'$ for all $y \in Y$. Set $W = \{w + (Y) \mid w \ll W'\}$. It is easy to see that W is uniformly recurrent and $I_A(W) = I(W')/(Y)$. It follows that $J = I_A(W)$. On the other hand let $W = \{u_i + (Y) \mid i \in \mathbb{Z}\}$ be a uniformly recurrent word in A. Then $W' = \{u_i \mid i \in \mathbb{Z}\}$ is a uniformly recurrent word in $F\langle Z\rangle$. Since W is nonzero, $w \ll W'$ implies that $w \notin (Y)$. It follows that $(Y) \subseteq I(W')$ and $I_A(W) = I(W')/(Y)$. Hence $A/I_A(W) \cong F\langle Z\rangle/I(W')$. Applying Lemma 14 (i), we conclude that $A/I_A(W)$ is monomially almost simple and so $I_A(W)$ is almost maximal. □

Proposition 17. *Let A be a monomial algebra. Then*

$$\mathrm{Nil}(A) = \cap\{J \mid J \textit{ is almost maximal ideal of } A\}$$

Proof. It follows from (1) that

$$\mathrm{Nil}(A) = \mathrm{Span}(\mathrm{Nil}(A) \cap (\{0\} \cup \mathcal{W})).$$

Therefore it is enough to show that for $u \in \mathcal{W}$ the condition $u \notin \mathrm{Nil}(A)$ is equivalent to the condition $u \notin J$ for some almost maximal ideal J. According to Proposition 6, $u \notin \mathrm{Nil}(A)$ if and only if u is almost dense in A. By Corollary 10, u is almost dense in A if and only if $u \ll W$ for some uniformly recurrent word W (or equivalently, $u \notin I_A(W)$). Applying Lemma 16, we complete the proof. □

Given a nonnegative integer m, we set

$$A_m = \mathrm{Span}(\{w \in \mathcal{W} \mid |w| = m\}).$$

Clearly $A_0 = F$ and $A = \oplus_{m=0}^{\infty} A_m$.

Lemma 18. *Let m be a natural number and* $f \in A_m$. *Then* $1 - f$ *is invertible if and only if f is nilpotent.*

Proof. It is enough to show that the invertibility of $1 - f$ implies that f is nilpotent. Let g be the inverse element of $1 - f$. Write $g = \sum_{i=0}^{k} g_i$ where $g_i \in A_i$. Clearly $I_r = \oplus_{i=r}^{\infty} A_i$ is an ideal of A for every nonnegative integer r. Set $J = I_{k+1}$. Obviously $f + J$ is a nilpotent element of A/J and so

$$\sum_{i=0}^{k}(g_i + J) = g + J = (1 - f)^{-1} + J = 1 + f + f^2 + \cdots + f^k + J.$$

Since $A/J = \oplus_{i=0}^{k}(A_i + J)$, we conclude that $g = 1 + f + f^2 + \cdots + f^k$ and so $1 = g(1 - f) = 1 - f^{k+1}$. Therefore, $f^{k+1} = 0$, and so f is nilpotent. □

A monomial algebra $A = F\langle Z\rangle/(Y)$ is called *finitely presented* if $\|Y\| < \infty$.

Theorem 19 (Gateva-Ivanova, [1], [2]). *Let A be a finitely presented monomial algebra. Then $J(A) = \text{Nil}(A)$.*

Theorem 20. *Let A be a monomially almost simple algebra. Then $J(A) = 0$.*

Proof. According to Theorem 15, there exists a uniformly recurrent word W in $F\langle Z\rangle$ such that $A = F\langle Z\rangle/I(W)$ and either A is almost simple or W is periodic. First suppose that W is periodic. Then $I(W)$ is finitely generated by Lemma 13. Applying Theorem 19 we conclude that $J(A) = \text{Nil}(A)$. It now follows from Corollary 12 that $J(A) = 0$.

Next assume that A is almost simple. Set $J = J(A)$. If $J = 0$, then there is nothing to prove. Suppose that $J \neq 0$. Then $(X + J)$ is a nilpotent ideal of A/J and so $X \subseteq J$ because $J(A/J) = 0$. Consider $x = x_1 + x_2 + \cdots + x_n$. Since $x \in J$, it follows from Lemma 18 that x is nilpotent. Let $x^m = 0$. It now follows from (1) that $w \in (Y)$ for any $w \in \langle Z\rangle$ with $|w| \geq m$. Therefore $J^m = 0$ and so $J = 0$ by Corollary 12. The proof is complete. □

Proof of Theorem 1. Given any almost maximal ideal I of A, we conclude from Theorem 20 that $J(A) \subseteq I$. Therefore

$$J(A) \subseteq \cap\{I \mid I \text{ is almost maximal ideal of } A\} = \text{Nil}(A)$$

by Proposition 17. As $J(A) \supseteq \text{Nil}(A)$, we obtain that $J(A) = \text{Nil}(A)$. In particular

$$J(A) = \cap\{I \mid I \text{ is almost maximal ideal of } A\}.$$

Since each almost maximal ideal is a monomial ideal by Lemma 16, we obtain that $J(A)$ is a monomial ideal as well. Again applying Lemma 16, we conclude that $J(A)$ is spanned by all monomials which are not subwords of any uniformly recurrent word in A. The proof is complete. □

References

[1] T. Gateva-Ivanova, Algorithmic determination of the Jacobson radical of monomial algebras, preprint, Sofia, 1987.

[2] T. Gateva-Ivanova and V. N. Latyshev, On recognizable properties of associative algebras, J. Symbolic Computation 6 (2-3) (1988), 371–388.

[3] R. C. Lyndon and M. P. Schützenberger, The equation $a^M = b^N c^P$ in a free group, Michigan Math. J. 9 (1962), 289–298.

[4] J. Okniński, Semigroup Algebras, Marcel Dekker, Inc., New York–Basel–Hong Kong 1991.

[5] V. A. Ufnarovskii, Combinatorial and asymptotic methods in algebra, in: Sovremennye problemy matematiki. Fundamental'nye napravlenija, Moscow, VINITI, 57 (1990), 5–177 (in Russian).

Department of Mechanics and Mathematics
Chair of Higher Algebra
Moscow State University
Moscow 119899, Russia

References

Lattice anti-isomorphisms of modules and the AB5* condition

*G. M. Brodskii**

All rings considered in this paper are associative with an identity element and all modules are unitary. Homomorphisms of right modules will be wirtten on the left hand side. We shall use the following notation: L^* is the dual of a partially ordered set L; $\mathrm{End}(X)$ and $L(X)$ are the endomorphism ring and the submodule lattice of a module X.

We recall some definitions and results. An element c of a complete lattice L is called (i) *compact* if for every directed subset $D \subseteq L$ with $c \leq \bigvee_{d\in D} d$, there exists $d_0 \in D$ such that $c \leq d_0$; (ii) *weakly compact* if for every directed subset $D \subseteq L$ with $c = \bigvee_{d\in D} d$, there exists $d_0 \in D$ such that $c = d_0$. A complete lattice L is called *(weakly) algebraic* if every element in L is a join of (weakly) compact elments. A complete lattice L is called *upper continuous* if $a \bigwedge \left(\bigvee_{d\in D} d\right) = \bigvee_{d\in D}(a \bigwedge d)$ for all elements $a \in L$ and for all directed subsets $D \subseteq L$.

The proofs of the following properties of an arbitrary complete lattice L may be found in [1]: 1) *every compact element in L is weakly compact;* 2) *if L is upper continuous, then every weakly compact element in L is compact;* 3) *if L is algebraic, then L is upper contiunous.* We have from these propositions

Lemma 1. *A complete lattice L is weakly algebraic and upper continuous if and only if L is algebraic.*

We now examine the lattices $L(X)$ and $L(X)^*$, where X is a module. The complete lattice $L(X)$ is algebraic. Thereto an element Y of $L(X)$ is compact if and only if Y is a finitely generated submodule of X. A submodule $Y \subseteq X$ is called *cofinite* if Y is a weakly compact element of the complete lattice $L(X)^*$. A module X is called *finitely cogenerated* if 0 is a cofinite submodule of X. It is easy to see that a submodule $Y \subseteq X$ is cofinite if and only if the factor module X/Y is finitely cogenerated. We say a module X satisfies the property AB5* if the complete lattice $L(X)^*$ is upper continuous. As is well-known ([2], Lemma 1.4), the complete lattice $L(X)^*$ is weakly algebraic. Therefore a module X satisfies the AB5* condition if and only if $L(X)^*$ is algebraic.

*Partially supported by the Russian Foundation for Fundamental Research.

A nonempty subset Q of an upper semilattice P is called an *ideal* of P if 1) $a, b \in Q$ implies that $a \bigvee b \in Q$; and 2) $a \in Q$ and $x \leq a$ imply that $x \in Q$. Throughout $I(P)$ is the lattice of ideals of an upper semilattice P with 0; $C(L)$ is the upper semilattice of compact elements of a complete lattice L; $\epsilon_L : L \to I(C(L))$ and $\delta_L : I(C(L)) \to L$ are the maps for which $\epsilon_L(x) = \{c \in C(L) \mid c \leq x\}$ and $\delta_L(Q) = \bigvee_{c \in Q} c$.

Lemma 2. *A complete lattice L is algebraic if and only if the maps* $\epsilon_L : L \to I(C(L))$ *and* $\delta_L : I(C(L)) \to L$ *are inverse order isomorphisms.*

Proof. Since $\epsilon_L(\delta_L(Q)) = \{c \in C(L) \mid c \leq \bigvee_{a \in Q} a\}$, we have $Q \subseteq \epsilon_L(\delta_L(Q))$. But for a compact element c of L the condition $c \leq \bigvee_{a \in Q} a$ implies that $c \leq \bigvee_{a \in F} a$, where F is a suitable finite subset of Q. Then $c \in Q$ because $Q \in I(C(L))$. Thus $\epsilon_L(\delta_L(Q)) = Q$ for all $Q \in I(C(L))$. Now consider $\delta_L(\epsilon_L(x)) = \bigvee_{x \geq c \in C(L)} c$. Clearly $\delta_L(\epsilon_L(x)) = x$ for every $x \in L$ if and only if L is algebraic. □

Let R be a ring, $\{Q_\lambda \mid \lambda \in \Lambda\}$ a minimal representing set of the simple right R-modules and E_λ the injective hull of Q_λ, $\lambda \in \Lambda$. Then $U = \bigoplus_{\lambda \in \Lambda} E_\lambda$ is a minimal cogenerator of the category of right R-modules. We identify the modules E_λ and the corresponding submodules of U, and consider the subring $S \subseteq \mathrm{End}(U)$ formed by endomorphisms φ satisfying the following condition: for every $\mu \in \Lambda$ there exists a finite subset $\Xi_{\varphi,\mu} \subseteq \Lambda$ such that $\varphi(E_\mu) \subseteq \sum_{\lambda \in \Xi_{\varphi,\mu}} E_\lambda$. We obtain the bimodule ${}_SU_R$. If M is a right R-module, let $D(M)$ denote the submodule of the left S-module $\mathrm{Hom}_R(M, U)$ formed by homomorphisms $h \in \mathrm{Hom}_R(M, U)$ for which $\mathrm{Im} h \subseteq \sum_{\lambda \in \Theta_h} E_\lambda$, where Θ_h is a suitable finite subset of Λ. Let we define the usual annihilator maps $\mathrm{Ann}_{D(M)} : L(M) \to L(D(M))$ and $\mathrm{Ann}_M : L(D(M)) \to L(M)$ by setting $\mathrm{Ann}_{D(M)}(A) = \{h \in D(M) \mid \ker h \supseteq A\}$ and $\mathrm{Ann}_M(J) = \bigcap_{h \in J} \ker h$ for all $A \in L(M)$ and $J \in L(D(M))$. This establishes a Galois connection between $L(M)$ and $L(D(M))$.

Lemma 3. *For any module* M_R *the maps* $\mathrm{Ann}_{D(M)} : L(M) \to \mathrm{Ann}_{D(M)}(L(M))$ *and* $\mathrm{Ann}_M : \mathrm{Ann}_{D(M)}(L(M)) \to L(M)$ *are inverse order antiisomorphisms.*

Proof. It is sufficient to show that for every $A \in L(M)$ there exists $J \in L(D(M))$ such that $A = \mathrm{Ann}_M(J)$. Since $M = \mathrm{Ann}_M(0)$, we examine only the case $A \neq M$. In view of Proposition 2.24 from [3], if $x \in M \setminus A$ then $g_x(x + A) \neq 0$ for suitable elements $\lambda(x) \in \Lambda$ and $g_x \in \mathrm{Hom}_R(M/A, E_{\lambda(x)})$. We put $h_x = i_{\lambda(x)} g_x \pi \in D(M)$ and $J = \sum_{x \in M \setminus A} Sh_x$, where $i_{\lambda(x)} : E_{\lambda(x)} \to U$ is the inclusion map and $\pi : M \to M/A$ is the canonical map. It is easy to check that $A = \bigcap_{x \in M \setminus A} \ker h_x = \mathrm{Ann}_M(J)$. This completes the proof. □

If M is a module, let $L_f(M)$ and $L^f(M)$ denote the subsets of $L(M)$, formed by all finitely generated submodules and by all cofinite submodules of M. We shall consider $L_f(M)$ and $L^f(M)$ as partially ordered sets under $\subseteq$.

Lemma 4. *For any module M_R the maps*

$$\mathrm{Ann}_{D(M)} : \mathrm{Ann}_M(L_f(D(M))) \to L_f(D(M))$$

and

$$\mathrm{Ann}_M : L_f(D(M)) \to \mathrm{Ann}_M(L_f(D(M))$$

are inverse order antiisomorphisms.

Proof. Let $J = \sum_{m=1}^{n} Sh_m$ be a finitely generated submodule of $D(M)$, U^n the direct product with injections i_m and projections π_m. Define a homomorphism $\gamma : M/\mathrm{Ann}_M(J) \to U^n$ as follows: $\pi_m \gamma \pi = h_m$, where $1 \leq m \leq n$ and $\pi : M \to M/\mathrm{Ann}_M(J)$ is the canonical map. For every $h \in \mathrm{Ann}_{D(M)}(\mathrm{Ann}_M(J))$ we choose a finite subset $\Theta_h \subseteq \Lambda$ for which $\mathrm{Im} h \subseteq V$, where $V = \sum_{\lambda \in \Theta_h} E_\lambda$. Then there exists a homomorphism $g : M/\mathrm{Ann}_M(J) \to V$ such that $h = \alpha g \pi$, where $\alpha : V \to U$ is the inclusion map. The exact diagram

$$\begin{array}{ccccc} 0 & \longrightarrow & M/\mathrm{Ann}_M(J) & \xrightarrow{\ \gamma\ } & U^n \\ & & \Big\downarrow g & & \\ & & V & & \end{array}$$

with injective V, can be extended commutatively by a homomorphism $\beta : U^n \to V$. Putting $\sigma_m = \alpha\beta i_m \in S$, $1 \leq m \leq n$, we obtain $h = \sum_{m=1}^{n} \sigma_m h_m \in J$. Therefore, $\mathrm{Ann}_{D(M)}(\mathrm{Ann}_M(J)) = J$ for all $J \in L_f(D(M))$. □

Lemma 5. *If M_R is a module then* $\mathrm{Ann}_M(L_f(D(M))) = L^f(M)$.

Proof. If $A \in \mathrm{Ann}_M(L_f(D(M)))$ then $A = \mathrm{Ann}_M(J)$ for a suitable $J = \sum_{m=1}^{n} Sh_m \in L_f(D(M))$. Since $A = \bigcap_{m=1}^{n} \ker h_m$, there exists a monomorphism of M/A in $\prod_{m=1}^{n} \mathrm{Im} h_m$. By [4] (Propositions 21.3(1), 14.7, 21.4(1)), the module M/A is finitely cogenerated. Thus $A \in L^f(M)$ and $\mathrm{Ann}_M(L_f(D(M))) \subseteq L^f(M)$.

Conversely, if $A \in L^f(M)$ then by Lemma 3 we have

$$A = \bigcap_{h \in \mathrm{Ann}_{D(M)}(A)} \ker h = \bigcap_{h \in F} \ker h$$

for some finite subset F of $\mathrm{Ann}_{D(M)}(A)$. It remains only to note that $A = \mathrm{Ann}_M(K) \in \mathrm{Ann}_M(L_f(D(M)))$, where $K = \sum_{h \in F} Sh \in L_f(D(M))$. □

Let M_R and ${}_TN$ be modules. If the lattices $L(M)$ and $L(N)$ are antiisomorphic, then the modules M and N are said to be *Baer dual* (to each other). Our main result is:

Theorem 6. *For a module M_R the following are equivalent:*

(1) *M is Baer dual to some module ${}_TN$;*

(2) *the maps* $\mathrm{Ann}_{D(M)} : L(M) \to L(D(M))$ *and* $\mathrm{Ann}_M : L(D(M)) \to L(M)$ *are inverse lattice antiisomorphisms;*

(3) *M satisfies the* AB5* *condition.*

Proof. (2) ⇒ (1) ⇒ (3) are trivial.

(3) ⇒ (2). Since the complete lattices $L(M)^*$ and $L(D(M))$ are algebraic, by Lemma 2 we obtain two pairs of inverse order isomorphisms:

$$L(M)^* \underset{\delta^*}{\overset{\epsilon^*}{\rightleftarrows}} I(L^f(M)^*)$$

and

$$L(D(M)) \underset{\delta}{\overset{\epsilon}{\rightleftarrows}} I(L^f(D(M))),$$

where $\epsilon^* = \epsilon_{L(M)^*}$; $\delta^* = \delta_{L(M)^*}$; $\epsilon = \epsilon_{L(D(M))}$ and $\delta = \delta_{L(D(M))}$. Thereto in view of Lemmas 4 and 5, we have the inverse order isomorphisms

$$L^f(M)^* \underset{\mathrm{Ann}_M}{\overset{\mathrm{Ann}_{D(M)}}{\rightleftarrows}} L_f(D(M)),$$

inducing the inverse lattice isomorphsims

$$I(L^f(M)^*) \underset{\sigma}{\overset{\rho}{\rightleftarrows}} I(L_f(D(M))).$$

It just remains to note that the diagram

$$\begin{array}{ccc}
L(M)^* & \underset{\mathrm{Ann}_M}{\overset{\mathrm{Ann}_{D(M)}}{\rightleftarrows}} & L(D(M)) \\
\epsilon^* \downarrow \uparrow \delta^* & & \epsilon \downarrow \uparrow \delta \\
I(L^f(M)^*) & \underset{\sigma}{\overset{\rho}{\rightleftarrows}} & I(L_f(D(M)))
\end{array}$$

is commutative in both directions. □

This theorem has been reported at Moscow University [5], and has been announced in [6]. Among consequences of the theorem there are "Existence Theorem" from [7], Theorem 6 and Corollary 1 from [8], Theorem 13 from [9] and other well-known results.

In conclusion, the author wishes to thank A. V. Mikhalev for carefully reading the manuscript and for his valuable suggestions.

References

[1] G. Birkhoff, Lattice Theory, Providence, RI 1967.

[2] G.M. Brodskii, The Hom functors and lattices of submodules, Trans. Moscow Math. Soc. 2 (1984), 169–192.

[3] D.W. Sharpe and P. Vamos, Injective Modules, Cambridge University Press, Cambridge 1972.

[4] R. Wisbauer, Foundations of Module and Ring Theory, Gordon and Breach Science Publishers, 1991.

[5] G.M. Brodskii, Dualisms in modules and the AB5* condition, Vestnik Moskov. Univ. Ser. I Mat. Mekh. 3 (1981), 77.

[6] G.M. Brodskii, Dualisms in modules and the AB5* condition, Russian Math. Surverys 38 (1983), 185.

[7] R. Baer, Linear Algebra and Projective Geometry, New York 1952.

[8] T. Onodera, Linearly compact modules and cogenerators, J. Fac. Sci. Hokkaido Univ. Ser. 1 22 (1972), 116–125.

[9] B. Lemonnier, AB5* et la dualité de Morita, C. R. Acad. Sci. Ser. A–B 289 (1979), A47–A50.

Department of Mechanics and Mathematics
Chair of Higher Algebra
Moscow State University
Moscow 119899, Russia

On certain subrings and ideals of prime rings

*M. A. Chebotar**

Abstract. Let R be a noncommutative prime ring of characteristic not 2 and let D be a nonzero derivation of R. A theorem of Bre\u0161har and Vukman asserts that U, the subring of R generated by all $[r^D, r]$, $r \in R$, contains a nonzero left ideal of R and a nonzero right ideal of R. We prove a more general result stating that U contains a nonzero two-sided ideal of R thus answering in a positive way on a question of Bre\u0161har and Vukman.

1. Preliminary results

We refer the reader to [2], [3] and [5] for the basic results of theory of prime rings with generalized polynomial identities. Throughout the paper R will denote a prime ring with extended centroid C and symmetric ring of quotients Q (see [3]). Given $q \in Q$, we denote by $\mathrm{ad}(q)$ the inner derivation of Q defined by q (i.e., $z^{\mathrm{ad}(q)} = [z, q]$ for all $z \in Q$). The Lie ring of all derivations of R is denoted by $\mathrm{Der}(R)$. It is well-known that every derivation of R has a unique extension up to a derivation of Q and $\mathrm{Der}(Q)$ is a right vector space over C. Therefore we may consider $\mathrm{Der}(R)$ as a Lie subring of $\mathrm{Der}(Q)$. We denote by $\mathrm{Der}(R)C$ the C-span of $\mathrm{Der}(R)$. Next we set

$$\mathrm{Inn}(R) = \{d \in \mathrm{Der}(R)C \mid d = \mathrm{ad}(q) \text{ for some } q \in Q\}.$$

Let M_0 be a C-basis of the right vector space $\mathrm{Inn}(R)$. We extended M to a right C-basis M of $\mathrm{Der}(R)C$. Setting $M_1 = M \setminus M$, we well-order M in such a way that $M_1 < M_0$. Now let S be a free semigroup with 1 on the set M. The elements of S will be called exponents, with 1 as empty exponent. Given $\Delta = \delta_1\delta_2 \dots \delta_n \in S$, $\delta_i \in M$, $i = 1, 2, \dots, n$, we set $|\Delta| = n$. We order S as follows: given $\Delta, \Delta' \in S$, we set $\Delta < \Delta'$ if either $|\Delta| < |\Delta'|$ or $|\Delta| = |\Delta'|$ and Δ is less then Δ' lexicographically. An exponent $\Delta \in S$ is called *regular* if either $\Delta = 1$ or $\Delta = \delta_1\delta_2 \dots \delta_n$ where

1. each $\delta_i \in M_1$,
2. $\delta_1 \leq \delta_2 \cdots \leq \delta_n$,
3. if $\mathrm{char}(R) = p$ and $\delta_{i+1} = \delta_{i+2} = \cdots = \delta_{i+s}$, then $s < p$ ([3]).

*Partially supported by the Russian Foundation for Fundamental Research.

Let X be an infinite set. We denote the Cartesian product of X and S by $X^S = \{x^\Delta \mid x \in X,\ \Delta \in S\}$. We then form the free product $Q\langle X^S\rangle$ of the C-algebra Q and the free C-algebra $C\langle X^S\rangle$ on the set X^S. An element $f \in Q\langle X^S\rangle$ is called regular if all the exponents involved in f are regular elements of S [3].

The following results will be used frequently in the sequel.

Theorem 1.1 ([6, Theorem 1]). *Let R be a prime ring with* $\operatorname{char}(R) \neq 2$ *and let d and h be derivations of R. If the composition dh is a derivation of R, then either $d = 0$ or $h = 0$.*

Theorem 1.2 ([1]). *Let R be a noncommutative prime ring with characteristic not equal to 2, and let D be a nonzero derivation of R. Then U, the subring of R generated by all $[r^D, r]$, $r \in R$, contains a nonzero left ideal of R and a nonzero right ideal of R.*

Lemma 1.3 ([4, Lemma 2]). *Let*

$$f = \sum_i \sum_{k=1}^{n_i} a_i^{(k)} x^{\Delta_i} b_i^{(k)} \in Q\langle X^S\rangle$$

where $a_i^{(k)}, b_i^{(k)} \in Q$ and $x \in X$. Suppose that f is a regular element with largest exponent Δ_1. If $\sum_k a_1^{(k)} \otimes b_1^{(k)} \neq 0$ in $Q \otimes_C Q$, then f is not an identity on R.

Lemma 1.4 ([3, Lemma 4A]). *Let $f(x) = \sum_i a_i x b_i \in Q\langle X\rangle$ where $a_i, b_i \in Q$ and $x \in X$. Then the following conditions are equivalent:*

1. $f(x) = 0$,
2. $\sum a_i \otimes_C b_i = 0$,
3. $f(I) = 0$ *for some nonzero ideal of R.*

Corollary 1.5. *Let $f(x) = \sum_{i=1}^{n} a_i x b_i \in Q\langle X\rangle$ where $a_i, b_i \in Q$ and $x \in X$. Let $f(x) = 0$ and $b_i \neq 0$ for some i. Then $a_i \in \sum_{j \neq i} C a_j$.*

Proof. By Lemma 1.4 (2), $\sum_{j=1}^n a_i \otimes b_i = 0$. Since $b_i \neq 0$, there exists a C-linear function $f : Q \to C$ such that $f(b_i) = 1$. It defines a linear mapping $F : Q \otimes_C Q$ given by the rule $F(a \otimes b) = af(b)$, $a, b \in Q$. We now have that

$$0 = F\left(\sum_{j=1}^{n} a_i \otimes b_i\right) = \sum_{j=1}^{n} a_j f(b_j)$$

and so $a_i = -\sum_{j \neq i} f(b_j) a_j$. The proof is complete. □

2. Main result

Theorem 2.1. *Let R be a noncommutative prime ring of characteristic not 2 and let D be a nonzero derivation of R. Then U, the subring of R generated by all $[r^D, r]$, $r \in R$, contains a nonzero two-sided ideal of R.*

Proof. Assume that U does not contain a nonzero ideal of R. Without loss of generality we may assume that $D \in M \subseteq S$. By Theorem 1.2, U contains a nonzero left ideal L and a nonzero right ideal P of R. Clearly LP, $LUP \subseteq U$ are ideals of R. Therefore $LP = 0 = LUP$. Since R is prime, $PL \neq 0$. Choose any $0 \neq a \in PL$. Then

$$a^2 = 0 \quad \text{and} \quad aUa \subseteq LUP = 0. \tag{1}$$

Recalling that $[r^D, r] \in U$ for all $r \in R$, we conclude that

$$a[x^D, x]a = 0 \tag{2}$$

is a generalized identity with derivations on R. Linearizing (2), we see that

$$ax^D ya - axy^D a + ay^D xa - ayx^D a = 0 \tag{3}$$

is an identity on R. There are two cases to consider.

Case 1. $D \in M_1$. Then substituting for y any fixed element $r \in R$, we infer from Lemma 1.3 that $a \otimes ra - ar \otimes a = 0$ in $Q \otimes_C Q$. Hence $ra \in Ca$ and so $ara \in aCa = Ca^2 = 0$ for all $r \in R$, a contradiction to the primeness of R.

Case 2. $D \in M_0$, i.e., $D = \mathrm{ad}(b)$ for some $b \in Q$. It follows from (4) that

$$\begin{aligned} 0 &= a(xb - bx)ya - ax(yb - by)a + a(yb - by)xa - ay(xb - bx)a \\ &= 2axbya - abxya - axyba + 2aybxa - abyxa - ayxba \end{aligned} \tag{4}$$

is a generalized polynomial identity on R. First suppose that $ab = ca$ for some $c \in C$. It follows from (4) that

$$ax(2bya - cya - yba) + (2ayb - cay)xa - ayxba = 0 \tag{5}$$

is a generalized polynomial identity on R. Since $ab = ca$ and $a^2 = 0$, we see that $aba = ca^2 = 0$. Therefore substitution of ya for y in (5) yields

$$0 = (2ayab - cay)xa - ayaxba = ayax(ca - ba)$$

for all $x, y \in R$. As R is prime, we conclude that $ba = ca$. Then (5) yields $ax(2bya - 2cya) + (2ayb - 2cay)xa = 0$ is a generalized polynomial identity on R. Fix any $r \in R$. Then $ax(2bra - 2cra) + (2arb - 2car)xa = 0$ and so by Lemma 1.4 we have that $2bra - 2cra = \lambda a = -(2arb - 2car)$ for some $\lambda = \lambda(r) \in C$, i.e., $bra - cra + arb - car = 0$ for all $r \in R$. It follows that

$$\begin{aligned} r^{\mathrm{ad}(b)\mathrm{ad}(a)} &= (rb - br)a - a(rb - br) = rba - bra - arb + abr \\ &= cra - bra - arb + car = 0 \end{aligned}$$

for all $r \in R$ and hence $\mathrm{ad}(b)\mathrm{ad}(a)$ is a derivation. By Theorem 1.1, either $b \in C$ or $a \in C$. Since $D = \mathrm{ad}(b) \neq 0$ and $a^2 = 0$, we obtain a contradiction in both cases. Therefore we may assume that a and ab are C-independent. Fix $r \in R$ such that $ara \neq 0$. Substituting xa for x and r for y in (4), we get that

$$ax(2abra - arba) - abxara - arxaba = 0 \tag{6}$$

is a generalized polynomial identity on R. Suppose that $aba = 0$. Since a and ab are C-independent, we obtain a contradiction to Lemma 1.4. Hence $aba \neq 0$ and so by Corollary 1.5 we have that $ar = \lambda a + \mu ab$ for some $\lambda = \lambda(r), \mu = \mu(r) \in C$. As $a^2 = 0$, we now infer from (6) that

$$\begin{aligned} 0 &= ax(2abra - \lambda aba - \mu ab^2 a) - \mu abxaba - \lambda axaba - \mu abxaba \\ &= ax(2abra - 2\lambda aba - \mu ab^2 a) - abx(2\mu aba) \end{aligned}$$

is a generalized polynomial identity on R. Applying Lemma 1.4, we conclude that $2\mu = 0 = 2abra - 2\lambda aba - \mu ab^2 a$. It follows that $ar = \lambda a$ and so $ara = 0$, a contradiction to the choice of r. Thus U contains a nonzero ideal of R. The proof is complete. □

This theorem answers in a positive way on a question of Brešar and Vukman [1].

Acknowledgement. I take this opportunity to express my gratitude to Professors K. I. Beidar, V. T. Markov and A. V. Mikhalev for their valuable comments.

References

[1] M. Brešar and J. Vukman, On certain subrings of prime rings with derivations, J. Austral. Math. Soc. Ser. A 54 (1993), 133–141.

[2] W. S. Martindale 3rd, Prime rings satisfying a generalized polynomial identity, J. Algebra 12 (1969), 576–584.

[3] W. S. Martindale 3rd, Lectures at Jilin University, Univ. of Massachusetts at Amherst, Department of Math. and Statistic, 1986.

[4] V. K. Kharchenko, Differential identities of prime rings Algebra and Logic 17 (2) (1978), 155–168.

[5] V. K. Kharchenko, Automorphisms and Derivations of Associative Rings, Mathematics and Its Applications (Soviet Series), Kluwer Academic Publishers, Dordrecht–Boston–London 1991.

[6] E. Posner, Derivations in prime rings, Proc. Amer. Math. Soc. 8 (1957), 79–92.

Department of Mechanics and Mathematics
Chair of Higher Algebra
Moscow State University
Moscow 119899, Russia

The induced linear maps on Minkowski algebras of convex sets

*Beifang Chen**

Abstract. The Minkowski algebras of closed convex sets and of relatively open convex sets are constructed in this paper by the Euler characteristics χ and $\bar{\chi}$, respectively. Various induced forward linear maps on these algebras are constructed in explicit formulas. The Sallee-Shephard map, which is an algebra isomorphism of Minkowski algebra of relatively open convex sets onto the Minkowski algebra of closed convex sets, is shown to be also an algebra isomorphism of Minkowski algebra of closed convex sets onto the Minkowski algebra of relatively open convex sets. This map is naturally obtained as an induced linear mapping from the identity map of the given space.

1. Introduction

Let $\mathbb{R}^n$ be the standard n-dimensional Euclidean space. By a convex body we mean a compact convex set of a Euclidean space. The *dilation* of a convex body K by a real number λ is the convex set

$$\lambda K = \{\lambda x \mid x \in K\},$$

and the *Minkowski sum* of two convex bodies, K_1 and K_2, is the set

$$K_1 + K_2 = \{x + y \mid x \in K_1, y \in K_2\}.$$

Given convex bodies $K_1, K_2, \ldots, K_m$ of $\mathbb{R}^n$ and nonnegative numbers $\lambda_1, \lambda_2, \ldots, \lambda_m$, it is well known [16, 17] that the volume of convex body $\lambda_1 K_1 + \cdots + \lambda_r K_m$, i.e., the n-dimensional Lebesgue measure, is a symmetric polynomial of variables $\lambda_1, \ldots, \lambda_m$. More specifically,

$$V(\lambda_1 K_1 + \cdots + \lambda_m K_m) = \sum_{i_1=1}^{m} \cdots \sum_{i_n=1}^{m} V(K_{i_1}, \ldots, K_{i_n}) \lambda_{i_1} \ldots \lambda_{i_n}, \qquad (1)$$

where $V(K_{i_1}, \ldots, K_{i_n})$ depends only on $K_1, \ldots, K_m$ and is independent of the order of $K_{i_1}, \ldots, K_{i_n}$.

*Research is partially supported by DAG93/94.SC09 and RGC94/95.

Since the time of Minkowski, people have been interested in extending formula (1) to more general sets or to a certain linear space on which the Minkowski sum can be well defined and the formula (1) is still valid. Hadwiger [9] considered a convexity ring which is the class of sets obtained by taking finite unions of convex bodies. Compact convex sets are also considered in the linear space of ordered pairs of convex bodies, of which two ordered pairs (A, B) and (C, D) are assumed equal if $A + D = B + C$. The addition, scalar multiplication, and difference for the pairs of convex bodies are defined by

$$(A, B) + (C, D) = (A + C, B + D),$$

$$a(A, B) = (aA, aB),$$

$$(A, B) - (C, D) = (A + D, B + C),$$

respectively. Here the convex body A may be identified with the pair $(A, \{o\})$. Groemer [8] considered the linear space generated by indicator functions of convex bodies, on which the Minkowski multiplication is defined as the convolution with respect to the Euler characteristic, which is viewed as a finitely additive measure on finite unions of convex bodies. The mixed volume is naturally extended to this linear space. The extension of Minkowski sum to unbounded closed convex sets and to relatively open convex sets were considered by Lawrence [12, 13]. The Minkowski product there is somewhat different from those considered in [4], but equivalent up to a constant. However, there are two Minkowski multiplications which were not considered in [12], due to the lack of the Euler characteristic $\bar{\chi}$. Przeslawski [20] extended the Minkowski sum to relatively open convex sets and to the so called *adhesive* closed convex sets [10, 20], but the method employed there cannot be applied to general unbounded closed convex sets. The Minkowski algebra in polyhedral case was considered by this author [2, 4] and it was applied to the incidence algebra [22] of polytopes. Morelli, Pukhlikov, and Khovanskii [18, 21] also considered the Minkowski algebra of polytopes and polyhedra for their cases. With the aid of the Euler characteristics χ [11, 23] and $\bar{\chi}$ [3, 4, 24], which are defined for both relatively open convex sets and for closed convex sets, the author [4] extended the Minkowski sum to both relatively open convex sets and to closed convex sets in an explicit way, and obtained *four* Minkowski algebras of convex sets. With the Minkowski multiplication and its dual form, the Geissinger multiplication, the mixed volume of convex bodies can be naturally expressed as the volume of Geissinger product of the indicator functions of the given convex bodies, up to a constant. All these authors independently studied Minkowski algebras from different angles for their own purposes. More recently, the Minkowski algebra has been applied to pattern recognition, computer vision, and mathematical morphology [7, 19].

It is worthwhile to notice that the Minkowski sum of two closed convex sets need not be a closed set. Moreover, the intersection of two convex curves may be a set of infinite singletons. For instance, the Minkowski sum of $\{(x, y) \mid y \geq e^{-x^2}, x \geq 0\}$ and $\{(x, y) \mid y = 0\}$ is the set $\{(x, y) \mid y > 0\}$ which is still convex, but it is not closed;

the intersection $\partial E_1 \cap \partial E_2$ of convex curves ∂E_1 and ∂E_2 is a discrete set of infinite singletons, where E_1 and E_2 are the convex hulls of the sequences $\{e^{i\theta_{2k}} \mid k \geq 0\}$ and $\{e^{i\theta_{2k+1}} \mid k \geq 0\}$ on the unit circle, respectively, and θ_k is a strictly increasing sequence of real numbers converging to 2π.

In this paper, first we shall systematically study the Minkowski algebra of relatively open convex sets and the Minkowski algebra of closed convex sets, then derive various induced linear maps, then consider the extension of the Minkowski sum to the linear spaces generated by indicator functions of relatively open convex sets and by that of closed convex sets. It should be pointed out that the convolution of indicator functions of a closed convex set and that of a relatively open convex set cannot be well-defined. So in general, we can not get Minkowski multiplications of indicator functions of a relatively open convex set and that of a closed convex set.

Throughout the paper, let us denote by V a finite dimensional affine space over the field $\mathbb{R}$ of real numbers. The indicator function 1_E of a subset E of V is the characteristic function of E, i.e., $1_E(x) = 1$ for $x \in E$ and $1_E(x) = 0$ otherwise. The closure of a convex set A is denoted by $\bar{A}$ and the relative interior of A is denoted by $\dot{A}$.

2. The Euler characteristic

Let r and s be real numbers such that $r + s = 1$. For a real-valued function $f(x)$ on $\mathbb{R}$, the *Euler integral* of f is defined as

$$\chi(f) = \int f(x)d\chi(x) = \sum_{x \in \mathbb{R} \cup \{\infty\}} \left[f(x) - rf(x^-) - sf(x^+)\right],$$

where $f(\infty) = 0$, $f(x^-)$ and $f(x^+)$ are left and right limits of f near $x \in \mathbb{R} \cup \{\infty\}$, respectively, and $f(\infty^-) = \lim_{x \to +\infty} f(x)$, $f(\infty^+) = \lim_{x \to -\infty} f(x)$. A function f is said to be *Euler integrable* if it is continuous except at finitely many points and has both left and right limits at those discontinuous points, including at ∞.

Let V be a d-dimensional real vector space and let $\boldsymbol{b} = \{v_1, \ldots, v_d\}$ be a basis of V. Then any function f on V can be represented as a function $f_{\boldsymbol{b}}(x_1, \ldots, x_d)$ of d coordinate variables $x = (x_1, \ldots, x_d)$. The *Euler integral* of f with respect to the basis $\boldsymbol{b}$ is defined as the iterated Euler integral

$$\chi(f) = \int_V f(x)d\chi(x) = \underbrace{\int \cdots \int}_{d} f_{\boldsymbol{b}}(x_1, \ldots, x_d)d\chi(x_1)\ldots d\chi(x_d).$$

We say that f is *Euler integrable with respect to the basis* $\boldsymbol{b}$ if each of its iterated Euler integrals is Euler integrable, and f is simply said to be *Euler integrable* if it is so under all bases.

It is clear that the set of Euler integrable functions on V is a real linear space. It can be shown that any indicator function of a relatively open convex set or a closed convex set is Euler integrable. Moreover, if U is a relatively open convex set and F is a closed convex set of V, it has been shown [3] that

$$\chi(U) = (-1)^{\dim(U)}, \tag{2}$$

$$\chi(F) = (-1)^{\dim(W)}, \quad \text{if } F = K \oplus W, \tag{3}$$

$$\chi(F) = 0, \quad \text{if } F = M \oplus W \oplus L, \tag{4}$$

where K is a compact convex set, W is a subspace, M is a manifold with or without boundary, and L is a closed half-line, $\oplus$ represents direct sum, i.e., $u \in E_1 \oplus E_2$ if and only if there exist unique $e_1 \in E_1$ and $e_2 \in E_2$ such that $u = e_1 + e_2$.

One can define another Euler characteristic $\bar{\chi}$ for unbounded closed polyconvex sets (finite unions of closed convex sets) and unbounded relatively open polyconvex sets (finite unions of relatively open convex sets). Let E be a relatively open polyconvex set or a closed polyconvex set. Let us define

$$\bar{\chi}(E) = \lim_{\rho \to +\infty} \chi(E \cap \overline{B(\rho)}),$$

where $B(\rho)$ is the open ball of radius ρ centered at the origin o. To avoid the situation of intersection of a closed convex set and a relatively open convex set (in such a situation the intersection may be a discrete set of infinite singletons), one can replace $B(\rho)$ by the open n-cube $I_\rho^n = (-\rho, \rho)^n$ if it is necessary. It has been shown [3] that if F is a closed convex set and U is a relatively open convex set, then

$$\bar{\chi}(F) = 1, \tag{5}$$

$$\bar{\chi}(U) = (-1)^{\dim(D)}, \quad \text{if } U = D \oplus W, \tag{6}$$

$$\bar{\chi}(U) = 0, \quad \text{if } U = \dot{M} \oplus W \oplus \dot{L}, \tag{7}$$

where D is a bounded relatively open convex set, W is a subspace, $\dot{M}$ is a manifold without boundary, and $\dot{L}$ is an open half-line.

The convex sets in the form of (4) or (7) are called *type* II sets. All the other convex sets in the form of (2), (3), (5) or (6) are called *type* I sets. Given a convex set E, we define the *recession cone* of E by

$$\operatorname{rec}(E) = \{v \in \mathbb{R}^n \mid v + E \subset E\}. \tag{8}$$

The recession cone of E can be characterized as the set of vectors v of $\mathbb{R}^n$ such that $p + \mathbb{R}_+ v = \{p + tv \mid t \geq 0\} \subset E$ for all $p \in E$.

Let $\mathcal{E}$ be a family of subsets of V. Let $S(V, \mathcal{E})$ denote the linear space of simple functions generated by indicator functions of members of $\mathcal{E}$. More precisely,

$$S(V, \mathcal{E}) = \left\{ \sum a_i 1_{E_i} \mid E_i \in \mathcal{E}, \text{the sum is finite} \right\}.$$

In this paper we shall consider such linear spaces of which $\mathcal{E}$ is one of the classes of subsets of V in the following.

$$\begin{aligned}
\mathcal{F} &= \text{the class of closed convex sets,}\\
\mathcal{O} &= \text{the class of relatively open convex sets,}\\
\mathcal{F}_b &= \mathcal{K} = \text{the class of bounded closed (compact) convex sets,}\\
\mathcal{O}_b &= \text{the class of bounded relatively open convex sets,}\\
\mathcal{P} &= \text{the class of polyhedra,}\\
\mathcal{P}_b &= \text{the class of bounded polyhedra.}
\end{aligned}$$

A *polyhedron* is meant to be a set of $\mathbb{R}^n$, which is obtained by taking unions, intersections, and complements of finitely many times of half-spaces. For the convenience of making statements, we adopt the following abbreviations:

$$\begin{aligned}
S(V) &= S(V, \mathcal{F} \cup \mathcal{O}),\\
S_b(V) &= S(V, \mathcal{F}_b \cup \mathcal{O}_b),\\
S_c(V) &= S(V, \mathcal{F}),\\
S_{bc}(V) &= S(V, \mathcal{F}_b),\\
S_o(V) &= S(V, \mathcal{O}),\\
S_{bo}(V) &= S(V, \mathcal{O}_b),\\
S_p(V) &= S(V, \mathcal{P}),\\
S_{bp}(V) &= S(V, \mathcal{P}_b).
\end{aligned}$$

3. Induced linear maps

Let V and W be finite dimensional real vector spaces and let ϕ be an affine linear map of V to W. It is trivial that ϕ induces a natural linear map ϕ^* of $S(W)$ to $S(V)$, which is defined as the pullback $\phi^*(f) = f \circ \phi$, $\forall f \in S(W)$. We are interested in the construction of push-forward maps by ϕ, since this cannot be usually done to general functions. In fact, there are four nontrivial push-forward linear maps ϕ_*^{χ}, $\bar{\phi}_*^{\chi}$, $\phi_*^{\bar{\chi}}$, $\bar{\phi}_*^{\bar{\chi}}$ of $S(V)$ to $S(W)$, which are induced by ϕ and the Euler characteristics χ and $\bar{\chi}$, respectively, and which are defined by

$$\phi_*^{\chi}(f)(w) = \lim_{\delta \to 0} \chi\left(f \cdot 1_{\phi^{-1}(\overline{B(w,\delta)})}\right), \tag{9}$$

$$\bar{\phi}_*^{\chi}(f)(w) = \lim_{\delta \to 0} \chi\left(f \cdot 1_{\phi^{-1}(B(w,\delta))}\right), \tag{10}$$

$$\phi_*^{\bar{\chi}}(f)(w) = \lim_{\delta \to 0} \bar{\chi}\left(f \cdot 1_{\phi^{-1}(\overline{B(w,\delta)})}\right), \tag{11}$$

$$\bar{\phi}_*^{\bar{\chi}}(f)(w) = \lim_{\delta \to 0} \bar{\chi}\left(f \cdot 1_{\phi^{-1}(B(w,\delta))}\right), \tag{12}$$

where $f \in S(V)$, $w \in W$, $B(w, \delta)$ is the open ball in W of radius ρ centered at w, and $\overline{B(w, \delta)}$ is the closure of $B(w, \delta)$.

Before we derive explicit formulas for ϕ_*^{χ}, $\bar{\phi}_*^{\chi}$, $\phi_*^{\bar{\chi}}$, $\bar{\phi}_*^{\bar{\chi}}$, let us state the following lemma which will be needed often in the proofs of later theorems.

Lemma 3.1. *Let ϕ be an affine linear map of a finite dimensional real vector space V to a real vector space W. Let E be a convex set of V. Then for all $w \in \phi(E)$, $\phi^{-1}(w) \cap E$ are simultaneously of type* I *sets or simultaneously of type* II *sets.*

Proof. It follows from the affine property that there exists a linear map f of V to W such that $\phi(v) = \phi(o) + f(v)$ for $v \in V$. Then

$$\phi^{-1}(w) \cap E = f^{-1}(w - \phi(o)) \cap E$$

are parallel convex sets for all $w \in \phi(E)$. Let $w_1, w_2 \in \phi(E)$. Assume that $\phi^{-1}(w_1) \cap E$ is of type II set, say $\phi^{-1}(w_1) \cap E = K_1 \oplus M_1 \oplus L_1$, where K_1 is a bounded convex set, M_1 is a manifold with or without boundary, and L_1 is a half-line. Let $v_1 \in \phi^{-1}(w_1) \cap E$ and $v_2 \in \phi^{-1}(w_2) \cap E$. Note that $\phi(v_1) = \phi(o) + f(v_1) = w_1$ and $\phi(v_1 + L_1) = \phi(o) + f(v_1 + L_1) = \phi(o) + \phi(v_1) + f(L_1) = w_1$. Then $f(L_1) = o$. Thus $\phi(v_2 + L_1) = \phi(o) + \phi(v_2) = \phi(v_2) = w_2$. This shows that $v_2 + L_1$ is contained in $\phi^{-1}(w_2) \cap E$. On the other hand, if $v_2 - L_1$ is contained in $\phi^{-1}(w_2) \cap E$, then with the same argument $v_1 - L_1$ must be contained in $\phi^{-1}(w_1) \cap E$, which is impossible. Therefore $\phi^{-1}(w_2) \cap E$ is a type II set. □

Let U be a relatively open convex set and let F be a closed convex set of V. It can be shown that

$$\phi_*^{\chi}(1_U)(w) = \lim_{\delta \to 0} \chi\left(U \cap \phi^{-1}(\overline{B(w,\delta)})\right) = \chi\left(U \cap \phi^{-1}(w)\right),$$

$$\phi_*^{\chi}(1_F)(w) = \lim_{\delta \to 0} \chi\left(F \cap \phi^{-1}(\overline{B(w,\delta)})\right) = \chi\left(F \cap \phi^{-1}(w)\right),$$

$$\phi_*^{\bar{\chi}}(1_U)(w) = \lim_{\delta \to 0} \bar{\chi}\left(U \cap \phi^{-1}(\overline{B(w,\delta)})\right) = \bar{\chi}\left(U \cap \phi^{-1}(w)\right).$$

Moreover, one can verify the following formulas by the affine property of ϕ.

$$\phi_*^{\chi}(1_U) = (-1)^{\dim U - \dim \phi(U)} 1_{\phi(U)}, \tag{13}$$

$$\bar{\phi}_*^{\chi}(1_U) = (-1)^{\dim U} 1_{\overline{\phi(U)}}, \tag{14}$$

$$\phi_*^{\bar{\chi}}(1_F) = 1_{\overline{\phi(F)}}, \tag{15}$$

$$\bar{\phi}_*^{\bar{\chi}}(1_F) = (-1)^{\dim \phi(F)} 1_{\phi(F)}. \tag{16}$$

If the fibers $F \cap \phi^{-1}(w)$ for one $w \in \phi(F)$ and $U \cap \phi^{-1}(w)$ for one $w \in \phi(U)$ are of type II sets, respectively, then it follows from definition of push-forward maps and (4) and (7) that

$$\phi_*^{\chi}(1_F) = \bar{\phi}_*^{\chi}(1_F) = \phi_*^{\bar{\chi}}(1_U) = \bar{\phi}_*^{\bar{\chi}}(1_U) = 0. \tag{17}$$

If $F \cap \phi^{-1}(w)$ for all $w \in \phi(F)$ and $U \cap \phi^{-1}(w)$ for all $w \in \phi(U)$ are of type I, respectively. It then can be verified by the affine property of ϕ that

$$\phi_*^{\chi}(1_F) = (-1)^{\dim H_1 - \dim \phi(H_1)} 1_{\phi(F)}, \tag{18}$$

$$\bar{\phi}_*^{\chi}(1_F) = (-1)^{\dim \phi(F) + \dim H_1 - \dim \phi(H_1)} 1_{\phi(F)}, \tag{19}$$

$$\phi_*^{\bar{\chi}}(1_U) = (-1)^{\dim U - \dim \phi(U) + \dim H_2 - \dim \phi(H_2)} 1_{\phi(U)}, \tag{20}$$

$$\bar{\phi}_*^{\bar{\chi}}(1_U) = (-1)^{\dim U + \dim H_2 - \dim \phi(H_2)} 1_{\overline{\phi(U)}}, \tag{21}$$

where H_1 is a maximal affine subspace of F and H_2 is a maximal subspace of U.

The above formulas (13–21) give the following theorem.

Theorem 3.2. *Let V and W be finite dimensional real vector spaces and let ϕ be an affine linear map of V to W. Then ϕ^* is a linear map of $S(W)$ to $S(V)$, and ϕ_*^{χ}, $\bar{\phi}_*^{\chi}$, $\phi_*^{\bar{\chi}}$, bar$\phi_*^{\bar{\chi}}$ are well-defined linear maps of $S(V)$ to $S(W)$, and*

(1) $\phi^*(S_o(W)) \subset S_o(V)$,

(2) $\phi^*(S_c(W)) \subset S_c(V)$,

(3) $\phi_*^{\chi}(S_o(V)) = S_o(\phi(V))$,

(4) $\phi_*^{\chi}(S_c(V)) = S_c(\phi(V))$,

(5) $\bar{\phi}_*^{\chi}(S_o(V)) = S_c(\phi(V))$,

(6) $\bar{\phi}_*^{\chi}(S_c(V)) = S_o(\phi(V))$,

(7) $\phi_*^{\bar{\chi}}(S_o(V)) = S_o(\phi(V))$,

(8) $\phi_*^{\bar{\chi}}(S_c(V)) = S_c(\phi(V))$,

(9) $\bar{\phi}_*^{\bar{\chi}}(S_o(V)) = S_c(\phi(V))$,

(10) $\bar{\phi}_*^{\bar{\chi}}(S_c(V)) = S_o(\phi(V))$. □

Let U be a relatively open convex set and let F be a closed convex set of V. It follows from the formulas (13–17) that

$$\chi(\phi_*^{\chi}(1_U)) = (-1)^{\dim U - \dim \phi(U)} \chi(\phi(U)) = \chi(U),$$

$$\chi(\bar{\phi}_*^{\bar{\chi}}(1_F)) = (-1)^{\dim \phi(F)} \chi(\phi(F)) = \bar{\chi}(F),$$

$$\bar{\chi}(\bar{\phi}_*^{\chi}(1_U)) = (-1)^{\dim U} \bar{\chi}(\overline{\phi(U)}) = \chi(U),$$

$$\bar{\chi}(\phi_*^{\bar{\chi}}(1_F)) = \bar{\chi}(\overline{\phi(F)}) = \bar{\chi}(F).$$

If the fibers $U \cap \phi^{-1}(w)$ for $w \in \phi(U)$ and the fibers $F \cap \phi^{-1}(w)$ for $w \in \phi(F)$ are of type II, then U and F also have to be of type II. It follows from (4) and (7) that

$$\chi(\phi_*^{\chi}(1_F)) = \chi(F) = \chi(\bar{\phi}_*^{\bar{\chi}}(1_U)) = \bar{\chi}(U) = 0,$$

$$\bar{\chi}(\bar{\phi}_*^{\chi}(1_F)) = \chi(F) = \bar{\chi}(\phi_*^{\bar{\chi}}(1_U)) = \bar{\chi}(U) = 0.$$

Otherwise, if $U \cap \phi^{-1}(w)$ for $w \in \phi(U)$ and $F \cap \phi^{-1}(w)$ for $w \in \phi(F)$ are of type I, we then have by (18–21)

$$\chi(\phi_*^\chi(1_F)) = (-1)^{\dim H - \dim \phi(U)} \chi(\phi(F)) = \chi(F),$$
$$\chi(\bar{\phi}_*^{\bar{\chi}}(1_U)) = (-1)^{\dim U - \dim H - \dim \phi(H)} \chi(\overline{\phi(U)}) = \bar{\chi}(U),$$
$$\bar{\chi}(\bar{\phi}_*^\chi(1_F)) = (-1)^{\dim \phi(F) + \dim H - \dim \phi(F)} \bar{\chi}(\phi(\dot{F})) = \chi(F),$$
$$\bar{\chi}(\phi_*^{\bar{\chi}}(1_U)) = (-1)^{\dim U - \dim \phi(U) + \dim H - \dim \phi(H)} \bar{\chi}(\phi(U)) = \bar{\chi}(U),$$

where H_1 is a maximal affine subspace of F, and H_2 is a maximal affine subspace of U. We thus obtain the following theorem.

Theorem 3.3. *Let* $\mathbf{V}$ *and* $\mathbf{W}$ *be finite dimensional real vector spaces and let* ϕ *be an affine linear map from* $\mathbf{V}$ *to* $\mathbf{W}$. *Then for any* f *of* $S(\mathbf{V})$ *and* g *of* $S(\mathbf{W})$,

$$\chi(\phi_*^\chi f) = \chi(f), \tag{22}$$
$$\chi(\bar{\phi}_*^{\bar{\chi}} f) = \bar{\chi}(f), \tag{23}$$
$$\bar{\chi}(\bar{\phi}_*^\chi f) = \chi(f); \tag{24}$$
$$\bar{\chi}(\phi_*^{\bar{\chi}} f) = \bar{\chi}(f). \tag{25}$$

□

The following theorem gives the relationships concerning tensor products of push-forward maps of affine linear maps.

Theorem 3.4. *Let* $\phi, \psi : V_i \longrightarrow \mathbf{W}_i$ *be affine maps of finite dimensional real vector spaces,* $i = 1, 2$. *Then*

$$(\phi \otimes \psi)_*^\chi = \phi_*^\chi \otimes \psi_*^\chi,$$
$$(\phi \otimes \psi)_*^{\bar{\chi}} = \phi_*^{\bar{\chi}} \otimes \psi_*^{\bar{\chi}},$$
$$(\overline{\phi \otimes \psi})_*^\chi = \bar{\phi}_*^\chi \otimes \bar{\psi}_*^\chi,$$
$$(\overline{\phi \otimes \psi})_*^{\bar{\chi}} = \bar{\phi}_*^{\bar{\chi}} \otimes \bar{\psi}_*^{\bar{\chi}}$$

on the tensor product $S(V_1) \otimes S(V_2) = S(V_1 \times V_2)$ *of spaces* $S(V_1)$ *and* $S(V_2)$, *where* $\phi \otimes \psi(v_1, v_2) = (\phi(v_1), \psi(v_2))$, $\forall (v_1, v_2) \in V_1 \times V_2$.

Proof. Let $A_i \subset V_i$ be closed convex sets, or relatively open sets. Let $B_i(w_i, \delta)$ be open balls of radius δ centered at w_i in $\mathbf{W}_i$, $i = 1, 2$. It is easy to check

$$\begin{aligned}(\phi \otimes \psi)_*^\chi(1_{A_1} \otimes 1_{A_2})(w_1, w_2) &= (\phi \otimes \psi)_*^\chi(1_{A_1 \times A_2})(w_1, w_2) \\ &= \lim_{\delta \to 0} \chi\left[(A_1 \times A_2) \cap (\phi \otimes \psi)^{-1}(\overline{B(w_1, \delta)} \times \overline{B(w_2, \delta)})\right] \\ &= \lim_{\delta \to 0} \chi\left[(A_1 \cap \phi^{-1}(\overline{B(w_1, \delta)})) \times (A_2 \cap \psi^{-1}(\overline{B(w_2, \delta)}))\right]\end{aligned}$$

$$
\begin{aligned}
&= \lim_{\delta\to 0} \chi\left[A_1 \cap \phi^{-1}(\overline{B(w_1,\delta)})\right] \lim_{\delta\to 0} \chi\left[A_2 \cap \psi^{-1}(\overline{B(w_2,\delta)})\right]\\
&= \phi_*^{\chi}(1_{A_1})(w_1)\psi_*^{\chi}(1_{A_2})(w_2)\\
&= \phi_*^{\chi}(1_{A_1}) \otimes \psi_*^{\chi}(1_{A_2})(w_1, w_2)\\
&= (\phi_*^{\chi} \otimes \psi_*^{\chi})(1_{A_1} \otimes 1_{A_2})(w_1, w_2);
\end{aligned}
$$

$$
\begin{aligned}
(\phi\otimes\psi)_*^{\bar\chi}(1_{A_1} \otimes 1_{A_2})(w_1, w_2) &= (\phi\otimes\psi)_*^{\bar\chi}(1_{A_1\otimes A_2})(w_1, w_2)\\
&= \lim_{\delta\to 0} \bar\chi\left[(A_1 \times A_2) \cap (\phi\otimes\psi)^{-1}(\overline{B(w_1,\delta)} \times \overline{B(w_2,\delta)})\right]\\
&= \lim_{\delta\to 0} \bar\chi\left[(A_1 \cap \phi^{-1}(\overline{B(w_1,\delta)})) \times (A_2 \cap \psi^{-1}(\overline{B(w_2,\delta)}))\right]\\
&= \lim_{\delta\to 0} \bar\chi\left[A_1 \cap \phi^{-1}(\overline{B(w_1,\delta)})\right] \lim_{\delta\to 0} \bar\chi\left[A_2 \cap \psi^{-1}(\overline{B(w_2,\delta)})\right]\\
&= \phi_*^{\bar\chi}(1_{A_1})(w_1)\psi_*^{\bar\chi}(1_{A_2})(w_2)\\
&= \phi_*^{\bar\chi}(1_{A_1}) \otimes \psi_*^{\bar\chi}(1_{A_2})(w_1, w_2)\\
&= (\phi_*^{\bar\chi} \otimes \psi_*^{\bar\chi})(1_{A_1} \otimes 1_{A_2})(w_1, w_2);
\end{aligned}
$$

$$
\begin{aligned}
(\overline{\phi\otimes\psi})_*^{\chi}(1_{A_1} \otimes 1_{A_2})(w_1, w_2) &= (\overline{\phi\otimes\psi})_*^{\chi}(1_{A_1\otimes A_2})(w_1, w_2)\\
&= \lim_{\delta\to 0} \chi\left[(A_1 \times A_2) \cap (\phi\otimes\psi)^{-1}(B(w_1,\delta) \times B(w_2,\delta))\right]\\
&= \lim_{\delta\to 0} \chi\left[(A_1 \cap \phi^{-1}(B(w_1,\delta))) \times (A_2 \cap \psi^{-1}(B(w_2,\delta)))\right]\\
&= \lim_{\delta\to 0} \chi\left[A_1 \cap \phi^{-1}(B(w_1,\delta))\right] \lim_{\delta\to 0} \chi\left[A_2 \cap \psi^{-1}(B(w_2,\delta))\right]\\
&= \bar\phi_*^{\chi}(1_{A_1})(w_1)\bar\psi_*^{\chi}(1_{A_2})(w_2)\\
&= (\bar\phi_*^{\chi} \otimes \bar\psi_*^{\chi})(1_{A_1} \otimes 1_{A_2})(w_1, w_2);
\end{aligned}
$$

$$
\begin{aligned}
(\overline{\phi\otimes\psi})_*^{\bar\chi}(1_{A_1} \otimes 1_{A_2})(w_1, w_2) &= (\overline{\phi\otimes\psi})_*^{\bar\chi}(1_{A_1\otimes A_2})(w_1, w_2)\\
&= \lim_{\delta\to 0} \bar\chi\left[(A_1 \times A_2) \cap (\phi\otimes\psi)^{-1}(B(w_1,\delta) \times B(w_2,\delta))\right]\\
&= \lim_{\delta\to 0} \bar\chi\left[(A_1 \cap \phi^{-1}(B(w_1,\delta))) \times (A_2 \cap \psi^{-1}(B(w_2,\delta)))\right]\\
&= \lim_{\delta\to 0} \bar\chi\left[A_1 \cap \phi^{-1}(B(w_1,\delta))\right] \lim_{\delta\to 0} \bar\chi\left[A_2 \cap \psi^{-1}(B(w_2,\delta))\right]\\
&= \bar\phi_*^{\bar\chi}(1_{A_1})(w_1)\bar\psi_*^{\bar\chi}(1_{A_2})(w_2)\\
&= (\bar\phi_*^{\bar\chi} \otimes \bar\psi_*^{\bar\chi})(1_{A_1} \otimes 1_{A_2})(w_1, w_2).
\end{aligned}
$$

□

Lemma 3.5. *Let $\phi : U \longrightarrow V$ and $\psi : V \longrightarrow W$ be affine linear maps of finite dimensional real vector spaces. Let E be a closed convex set, or a relatively open* *convex set of U. Then, for any $w \in \psi\phi(E)$, $(\phi\psi)^{-1}(w) \cap E$ is of type* II *if and only if that $\phi^{-1}(v) \cap E$ is of type* II *for all $v \in \psi^{-1}(w) \cap \phi(E)$, or $\psi^{-1}(w) \cap \phi(E)$ is of type* II.

Proof. It can be directly verified that $\phi^{-1}(v) \cap E$ is non-empty for all $v \in \psi^{-1}(w) \cap \phi(E)$.

$$(\phi\psi)^{-1}(w) \cap E = \bigcup_{v \in \psi^{-1}(w) \cap \phi(E)} \phi^{-1}(v) \cap E.$$

Then the necessary and sufficient condition for $(\phi\psi)^{-1}(w) \cap E$ to be of type II is that either $\phi^{-1}(v) \cap E$ is of type II for all $v \in \psi^{-1}(w) \cap \phi(E)$ or $\psi^{-1}(w) \cap \phi(E)$ itself is of type II. □

Theorem 3.6. *Let $\phi : U \longrightarrow V$ and $\psi : V \longrightarrow W$ be affine linear maps of finite dimensional real vector spaces. Then*

(1) $(\psi\phi)^{\chi}_{*} = \psi^{\chi}_{*}\phi^{\chi}_{*} = \bar{\psi}^{\bar{\chi}}_{*}\bar{\phi}^{\chi}_{*}$,

(2) $(\psi\phi)^{\bar{\chi}}_{*} = \psi^{\bar{\chi}}_{*}\phi^{\bar{\chi}}_{*} = \bar{\psi}^{\chi}_{*}\bar{\phi}^{\bar{\chi}}_{*}$,

(3) $(\overline{\psi\phi})^{\chi}_{*} = \bar{\psi}^{\chi}_{*}\phi^{\chi}_{*} = \psi^{\bar{\chi}}_{*}\bar{\phi}^{\chi}_{*}$,

(4) $(\overline{\psi\phi})^{\bar{\chi}}_{*} = \bar{\psi}^{\bar{\chi}}_{*}\phi^{\bar{\chi}}_{*} = \psi^{\chi}_{*}\bar{\phi}^{\bar{\chi}}_{*}$,

(5) $\psi^{\bar{\chi}}_{*}\phi^{\chi}_{*} = \bar{\psi}^{\chi}_{*}\bar{\phi}^{\chi}_{*}$,

(6) $\psi^{\chi}_{*}\phi^{\bar{\chi}}_{*} = \bar{\psi}^{\bar{\chi}}_{*}\bar{\phi}^{\bar{\chi}}_{*}$,

(7) $\psi^{\chi}_{*}\bar{\phi}^{\chi}_{*} = \bar{\psi}^{\bar{\chi}}_{*}\phi^{\chi}_{*}$,

(8) $\psi^{\bar{\chi}}_{*}\bar{\phi}^{\bar{\chi}}_{*} = \bar{\psi}^{\chi}_{*}\phi^{\bar{\chi}}_{*}$.

Proof. Let U be a relatively open convex set and let F be a closed convex set of U. It follows from (13-16) that

$$(\psi\phi)^{\chi}_{*}(1_U) = \psi^{\chi}_{*}\phi^{\chi}_{*}(1_U) = \bar{\psi}^{\bar{\chi}}_{*}\bar{\phi}^{\chi}_{*}(1_U) = (-1)^{\dim U - \dim \psi\phi(U)} 1_{\psi\phi(U)},$$

$$(\psi\phi)^{\bar{\chi}}_{*}(1_F) = \psi^{\bar{\chi}}_{*}\phi^{\bar{\chi}}_{*}(1_F) = \bar{\psi}^{\chi}_{*}\bar{\phi}^{\bar{\chi}}_{*}(1_F) = 1_{\overline{\psi\phi(F)}},$$

$$(\overline{\psi\phi})^{\chi}_{*}(1_U) = \bar{\psi}^{\chi}_{*}\phi^{\chi}_{*}(1_U) = \psi^{\bar{\chi}}_{*}\bar{\phi}^{\chi}_{*}(1_U) = (-1)^{\dim U} 1_{\overline{\psi\phi(U)}},$$

$$(\overline{\psi\phi})^{\bar{\chi}}_{*}(1_F) = \bar{\psi}^{\bar{\chi}}_{*}\phi^{\bar{\chi}}_{*}(1_F) = \psi^{\chi}_{*}\bar{\phi}^{\bar{\chi}}_{*}(1_F) = (-1)^{\dim \psi\phi(F)} 1_{\psi\phi(F)}.$$

If the fiber $F \cap (\psi\phi)^{-1}(w)$ is of type I for all $w \in \psi\phi(F)$, it can be seen by (18), (19), and (21) that

$$\begin{aligned}(\psi\phi)^{\chi}_{*}(1_F) &= \psi^{\chi}_{*}\phi^{\chi}_{*}(1_F) = \bar{\psi}^{\bar{\chi}}_{*}\bar{\phi}^{\chi}_{*}(1_F) \\ &= (-1)^{\dim H - \dim \psi\phi(H)} 1_{\overline{\psi\phi(F)}},\end{aligned}$$

$$\begin{aligned}(\overline{\psi\phi})_*^{\chi}(1_F) &= \bar{\psi}_*^{\chi}\phi_*^{\chi}(1_F) = \psi_*^{\bar{\chi}}\bar{\phi}_*^{\chi}(1_F)\\ &= (-1)^{\dim\psi\phi(F)+\dim H-\dim\psi\phi(H)}1_{\psi\phi(F)},\end{aligned}$$

where H is a maximal affine subspace of F. While if $F \cap (\psi\phi)^{-1}(w)$ is of type II for all $w \in \psi\phi(F)$. It can then be shown by (17) and Lemma 3.5 that

$$(\psi\phi)_*^{\chi}(1_F) = \psi_*^{\chi}\phi_*^{\chi}(1_F) = \bar{\psi}_*^{\bar{\chi}}\bar{\phi}_*^{\chi}(1_F) = 0,$$

$$(\overline{\psi\phi})_*^{\chi}(1_F) = \bar{\psi}_*^{\chi}\phi_*^{\chi}(1_F) = \psi_*^{\bar{\chi}}\bar{\phi}_*^{\chi}(1_F) = 0.$$

If the fiber $U \cap (\psi\phi)^{-1}(w)$ is of type I for all $w \in \psi\phi(U)$, it can be shown by (18), (20), and (21) that

$$\begin{aligned}(\psi\phi)_*^{\bar{\chi}}(1_U) &= \psi_*^{\bar{\chi}}\phi_*^{\bar{\chi}}(1_U) = \bar{\psi}_*^{\chi}\bar{\phi}_*^{\bar{\chi}}(1_U)\\ &= (-1)^{\dim U-\dim\psi\phi(U)+\dim H-\dim\psi\phi(H)}1_{\psi\phi(U)},\end{aligned}$$

$$\begin{aligned}(\overline{\psi\phi})_*^{\bar{\chi}}(1_U) &= \bar{\psi}_*^{\bar{\chi}}\phi_*^{\bar{\chi}}(1_U) = \psi_*^{\chi}\bar{\phi}_*^{\bar{\chi}}(1_U)\\ &= (-1)^{\dim U+\dim H-\dim\psi\phi(H)}1_{\overline{\psi\phi(U)}},\end{aligned}$$

where H is a maximal affine subspace of U. On the other hand, if $U \cap (\psi\phi)^{-1}(w)$ is of type II for all $w \in \psi\phi(U)$, similarly, by (17) and Lemma 3.5, we have

$$(\psi\phi)_*^{\bar{\chi}}(1_U) = \psi_*^{\bar{\chi}}\phi_*^{\bar{\chi}}(1_U) = \bar{\psi}_*^{\chi}\bar{\phi}_*^{\bar{\chi}}(1_U) = 0,$$

$$(\overline{\psi\phi})_*^{\bar{\chi}}(1_U) = \bar{\psi}_*^{\bar{\chi}}\phi_*^{\bar{\chi}}(1_U) = \psi_*^{\chi}\bar{\phi}_*^{\bar{\chi}}(1_U) = 0.$$

Whenever $\phi(U) \cap \psi^{-1}(w)$ is of type I set for all $w \in \psi\phi(U)$, by (13), (14), and (18–21)

$$\psi_*^{\bar{\chi}}\phi_*^{\chi}(1_U) = \bar{\psi}_*^{\chi}\bar{\phi}_*^{\chi}(1_U) = (-1)^{\dim U-\dim\psi\phi(U)+\dim\phi(H)-\dim\psi\phi(H)}1_{\psi\phi(U)},$$

$$\psi_*^{\chi}\bar{\phi}_*^{\chi}(1_U) = \bar{\psi}_*^{\bar{\chi}}\phi_*^{\chi}(1_U) = (-1)^{\dim U+\dim\phi(H)-\dim\psi\phi(H)}1_{\overline{\psi\phi(U)}}.$$

Otherwise, when $\phi(U) \cap \psi^{-1}(w)$ is of type II set for all $w \in \psi\phi(U)$, we have

$$\psi_*^{\bar{\chi}}\phi_*^{\chi}(1_U) = \bar{\psi}_*^{\chi}\bar{\phi}_*^{\chi}(1_U) = 0,$$

$$\psi_*^{\chi}\bar{\phi}_*^{\chi}(1_U) = \bar{\psi}_*^{\bar{\chi}}\phi_*^{\chi}(1_U) = 0.$$

If $U \cap \phi^{-1}(w)$ is of type I for all $v \in \phi(U)$, then by (13), (15), (16), (18), (20), and (21)

$$\psi_*^{\chi}\phi_*^{\bar{\chi}}(1_U) = \bar{\psi}_*^{\bar{\chi}}\bar{\phi}_*^{\bar{\chi}}(1_U) = (-1)^{\dim U-\psi\phi(U)+\dim H-\phi(H)}1_{\psi\phi(U)}.$$

$$\psi_*^{\bar{\chi}}\bar{\phi}_*^{\bar{\chi}}(1_U) = \bar{\psi}_*^{\chi}\phi_*^{\bar{\chi}}(1_U) = (-1)^{\dim U+\dim H-\dim\phi(H)}1_{\overline{\psi\phi(U)}}.$$

If $U \cap \phi^{-1}(w)$ is of type II for all $v \in \phi(U)$, then by (17)

$$\psi_*^{\chi}\phi_*^{\bar{\chi}}(1_U) = \bar{\psi}_*^{\bar{\chi}}\bar{\phi}_*^{\bar{\chi}}(1_U) = 0,$$

$$\psi_*^{\bar{\chi}}\bar{\phi}_*^{\bar{\chi}}(1_U) = \bar{\psi}_*^{\chi}\phi_*^{\bar{\chi}}(1_U) = 0.$$

If $F \cap \phi^{-1}(w)$ is of type I for all $w \in \phi(F)$, then it is shown that

$$\psi_*^{\bar{\chi}}\phi_*^{\chi}(1_F) = \bar{\psi}_*^{\chi}\bar{\phi}_*^{\chi}(1_F) = (-1)^{\dim H - \dim \phi(H)} 1_{\overline{\psi\phi(F)}},$$

$$\psi_*^{\chi}\bar{\phi}_*^{\chi}(1_F) = \bar{\psi}_*^{\bar{\chi}}\phi_*^{\chi}(1_F) = (-1)^{\dim H - \dim \phi(H) - \dim \psi\phi(F)} 1_{\psi\dot{\phi}(F)}.$$

Otherwise, whenever $F \cap \phi^{-1}(w)$ is of type II for all $w \in \phi(F)$, then

$$\psi_*^{\bar{\chi}}\phi_*^{\chi}(1_F) = \bar{\psi}_*^{\chi}\bar{\phi}_*^{\chi}(1_F) = 0,$$

$$\psi_*^{\chi}\bar{\phi}_*^{\chi}(1_F) = \bar{\psi}_*^{\bar{\chi}}\phi_*^{\chi}(1_F) = 0.$$

Finally, when $\phi(F) \cap \psi^{-1}(w)$ is of type I for all $w \in \psi\phi(F)$, it can be shown that

$$\psi_*^{\bar{\chi}}\bar{\phi}_*^{\bar{\chi}}(1_F) = \bar{\psi}_*^{\chi}\phi_*^{\bar{\chi}}(1_F) = (-1)^{\dim \psi\phi(F) + \dim \phi(H) - \dim \psi\phi(H)} 1_{\psi\dot{\phi}(F)}.$$

$$\psi_*^{\chi}\phi_*^{\bar{\chi}}(1_F) = \bar{\psi}_*^{\bar{\chi}}\bar{\phi}_*^{\bar{\chi}}(1_F) = (-1)^{\dim \phi(H) - \dim \psi\phi(H)} 1_{\psi\phi(F)}.$$

When $\phi(F) \cap \psi^{-1}(w)$ is of type I for all $w \in \psi\phi(F)$, therefore

$$\psi_*^{\bar{\chi}}\bar{\phi}_*^{\bar{\chi}}(1_F) = \bar{\psi}_*^{\chi}\phi_*^{\bar{\chi}}(1_F) = 0,$$

$$\psi_*^{\chi}\phi_*^{\bar{\chi}}(1_F) = \bar{\psi}_*^{\bar{\chi}}\bar{\phi}_*^{\bar{\chi}}(1_F) = 0.$$

□

4. Convolutions

Let V be a finite dimensional real vector space. Let f and g be simple functions both in $S_o(V)$ or both in $S_c(V)$. The *convolutions*, $f * g$ and $f \mathbin{\bar{*}} g$, of f and g with respect to the Euler characteristics χ and $\bar{\chi}$ are defined by

$$(f * g)(y) = \int_V f(x)g(y-x)d\chi(x), \forall y \in V$$

$$(f \mathbin{\bar{*}} g)(y) = \int_V f(x)g(y-x)d\bar{\chi}(x), \forall y \in V.$$

It can be easily seen that for compact convex sets K_1 and K_2,

$$1_{K_1} * 1_{K_2} = 1_{K_1} \mathbin{\bar{*}} 1_{K_2} = 1_{K_1+K_2}.$$

This is why the convolutions $*$ and $\bar{*}$ are also called *Minkowski multiplications* with respect to the Euler characteristics χ and $\bar{\chi}$.

Let A and B be both relatively open convex sets or both closed convex sets of V. For any fixed $v \in V$,

$$1_A(x)1_B(v-x) = 1 \Leftrightarrow x \in A \cap (v - B).$$

Let $\alpha : V \times V \longrightarrow V$ denote the vector addition, i.e., $\alpha(x, y) = x + y$. It is easily verified that the projection $\beta : V \times V \longrightarrow V$ onto its first component, i.e., $\beta(x, y) = x$, is an affine one-to-one correspondence between $(A \times B) \cap \alpha^{-1}(v)$ and $A \cap (v - B)$ for each $v \in V$. Then

$$(1_A * 1_B)(v) = \chi(A \cap (v - B)) = \chi((A \times B) \cap \alpha^{-1}(v)) = \alpha_*^{\chi}(1_{A \times B})(v),$$

$$(1_A \mathbin{\bar{*}} 1_B)(v) = \bar{\chi}(A \cap (v - B)) = \bar{\chi}((A \times B) \cap \alpha^{-1}(v)) = \alpha_*^{\bar{\chi}}(1_A \times 1_B)(v).$$

Thus for any $f, g \in S_o(V)$, or $f, g \in S_c(V)$, the convolutions can be represented as

$$f * g = \alpha_*^{\chi}(f \otimes g),$$

$$f \mathbin{\bar{*}} g = \alpha_*^{\bar{\chi}}(f \otimes g),$$

where $f \otimes g$ is the function on $V \times V$ defined by $(f \otimes g)(x, y) = f(x)g(y)$. Therefore $*$ and $\bar{*}$ are bilinear operations on both $S_o(V)$ and $S_c(V)$. The bilinearity of $*$ and $\bar{*}$ follows from the linearity of α_*^{χ} and $\alpha_*^{\bar{\chi}}$ and the bilinearity of $\otimes$. Moreover, for relatively open convex sets U_1 and U_2 and closed convex sets F_1 and F_2, we have the following explicit formulas

$$1_{U_1} * 1_{U_2} = (-1)^{\dim U_1 + \dim U_2 - \dim(U_1 + U_2)} 1_{U_1 + U_2},$$

$$1_{F_1} * 1_{F_2} = \chi(\operatorname{rec}(F_1) \cap \operatorname{rec}(-F_2)) 1_{\overline{F_1 + F_2}},$$

$$1_{U_1} \mathbin{\bar{*}} 1_{U_2} = (-1)^{\dim U_1 + \dim U_2 - \dim(U_1 + U_2)} \chi(\operatorname{rec}(U_1) \cap \operatorname{rec}(-U_2)) 1_{U_1 + U_2},$$

$$1_{F_1} \mathbin{\bar{*}} 1_{F_2} = 1_{\overline{F_1 + F_2}},$$

If $F_1 \cap (v - F_2)$ is of type II set for one $v \in F_1 + F_2$, then $1_{F_1} * 1_{F_2} = 0$. Otherwise,

$$1_{F_1} * 1_{F_2} = (-1)^{\dim H_1 + \dim H_2 - \dim(H_1 + H_2)} 1_{\overline{F_1 + F_2}},$$

where H_1 is a maximal affine subspace of F_1 and H_2 is a maximal affine subspace of F_2. Similarly, whenever $U_1 \cap (v - U_2)$ is of type II, then $1_{U_1} \mathbin{\bar{*}} 1_{U_2} = 0$. Whenever $U_1 \cap (v - U_2)$ is of type I, then

$$1_{U_1} \mathbin{\bar{*}} 1_{U_2} = (-1)^{\dim U_1 + \dim U_2 - \dim(U_1 + U_2) + \dim H_1 + \dim H_2 - \dim(H_1 + H_2)} 1_{U_1 + U_2}.$$

Let A, B, C be subsets of V. For any fixed $v \in V$, it can be checked that

$$(A \times B) \cap \alpha^{-1}(v - C) = \{(x, y) \mid 1_A(x)1_B(y)1_C(v - x - y) = 1\}.$$

Let $\gamma : V \times V \longrightarrow V \times V$ be the linear map which is defined by

$$\gamma(x, y) = (x, x + y), \ \forall (x, y) \in V \times V.$$

It can be verified that

$$\gamma[(A \times B) \cap \alpha^{-1}(v - C)] = \{(x, y) \mid 1_A(x)1_B(y - x)1_C(v - y) = 1\}.$$

By the definition of convolution $*$, for each $v \in \boldsymbol{V}$, it follows that

$$\begin{aligned} 1_A * (1_B * 1_C)(v) &= \int 1_A(x)(1_B * 1_C)(v - x)d\chi(x) \\ &= \int\int 1_A(x)1_B(y)1_C(v - x - y)d\chi(y)d\chi(x) \\ &= \int\int 1_A(x)1_B(y)1_C(v - x - y)d\chi(x, y) \\ &= \chi[(A \times B) \cap \alpha^{-1}(v - C)], \end{aligned}$$

$$\begin{aligned} (1_A * 1_B) * 1_C(v) &= \int (1_A * 1_B)(y)1_C(v - y)d\chi(y) \\ &= \int\int 1_A(x)1_B(y - x)1_C(v - y)d\chi(x)d\chi(y) \\ &= \int\int 1_A(x)1_B(y - x)1_C(v - y)d\chi(x, y) \\ &= \chi[\gamma((A \times B) \cap \alpha^{-1}(v - C))] \\ &= \chi[(A \times B) \cap \alpha^{-1}(v - C)]. \end{aligned}$$

Thus

$$(1_A * 1_B) * 1_C = 1_A * (1_B * 1_C).$$

Therefore for functions $f, g, h \in S_o(\boldsymbol{V})$, or for $f, g, h \in S_c(\boldsymbol{V})$,

$$(f * g) * h = f * (g * h).$$

The associativity of the Minkowski multiplication $\bar{*}$ can be similarly obtained. It follows from Theorem 3.3 that

$$\begin{aligned} \chi(f * g) &= \chi[\alpha_*^{\chi}(f \otimes g)] \\ &= \chi(f \otimes g) \\ &= \chi(f)\chi(g), \end{aligned}$$

$$\begin{aligned} \bar{\chi}(f \bar{*} g) &= \bar{\chi}[\alpha_*^{\bar{\chi}}(f \otimes g)] \\ &= \bar{\chi}(f \otimes g) \\ &= \bar{\chi}(f)\bar{\chi}(g). \end{aligned}$$

We summerize the above observations in the following theorem.

Theorem 4.1. *For any finite dimensional real vector space* $\boldsymbol{V}$, $S_o(\boldsymbol{V})$ *and* $S_c(\boldsymbol{V})$ *are commutative algebras with identity* $1_{\{o\}}$ *under the multiplications* $*$ *and* $\bar{*}$, χ *and* $\bar{\chi}$

are algebra homomorphisms of the Minkowski algebras $(S_o(V,*)$, $(S_c(V,*)$, *and* $(S_o(V,\bar{*})$, $(S_c(V,\bar{*})$ *to* $\mathbb{R}$, *respectively.* □

Let ϕ be a linear map from a vector space V to a vector space W. It is easily verified that the following diagram is commutative.

$$\begin{array}{ccc} V \times V & \xrightarrow{\alpha} & V \\ {\scriptstyle \phi\otimes\phi}\downarrow & & \downarrow{\scriptstyle \phi} \\ W \times W & \xrightarrow{\alpha} & W \end{array}$$

where $(\phi \otimes \phi)(x, y) = (\phi(x), \phi(y)), \forall (x, y) \in V \times V$. It follows from Theorem 3.6 that

$$\begin{aligned} \phi_*^X(f * g) &= \phi_*^X \alpha_*^X (f \otimes g) \\ &= (\phi\alpha)_*^X (f \otimes g) \\ &= (\alpha(\phi \otimes \phi))_*^X (f \otimes g) \\ &= \alpha_*^X (\phi \otimes \phi)_*^X (f \otimes g) \\ &= \alpha_*^X (\phi_*^X \otimes \phi_*^X)(f \otimes g) \\ &= \alpha_*^X (\phi_*^X(f) \otimes \phi_*^X(g)) \\ &= \phi_*^X(f) * \phi_*^X(g). \end{aligned}$$

$$\begin{aligned} \phi_*^{\bar{X}}(f \bar{*} g) &= \phi_*^{\bar{X}} \alpha_*^{\bar{X}} (f \otimes g) \\ &= (\phi\alpha)_*^{\bar{X}} (f \otimes g) \\ &= (\alpha(\phi \otimes \phi))_*^{\bar{X}} (f \otimes g) \\ &= \alpha_*^{\bar{X}} (\phi \otimes \phi)_*^{\bar{X}} (f \otimes g) \\ &= \alpha_*^{\bar{X}} (\phi_*^{\bar{X}} \otimes \phi_*^{\bar{X}})(f \otimes g) \\ &= \alpha_*^{\bar{X}} (\phi_*^{\bar{X}}(f) \otimes \phi_*^{\bar{X}}(g)) \\ &= \phi_*^{\bar{X}}(f) \bar{*} \phi_*^{\bar{X}}(g). \end{aligned}$$

$$\begin{aligned} \bar{\phi}_*^X(f * g) &= \bar{\phi}_*^X \alpha_*^X (f \otimes g) \\ &= (\overline{\phi\alpha})_*^X (f \otimes g) \\ &= (\overline{\alpha(\phi \otimes \phi)})_*^X (f \otimes g) \\ &= \alpha_*^{\bar{X}} (\overline{\phi \otimes \phi})_*^X (f \otimes g) \\ &= \alpha_*^{\bar{X}} (\bar{\phi}_*^X \otimes \bar{\phi}_*^X)(f \otimes g) \\ &= \alpha_*^{\bar{X}} (\bar{\phi}_*^X(f) \otimes \bar{\phi}_*^X(g)) \\ &= \bar{\phi}_*^X(f) \bar{*} \bar{\phi}_*^X(g). \end{aligned}$$

$$
\begin{aligned}
\bar{\phi}_*^{\bar{\chi}}(f \bar{*} g) &= \bar{\phi}_*^{\bar{\chi}} \alpha_*^{\bar{\chi}} (f \otimes g) \\
&= (\overline{\phi\alpha})_*^{\bar{\chi}} (f \otimes g) \\
&= (\overline{\alpha(\phi \otimes \phi)})_*^{\bar{\chi}} (f \otimes g) \\
&= \alpha_*^{\bar{\chi}} (\overline{\phi \otimes \phi})_*^{\bar{\chi}} (f \otimes g) \\
&= \alpha_*^{\bar{\chi}} (\phi_*^{\bar{\chi}} \otimes \phi_*^{\bar{\chi}})(f \otimes g) \\
&= \alpha_*^{\bar{\chi}} (\bar{\phi}_*^{\bar{\chi}}(f) \otimes \bar{\phi}_*^{\bar{\chi}}(g)) \\
&= \bar{\phi}_*^{\bar{\chi}}(f) * \bar{\phi}_*^{\bar{\chi}}(g).
\end{aligned}
$$

Theorem 4.2. *Let ϕ be a linear map of real vector spaces $\mathbf{V}$ to $\mathbf{W}$. Then ϕ_*^{χ} is an algebra homomorphism of $S_o(\mathbf{V}, *)$ to $S_o(\mathbf{W}, *)$, of $S_c(\mathbf{V}, *)$ to $S_c(\mathbf{W}, *)$; $\phi_*^{\bar{\chi}}$ is an algebra homomorphism of $S_o(\mathbf{V}, \bar{*})$ to $S_o(\mathbf{W}, \bar{*})$, of $S_c(\mathbf{V}, \bar{*})$ to $S_c(\mathbf{W}, \bar{*})$; $\bar{\phi}_*^{\chi}$ is an homomorphism of $S_o(\mathbf{V}, *)$ to $S_c(\mathbf{W}, \bar{*})$, of $S_c(\mathbf{V}, *)$ to $S_o(\mathbf{W}, \bar{*})$; and $\bar{\phi}_*^{\bar{\chi}}$ is an algebra homomorphism of $S_o(\mathbf{V}, \bar{*})$ to $S_c(\mathbf{W}, *)$, of $S_c(\mathbf{V}, \bar{*})$ to $S_o(\mathbf{W}, *)$.* □

Let ϕ be the identity map ι of $\mathbf{V}$ to itself. It is obvious by definition that $\iota_*^{\chi} = \iota_*^{\bar{\chi}}$ and $\bar{\iota}_*^{\chi} = \bar{\iota}_*^{\bar{\chi}}$. So we denote ι_*^{χ} and $\iota_*^{\bar{\chi}}$ by ι_* and denote $\bar{\iota}_*^{\chi}$ and $\bar{\iota}_*^{\bar{\chi}}$ by $\bar{\iota}_*$. It can be easily seen by (13), (15), (18), and (20) that ι_* is the identity map while $\bar{\iota}_*$ is a linear automorphism. Moreover, it follows from (1), (2), (5), and (6) of Theorem 3.6 that

$$\bar{\iota}_*^2 = \bar{\iota}_* \bar{\iota}_* = \iota_* \iota_* = \text{identity}.$$

This indicates that $\bar{\iota}_*$ is a linear involution. The algebra isomorphism $\bar{\iota}_*$ from $S_o(\mathbf{V}, *)$ onto $S_c(\mathbf{V}, \bar{*})$ is called *Sallee-Shephard mapping* [12, 13].

Theorem 4.3. *$\bar{\iota}_*$ is a linear involution on $S(\mathbf{V})$, and it is an algebra isomorphism of $S_o(\mathbf{V}, *)$ onto $S_c(\mathbf{V}, \bar{*})$, of $S_o(\mathbf{V}, \bar{*})$ onto $S_c(\mathbf{V}, *)$, of $S_c(\mathbf{V}, *)$ onto $S_o(\mathbf{V}, \bar{*})$, and of $S_c(\mathbf{V}, \bar{*})$ onto $S_o(\mathbf{V}, *)$.* □

5. Remarks

(1) The associativity of $\bar{*}$ in $S_o(\mathbf{V})$ and $S_c(\mathbf{V})$ can be derived from the associativity of $*$ in $S_c(\mathbf{V})$ and $S_o(\mathbf{V})$ by the isomorphism $\bar{\iota}_*$, respectively.

(2) By definition of χ and $\bar{\chi}$, $*$ and $\bar{*}$ are the same on $S_{bo}(\mathbf{V})$ and $S_{bc}(\mathbf{V})$.

(3) It is clear that $S_p(\mathbf{V}) \subset S_o(\mathbf{V}) \cap S_c(\mathbf{V})$. We infer that $S_o(\mathbf{V}) \cap S_c(\mathbf{V})$ is the same as $S_p(\mathbf{V})$.

(4) Let t be any real number and let ϕ_t denote the dilation by t. Define

$$t_*(f) = \begin{cases} (\phi_t)_*(f), & t \geq 0, \\ (\bar{\phi}_t)_*(f), & t < 0. \end{cases}$$

If f is a unit in $S_{bp}(V)$, i.e., f is invertible with respect to $*$ or $\bar{*}$, then $t_*(f)$ is a one parameter subgroup of $S_{bp}(V)$.

(5) The mixed volume of convex bodies $K_1, \dots, K_n$ of $\mathbb{R}^n$ can be represented as

$$V(K_1, \dots, K_n) = \frac{(-1)^{n-1}}{n!} V(1_{K_1} \circ \cdots \circ 1_{K_n}),$$

where $\circ$ is the Geissinger multiplication [1, 5, 6].

(6) The valuation rings of lattices of finite unions of relatively open convex sets and that of closed convex sets are isomorphic to $S_o(V)$ and $S_c(V)$, respectively.

(7) It is difficult to define Minkowski multiplications on $S(V)$ so that $S_o(V)$ and $S_c(V)$ are subalgebras.

(8) One may also consider convex algebra like the polytope algebra [15] introduced by McMullen. More related references can be found in [14].

References

[1] M. Barnabei, A. Brini, and G.-C. Rota, Theory of Möbius functions, Russian Math. Surveys 41 (1986), 135–188.

[2] B. F. Chen, The incidence algebra of polyhedra and Minkowski algebra, Adv. Math., to appear.

[3] B. F. Chen, On the Euler characteristics of finite unions of convex sets, Discrete Comput. Geom. 10 (1993), 79–93.

[4] B. F. Chen, The Minkowski algebra of convex sets, preprint of MIT, May 1992.

[5] B. F. Chen, The mixed volumes and Geissinger multiplications of convex sets, Stud. Appl. Math. 91 (1994), 39–50.

[6] L. Geissinger, Valuations on distributive lattices I, II, III, Arch. Math. 24 (1973), 230–239, 337–345, 475–481.

[7] P. K. Ghosh, Vision, geometry, and Minkowski operators, Contemp. Math. 119 (1991), 63–83.

[8] H. Groemer, Minkowski addition and mixed volumes, Geom. Dedicata 6 (1977), 141–163.

[9] H. Hadwiger, Vorlesungen über Inhalt, Oberfläche und Isoperimetrie, Springer-Verlag, Berlin 1957.

[10] V. L. Klee, Asymptotes and projections of convex bodies, Math. Scand. 8 (1960), 356–362.

[11] V. L. Klee, The Euler characteristic in combinatorial geometry, Amer. Math. Monthly. 70 (1963), 119–127.

[12] J. Lawrence, Minkowski rings, preprint, September 1985.

[13] J. Lawrence, Valuations and Polarity, Discrete Comput. Geom. 3 (1988), 307–324.

[14] P. McMullen and R. Schneider, Valuations of convex bodies, in: Convexity and Its Applications (ed. by P. M. Gruber and J. M. Wills), 170–247, Birkhäuser, Boston 1983.

[15] P. McMullen, The polytope algebra, Adv. Math. 78 (1989), 76–130.

[16] H. Minkowski, Theorie der konvexen Körper, insbesondere Begründung ihres Oberflächenbegriffs, Ges. Abh., Leipzig–Berlin 2 (1911), 131–229.

[17] H. Minkowski, Volumen und Oberfläche, Math. Ann. 57 (1903), 447–495, Ges. Abh. 2, 230–276.

[18] R. Morelli, A theory of polyhedra, Adv. Math. 97 (1993), 1–73.

[19] D. M. Mount and R. Silverman, Combinatorial and computational aspects of Minkowski decompositions, Contemp. Math. 119 (1991), 107–124.

[20] K. Przeslawski, Linear algebra of convex sets and the Euler characteristic, Linear and Multilinear Algebra 31 (1992), 143–191.

[21] A. V. Pukhlikov and A. G. Khovanskii, Finitely additive measures of virtual polytopes, St. Petersburg Math. J 4 (1993), 337–356.

[22] G.-C. Rota, On the foundations of combinatorial theory I. Theory of Möbius functions, Z. Wahrscheinlichkeitstheorie 2 (1964), 340–368.

[23] G.-C. Rota, On the combinatorics of the Euler characteristic, Studies in Pure Mathematics (Papers presented to Richard Rado), 221–233, Academic Press, London 1971.

[24] S. Schanuel, Negative sets have Euler characteristic and dimension, Lecture Notes in Math. 1488, Springer-Verlag, New York 1991.

Department of Mathematics
Hong Kong University of Science and Technology
Clear Water Bay
Kowloon, Hong Kong

Superconformal algebras and affine algebras with extended symmetry

S.-J. Cheng

Abstract. We explain which of the super-symmetries have super-affine backgrounds and give explicit formulas for those $N = 2$-symmetries which can be realized on the Fock spaces related to affine algebras with $2 + k$ extended symmetry, where $k \geq 1$.

1. Introduction

$N = 2$-symmetry is of interest today mainly because of its intimate connection to the so-called Mirror symmetry. In this short paper we will explain an interesting connection between the representations of affine algebras with extended symmetry and the representations of the so-called superconformal algebras. For a more detailed discussion the reader is referred to [1] and [2]. Superconformal algebras are in some sense super-extensions of the Virasoro algebra, which can be represented as the universal central extension of the Lie algebra of first order differential operators with polynomial coefficients on the circle. To be precise the Virasoro algebra Vir is generated over $\mathbb{C}$ by elements $\{L_m\}_{m\in\mathbb{Z}}$ and a central element z satisfying the commutation relations

$$[L_m, L_n] = (m-n)L_{m+n} + \left(\frac{m^3 - m}{12}\right)\delta_{m,-n}z.$$

$\mathrm{Vir}/\mathbb{C}z$ can be regarded as the centerless $N = 0$ superconformal algebra.

The Sugawara construction [8] allows one to construct on the representation space of a unitary highest weight representation of an affine Kac–Moody algebra a unitary representation of the Virasoro algebra. We will investigate if this construction carries over to higher order symmetries. To formulate our question precisely: Is it possible to construct on a representation space of an affine algebra with $N+k$-extended symmetry a representation of an N-superconformal algebra? The question was answered in the positive in the special case when $N = 1$ and $k = 0$ by Kac and Todorov [4]. We will answer this question in general.

2. Superconformal algebras and super-affine algebras

Let us define the objects of our study, namely superconformal algebras and affine algebras with extended symmetry. Following [6] we give a purely algebraic definition of the N-superconformal algebra. Roughly speaking, a superconformal algebra can be regarded as infinitessimal generators of super-analytic transformations satisfying certain properties. Let t be a complex variable and $\theta_1, \dots, \theta_N$ be odd Grassmann variables. We let $\mathbb{C}[t, t^{-1}]$ be the algebra of Laurent polynomials and $\Lambda(N)$ be the Grassmann superalgebra in the odd variables $\theta_1, \dots, \theta_N$ for $i = 1, \dots, N$. On the associative superalgebra $\mathbb{C}[t, t^{-1}] \otimes \Lambda(N)$ we define derivations $D_i := \theta_i \frac{\partial}{\partial t} + \frac{\partial}{\partial \theta_i}$. One checks immediately that

$$[D_i, D_j] = 2\delta_{i,j} \frac{\partial}{\partial t}.$$

We write $W(1, N)$ for the superalgebra of derivations of $\mathbb{C}[t, t^{-1}] \otimes \Lambda(N)$ over $\mathbb{C}$. Let $u \in \mathbb{C}[t, t^{-1}] \otimes \Lambda(N)$ be a homogeneous (in the $\mathbb{Z}_2$-grading) element. Define $L(u)$ in $W(1, N)$ to be

$$L(u) = u \frac{\partial}{\partial t} + \frac{1}{2} \sum_{j=1}^{N} (-1)^{\deg u} D_j(u) D_j.$$

The N-superconformal algebra is the subalgebra of $W(1, N)$ consisting precisely of elements of the above form.

Next we define affine algebras with extended symmetry. Here we just generalize the notion of an affine Kac–Moody algebra. Let S be a finite dimensional simple complex Lie algebra equipped with an invariant symmetric nondegenerate bilinear form $(\cdot, \cdot)_S$. We form the super-loop algebra $\text{Ł}_N = S \otimes \mathbb{C}[t, t^{-1}] \otimes \Lambda(N)$. Let α be a nontrivial 2-cocycle of $S \otimes \mathbb{C}[t, t^{-1}] \otimes \Lambda(N)$. α give rise to a central extension of $S \otimes \mathbb{C}[t, t^{-1}] \otimes \Lambda(N)$. We call such a central extension of the super-loop algebra an affine algebra with N-extended symmetry. If $N = 0$, then α is unique up to a constant and in this case we obtain the usual affine algebra. If $N \geq 0$, the situation changes drastically. In this case α is by no means unique. However, we can describe all 2-cocycles as follows: Let $\lambda \in \mathbb{C}[t, t^{-1}] \otimes \Lambda(N)$. We denote by $\oint \lambda$ the coefficient of $t^{-1} \otimes \lambda_1 \wedge \cdots \wedge \lambda_N$. Define on Ł_N an invariant super-symmetric nondegenerate bilinear form by

$$(s_1 \otimes \lambda_1, s_2 \otimes \lambda_2) := (s_1, s_2)_S \oint \lambda_1 \wedge \lambda_2.$$

It can be proved that the outer derivations of Ł_N is precisely the superalgebra of derivations of $\mathbb{C}[t, t^{-1}] \otimes \Lambda(N)$. Let

$$D = a_0 \frac{\partial}{\partial t} + \sum_{i=1}^{N} a_i \frac{\partial}{\partial \theta_i},$$

with $\frac{\partial}{\partial t}a_0+(-1)^{\deg a_i}\frac{\partial}{\partial\theta_i}a_i := \mathrm{div}D = 0$. Then $\alpha_D(x, y) := (D(x), y)$ is a 2-cocycle. And conversely all 2-cocycles are of this form [1] [6].

Before going to our main results let us give some remarks on fields. A field $\phi(z)$ is a formal power series in z, where the coefficients are elements of some Lie superalgebra. We write

$$\phi(z) = \sum_j \phi_j z^{-j-\Delta},$$

where Δ is the conformal dimension. Given another field $\psi(z)$ we denote their normally ordered product by $:\phi(z)\psi(z):$. The derivative of a field $\phi(z)$ will be denoted by $\partial\phi(z)$. The singular terms of the operator product expansion (OPE) will written as $\phi(z)\cdot\psi(w)$. Since all our fields are local with respect to one another, the OPE determines all commutation relations completely.

3. Compatibility

As we have seen above the centerless N-superconformal algebra is a subalgebra of $W(1, N)$. A 2-cocycle α_D of the N-super-loop algebra is determined by an element $D \in W(1, N)$ with $\mathrm{div}D = 0$. α_D determines a super-symmetric bilinear form on $\mathbb{C}[t, t^{-1}] \otimes \Lambda(N)$. We will call an element $L \in W(1, N)$ compatible with α_D if L is super-skewsymmetric with respect to α_D. Let's denote by $W(1, N)^\alpha$ the subalgebra of $W(1, N)$ consisting of all such elements. It turns out that we obtain a representation of the N-superconformal algebra on a representation space of $\hat{Ł}_N$, the central extension of $Ł_N$, if $W(1, N)^\alpha$ contains the N-superconformal algebra. In the case when $N = 0$ one obtains the Sugawara construction and in the case when $N = 1$ this gives rise to the Kac–Todorov construction. For $N = 2$ the situation is more complicated. We have to consider a more general situation: Certainly the N-superconformal algebra can be embedded into $W(1, N+k)$, $k \in \mathbb{N}$. In $\mathbb{C}[t, t^{-1}] \otimes \Lambda(N+k)$ we can consider a divergence free derivation D which in turn gives rise to a 2-cocycle α. We can ask for which α does $W(1, N+k)^\alpha$ contain the N-superconformal algebra. We have the following results [3]:

Proposition 1. *If $N \geq 3$ and α is compatible with the N-superconformal algebra, then $\alpha = 0$.*

Proposition 2. *D is compatible with the $N = 2$-superconformal algebra if and only if $D = \sum_{i=3}^{2+k} \alpha_i \frac{\partial}{\partial\theta_i}$, where $\alpha_i \in \Lambda(\theta_3, \ldots, \theta_{2+k})$.*

So it turns out that there are only finitely many 2-cocycles that are compatible with the $N = 2$-symmetry for any fixed k. If $N \geq 3$, there aren't any compatible 2-cocycles at all.

4. Construction of representations

We can expect to construct $N = 2$-symmetry on a representation space of an affine algebra with $2+k$-extended symmetry, where $k \geq 1$. Let $D = \sum_{i=3}^{2+k} \alpha_i \frac{\partial}{\partial \theta_i}$, where $\alpha_i \in \Lambda(\theta_3, \ldots, \theta_{2+k})$ and $\mathrm{div} D = 0$. We have to restrict ourselves to the maximal subspace $\bar{K}$ of $\mathbb{C}[t, t^{-1}] \otimes \Lambda(N+k)$ on which α_D is nondegenerate. This is essential and the explanation is given in [2], [3]. It follows that we would have to use a central extension of $S \otimes \bar{K}$ to construct our $N = 2$-symmetry. However, $S \otimes \bar{K}$ may not even be a Lie superalgebra in general. This problem forces us to consider only special Lie algebras S for which $S \otimes \bar{K}$ is a Lie superalgebra. It is not essential here that S is simple as long as S carries an invariant symmetric nondegenerate bilinear form. If S is abelian, then certainly S satisfies the condition. From now on we let S be the one-dimensional Lie algebra equipped with such a form. Let D be such a compatible derivation. Consider D as a derivation of $\Lambda(\theta_3, \ldots, \theta_{2+k})$. Let K be the subspace of $\Lambda(\theta_3, \ldots, \theta_{2+k})$ on which α_D is nondegenerate. α_D is symmetric on the odd part $K_{\bar{1}}$ and skew-symmetric on the even part $K_{\bar{0}}$. Choose an orthonormal basis $\{\lambda_1, \ldots, \lambda_l\}$ of $K_{\bar{1}}$ and a basis $\{\mu_1, \ldots, \mu_m, \mu_1^*, \ldots, \mu_m^*\}$ of $K_{\bar{0}}$ such that $\alpha_D(\mu_i, \mu_j^*) = \delta_{ij}$. Of course $r = 2m + l$ is the rank of D.

Let $\mathfrak{a}_i$ be the Lie superalgebra generated by $\{t^n \otimes \theta_1 \wedge \lambda_i, t^n \otimes \theta_2 \wedge \lambda_i, t^n \otimes \theta_1 \wedge \theta_2 \wedge \lambda_i, t^n \otimes \lambda_i, z_i; n \in \mathbb{Z}\}$ with commutation relations defined by

$$[t^n \otimes v, t^m \otimes v'] = \delta_{n,-m} \alpha_D(v, v') z_i$$

Here z_i is a central element. From this Lie superalgebra one constructs 4 free fields [1]:

$$\alpha^+(z) = \sum_{r \in \frac{1}{2}+\mathbb{Z}} \alpha_r^+ z^{-r-\frac{1}{2}}, \alpha_r^+ = t^{r-\frac{1}{2}} \otimes \lambda_i \wedge \theta_1$$

$$\alpha^-(z) = \sum_{r \in \frac{1}{2}+\mathbb{Z}} \alpha_r^- z^{-r-\frac{1}{2}}, \alpha_r^- = t^{r-\frac{1}{2}} \otimes \theta_2 \wedge \lambda_i$$

$$\psi^+(z) = \sum_{j \in \mathbb{Z}} \psi_j^+ z^{-j-1}, \psi_j^+ = t^j \otimes \lambda_i$$

$$\psi^-(z) = \sum_{j \in \mathbb{Z}} \psi_j^- z^{-j-1}, \psi_j^- = -j t^{j-1} \otimes \theta_1 \wedge \theta_2 \wedge \lambda_i.$$

They satisfy the following nontrivial OPE's:

$$\alpha^+(z) \cdot \alpha^-(w) \sim \frac{k}{(z-w)}, \quad \psi^+(z) \cdot \psi^-(w) \sim \frac{k}{(z-w)^2}.$$

The first two are even fields of conformal dimension $\frac{1}{2}$, and the last two are odd of dimension 1. k is the scalar with which z_i acts on the Fock space on these free fields.

Proposition 3 ([2]). *On the Fock space of* $\mathfrak{a}_i$ *one has an* $N = 2$*-symmetry of central charge* -1. *The representation is given by*

$$L(z) = \frac{1}{2k}(:\partial\alpha^-(z)\alpha^+(z): - :\partial\alpha^+(z)\alpha^-(z):) - \frac{1}{k} :\psi^+(z)\psi^-(z):$$

$$T(z) = \frac{1}{2k}(:\alpha^-(z)\alpha^-(z): - :\alpha^+(z)\alpha^+(z):)$$

$$G^+(z) = \frac{1}{2k}(:\alpha^-(z)\psi^+(z): + :\alpha^+(z)\psi^-(z): - :\alpha^-(z)\psi^-(z): - :\alpha^+(z)\psi^+(z):)$$

$$G^-(z) = \frac{1}{2k}(:\alpha^-(z)\psi^+(z): + :\alpha^+(z)\psi^-(z): + :\alpha^-(z)\psi^-(z): + :\alpha^+(z)\psi^+(z):)$$

Now define $\mathfrak{b}_i$ to be the Lie superalgebra generated by $\{t^n\otimes\theta_1\wedge\mu_i, t^n\otimes\theta_1\wedge\mu_i^*, t^n\otimes\theta_2\wedge\mu_i, t^n\otimes\theta_2\wedge\mu_i^*, t^n\otimes\theta_1\wedge\theta_2\wedge\mu_i, t^n\otimes\theta_1\wedge\theta_2\wedge\mu_i^*, t^n\otimes\mu_i, t^n\otimes\mu_i^*, \tilde{z}_i;\ n\in\mathbb{Z}\}$ where $\tilde{z}_i$ is central and commutation relations are defined by α_D. Form $\mathfrak{b}_i$ one obtains 8 free fields: The odd fields are

$$a^+(z) = \sum_{r\in\frac{1}{2}+\mathbf{Z}} a_r^+ z^{-r-\frac{1}{2}},\ a_r^+ = -t^{r-\frac{1}{2}}\otimes\theta_1\wedge\mu_i$$

$$a^-(z) = \sum_{r\in\frac{1}{2}+\mathbf{Z}} a_r^- z^{-r-\frac{1}{2}},\ a_r^- = t^{r-\frac{1}{2}}\otimes\theta_2\wedge\mu_i^*$$

$$b^+(z) = \sum_{r\in\frac{1}{2}+\mathbf{Z}} b_r^+ z^{-r-\frac{1}{2}},\ b_r^+ = t^{r-\frac{1}{2}}\otimes\theta_2\wedge\mu_i$$

$$b^-(z) = \sum_{r\in\frac{1}{2}+\mathbf{Z}} b_r^- z^{-r-\frac{1}{2}},\ b_r^- = -t^{r-\frac{1}{2}}\otimes\theta_1\wedge\mu_i^*$$

The even fields are

$$A^+(z) = \sum_{n\in\mathbf{Z}} A_n^+ z^{-n-1},\ A_n^+ = t^n\otimes\mu_i$$

$$A^-(z) = \sum_{n\in\mathbf{Z}} A_n^- z^{-n-1},\ A_n^- = -nt^{n-1}\otimes\theta_1\wedge\theta_2\wedge\mu_i^*$$

$$B^+(z) = \sum_{n\in\mathbf{Z}} B_n^+ z^{-n-1},\ B_n^+ = -nt^{n-1}\otimes\theta_1\wedge\theta_2\wedge\mu_i$$

$$B^-(z) = \sum_{n\in\mathbf{Z}} B_n^- z^{-n-1},\ B_n^- = t^n\otimes\mu_i^*.$$

They satisfy the following nontrivial operator product expansions:

$$a^+(z)\cdot a^-(w) \sim \frac{-k}{(z-w)} \qquad A^+(z)\cdot A^-(w) \sim \frac{k}{(z-w)^2}$$

$$b^+(z)\cdot b^-(w) \sim \frac{k}{(z-w)} \qquad B^+(z)\cdot B^-(w) \sim \frac{-k}{(z-w)^2}$$

Here $k \in \mathbb{C}^*$ is the scalar with which z acts on the Fock space of $\mathfrak{b}_i$. We have [3]

Proposition 4. *On the Fock space of the Lie superalgebra $\mathfrak{b}_i$ we have an action of the $N = 2$-superconformal algebra. The action is given as follows:*

$$L(z) = \frac{1}{2k}(: a^+(z)\partial a^-(z) : + : a^-(z)\partial a^+(z) : + : \partial b^+(z)b^-(z) :$$
$$+ : \partial b^-(z)b^+(z) : +2 : A^+(z)A^-(z) : +2 : B^+(z)B^-(z) :)$$
$$T(z) = \frac{-i}{k}(: a^-(z)b^+(z) : - : a^+(z)b^-(z) :)$$
$$G^+(z) = \frac{1}{2k}(: A^+(z)a^-(z) : - : A^-(z)a^+(z) : +B^+(z)b^-(z) : - : B^-(z)b^+(z) :)$$
$$+ \frac{i}{2k}(: A^+(z)b^-(z) : + : A^-(z)b^+(z) : -B^+(z)a^-(z) : -B^-(z)a^+(z) :)$$
$$G^-(z) = \frac{1}{2k}(: A^+(z)a^-(z) : - : A^-(z)a^+(z) : +B^+(z)b^-(z) : - : B^-(z)b^+(z) :)$$
$$- \frac{i}{2k}(: A^+(z)b^-(z) : + : A^-(z)b^+(z) : -B^+(z)a^-(z) : -B^-(z)a^+(z) :)$$

This representation has central charge 2.

Since the central extension of $S \otimes \bar{K}$ is isomorphic to l copies of $\mathfrak{a}_i$ and m copies of $\mathfrak{b}_i$ with identified central elements, we see that the $N = 2$ acts on the Fock space of the central extension of $S \otimes \bar{K}$ with central charge $2m - l$.

We conclude with 2 remarks:

Remark 1. The representation in Proposition 3 coincides with a special case of the so-called (b, c, β, γ)-model. [2] [9]

Remark 2. The $N = 2$-symmetry in Proposition 4 coincides with the Manin triple model [7] where one takes as the underlying Manin triple the four-dimensional abelian Lie algebra.

References

[1] Cheng, S.-J., Representations of Central Extensions of Differentiable Simple Lie Superalgebras, Comm. Math. Phys. 154 (1993), 555–568.

[2] Cheng, S.-J., Construction of the $N = 2$ Superconformal Algebra from Affine Algebras with Extended Symmetry I, Lett. Math. Phys., to appear.

[3] Cheng, S.-J., Construction of the $N = 2$ Superconformal Algebra from Affine Algebras with Extended Symmetry II, to appear.

[4] Kac, V., Todorov, I., Superconformal Current Algebras and Their Unitary Representations, Comm. Math. Phys. 102 (1985), 337–347.

[5] Thielemans, K., A Mathematica Package for Computing Operator Product Expansions, Internat. J. Modern Phys. C 2, No.3 (1991), 787–798.

[6] Coquereuax, R., Frappat, L., Ragoucy, E., Sorba, P., Extended Super Kač-Moody Algebras and Their Super-Derivations Algebras, Comm. Math. Phys. 133 (1990), 1–35.

[7] Getzler, E., Manin Pairs and Topological Field Theory, hep-th. 9309057 (1993).

[8] Sugawara, H., A Field Theory of Currents, Phys. Rev. 170 (1968), 1659–1662.

[9] Fre, P., Girardello, L., Lerda, A., Soriani, P., Topological First-order systems with Landau–Ginzburg interactions, Nuclear Phys. B387 (1992), 333–372.

Department of Mathematics,
National Cheng Kung University
Tainan
Taiwan 701, ROC

Orbits of the coadjoint representation and Yang–Baxter equation

M. I. Gekhtman[1,2] *and A. Stolin*[3]

1991 Mathematics Subject Classification: 17B37, 81R50

0. Introduction

In this paper we present a quantization procedure for the Lie–Kirillov–Kostant bracket on a wide class of orbits of the coadjoint representation.

Let $\mathcal{L}$ be a real finite dimensional Lie algebra and $\mathcal{L}^*$ be its dual. Then the algebra of smooth functions $C^\infty(\mathcal{L}^*)$ is endowed with the Lie–Kirillov–Kostant bracket $\{,\}_L$: for two linear functions ℓ_1, ℓ_2 on $\mathcal{L}^*$, $\{\ell_1, \ell_2\}_L = [\ell_1, \ell_2]$, where ℓ_i is identified with elements of $\mathcal{L}$. It is well known that this bracket is degenerate and its symplectic leaves coincide with orbits of the coadjoint representation of G in $\mathcal{L}^*$, where G is a Lie group with the algebra $\mathcal{L}$.

The problem arising from quantum mechanics is to quantize this bracket, that is to find a new operation $*$ on $C^\infty(\mathcal{L}^*)$ or $C^\infty(\theta)$, where θ is an orbit in $\mathcal{L}^*$, such that

1) $*$ depends on a formal parameter h,

2) $f * g = fg + \frac{h}{2}\{f, g\} + h^2 F_2(f, g) + \cdots + h^n F_n(f, g) + \cdots$, where F_i are bidifferential operators on $C^\infty(\mathcal{L}^*)$ or $C^\infty(\theta)$,

3) $*$ is an associative operation,

4) $(f_1 + f_2) * g = f_1 * g + f_2 * g$,

5) $\alpha * f = f * \alpha = \alpha f$, where $\alpha \in \mathbb{R}$ is considered as a constant function.

The operation that satisfies conditions 1)–5) is called a *star-product* (see, for example, [2], [10]).

In [2] the general construction of the quantization was proposed for $\mathcal{L}^*$ and the so-called regular orbits. We should note that the $*$-operation is not unique. One of

[1]Recipient of a Dov Biegun Postdoctoral Fellowship.

[2]Partially supported by Swedish Natural Science Research Council.

[3]Supported by Magnusson Foundation of Swedish Academy of Science.

the ways to introduce a star-product is to use the Fourier transform and the Campbell–Hausdorff formula (see [1]). For this star-product we shall show how to compute bidifferential operators F_i.

The idea of this work is to use a decomposition of the Lie algebra $\mathcal{L}$ into a direct sum of two subalgebras, one of which is the stabilizer of an orbit in $\mathcal{L}^*$. It turns out that the subalgebra complement to the stabilizer is a so-called Frobenius Lie algebra. It is well-known that one can naturally assign to a Frobenius Lie algebra the solution of the classical Yang–Baxter equation that defines a Poisson structure on the corresponding Lie group. Moreover, there is a local Poisson diffeomorphism between this group and the orbit associated with the decomposition of $\mathcal{L}$.

Therefore, to quantize the Lie–Kirillov–Kostant bracket in this case, it is sufficient to quantize the corresponding Poisson structure on the group. In the Frobenius case we present a construction that seems to be more simple than the general case of [5]. We list several examples of orbits satisfying our assumptions, including Hermitian symmetric spaces of noncompact type.

We should remark that the obtained decompositions can be used to construct new rational solutions of the classical Yang–Baxter equation. One can also apply the Adler–Kostant–Symes scheme to these decompositions. We discuss these applications in Appendices.

We are grateful to Professor L. Takhtajan for valuable discussions.

I. On bidifferential operators defining star-product

1. In what follows we denote elements of $\mathcal{L}^*$ by $a, b, x, y, \ldots$ and elements of $\mathcal{L}$ by $\alpha, \beta, \xi, \eta, \ldots$. We choose dual bases $e_i, \varepsilon^i, i = 1, \ldots, n$ in $\mathcal{L}^*$ and $\mathcal{L}$ respectively, satisfying the conditions

$$\langle e_i, \varepsilon^j \rangle = \delta_i^j, \quad [\varepsilon^i, \varepsilon^j] = f_k^{ij} \varepsilon^k.$$

The vector space $C^\infty(\mathcal{L}^*)$ is a commutative cocommutative Hopf algebra with an ordinary multiplication and the comultiplication Δ_0, counit ε and antipode S defined as follows:

$$(\Delta_0 f)(x, y) = f(x + y), \quad \varepsilon(f) = f(0), (Sf)(x) = f(-x).$$

Similarly, $C^\infty(\mathcal{L})$ is endowed with the structure of a commutative cocommutative Hopf algebra.

It is convenient to use in our considerations the special functions $f_\xi \in C^\infty(\mathcal{L}^*)$, $g_x \in C^\infty(\mathcal{L})$ given by

$$f_\xi(x) = g_x(\xi) = e^{i\langle x, \xi \rangle}.$$

Let $\hat{f}(\xi) = \int f(x)e^{-i\langle x,\xi\rangle}dx$, $\hat{g}(x) = \int g(\xi)e^{-i\langle x,\xi\rangle}d\xi$ be Fourier transforms of functions $f \in C^\infty(\mathcal{L}^*)$, $g \in C^\infty(\mathcal{L})$. Note that whenever it is necessary, we understand the Fourier transform in the sense of distributions.

Lemma 1. *The formula*

$$(f, g) := \int f(x)\hat{g}(x)dx = \int \hat{f}(\xi)g(\xi)d\xi \tag{1}$$

defines a Hopf algebra pairing between $C^\infty(\mathcal{L}^*)$ *and* $C^\infty(\mathcal{L})$. *In particular,*

$$\begin{aligned}
(f, g_x) &= f(x),\\
(f_\xi, g) &= g(\xi),\\
(f_\xi, g_x) &= f_\xi(x) = g_x(\xi) = e^{i\langle x,\xi\rangle},\\
(f, 1) &= (f, g_0) = \varepsilon(f).
\end{aligned}$$

Remark. Clearly, the pairing (1) is not always well-defined. However, in all the situations occurring below we can make it meaningful by considering as the pairing either between the space of Schwartz functions and the corresponding space of distributions or between the smooth functions and the distributions with compact support. In particular, identifying the basis vectors e_i, ε^j with linear functions on $\mathcal{L}$ and $\mathcal{L}^*$ respectively, we obtain the relations

$$(\varepsilon^j, e_i) = \delta_i^j.$$

We define the action of $\mathcal{L}$ on $C^\infty(\mathcal{L}^*)$ and $C^\infty(\mathcal{L})$ as follows:

$$\begin{aligned}
(\varphi_{\mathcal{L}}(\eta)f)(x) &= \{f, \eta\}_L(x),\\
(\pi_{\mathcal{L}}(\eta)g)(\xi) &= (\{\eta, f_\xi\}_L, g).
\end{aligned}$$

Here the Poisson bracket $\{\cdot,\cdot\}_L$ for smooth functions on $\mathcal{L}^*$ is the Lie–Kirillov–Kostant bracket

$$\{f_1, f_2\}_L(x) = \langle x, [\nabla f_1, \nabla f_2]\rangle,$$

and elements $\eta \in \mathcal{L}$ are identified with linear functions on g.

Our next aim is to introduce an associative operation $*$ on $C^\infty(\mathcal{L}^*)$ and a co-associative operation Δ on $C^\infty(\mathcal{L})$, that depend on the parameter h and are dual with respect to the pairing (1). The idea of introducing the operation $*$ goes back to Berezin's work [3]. In the form presented below it first appeared in [1].

Let $C_h(\xi, \eta)$ be the Campbell–Hausdorff series, uniquely defined by the relation

$$e^{h\xi} \cdot e^{h\eta} = e^{hC_h(\xi,\eta)},$$

where ξ and η are elements of the Lie algebra $\mathcal{L}$. $C_h(\xi, \eta)$ has a form

$$C_h(\xi, \eta) = \xi + \eta + \frac{h}{2}[\xi, \eta] + \frac{h^2}{12}([\xi, [\xi, \eta]] + [\eta, [\eta, \xi]]) + \cdots$$

and satisfies the following associativity condition

$$C_h(C_h(\xi, \eta), \zeta) = C_h(\xi, C_h(\eta, \zeta)). \tag{2}$$

Definition. Let $f_1, f_2 \in C^\infty(\mathcal{L}^*)$ and $g \in C^\infty(\mathcal{L})$. Then
1) $(f_1 * f_2)(x) = \int \hat{f}_1(\xi)\hat{f}_2(\eta)e^{i\langle x, C_h(\xi,\eta)\rangle}d\xi d\eta$
2) $(\Delta g)(\xi, \eta) = g(C_h(\xi, \eta))$.

It follows immediately from (2) that $\Delta : C^\infty(\mathcal{L}) \to C^\infty(\mathcal{L})^{\otimes 2} = C^\infty(\mathcal{L} \times \mathcal{L})$ is a co-associative operation.

Lemma 2.
1) $*$ *is an associative operation on* $C^\infty(\mathcal{L}^*)$.
2) $1 * f = f * 1 = f$.
3) $(f_1 * f_2, g) = (f_1 \otimes f_2, \Delta g)$.
4) $*$ *is* Ad**-invariant.*

Proof. Consider

$$(f_1 * f_2) * f_3 = \int \widehat{f_1 * f_2}(\xi) f_3(\eta) e^{i\langle x, C_h(\xi,\eta)\rangle} d\xi d\eta.$$

It is easy to see that

$$\widehat{f_1 * f_2}(\xi) = \int \hat{f}_1(l)\hat{f}_2(\mu)\delta(\xi - C_h(l, \mu))dl d\mu \tag{3}$$

and, therefore,

$$(f_1 * f_2) * f_3 = \int \hat{f}_1(l)\hat{f}_2(\mu)\hat{f}_3(\eta)e^{i\langle x, C_h(C_h(l,\mu),\eta)\rangle}dl d\mu d\eta.$$

Similarly,

$$f_1 * (f_2 * f_3) = \int \hat{f}_1(l)\hat{f}_2(\mu)\hat{f}_3(\eta)e^{i\langle x, C_h(l, C_h(\mu,\eta))\rangle}dl d\mu d\eta.$$

Thus, the associativity of $*$ follows from (2).

The second statement is the consequence of the properties of the pairing (1). Using (3) again, we get

$$\begin{aligned}(f_1 * f_2, g) &= \int \widehat{f_1 * f_2}(\xi) g(\xi) d\xi \\ &= \int \hat{f}_1(l)\hat{f}_2(\mu)\delta(\xi - C_h(l,\mu))g(\xi)\,dl\,d\mu\,d\xi \\ &= \int \hat{f}_1(l)\hat{f}_2(\mu)g(C_h(l,\mu))\,dl\,d\mu = (f_1 * f_2, \Delta g),\end{aligned}$$

which proves 3). The statement 4) is an immediate consequence of the Ad*-invariance of $C_h(\xi,\eta)$. □

Lemma 2 implies that endowed with the operation $*$, $C^\infty(\mathcal{L}^*)$ becomes the non-commutative cocommutative Hopf algebra with comultiplication Δ_0, antipode S and counit ε defined as above, whereas comultiplication Δ transforms $C^\infty(\mathcal{L})$ into the commutative, but non-cocommutative Hopf algebra. We shall denote these new Hopf algebras by ${C^\infty}_h(\mathcal{L}^*)$ and ${C^\infty}_h(\mathcal{L})$ respectively. Statement 3) of Lemma 2 shows that the pairing (1) is a Hopf algebra pairing between ${C^\infty}_h(\mathcal{L}^*)$ and ${C^\infty}_h(\mathcal{L})$.

Remark. The operation $*$ looks especially simple, when applied to functions f_ξ and f_η:

$$(f_\xi * f_\eta)(x) = f_{C_h(\xi,\eta)}(x) = g_x(C_h(\xi,\eta)) = (\Delta g_x)(\xi,\eta). \tag{4}$$

Proposition 1. *The universal enveloping algebra $\mathfrak{U}(\mathcal{L})$ is isomorphic to the Hopf subalgebra of polynomial functions in ${C^\infty}_h(\mathcal{L})$.*

Proof. Consider the mapping $\eta \mapsto \eta(x) = \langle x, \eta\rangle \in {C^\infty}_h(\mathcal{L}^*)$. Since $\hat{\eta}(\xi) = i\langle\nabla\delta(\xi), \eta\rangle$, we have

$$\begin{aligned}(\eta_1 * \eta_2)(x) &= -\int \langle\nabla\delta(\xi),\eta_1\rangle\langle\nabla\delta(\eta),\eta_2\rangle e^{i\langle x, C_h(\xi,\eta)\rangle} d\xi\, d\eta \\ &= \eta_{1,i}\eta_{2,j}\Big\langle x, \frac{\partial^2}{\partial\xi_i\partial\eta_j} C_h(\xi,\eta)\Big|_{\xi=\eta=0}\Big\rangle \\ &\quad + \eta_{1,i}\eta_{2,j}\Big\langle x, \frac{\partial}{\partial\xi_i} C_h(\xi,\eta)\Big|_{\xi=\eta=0}\Big\rangle \Big\langle x, \frac{\partial}{\partial\eta_j} C_h(\xi,\eta)\Big|_{\xi=\eta=0}\Big\rangle \\ &= \frac{h}{2}\langle x, [\eta,\eta_2]\rangle + \langle x,\eta_1\rangle\,\langle x,\eta_2\rangle = \eta_1(x)\eta_2(x) + \frac{1}{2}\,\{\eta_1,\eta_2\}_L(x).\end{aligned}$$

Therefore,

$$(\eta_1 * \eta_2)(x) - (\eta_2 * \eta_1)(x) = h[\eta_1,\eta_2](x).$$

Moreover,

$$(\Delta_0\eta)(x) = \eta(x+y) = \langle x,\eta\rangle + \langle y,\eta\rangle = (\eta\otimes 1 + 1\otimes\eta)(x,y),$$

which completes the proof. □

2. Let us consider the space $\mathrm{Diff}(\mathcal{L}^*)$ of differential operators $\sum_\alpha f_\alpha(x)D^\alpha$ on $C^\infty(\mathcal{L}^*)$ with polynomial coefficients. We denote by $\mathrm{Diff}_h(\mathcal{L}^*)$ the algebra of formal power series $\Psi_h = \Psi_0 + h\Psi_1 + \cdots + h^n\Psi_n + \cdots$, where $\Psi_0, \Psi_1, \ldots \in \mathrm{Diff}(\mathcal{L}^*)$.

We are looking for an element F_h from $\mathrm{Diff}_h(\mathcal{L}^*)^{\otimes 2}$ which represents the associative operation $*$ in the following sense:

$$(mF_h)(f_1 \otimes f_2) = f_1 * f_2, \tag{5}$$

where $m : C^\infty(\mathcal{L}^*)^{\otimes 2} \to C^\infty(\mathcal{L}^*)$ is the multiplication map

$$m(f_1 \otimes f_2) = f_1 f_2.$$

Proposition 2. *There exists a (not unique) element from F_h from* $\mathrm{Diff}_h(\mathcal{L}^*)^{\otimes 2}$ *satisfying* (5).

Proof. The idea of construction of F_h is close to the considerations in [5]. Namely, let us consider the formal series expansion for $e^{i\langle x, C_h(\xi,\eta)-\xi-\eta\rangle}$:

$$e^{i\langle x, C_h(\xi,\eta)-\xi-\eta\rangle} = 1 + \frac{ih}{2}\langle x, [\xi,\eta]\rangle + \cdots + h^n P_n(x;\xi,\eta) + \cdots,$$

where P_n is a polynomial function of x, ξ, η. Then

$$f_1 * f_2 =$$
$$\int \hat{f}_1(\xi)\hat{f}_2(\eta)(1 + \frac{ih}{2}\langle x, [\xi,\eta]\rangle + \cdots + h^n P_n(x;\xi,\eta)) + \cdots)e^{i\langle x,\xi\rangle}e^{i\langle x,h\rangle}d\xi d\eta.$$

Let us define F_h as follows:

$$F_h(f_1 \otimes f_2)(x,y) =$$
$$\int \hat{f}_1(\xi)\hat{f}_2(\eta)(1 + \frac{ih}{2}\langle x, [\xi,\eta]\rangle + \cdots + h^n P_n(x;\xi,\eta) + \cdots)e^{i\langle x,\xi\rangle}e^{i\langle y,\eta\rangle}d\xi d\eta. \tag{6}$$

Using the properties of the inverse Fourier transform, one can see that F_h is a formal series of bi-differential operators

$$F_h = 1 + \frac{h}{2}F_0 + \cdots + h^n Q_n(x; \nabla_x, \nabla_y) + \cdots, \tag{7}$$

where Q_n is a polynomial function.

Clearly, F_h satisfies (5). Non-uniqueness of the choice of F_h is connected with the fact that if some fixed monomial term in $Q_n(x;\nabla_x,\nabla_y)$ contains x^k, then we can substitute x^k by y^k without affecting identity (5).

Direct calculation shows that the operator F_0 in formula (7) has the following form:

$$F_0 = \varphi_{\mathcal{L}}(\varepsilon^i) \otimes \frac{\partial}{\partial x^i}, \tag{8}$$

where

$$\varphi_{\mathcal{L}}(\varepsilon^i) = f_k^{\alpha i} x^k \frac{\partial}{\partial x^\alpha}$$

is the operator that was introduced in Section 1. It is easy to see from the formula $F_0(f_1 \otimes f_2)(x, y) = \{f_1, \varepsilon^i\}_L(x) \cdot \frac{\partial}{\partial y^i} f_2(y)$ that

$$(mF_0)(f_1 \otimes f_2) = \{f_1, f_2\}_L. \tag{9}$$

Relations (9) and (5) together with statements 1), 2) of Lemma 2 imply that the operation $*$ is a star-product. Of course, the operator (8) is not the only operator satisfying (9). However the following result holds true for every operator F_0 that can be obtained from (6) after substitution of x by the linear function $l(x, y)$ such that $l(x, x) = x$ in the term $\frac{ih}{2}\langle x, [\xi, \eta]\rangle$.

Proposition 3. *Let S be the operator in $C^\infty(\mathcal{L}^*)^{\otimes 2}$ interchanging the tensor factors. Then the operator*

$$r = S[F_0, S] = f_k^{ij}(x^k + y^k)\frac{\partial}{\partial x^i}\frac{\partial}{\partial y^j}$$

satisfies the modified classical Yang–Baxter equation.

Proof. Let us consider a semidirect sum $\mathcal{L}^* + \mathcal{L}$ of the linear space $\mathcal{L}^*$ considered as a commutative Lie algebra and the Lie algebra $\mathcal{L}$. The commutator between e_i, ε^j in $\mathcal{L}^* + \mathcal{L}$ is given by

$$[e_i, \varepsilon^j] = f_i^{\alpha j} e_\alpha.$$

The Lie algebra $\mathcal{L}^* + \mathcal{L}$ can be mapped homomorphically into $\mathrm{Diff}_h(\mathcal{L}^*)^{\otimes 2}$:

$$e_i \overset{\rho}{\mapsto} \frac{\partial}{\partial x^i}, \ \varepsilon^j \overset{\rho}{\mapsto} \varphi_{\mathcal{L}}(\varepsilon^j). \tag{10}$$

Then we can identify the operator r with the element $\varepsilon^i \otimes e_i - e_i \otimes \varepsilon^i$ in $\mathcal{U}(\mathcal{L}^* + \mathcal{L})^{\otimes 2}$ which is known to satisfy the modified classical Yang–Baxter equation [7]. □

Proposition 3 shows that we can choose the element F_h satisfying (6) to be equal to

$$F_h = 1 + \frac{h}{2} r + \cdots.$$

Moreover, using the structure of the Campbell–Hausdorff series and substituting if necessary x^k by y^k, one can show inductively that the following statement is true.

Lemma 3. *The element F_h satisfying* (5) *can be chosen in the form*

$$F_h = 1 + \frac{h}{2} r + \cdots + h^n \tilde{Q}_n(x, y; \nabla_x, \nabla_y) + \cdots, \tag{11}$$

where for every n the bidifferential operator $\tilde{Q}_n(x, y; \nabla_x, \nabla_y)$ *belongs to the image* $\mathfrak{U}(\mathfrak{L}^* + \mathfrak{L})^{\otimes 2}$ *under the mapping* (10).

Then it is not hard to check, that in view of the associativity of the operation $*$ the operator F_h satisfies the identity

$$m(\Delta_0 \otimes \mathrm{id})F_h \circ (F_h)_{12} = m(\mathrm{id} \otimes \Delta_0)F_h \circ (F_h)_{23}, \tag{12}$$

where Δ_0 is the natural cocommutative comultiplication on $\rho(\mathfrak{U}(\mathfrak{L}^* + \mathfrak{L}))$.

Remark. Identity (12) does not imply the equality

$$(\Delta_0 \otimes \mathrm{id})F_h \circ (F_h)_{12} = (\mathrm{id} \otimes \Delta_0)F_h \circ (F_h)_{23}, \tag{13}$$

which would have been used to introduce the structure of the noncocommutative Hopf algebra structure on $\rho(\mathfrak{U}(\mathfrak{L}^* + \mathfrak{L}))$ (see, e.g. [7]). Moreover, using the arguments from [5], one can show that among elements F_h satisfying (5) there is no one obeying (13). Indeed, if such F_h exists, then the operator $r = S[F_0, S]$ gives a solution of the classical Yang–Baxter equation, which is not the case.

However one can define an element $\Phi \in \mathrm{Diff}_h(\mathfrak{L}^*)^{\otimes 3}$ by

$$(\Delta_0 \otimes \mathrm{id})F_h \circ (F_h)_{12} \circ \Phi = (\mathrm{id} \otimes \Delta_0)F_h \circ (F_h)_{23}.$$

Proposition 4.

1) $\Phi \equiv 1 \pmod{h^2}$.
2) $f_1 * f_2 * f_3 = \sum \Phi^{(1)} f_1 * \Phi^{(2)} f_2 * \Phi^{(3)} f_3$.
3) *The operator* Φ *satisfies the pentagon equation.*
4) *An operation* $\Delta(\cdot) = F_h^{-1}\Delta_0(\cdot)F_h$ *endows* $\rho(\mathfrak{U}(\mathfrak{L}^* + \mathfrak{L}))$ *with a quasi-Hopf algebra structure.*
5) *The operator* $R = S(F_h^{-1})F_h$ *satisfies the hexagon equation*

$$(\Delta \otimes 1)R = \Phi_{231}^{-1} R_{13} \Phi_{132} R_{23} \Phi_{123}^{-1}.$$

The first two statements are due to Lemma 3 (11), (12), respectively, while the remaining three can be derived directly from the identity (13) (cf. [7], [10]).

II. Quantization of the Lie–Kirillov–Kostant bracket and Frobenius Lie algebras

Let $r \in \mathfrak{L}^{\otimes 2}$ be a solution of the classical Yang–Baxter equation. We write r in the form $r = r_{ij}\varepsilon^i \otimes \varepsilon^j$, where $\{\varepsilon^i\}$ is a basis of $\mathfrak{L}$. Let ∂_i be the left-invariant vector field corresponding to ε^i. Then $\{f, g\}_r = r_{ij}\partial_i f \cdot \partial_j g$ is a Poisson bracket on $C^\infty(G)$, which we call *r-matrix bracket* (see [4]). Let us remind that r is supposed to be a non-degenerate element of $\mathcal{L}^{\otimes 2}$, i.e., $\det r_{ij} \neq 0$. Moreover, being an element of $(\mathcal{L}^*)^{\otimes 2}$, the matrix inverse to r, is a 2-cocycle on $\mathcal{L}$. If this cocycle is a coboundary, we call $\mathcal{L}$ a Frobenius Lie algebra. In other words, $\mathcal{L}$ is Frobenius if there exists $f \in \mathcal{L}^*$ such that the bilinear form $f([x, y])$ is non-degenerate.

Let $\mathcal{L}$ be a Lie algebra and $f \in \mathcal{L}^*$. Consider the subspace $\mathcal{L}_f \subset \mathcal{L}$, where $\mathcal{L}_f = \{a \in \mathcal{L} : f([a, x]) = 0 \text{ for any } x \in \mathcal{L}\}$. Clearly $\mathcal{L}_f$ is a subalgebra of $\mathcal{L}$.

Proposition 5. *Let A be a subalgebra of $\mathcal{L}$ such that $\mathcal{L}$ is a vector space direct sum of A and $\mathcal{L}_f$. Then A is a Frobenius Lie algebra with the coboundary given by f.*

Proof. We have to prove that $f([x, y])$ is a non-degenerate 2-form on A. Let $x_0 \in A$ be such that $f([x_0, y]) = 0$ for any $y \in A$. Prove that $f([x_0, a]) = 0$ for any $a \in \mathcal{L}$. Indeed, since $A \oplus \mathcal{L}_f = \mathcal{L}$, we have $a = a_1 + a_2$, where $a_1 \in A$, $a_2 \in \mathcal{L}_f$. Then $f([x_0, a]) = f([x_0, a_1]) + f([x_0, a_2])$. The first summand is equal to zero by the definition of x_0, and the second is zero since $a_2 \in \mathcal{L}_f$. Therefore, $x_0 \in \mathcal{L}_f$, which contradicts to the fact that $\mathcal{L}_f \cap A = 0$. □

Corollary. *Let $\mathcal{L}$ be a real simple Lie algebra of noncompact type with the Lie group G and ω be an inner automorphism of order 2. Let $\mathcal{L}^\omega$ (G^ω) be the subalgebra (subgroup) of fixed points of ω. Suppose G/G^ω is a noncompact Hermitian symmetric space. Let $\mathcal{L} = I \oplus \mathcal{L}^\omega$ be the corresponding Iwasawa decomposition. Then I is a Frobenius Lie algebra.*

Proof. Since $\mathcal{L}$ is simple, every element of $\mathcal{L}^*$ may be given in the form $K(a, \cdot)$, where K is the Killing form. Since ω is an inner automorphism, $\omega = \exp t_0$ for some $t_0 \in \mathcal{L}$. Obviously, $\mathcal{L}^\omega = \mathcal{L}_{t_0}$ and hence Proposition 5 implies the statement. □

Now let $f \in \mathcal{L}^*$ and $A \subset \mathcal{L}$ satisfy the conditions of Proposition 5. Let $G(A)$ be the subgroup of G corresponding to the subalgebra A and $G(\mathcal{L}_f)$ be the subgroup corresponding to $\mathcal{L}_f$. Let also $\theta_f \subset \mathcal{L}^*$ be the orbit through f. First of all, $\theta_f \cong G/G(\mathcal{L}_f)$. On the other hand, the formula $A \oplus \mathcal{L}_f = \mathcal{L}$ shows that $G(A)$ is locally diffeomorphic to $G/G(\mathcal{L}_f) \simeq \theta_f$. Moreover, the set $\mathrm{Ad}_k^* f$, where $k \in G(A)$, is open and dense in θ_f. We already know that A is a Frobenius Lie algebra and therefore there exists the r-matrix Poisson bracket $\{\cdot, \cdot\}_r$ on $G(A)$. We would like to compare it with the Lie–Kirillov–Kostant bracket $\{\cdot, \cdot\}_L$ on θ_f.

Proposition 6. *Let s be the local diffeomorphism between $G(A)$ and θ_f given by the formula $s(k) = \mathrm{Ad}_k^* f$. Then s is a Poisson map with respect to $\{\cdot, \cdot\}_r$ and $-\{\cdot, \cdot\}_L$.*

Proof. Clearly, $C^\infty(\theta_f)$ is generated by functions $\lambda_a(\mathrm{Ad}_g^* f) = (\mathrm{Ad}_g^* f)(a)$, where $g \in G$, $a \in \mathcal{L}$. We may consider the factorization $g = k \cdot g_1$, where $k \in G(A)$, $g_1 \in G(\mathcal{L}_f)$. Let $\{\varepsilon^i\}$ be a basis of A, ∂_i be the corresponding left-invariant vector fields. Since $\mathrm{Ad}^*_{g_1} f = f$, all we need is to prove the identity

$$r_{ij}\partial_i(\mathrm{Ad}^*_k f(a))\partial_j(\mathrm{Ad}^*_k f(b)) = (\mathrm{Ad}^*_k f)[b, a],$$

where $a, b \in \mathcal{L}$. We have $\partial_i(\mathrm{Ad}^*_k f(a)) = f([\varepsilon^i, \mathrm{Ad}_k(a)])$ and hence one should prove that $r_{ij} f([\varepsilon^i, x]) \cdot f([\varepsilon^j, y]) = f([y, x])$. Let us suppose for a moment that $x, y \in A$ and recall that the matrix (r_{ij}) is inverse to $(f([\varepsilon^i, \varepsilon^j]))$. The latter means that $r_{ij} f([\varepsilon^j, \varepsilon^m]) = \delta_{im}$, where δ_{im} is the Kronecker symbol. Then $r_{ij} f([\varepsilon^i, x]) \cdot f([\varepsilon^j, \varepsilon^m]) = f([\varepsilon^m, x])$ for all m. Therefore, $r_{ij} f([\varepsilon^i, x]) \cdot f([\varepsilon^j, y]) = f([y, x])$ for $x, y \in A$. Since $f([x_0, x]) = 0$ if $x_0 \in \mathcal{L}_f$ and $A \oplus \mathcal{L}_f = \mathcal{L}$, the general case follows from this. □

Proposition 6 reduces the quantization of $C^\infty(\theta_f)$ to the quantization of $C^\infty(G(A))$ with respect to the r-matrix bracket. The case of general r-matrix was treated in [5]. However, our r-matrix comes from a Frobenius Lie algebra which makes it possible to apply considerations from part I.

Let us first recall that if A is a Frobenius Lie algebra, then $G(A)$ acts locally transitively on A^*, and hence A^* and $G(A)$ are locally diffeomorphic (see [6]). Then Proposition 6 implies the following.

Proposition 7. *Let A be a Frobenius Lie algebra with a given coboundary f. Let r be the corresponding solution of the classical Yang–Baxter equation, and $\{\cdot,\cdot\}_r$ be the corresponding r-matrix Poisson bracket on $C^\infty(G(A))$. Let q be the local diffeomorphism between $G(A)$ and A^* given by the formula $q(k) = \mathrm{Ad}^*_k f$. Then q is a Poisson map with respect to $\{\cdot,\cdot\}_r$ and $-\{\cdot,\cdot\}_L$, where $\{\cdot,\cdot\}_L$ is the Lie–Kirillov–Kostant bracket on A^*.*

Now we can carry the star-operation from $C^\infty(A^*)$ defined in part I to $C^\infty(G(A))$. Since by Lemma 2, the operation $*$ is Ad^*-invariant on $C^\infty(A^*)$, Theorem 1 from [5] shows that the corresponding star-operation on $C^\infty(G(A))$ can be uniquely written in the form

$$(mQ_h)(\phi_1 \otimes \phi_2) = \phi_1 * \phi_2,$$

where $Q_h = 1 + hQ_1 + \cdots + h^n Q_n + \cdots$ and Q_n are left-invariant bidifferential operators on $C^\infty(G(A))$ and thus are identified with elements of $\mathfrak{U}(A) \otimes \mathfrak{U}(A)$. Moreover, by Drinfeld's theorem $S(Q_h)^{-1} Q_h$ satisfies the quantum Yang–Baxter equation, where S interchanges factors in the tensor product.

Proposition 8. *Let A be a Frobenius Lie algebra and let $f \in A^*$ be such that $f([x, y])$ is a non-degenerate bilinear form on A. Let $r \in A \otimes A$ be the corresponding solution of the classical Yang–Baxter equation. Then the operator Q_h defined above can be written in the form*

$$Q_h = 1 + \frac{h}{2} r + h^2\left(\frac{1}{12} r_{js} r_{lp} f_k^{lj} (\varepsilon^k \varepsilon^s \otimes \varepsilon^p + \varepsilon^p \otimes \varepsilon^k \varepsilon^s) + \frac{1}{8} r_{qs} r_{kp} \varepsilon^k \varepsilon^s \otimes \varepsilon^q \varepsilon^p\right) + o(h^2). \quad (14)$$

Proof. By Lemma 3, the operator F_h, that defines the star-operation on A^*, can be expressed through $\varphi_A(\varepsilon^i)$ and $\frac{\partial}{\partial x^i}$. The local diffeomorphism between A^* and $G(A)$ transforms $\varphi_A(\varepsilon^i)$ into the left-invariant vector field, generated by $\varepsilon^i \in A$. Since A is a Frobenius Lie algebra, $\varphi_A(\varepsilon^i)$ form a basis of the tangent space to A^* at the point f. One can express $\frac{\partial}{\partial x^i}$ at the point f through $\varphi_A(\varepsilon^i)$: $\frac{\partial}{\partial x^i} = r_{ik}\varphi_A(\varepsilon^k)$ at f. Substituting the obtained expression in the formula for F_h, and using the local diffeomorphism between A^* and $G(A)$, we get (14). □

Examples

1. The corollary to Proposition 5 enables us to quantize the Lie–Kirillov–Kostant bracket on $C^\infty(H)$, where H is a Hermitian symmetric space of noncompact type.

2. Let $\mathfrak{L} = \mathfrak{sl}(n)$. Consider $f = x_\alpha$, where α is the maximal root. Then A can be chosen as the following subalgebra of $\mathfrak{sl}(n)$:

$$A = \left\{ \begin{pmatrix} a & 0 & \cdots & 0 \\ * & 0 & \cdots & 0 \\ \vdots & 0 & \cdots & 0 \\ * & * & \cdots & -a \end{pmatrix} \right\}.$$

Here A consists of matrices with non-zero elements just in the first column and in the last row.

3. $\mathfrak{L} = \mathfrak{sl}(n)$, $f = \{$the nilpotent Jordan block$\}$. Let $P_1 \subset \mathfrak{sl}(n)$ be a parabolic subalgebra of the form

$$P_1 = \left\{ \begin{pmatrix} * & \cdots & * & 0 \\ * & \cdots & * & 0 \\ \vdots & & & \vdots \\ * & \cdots & * & 0 \\ * & \cdots & * & * \end{pmatrix} \right\}.$$

Then one can show that $P_1 \oplus \mathfrak{L}_f = \mathfrak{sl}(n)$ and we may put $A = P_1$.

4. $\mathfrak{L} = \mathfrak{sl}(n)$, f has distinct eigenvalues. This case was treated in [2]. We propose another way employing the decomposition $\mathfrak{sl}(n) = h \oplus T^{-1} P_1 T$, where P_1 is defined above, h is the diagonal subalgebra and $T = \mathrm{id} + E_{1n} + E_{2n} + \cdots + E_{n-1,n} \in \mathrm{GL}(n)$. Here E_{ij} is the matrix which has 1 at $\{ij\}$-entry and 0 elsewhere.

5. $\mathcal{L} = \mathfrak{sl}(n)$, $f = \mathrm{diag}(1, \ldots, 1, 1-n)$. The corresponding decomposition is

$$\mathfrak{sl}(n) = \left\{\begin{pmatrix} * & \cdots & * & 0 \\ * & \cdots & * & 0 \\ \vdots & & \vdots & \vdots \\ * & \cdots & * & 0 \\ 0 & \cdots & 0 & * \end{pmatrix}\right\} + \left\{T\begin{pmatrix} * & 0 & \cdots & 0 & * \\ 0 & * & 0 & \cdots & * \\ \vdots & & \ddots & & \vdots \\ 0 & \cdots & 0 & * & * \\ 0 & \cdots & & 0 & * \end{pmatrix}T^{-1}\right\},$$

where T is the same as in the previous example.

6. $\mathcal{L} = \mathfrak{sl}(3)$. Using the Jordan normal form, consider the following cases:

1) f has distinct eigenvalues. This case was treated in [2] and above.

2) $f = \begin{pmatrix} 0 & 1 & 0 \\ 0 & 0 & 1 \\ 0 & 0 & 0 \end{pmatrix}$. This is a particular case of Example 3.

3) $f = \begin{pmatrix} 0 & 1 & 0 \\ 0 & 0 & 0 \\ 0 & 0 & 0 \end{pmatrix}$. In this case f is conjugated with $f_1 = \begin{pmatrix} 0 & 0 & 1 \\ 0 & 0 & 0 \\ 0 & 0 & 0 \end{pmatrix}$, and hence, this is a particular case of Example 2. Note that, due to results of [S], there exists a family of subalgebras satisfying the conditions of Proposition 5. They are of the form

$$A_{\lambda_1,\lambda_2,\lambda_3} = \left\{\begin{pmatrix} \lambda_1 t & 0 & 0 \\ * & \lambda_2 t & 0 \\ * & * & \lambda_3 t \end{pmatrix}\right\} \quad \text{where } \lambda_1 + \lambda_2 + \lambda_3 = 0.$$

4) $f = \begin{pmatrix} 1 & 0 & 0 \\ 0 & 1 & 0 \\ 0 & 0 & -2 \end{pmatrix}$. In this case, it is possible to choose A to be of the form $TA_{1,0,-1}T^{-1}$, where $A_{1,0,-1}$ is determined in (3) and $T = \begin{pmatrix} 1 & 0 & -1 \\ 0 & 1 & 0 \\ 0 & 0 & 1 \end{pmatrix}$.

5) $f = \begin{pmatrix} \lambda & 1 & 0 \\ 0 & \lambda & 0 \\ 0 & 0 & -2\lambda \end{pmatrix}$. In this case A may be chosen to be of the form

$$A = \left\{\begin{pmatrix} 1 & 0 & 1 \\ 0 & 1 & 0 \\ 0 & 0 & 1 \end{pmatrix}\begin{pmatrix} * & 0 & 0 \\ * & * & * \\ * & * & * \end{pmatrix}\begin{pmatrix} 1 & 0 & -1 \\ 0 & 1 & 0 \\ 0 & 0 & 1 \end{pmatrix}\right\}.$$

Appendix A

We would like to describe a relation between rational solutions of the classical Yang–Baxter equation (CYBE) and decompositions of $\mathfrak{sl}(n)$ obtained above. It is surprising that the decompositions of $\mathfrak{sl}(3)$ lead to rational solutions in $\mathfrak{o}(8)$ and $\mathfrak{e}_6$. First we remind some general facts concerning rational solutions.

Definition. Let $X(u, v) = \frac{c_2}{u-v} + r(u, v)$ be a function from $\mathbb{C}^2$ to $\mathfrak{g} \otimes \mathfrak{g}$, where $\mathfrak{g}$ is a simple Lie algebra over $\mathbb{C}$. We say that X satisfies the CYBE if:

1) c_2 is the quadratic Casimir element of $\mathfrak{g} \otimes \mathfrak{g}$ ($c_2 = \sum_\mu I_\mu \otimes I_\mu$, where I_μ is an orthogonal basis of $\mathfrak{g}$ with respect to the Killing form);
2) $r(u, v)$ is a polynomial in u, v;
3) $X(u, v) = -X(v, u)^\sigma$, where σ interchanges factors in $\mathfrak{g} \otimes \mathfrak{g}$;
4) $[X^{12}(u_1, u_2), X^{13}(u_1, u_3)] + [X^{12}(u_1, u_2), X^{23}(u_2, u_3)] + [X^{13}(u_1, u_3), X^{23}(u_2, u_3)] = 0$. Here $[a^{12}, a^{13}]$, where $a = \sum p_i \otimes q_i \in \mathfrak{g} \otimes \mathfrak{g}$, is defined as follows: Let $a^{12} = \sum p_i \otimes q_i \otimes 1 \in U(\mathfrak{g})^{\otimes 3}$ and $a^{13} = \sum p_i \otimes 1 \otimes q_i \in U(\mathfrak{g})^{\otimes 3}$. Then $[a^{12}, a^{13}]$ is the usual commutator in the associative algebra $U(\mathfrak{g})^{\otimes 3}$. It is easy to check that in fact $[a^{12}, a^{13}] \in \mathfrak{g}^{\otimes 3}$. The other two summands are defined in the same way.

Now we need a fact proved in [8]. Set $\mathbb{O} = \mathbb{C}[[u^{-1}]]$, $K = \mathbb{C}((u^{-1}))$, $P = \mathbb{C}[u]$.

Theorem 1. *There is a natural one-to-one correspondence between rational solutions of the CYBE and subspaces $W \subset \mathfrak{g} \otimes K$, such that*

1) *W is a subalgebra in $\mathfrak{g} \otimes K$;*
2) *$W \supset u^{-N} \cdot (\mathfrak{g} \otimes \mathbb{O})$ for some N;*
3) *$W \cap (\mathfrak{g} \otimes P) = \{0\}$;*
4) *$W \oplus (\mathfrak{g} \otimes P) = \mathfrak{g} \otimes K$;*
5) *$W = W^\perp$ with respect to the inner product $(x, y) = \operatorname{Res} \operatorname{tr}(\operatorname{ad} x \cdot \operatorname{ad} y)$ on $\mathfrak{g} \otimes K$.*

The proof is based on the following observation: Let $X(u, v) = \frac{c_2}{u-v} + r(u, v)$ be a rational solution of the CYBE. Let $f_{ms} = I_m u^{-s-1} \in u^{-1}(\mathfrak{g} \otimes \mathbb{O})$, where $\{I_m\}$ is the orthonormal basis of $\mathfrak{g}$ with respect to the Killing form. Let also $\ell_{mk} = I_m u^k \in \mathfrak{g} \otimes p$. Write $r(u, v)$ in the form $r(u, v) = \sum a_k u^k \otimes b_j v^j$, with a_k, b_j from $\mathfrak{g}$. Then W is generated by $\{f_{ms} + \sum (b_j v^j, f_{ms}) \cdot a_k u^k\} \in \mathfrak{g} \otimes K$. It turns out that the CYBE for $X(u, v)$ becomes the Jacobi identity for these elements, the property $W = W^\perp$ is equivalent to the skew-symmetry of X and the property $W \supset u^{-N} \cdot (\mathfrak{g} \otimes \mathbb{O})$ reflects the fact that r is a polynomial. Conversely, one can show that W, satisfying the conditions of Theorem 1, is of the form $W = \{f + \varsigma(f)\}$, where $f \in u^{-1}(\mathfrak{g} \otimes \mathbb{O})$ and $\varsigma : u^{-1}(\mathfrak{g} \otimes \mathbb{O}) \to \mathfrak{g} \otimes P$ is a linear map. Then $X = \sum_{m,s} (f_{ms} + \varsigma(f_{ms})) \otimes \ell_{ms}$.

Therefore, we can consider decompositions of $\mathfrak{g} \otimes K$ satisfying the conditions of Theorem 1 instead of rational solutions.

Developing this method, it is possible to construct infinitely many rational solutions in the case $\mathfrak{g} = \mathfrak{sl}(n)$. For other simple Lie algebras, much less is known. In [9] there were constructed rational solutions of the CYBE for $\mathfrak{o}(5)$, $\mathfrak{o}(7)$, $\mathfrak{o}(10)$, $\mathfrak{o}(14)$ and $\mathfrak{g}_2$.

Now we will construct a solution for $\mathfrak{g} = \mathfrak{o}(8)$ using an $\mathfrak{sl}(3)$-decomposition. First we need the following matrix realization of $\mathfrak{o}(n)$: $\mathfrak{o}(n) = \{$set of $n \times n$ matrices skew-symmetric with respect to the antidiagonal$\}$. Let also $p(n)$ be the maximal parabolic subalgebra of $\mathfrak{o}(n)$ with a common eigen-vector $e_n = \mathrm{col}[0, \dots, 0, 1] \in \mathbb{C}^n$. Put $H = \mathrm{diag}(u^{-1}, 1, \dots, 1, u) \in SO(n, K)$.

Theorem 2. *Let L be a subalgebra of $\mathfrak{o}(n)$ satisfying the condition $L \oplus p(n) = \mathfrak{o}(n)$. Then the subalgebra $W(L) = H^{-1}(L + u^{-1}L^{\perp} + u^{-2}(\mathfrak{o}(n) \otimes \mathbb{O}))H$ satisfies all the conditions of Theorem 1, and hence provides a rational solution of the CYBE. Here $L^{\perp}$ is the orthogonal complement to L with respect to the Killing form.*

Proof. It suffices to prove that $(L + u^{-1}L^{\perp} + u^{-2}(\mathfrak{o}(n) \otimes \mathbb{O})) \oplus H(\mathfrak{o}(n) \otimes P)H^{-1} = \mathfrak{o}(n) \otimes K$. The latter is equivalent to $(L + u^{-1}L^{\perp} + u^{-2}(\mathfrak{o}(n) \otimes \mathbb{O})) \oplus (H(\mathfrak{o}(n) \otimes P)H^{-1} \cap (\mathfrak{o}(n) \otimes \mathbb{O})) = \mathfrak{o}(n) \otimes \mathbb{O}$.

It is easily seen that $(H(\mathfrak{o}(n) \otimes P)H^{-1}) \cap (\mathfrak{o}(n) \otimes \mathbb{O}) = p(n) \oplus u^{-1}p(n)^{\perp}$. The condition $L \oplus p(n) = \mathfrak{o}(n)$ implies that $L^{\perp} \oplus p(n)^{\perp} = \mathfrak{o}(n)$, and therefore $W(L) \oplus (\mathfrak{o}(n) \otimes P) = \mathfrak{o}(n) \otimes K$.

Clearly, $W(L)$ is a subalgebra of $\mathfrak{o}(n) \otimes K$, which contains $u^{-3}(\mathfrak{o}(n) \otimes \mathbb{O})$. We have verified conditions 1–4 in Theorem 1. The last condition for $W(L)$ in Theorem 1 follows from the construction of $W(L)$. □

Let $\mathbb{C}^n$ be endowed with the symmetric bilinear non-degenerate form given by the formula: $B(x, y) = x_1 y_n + x_2 y_{n-1} + \cdots + x_n y_1$. Let G_1^n be the Grassman manifold of isotropic lines with respect to B. Then $G_1^n \cong SO(n, \mathbb{C})/P(n)$, where $P(n)$ is the parabolic subgroup of $SO(n, \mathbb{C})$ corresponding to the Lie subalgebra $p(n)$.

Corollary. *To construct a rational solution of the CYBE for $\mathfrak{o}(n, \mathbb{C})$, it is sufficient to find a subgroup $G \subset SO(n, \mathbb{C})$, which acts locally transitively on G_1^n without stationary subgroup.*

Now we are ready to describe the rational solution of the CYBE for $\mathfrak{o}(8)$, which arises from the following orbital decomposition of $\mathfrak{sl}(3) = P_1 \oplus \begin{pmatrix} 0 & a & b \\ 0 & 0 & a \\ 0 & 0 & 0 \end{pmatrix}$, where

$$P_1 = \begin{pmatrix} * & * & 0 \\ * & * & 0 \\ * & * & * \end{pmatrix}.$$

Theorem 3. *Consider the* Ad*-representation of* SL(3, ℂ) *on* $\mathfrak{sl}(3,\mathbb{C}) \cong \mathbb{C}^8$, *equipped with the Killing form. Then* $G(P_1)$ *acts locally transitively on* G_1^8 *without stationary subgroup, where* $G(P_1)$ *is the parabolic subgroup of* SL(3) *corresponding to* P_1.

Proof. Consider the point $s = \begin{pmatrix} 0 & 1 & 0 \\ 0 & 0 & 1 \\ 0 & 0 & 0 \end{pmatrix} \in \mathfrak{sl}(3)$. The line through this point is isotropic with respect to the Killing form. Therefore, the orbit of SL(3) through S is contained in G_1^8. Since the stationary subalgebra of S is isomorphic to $\begin{pmatrix} 0 & a & b \\ 0 & 0 & a \\ 0 & 0 & 0 \end{pmatrix}$, the orbit has dimension 6. Taking into account that dim $G_1^8 = 6$, we see that SL(3) acts locally transitively. The decomposition $\mathfrak{sl}(3) = P_1 \oplus \begin{pmatrix} 0 & a & b \\ 0 & 0 & a \\ 0 & 0 & 0 \end{pmatrix}$ shows that $G(P_1)$ acts also locally transitively and without stationary subgroup. □

It is possible to show that all the decompositions of $\mathfrak{sl}(3)$ considered above lead to rational solutions of the CYBE for $\mathfrak{e}_6$. The reduction in this case is more complicated, and we hope to discuss this in future publications.

Appendix B

It is well-known (see, for example, [Ko]) how to construct nonlinear equations associated to the given decomposition of the Lie algebra $\mathcal{L}$ into the direct sum of two subalgebras $\mathcal{L} = \mathcal{L}_1 + \mathcal{L}_2$. The procedure which is called the Adler–Kostant–Symes scheme is as follows.

Let $B(\cdot,\cdot)$ be an invariant nondegenerate form on $\mathcal{L}$ and $\mathcal{L} = \mathcal{L}_1^\perp + \mathcal{L}_2^\perp$ be a decomposition of $\mathcal{L}$ into into the direct sum of linear subspaces orthogonal to $\mathcal{L}_1$ and $\mathcal{L}_2$ w.r.t B. Then the pull-back of the Lie–Kirillov–Kostant Poisson structure on $\mathcal{L}_1^*$ and $\mathcal{L}_2^\perp$ gives one the following formula for the Poisson bracket of functions $f_1, f_2 \in C^\infty(\mathcal{L}_2^\perp)$:

$$\{f_1, f_2\}(x) = B(x, [\pi_1(\nabla f_1), \pi_1(\nabla f_2)]), \tag{15}$$

where π_1 is the projection onto $\mathcal{L}_1$ along $\mathcal{L}_2$. Then, by the Adler–Kostant–Symes theorem, restrictions to $\mathcal{L}_2^\perp$ of any two Ad-invariant functions are in involution w.r.t. (15) and the equation of motion that corresponds to the Ad-invariant Hamiltonian H is given by

$$\dot{x} = [x, \pi_1(\nabla H)]. \tag{16}$$

We may apply the Adler–Kostant–Symes scheme to decompositions considered in the end of part II. For example, for the decomposition 5,

$$\mathfrak{sl}(n) = \left\{ \begin{pmatrix} * & \cdots & * & 0 \\ * & \cdots & * & 0 \\ \vdots & & \vdots & \vdots \\ * & \cdots & * & 0 \\ 0 & \cdots & 0 & * \end{pmatrix} \right\} + \left\{ T \begin{pmatrix} * & 0 & \cdots & 0 & * \\ 0 & * & 0 & \cdots & * \\ \vdots & & \ddots & & \vdots \\ 0 & \cdots & 0 & * & * \\ 0 & \cdots & & 0 & * \end{pmatrix} T^{-1} \right\},$$

where $T = \mathrm{id} + E_{1n} + E_{2n} + \cdots + E_{n-1,n}$. We have

$$\mathcal{L}_2^{\perp} = \left\{ T \begin{pmatrix} L & y \\ 0 & 0 \end{pmatrix} T^{-1} \right\},$$

where $L = (l_{jk})$ is a $(n-1) \times (n-1)$ off-diagonal matrix and $y = (y_i)$ is an $n-1$-dimensional vector-column. Here we assume $B(X, Y)$ to be equal to $\mathrm{tr}(XY)$. In this case the Poisson brackets (15) for variables l_{jk}, y_i has the form

$$\{y_i, y_j\} = 0, \{y_i, l_{jk}\} = (\delta_i^k - \delta_i^j)(l_{jk} - y_j)$$
$$\{l_{jk}, l_{ps}\} = \delta_j^s l_{pk} - \delta_p^k l_{js} + (\delta_p^j - \delta_p^k) l_{jk} - (\delta_p^j - \delta_s^j) l_{ps}.$$

Equation (16) that corresponds to the simplest nontrivial Ad-invariant Hamiltonian $H = \mathrm{tr}(L^2)$ is equivalent to the nonlinear system

$$\dot{l}_{jk} = l_{jk} \sum_{i=1}^{n-1} (l_{ik} - l_{ij}),$$

$$\dot{y}_i = \sum_{j=1}^{n-1} l_{ij} y_j - \sum_{k=1}^{n-1} l_{ki} y_i.$$

References

[1] L. Abellanos and Alonso L. Martinez, Quantization from the algebraic viewpoint, J. Math. Phys. 17 (1976), 1363–1365.

[2] F. Bayen, M. Flato, C. Fronsdal, A. Lichnerowicz and D. Sternheimer, Deformation theory and quantization. I, Deformations of symplectic structures, Ann. Physics 110 (1978), 67–110.

[3] F. A. Berezin, Several remarks on the associative hull of a Lie algebra, Functional. Anal. Appl. 1 (1967), 1–14.

[4] V. G. Drinfeld, Hamiltonian structures on Lie groups, Lie bialgebras and the geometrical meaning of the classical Yang–Baxter equations, Soviet Math. Dokl. 27 (1983), 68–71.

[5] V. G. Drinfeld, On constant, quasiclassical solutions of the Yang–Baxter quantum equation, Soviet Math. Dokl. 28 (1983), 667–671.

[6] A. Elashvili, Frobenius Lie algebras, Functional Anal. Appl. 16 (1982), 94–95.

[7] S. Majid, Quasitriangular Hopf algebras and Yang–Baxter equations, Internat. J. Modern Phys. 5 (1990), 1–91.

[8] Stolin, A., On rational solutions of Yang–Baxter equation for $sl(n)$, Math. Scand. 69 (1991), 56–80.

[9] A. Stolin, On rational solutions of Yang–Baxter equations, Maximal orders in loop algebras, Commun. Math. Phys. 141 (1991), 533–548.

[10] L. A. Takhtajan, Introduction to quantum groups, in: Quantum groups (Clausthal, 1989), Lecture Notes in Phys. 370, 1990, 3–28.

Department of Theoretical Mathematics
The Weizmann Institute of Science
Rehovot 76100, Israel

Current address:
Department of Mathematics
University of Michigan
Ann Arbour, MI 48109-109, USA
e-mail: `gekhtman@math.lsa.umich.edu`

Department of Mathematics
Royal Institute of Technology
10044 Stockholm, Sweden
e-mail: `astolin@math.kth.se`

On the finite basis property of T-spaces over a field of finite characteristic

*A. V. Grishin**

Let $F = k\langle x_1, \ldots, x_i, \ldots \rangle$ be the free countably generated associative algebra over an infinite field k, and let T be the semigroup of endomorphisms (substitutions) of the algebra F.

The endomorphism τ of the algebra F defined by the correspondence $x_i \longmapsto g_i$, where $g_i \in F$, is called a *substitution of type* $(x_1, \ldots, x_i, \ldots) \mapsto (g_1, \ldots, g_i, \ldots)$. We indicate by f^τ the image of the polynomial f with respect to the substitution τ.

The substitution τ is called *regular* if τ is of type $(x_1, \ldots, x_i, \ldots) \mapsto (g_1 x_1 h_1, \ldots, g_i x_i h_i, \ldots)$, where $g_i, h_i \in k\langle 1, x_1, \ldots, x_i, \ldots \rangle$. It is clear that the set T^* of regular substitutions is a subsemigroup of the semigroup T.

Following [1], we say that a linear subspace V of the algebra F is a T*-space* of F if V is closed under substitutions, i.e, if $f \in V$, $\tau \in T$, then $f^\tau \in V$. Denote by S^T the T-space generated by some subset S of some T-space.

Analogously we can define a T^*-space of the algebra F and denote by S^{T^*} the T^*-space generated by S.

In particular a T*-ideal* (T^**-ideal*) of the algebra F is an ideal which is also a T-space (T^*-space).

We can define in an obvious way the right action of the semigroups T and T^* on any relatively free algebras. Therefore, we can define a T-space (T^*-space) in this case.

Denote by $k[x_1, \ldots, x_i, \ldots] = k\langle x_1, \ldots, x_i, \ldots \rangle / [F, F]$ the algebra of commutative polynomials (the image of x_i in $F/[F, F]$ is denoted for simplicity by the same letter).

The T-space (T^*-space) V is called *finitely based* if $V = \{S\}^T$, where S is finite subset of V.

Define the Capelli polynomial

$$c_m = c_m(x_0, \ldots, x_m, y_0, \ldots, y_m) = \sum_{G' \in \mathrm{Sym}(m)} (-1)^\sigma y_0 x_{\sigma(1)} y_1 \ldots x_{\sigma(m)} y_m.$$

*Partially supported by the Russian Foundation for Fundamental Research.

If $\operatorname{char} k = 0$, then every T-space of the algebra $F/\{c_m\}^T$ is finitely based (see [2]).

In the case of characteristic $p > 0$ the situation is not clear enough (I think that there are many counter-examples).

But in the commutative case we have the following result.

Proposition 1. *Let k be a field of characteristic $p > 0$. Then*

1) *every T^*-space of $k[x_1, \dots, x_i, \dots]$ generated by polynomials of the algebra $k[x_1, \dots, x_d]$ is finitely based;*
2) *every T-space of $k[x_1, \dots, x_i, \dots]$ is finitely based.*

Proof. As k is infinite we can assume that given a T-space (T^*-space) is generated by monomials (of course, not necessarily multilinear).

1) We consider the following action on monomials $f = x_1^{m_1} \dots x_d^{m_d}$ (the particular case of construction is contained in [1]). Let $t_1 = x_{d+1}, \dots, t_d = x_{2d}$ and let $n_1, \dots, n_d$ be any positive integers. Denote by $f^{(n_1,\dots,n_d)}$ the $(n_1, \dots, n_d)$-homogeneous component of the polynomial

$$(x_1(1+t_1))^{m_1} \dots (x_d(1+t_d))^{m_d}.$$

It is clear that $f^{(n_1,\dots,n_d)} \in \{f\}^{T^*}$ and

$$f^{(n_1,\dots,n_d)} = \left(\binom{m_1}{n_1}\right) \dots \left(\binom{m_d}{n_d}\right) t_1^{n_1} \dots t_d^{n_d} f.$$

Let $m_i = p^{s_i} l_i$ where $(l_i, p) = 1$. It is easy to show that $\left(\binom{m_i}{p_i^{s_i}}\right)$ is not divisible by p. Indeed

$$\left(\binom{m_i}{p^{s_i}}\right) = \frac{p^{s_i} l_i (p^{s_i} l_i - 1) \dots (p^{s_i} l_i - p^{s_i} + 1)}{p^{s_i} (p^{s_i} - 1) \dots (p^{s_i} - p^{s_i} + 1)}$$

where the numbers $p^{s_i} l_i - p^r a$ and numbers $p^{s_i} - p^r a$ (where $1 \le r \le s_i - 1$, $p^{s_i} < p^{s_i}$, $(a, p) = 1$) contain p in the same powers r, and the other members of this expression are not divisible by p. Therefore, $f^* = t_1^{p_1^{s_1}} \dots t_d^{p_d^{s_d}} f \in \{f\}^{T^*}$.

It is clear that every monomial

$$(t_1 x_1^{n_1})^{p^{r_1}} \dots (t_d x_d^{n_d})^{p^{r_d}},$$

where $l_i \le n_i$ and $s_i \le r_i$ for all $i = 1, \dots, d$, may be obtained from f^* by using the regular substitutions. It is a well-known fact that in every set of vectors $(n_1, r_1, \dots, n_d, r_d)$, where $n_i, r_i \in \mathbb{Z}$ and $n_i \ge 0$ and $r_i \ge 0$, there exists a finite set of minimal elements (in the sense of $(n_1, r_1, \dots, n_d, r_d) \prec (n_1', r_1', \dots, n_d', r_d') \Leftrightarrow n_i \le n_i', r_i \le r_i'$). Therefore every set S of monomials f^* generates a finitely based T^*-space.

Finally, note that if the monomials $f_1, \dots, f_q$ do not contain the elements $t_1, \dots, t_d$ and $f^* \in \{f_1, \dots, f_q\}^{T^*}$, then $f = f^*|_{t_1=\dots=t_d=1} \in \{f_1, \dots, f_q\}^{T^*}$.

2) Let a set of monomials depend on an infinite set of variables. Choosing the names of variables suitably (the substitution is not regular) we can assume that $f = x_1^{m_1} \dots x_d^{m_d}$, where $m_1 \geq \dots \geq m_d$, and carry out the proof by analogy with 1). The proposition is proved. □

References

[1] A. V. Grishin, On the finite basis property for systems of generalized polynomials, Izv. Akad. Nauk SSSR Ser. Mat. 54 (1990), 859–927 (in Russian).

[2] A. V. Grishin, On the finite basis property of abstract T-space, Third International Algebraic Conf., Krasnojarsk, 1993 (in Russian).

Department of Mechanics and Mathematics
Chair of Higher Algebra
Moscow State University
Moscow 119899, Russia

On the conjecture of the resolution of a finite set of points in projective space

Ngau Lam

1. Introduction

In this paper, we will fix the following notations. Let $X \subseteq \boldsymbol{P}^r$ be a finite set of distinct points, not contained in any hyperplane, with reduced structure. Let I_X be the homogeneous ideal of X and $\mathcal{I}_X$ be the ideal sheaf of X. We work over an algebraically closed field K of any characteristic.

A special case of a conjecture of M. Green and R. Lazarsfeld (see [GL2]) says that if $X \subseteq \boldsymbol{P}^r$ with $|X| \leq 2r$ and if for all $1 \leq s \leq r$, no $2s+1$ points of X lie in a projective s-plane, then I_X is generated by quadrics. We will prove this conjecture, see Theorem 3 in Section 2, by using a result in Matroid Theory due to J. Edmonds [Ed] and A. Horn [Ho]. Using Matroid Theory, D. Eisenbud and J. Koh [EK] have shown: If $X \subseteq \boldsymbol{P}^r$ with $|X| \leq 2r$ and for all $1 \leq s \leq r$, no $2s+1$ points of X lie in a projective s-plane, then X is scheme theoretically cut out by quadrics.

As a consequence of our theorem, we can give a new proof of a theorem due to Fujita [Fji] and St. Donat [St. D], strengthening earlier work of Mumford [M] which deals with line bundles of slightly larger degree: Let C be a smooth curve of g. If C is embedded into a projective space by a complete linear system of a line bundle with degree at least $2g+2$, then the homogeneous ideal of C is generated by quadrics. (Also see [EKS], [G], [GL1], [GL2] and [KS].)

Let $E_\bullet$ be a minimal graded free resolution of I_X over the homogeneous co-ordinate ring S of $\boldsymbol{P}^r$:

$$0 \to E_r \to \cdots \to E_2 \to E_1 \to I_X \to 0,$$

where $E_i = \bigoplus S(-a_{ij})$.

For a given integer $p \geq 1$, M. Green and R. Lazarsfeld [GL2] asked whether X satisfies the following property:

$$(N_p) \qquad E_i = \bigoplus S(-i-1) \qquad \text{(that is } a_{ij} = i+1 \text{ for all j)} \quad \text{for } 1 \leq i \leq p.$$

In particular, X satisfies N_1 iff I_X is generated by quadrics.

Their main result is:

Theorem ([GL2]). *If $X \subseteq \mathbf{P}^r$ consists of $\leq 2r + 1 - p$ points in linear general position, then I_X satisfies property (N_p).*

Also they make the following conjecture giving a sharp condition for the property (N_p):

Conjecture ([GL2]). If $X \subseteq \mathbf{P}^r$ fails to satisfies (N_p), then there is an integer $s \leq r$ and a subset $Y \subseteq X$ consisting of at least $2s + 2 - p$ points such that Y is contained in a linear subspace $\mathbf{P}^s \subseteq \mathbf{P}^r$ such that (N_p) fails for $Y \subseteq \mathbf{P}^r$. In particular, if $X \subseteq \mathbf{P}^r$ with $|X| \leq 2r$ and for all $1 \leq s \leq r$, no $2s + 1$ points of X lie in a projective s-plane, then I_X is generated by quadrics.

In fact, see Proposition 4 below, $X = H \cap C$, where H is a general hyperplane section of a smooth curve C of genus g which is embedded into $\mathbf{P}^{r+1}$ by the complete linear system of a line bundle with degree at least $2g + 1 + p$, will satisfy for all $p + 1 \leq s \leq r$, no $2s + 2 - p$ points of X lie in a s-plane and for all $0 \leq s \leq p$, no $p + 2$ points of X lie in a s-plane. So their conjecture will give a stronger statement about the minimal graded free resolution of the homogeneous ideal I_X of a curve C which is embedded into $\mathbf{P}^r$ by the complete linear system of a line bundle with degree at least $2g + 1 + p$.

The author is grateful to Professor David Eisenbud for his extremely helpful discussions and encouragement.

2. Small numbers of points in projective space

Theorem 1 ([Ed], [Ho]). *Let X be a finite set of zero-cycle in $\mathbf{P}^r$. If no k-plane in $\mathbf{P}^r$ contains $> 2k + 2$ points of X counting with multiplicity, then $X = A \cup B$ where A and B are linearly independent sets.*

Theorem 1 is the main tool from Matroid Theory that we will use to prove our theorems. First we would like to give a new proof of a theorem in [EK1] here to motivate how to use Theorem 1 to prove our main result.

Theorem 2. *Let $X \subseteq \mathbf{P}^r$ be a finite set of distinct points with reduced structure and suppose that X is not contained in any hyperplane. If for all $1 \leq s \leq r$, no $2s + 2$ points of X lie in a s-plane, then X imposes independent conditions on quadrics.*

Proof. For a fixed point $p \in X$, let $Y_p = X \cup \{p\}$ (Y_p is really a 0-cycle where the point p has multiplicity 2.). By Theorem 1, we can divide Y_p into two sets A_p and B_p

such that each of A_p and B_p consists of linearly independent points. Since p occurs twice, we know that $p \in A_p$ and $p \in B_p$. Since A_p is a linearly independent set, so we can choose a linear form x_p such that $x_p(A_p \backslash \{p\}) = 0$ such that x_p does not vanish on p. Similarly we can choose a linear form y_p such that $y_p(B_p \backslash \{p\}) = 0$ and y_p does not vanish on p. Therefore the quadric $x_p y_p$ vanishes on all the points of X except p. So for any point $p \in X$, we can find a 2-form h such that h vanishes on all the points of X except p. Hence X imposes independent conditions on quadrics. □

Theorem 3. *Let $X \subseteq \boldsymbol{P}^r$ be a finite set of distinct points with reduced structure and suppose that X is not contained in any hyperplane. If for all $1 \leq s \leq r$, no $2s+1$ points of X lie in a s-plane, then I_X is generated by quadrics.*

Proof. By Theorem 2, we know that X imposes independent conditions on quadrics; that is the natural map $H^0(\boldsymbol{P}^r, \mathcal{O}_{\boldsymbol{P}^r}(2)) \to H^0(\boldsymbol{P}^r, \mathcal{O}_X(2))$ is surjective. It follows that $H^i(\boldsymbol{P}^r, \mathcal{I}_X(3-i)) = 0$ for $i \geq 1$; that is the ideal sheaf $\mathcal{I}_X$ is 3-regular in the sense of Mumford [M1]. Since $\mathcal{I}_X$ is 3-regular, by a theorem of Mumford [M1] we have I_X is generated by elements with degree less than or equal to 3. Therefore to show I_X is generated by quadrics is equivalent to show $\mathrm{Tor}^S_{0,3}(I_X, K) = 0$ where S is the coordinate ring of $\boldsymbol{P}^r$ and K is the ground field. But $\mathrm{Tor}^S_{0,3}(I_X, K) = \mathrm{Tor}^S_{1,3}(S/I_X, K)$ and $S/I_X = K \oplus H^0(\boldsymbol{P}^r, \mathcal{O}_{\boldsymbol{P}^r}(1)) \oplus_{n\geq 2} H^0(\boldsymbol{P}^r, \mathcal{O}_X(n))$ by Theorem 2, therefore it is sufficient to show that the following sequence is exact.

$$\bigwedge^2 V \otimes V \xrightarrow{\alpha} V \otimes H^0(\boldsymbol{P}^r, \mathcal{O}_X(2)) \xrightarrow{\beta} H^0(\boldsymbol{P}^r, \mathcal{O}_X(3)),$$

where $V = H^0(\boldsymbol{P}^r, \mathcal{O}_{\boldsymbol{P}^r}(1))$, α and β are Koszul maps.

For each $p \in X$, we fix an isomorphism between $H^0(\boldsymbol{P}^r, \mathcal{O}_p(1))$ and K. Since $H^0(\boldsymbol{P}^r, \mathcal{O}_X(1)) = \bigoplus_{p\in X} H^0(\boldsymbol{P}^r, \mathcal{O}_p(1))$, so we can identify a homogeneous form as a function on X by the isomorphism between $H^0(\boldsymbol{P}^r, \mathcal{O}_p(1))$ and K for all point $p \in X$. For each $p \in X$, we define $V_p = \{x \in V | x(p) = 0\}$ and let e_p be the element in $H^0(\boldsymbol{P}^r, \mathcal{O}_X(2))$ supported at p such that $e_p(p) = 1$. Because $H^0(\boldsymbol{P}^r, \mathcal{O}_X(1)) = \bigoplus_{p\in X} H^0(\boldsymbol{P}^r, \mathcal{O}_p(1))$, we have $\mathrm{Ker}\,\beta = \bigoplus_{p\in X} V_p \otimes e_p$. Thus it enough to show that for each $p \in X$, $V_p \otimes e_p$ is contained in $\mathrm{Im}\,\alpha$.

Now we fix $p \in X$. For any $q \in \boldsymbol{P}^r$ such that $q \neq p$, let $Y_q = X \cup \{p, q\}$ (Y_q is really a 0-cycle where p and possibly q have multiplicity 2.). By Theorem 1, we can divide Y_q into two sets A_q and B_q such that each of A_q and B_q consists of linearly independent points. Relabeling it if necessary, we can assume $q \in A_q$ and since p occurs twice, we know that $p \in A_q$ and $p \in B_q$. For any $a \in A_q$, we choose a linear form $x_{q,a}$ such that $x_{q,a}(A_q \backslash \{a\}) = 0$ and $x_{q,a}(a) = 1$. For any $b \in B_q$, we choose a fixed linear form $y_{q,b}$ such that $y_{q,b}(B_q \backslash \{b\}) = 0$ and $y_{q,b}(b) = 1$. We have

$$\alpha(x_{q,p} \wedge x_{q,q} \otimes y_{q,p}) = x_{q,p} \otimes f_*(x_{q,q} y_{q,p}) - x_{q,q} \otimes f_*(x_{q,p} y_{q,p}) = x_{q,q} \otimes e_p$$

because one of $x_{q,q}$ and $y_{q,p}$ vanishes at each points of X, where

$$f_* : H^0(P^r, \mathcal{O}_{P^r}(2)) \to H^0(P^r, \mathcal{O}_X(2))$$

is the natural map. This shows $x_{q,q} \otimes e_p \in \operatorname{Im} \alpha$.

Define $W = \{x_{q,q} \in V | q \in P^r, q \neq p\}$. We will show $W = V_p$. Suppose $W \neq V_p$, we can find an element c in the zero set of W such that $c \neq p$ since $W \subset V_p$ and $W \neq V_p$. Thus we have $Y_c = A_c \cup B_c$ and $x_{c,c} \in W$ such that $x_{c,c}(c) \neq 0$. We get a contradiction to c belongs to the zero set of W. Hence we have $W = V_p$. This shows $V_p \otimes e_p \subset \operatorname{Im} \alpha$ and completes the proof of the theorem. □

3. Applications to algebraic curves

Proposition 4. *Let C be a smooth curve of genus g. Suppose that C is embedded into P^{r+1} by a complete linear system of a line bundle L and H is a hyperplane section such that $H \cap C$ is a reduced scheme. If $\deg L = 2g + 1 + p$ with $p \geq 0$, then for all $p + 1 \leq s \leq r$, no $2s + 2 - p$ points of X lie in a s-plane and for all $0 \leq s \leq p$, no $p + 2$ points of X lie in a s-plane.*

Proof. First we show that for all $0 \leq s \leq p$, no $s + 2$ points of X lie in a s-plane. Otherwise let $Y \subset X$ such that Y spans a s-plane and $|Y| > s + 1$. We choose a subset Z of Y such that $|Z| = s + 2$ and Z spans the s-plane. By the Riemann–Roch Theorem, we have $h^0(C, L(-Z)) = r + 1 - s$. On the other hand, by the Riemann–Roch Theorem we have

$$h^0(C, L(-Z)) = \deg L - |Z| + 1 - g = r + 2 - |Z| = r - s$$

since $\deg L(-Z) = 2g + 1 + p - (s + 2) \geq 2g - 1$. We get a contradiction.

Now we are going to show that for all $p+1 \leq s \leq r$, no $2s+2-p$ points of X lie in a s-plane. Otherwise let $Y \subset X$ such that Y spans a s-plane and $|Y| \geq 2s+2-p$ with $p+1 \leq s \leq r$. By the Riemann–Roch Theorem, we have $h^0(C, L(-Y)) = r+1-s$. Therefore

$$\begin{aligned} \deg L(-Y) - 2(h^0(C, L(-Y)) - 1) &\leq \deg L - 2s - 2 + p - 2(r + 1 - s - 1) \\ &= \deg L - 2r - 2 + p \\ &= \deg L - 2 + p - 2(\deg L - g - 1) \\ &= 2g + p - \deg L \\ &< 0. \end{aligned}$$

By Clifford's Theorem, we have $h^1(C, L(-Y)) = 0$. Hence

$$\begin{aligned} h^0(C, L(-Y)) - h^1(C, L(-Y)) - \deg L(-Y) - 1 + g & \\ &= r + 1 - s - \deg L + |Y| - 1 + g \\ &> r + 1 - s - \deg L + (2s + 1 - p) - 1 + g \\ &= r + 1 + s - p - (r + g) + g \\ &= s - p - 1 \\ &\geq 0. \end{aligned}$$

We get $h^0(C, L(-Y)) - h^1(C, L(-Y)) > \deg L(-Y) + 1 - g$ which contradicts to the Riemann–Roch Theorem. So for all $p + 1 \leq s \leq r$, no $2s + 2 - p$ points of X lie in a s-plane and completes the proof of the proposition. □

From this we get a new proof of following two results. First we start to prove a well-known result of Castelnuovo [C], Mattuck [Mat] and Mumford [M]. Readers can find a sharper result of the following corollary in [GL1] for characteristic zero and in [KS] for any characteristic.

Corollary 5. *Let C be a smooth curve of genus g. If C is embedded into a projective space by a complete linear system of a line bundle with degree at least $2g + 1$, then C is projectively normal.*

Proof. Suppose that C is embedded into a projective space $\mathbf{P}^{r+1}$ by the complete linear system of L with degree of $L \geq 2g + 1$. Let H be a fixed general hyperplane such that $H \cap C$ is a reduced scheme. Let $X = H \cap C$. By Proposition 4 and Theorem 2, we know that X imposes independent conditions on quadrics; that is the natural map

$$H^0(\mathbf{P}^r, \mathcal{O}_{\mathbf{P}^r}(2)) \to H^0(\mathbf{P}^r, \mathcal{O}_X(2))$$

is surjective. Therefore the natural maps $H^0(\mathbf{P}^r, \mathcal{O}_{\mathbf{P}^r}(n)) \to H^0(\mathbf{P}^r, \mathcal{O}_X(n))$ are surjective for $n \geq 2$. Thus the natural maps $H^0(\mathbf{P}^{r+1}, \mathcal{O}_C(n)) \to H^0(\mathbf{P}^{r+1}, \mathcal{O}_X(n))$ is also surjective for $n \geq 2$.

Consider the following commutative diagram:

$$\begin{array}{ccccccccc} 0 & \to & H^0(\mathcal{O}_{\mathbf{P}^{r+1}}(n-1)) & \to & H^0(\mathcal{O}_{\mathbf{P}^{r+1}}(n)) & \to & H^0(\mathcal{O}_{\mathbf{P}^r}(n)) & \to & 0 \\ & & \downarrow & & \downarrow & & \downarrow & & \\ 0 & \to & H^0(\mathcal{O}_C(n-1)) & \to & H^0(\mathcal{O}_C(n)) & \to & H^0(\mathcal{O}_X(n)) & \to & 0 \end{array}$$

Let φ_{n*} be the map $H^0(\mathbf{P}^{r+1}, \mathcal{O}_{\mathbf{P}^r}(n)) \to H^0(\mathbf{P}^{r+1}, \mathcal{O}_C(n))$. Row sequences are exact for $n \geq 2$ and the right vertical map is surjective for $n \geq 2$ in the commutative diagram. For fixed $n \geq 2$, if the left vertical map $\varphi_{(n-1)*}$ is surjective, then the middle map φ_{n*} is also surjective. Since the map φ_{1*} is surjective, by induction we know that the maps φ_{n*} are surjective for $n \geq 1$. Hence C is projective normal in $\mathbf{P}^{r+1}$. □

The following corollary is due to Fujita [Fji] and St. Donat [St. D], strengthening earlier work of Mumford [M] which deals with line bundles of slightly larger degree. Some conjectures related to the following theorem and Clifford index will refer to [GL1].

Corollary 6. *Let C be a smooth curve of g. If C is embedded into a projective space by a complete linear system of a line bundle with degree at least* $2g+2$*, then the homogeneous ideal of C is generated by quadrics.*

Proof. Suppose that C is embedded into a projective space P^{r+1} by the complete linear system of L with degree of $L \geq 2g+2$. Let H be a fixed general hyperplane such that $H \cap C$ is a reduced scheme. C is projectively normal in P^{r+1} by Corollary 5. Thus the ideal of C is generated by quadrics iff the ideal of $H \cap C$ in P^r is generated by quadrics.

Let $X = H \cap C$ and I_X be the homogeneous ideal of X in P^r. By Proposition 4, we know that for all $1 \leq s \leq r$, no $2s+1$ points of X lie in a s-plane. By Theorem 3, we have I_X is generated by quadrics and completes the proof of the corollary. □

References

[ACGH] E. Arbarello, M. Cornalba, P. Griffiths and J. Harris, Geometry of Algebraic Curves, Springer Verlag, New York 1984.

[BS] D. Bayer and M. Stillman, Macaulay, a computer algebra program, (1982–1990).

[C] G. Castelnuovo, Sui Multipli di uni serie lineare di gruppi di punti appartenente ad una curva algebrica, Rend. Circ. Mat. Palermo 7 (1893), 99–119.

[EK] D. Eisenbud and J. Koh, Remarks on points in projective space in: Commutative Algebra, Conf. Proc. of the 1987 conference in Berkeley, ed. M. Hochster, C. Huneke and J. D. Sally, Math. Sciences Res. Inst. Publ. 15 (1989), 157–172.

[EKS] D. Eisenbud, J. Koh and M. Stillman, Determinantal equations for curves of high degree, Amer. J. Math 110 (1988), 513–539.

[Fji] T. Fujita, Defining equations for certain types of polarized Varieties, Complex Analysis and Algebraic Geometry, Cambridge Univ. Press, 1977, 361–366.

[G] M. Green, Koszul cohomology and the geometry of projective varieties, J. Differential Geometry 19 (1984), 125–171.

[GL1] M. Green and R. Lazarsfeld, On the projective normality of complete linear series on an algebraic curve, Invent. Math. 83 (1986), 73–90.

[GL2] M. Green and R. Lazarsfeld, Some results on the syzygies of finite sets and algebraic curves, Compositio Math. 67 (1988), 301–314.

[Ho] A. Horn, A characterization of unions of linearly independent sets, J. London Math. Soc. 30 (1955), 494–496.

[Ha] R. Hartshorne, Algebraic Geometry, Springer-Verlag, New York 1977.

[KS] J. Koh and M. Stillman, Linear sygzgies and line bundles on an algebraic curve, J. Algebra 125 (1989), 120–132.

[M] D. Mumford, Varieties defined by quadratic equations, Corso C.I.M.E., 1969, in questions on algebraic varieties. Rome, Cremonese, 30–100, 1970.

[M1] D. Mumford, Lectures on curves on an algebraic surface, Ann. Math. Stud. 59 (1966).

[Mat] A. Mattuck, Symmetric products an Jacobians, Amer. J. Math. 83 (1961), 189–206.

[St. D] B. St.-Donat, Sur les equations definissant une courbe algebrique, C.R. Acad. Sci. Paris, Ser. A 274 (1972), 324–327.

Department of Mathematics
National Cheng Kung University
Tainan
Taiwan 701, ROC

Lie nilpotency: recognition and word problem

*V. N. Latyshev**

In this note we deal with the Lie nilpotent associative algebras and emphasize, that the related algorithms are along the line of Gröbner basis theory.

Let $K\langle X\rangle$ denotes free associative algebra over field K with free generators $X = \{x_1, \ldots, x_n\}$. The set X also generates under bracket operation $[y_1, y_2] = y_1y_2 - y_2y_1$ the free associative algebra $K_0\langle X\rangle^{(-)}$ over the field K. We consider the empty word 1 as belonging to $\langle X\rangle$. The "long" commutator (right-normal product) is defined by induction: $[y_1, \ldots, y_r] = [[y_1, \ldots, y_{r-1}], y_r]$. It may be regarded as a "non-associative monomials". The elements of the Lie algebra $K_0\langle X\rangle^{(-)}$ are called Lie elements of the free associative algebra $K\langle X\rangle$. The elements of the linear span of the products of the long commutators are called eigen elements. For example, the Lie elements are eigen. It is well known fact that any total characterestric ideal (T-ideal) is generated by its eigen elements. (In what follows we suppose, that char $K = 0$.) The reader is basically assumed to be aquainted with the notion of Gröbner basis. The necessary background is available from the textbooks and monogfraphs on computer algebra. The algebra A is said to be s.f.p. (standard finitely presented) algebra if its ideal of defined relations is generated by a Gröbner basis with finite number of members. Obviously this involves the solvability of the word problem in A. The property P of algebra is said to be recognizable, if there is an algorithm for checking whether an arbitrary s.f.p. algebra A satisfies P or not. The property of Lie nilpotency of algebra A implies the polynomial identity $[y_1, \ldots, y_r] = 0$ holds in it for some integer r. The commutator ideal of such algebra is nilpotent of some index $q \leq q(r, n)$, where n is a number of generators of A, and the function $q(r, n)$ may be computed algorithmically.

Theorem 1. *Lie nilpotency of a fixed degree r is a recognizable property.*

Proof. Let us consider the following types of eigen elements belonging to the T-ideal $T^{(r)} \lhd K\langle X\rangle$ defined by commutator $[y_1, \ldots, y_r]$ (as above, q is the index of nilpotency of the commutator ideal in the algebra $K\langle X\rangle/T^{(r)}$):

(i) $[x_{i_1}, \ldots, x_{i_r}]$;

(ii) $[x_{i_1}, \ldots, x_{i_{s_1}}] \ldots [x_{j_1}, \ldots, x_{j_{s_q}}]$; (1)

(iii) $\{f_1, \ldots, f_t\}$ is the basis of the linear subspace in $T^{(r)}$ generated by eigen elements of degree $\leq D = \max\{q(r-1), r\}$.

*Partially supported by the Russian Foundation for Fundamental Research.

It is not so difficult to verify that these elements generate $T^{(r)}$ as an ideal. Thus, an algebra $A = K\langle a_1, \ldots, a_n\rangle$ with n generators a_i satisfies the polynomial identity $[y_1, \ldots, y_r] = 0$ iff the eigen polynomials (1) vanish by all substitutions of x_i by a_j. This yields the needed algorithm or recognition. □

The variety $V^{(r)}$ of Lie nilpotent algebras of a fixed degree r is finitely approximable and then the word problem is solvable in it. But we intend to indicate that this problem can be solved by means of an algorithm using the notion of Gröbner bases.

Of course, the problem under consideration is equivalent to the membership (to ideals) problem in free algebra A of the variety $V^{(r)}$, which is given as the factor algebra $A \cong K\langle X\rangle/T^{(r)}$, i.e. it is finitely presented, and the noncommutative polynomials (1) are its defined relations. Let $\varphi : K\langle X\rangle \to K\langle X\rangle/T^{(r)}$ be the natural epimorphism. The image $L = \varphi(k_0\langle X\rangle^{(-)})$ of the free Lie algebra $k_0\langle X\rangle^{(-)}$ in A is finite dimensional, because all homogeneous Lie elements of degree $\geq r$ belong to the T-ideal $T^{(r)}$. If $\xi_1 = \varphi(g_1), \ldots, \xi_m = \varphi(g_m) \in L$ is a basis of Lie algebra L, $g_i \in k_0\langle X\rangle^{(-)}$, $m = \dim L$, then the elements g_i may be chosen in such a way that $g_1 = x_1, \ldots, g_n = x_n$ and $g_{n+1}, \ldots, g_m$ are homogeneous Lie elements of degree $< r$. Taking into account the homogenity of the ideal $T^{(r)}$, the elements g_i can be founded by means of the "linear procedures". In the same time we express all ξ_i's as some (Lie) polynomials in $\xi_1, \ldots, \xi_n$ and compute the structural constants of L in this basis.

Let U_L be the universal enveloping algebra of the Lie algebra L, and $e_1, \ldots, e_n$ be copies in U_L of $\xi_1, \ldots, \xi_n$ respectively. Of course, the elements $e_1, \ldots, e_n$ generate U_L. There are two canonical epimorphisms associated with U_L. Firstly, it is the epimorphism of algebras $\psi : k\langle X\rangle \to U_L$, which extends the map of sets $\psi : X \to U_L$, $\psi(x_i) = e_i$, $i = 1, \ldots, n$. Secondly, it is the epimorphism of algebras $\theta : U_L \to A$, by which $\theta(e_i) = \psi_i$, $i = 1, \ldots, n$. The diagram of epimorphisms

$$\begin{array}{ccc} k\langle X\rangle & \xrightarrow{\ \psi\ } & U_L \\ & \overset{\varphi}{\searrow} & \downarrow \\ & & A \cong k\langle X\rangle/T^{(r)} \end{array}$$

is commutative and that's why the ideal $\ker\theta \lhd U_L$ is generated by the elements of the form $h_j(e_1, \ldots, e_n)$, where $h_j(e_1, \ldots, e_n)$ is one of the polynomials (1). Thus, there are isomorphisms:

$$A \cong k\langle X\rangle/T^{(r)} \cong U_L/I.$$

Theorem 2. *There is an algorithm based on the notion of Gröbner bases which solves the membership problem in the variety of the Lie nilpotent algebras of a fixed index.*

Proof. We shall use all techniques introduced above. Let us consider the problem of membership to ideal $B \lhd A$ generated by the finitely many elements $b_i(\xi_1, \ldots, \xi_n) = \varphi(b_i(x_1, \ldots, x_n))$, $b_i(x_1, \ldots, x_n) \in K\langle X\rangle$. This problem can be translated into one of membership to the ideal $J \lhd U_L$ generated by the elements $b_i(e_1, \dots, e_n) = \psi(b_i(x_1, \dots, x_n))$ and $h_i(e_1, \dots, e_n)$, $I \subseteq J$. In order to solve the latter problem it is enough to construct the Gröbner basis of the ideal J and to apply the reduction process whith respect to it [2]. □

The crude form of the algorithm mentioned above in the proof of the Theorem 2 can be refined in the routine way.

See the abstract on this theme in [2].

References

[1] V. N. Latyshev, Combinatorial Ring Theory, Standard Basis, Moscow Univ. Issue, 1988 (in Russian).

[2] V. N. Latyshev, About an equality algorithm in Lie nilpotent associative algebra, Kiev Univ. Issue Math. & Mech. 27 (1985), Kiev: Higher School, 67–69 (in Ukrainian).

Department of Mechanics and Mathematics
Chair of Higher Algebra
Moscow State University
Moscow 119899, Russia

Annihilators of power values of a derivation

Tsiu-Kwen Lee and Tsong-Cherng Lee

Abstract. Let R be a prime ring without nonzero nil right ideals and let U be a noncommutative Lie ideal of R. If d is a derivation of R and $a \in R$ such that $ad(u)^{n(u)} = 0$ for all $u \in U$, where $n(u)$ is a positive integer depending on u, then either $a = 0$ or $d = 0$.

1991 Mathematics Subject Classification: Primary 16W25.
Key words and phrases: Prime rings, derivations, Lie ideals, annihilators.

In [1] M. Brešar proved the theorem: Let R be a semiprime ring with a derivation d and $a \in R$. If $ad(x)^n = 0$ for all $x \in R$, where n is a fixed integer, then $ad(R) = 0$ when R is an $(n-1)!$-torsion free ring. In a recent paper [7] Lee and Lin proved Brešar's theorem without the assumption of $(n-1)!$-torsion free on R. In fact, they studied the Lie ideal case as given by C. Lanski [6] and then obtained Brešar's result as the corollary to the main result of [7]. More precisely, for the prime case the following theorem was proved.

Theorem. *Let R be a prime ring with a derivation d and let U be a Lie ideal of R, $a \in R$. Suppose that $ad(u)^n = 0$ for all $u \in U$, where n is a fixed integer. Then*

(i) $ad(U) = 0$ *unless* $\operatorname{char} R = 2$ *and* $\dim_C RC = 4$, *where C stands for the extended centroid of R,*

(ii) $ad(R) = 0$ *if* $[U, U] \neq 0$.

Our purpose in this paper is to remove the fixed bound n in the theorem above. Of course, to avoid the Köthe conjecture the ring R must be assumed to have no nonzero nil right ideals. We shall prove the following result in this paper.

Main Theorem. *Let R be a prime ring without nonzero nil right ideals, U a noncommutative Lie ideal of R and $a \in R$. If d is a derivation of R such that $ad(u)^{n(u)} = 0$ for all $u \in U$, where $n(u)$ depends on u, then either $a = 0$ or $d = 0$.*

Throughout this paper let R be a prime ring with center Z, Q the two-sided Martindale quotient ring of R and C the extended centroid of R. Given two elements $x, y \in R$, $[x, y]$ stands for the element $xy - yx$; also, for two subsets A, B of R, $[A, B]$ is then the additive subgroup of R generated by all elements $[x, y]$ with

$x \in A,\ y \in B$. For any subset S of R, denote by $r_R(S)$ the right annihilator of S in R, that is, $r_R(S) = \{x \in R \mid Sx = 0\}$ and $\ell_R(S)$ is defined similarly. We proceed with a series of lemmas to compose the proof of the Main Theorem.

In what follows we always assume the following situation. *Let R be a prime ring without nonzero nil right ideals, U a noncommutative Lie ideal of R, $a \in R$ and d a derivation of R such that $ad(u)^{n(u)} = 0$ for all $u \in U$, where $n(u)$ depends on u.*

Denote by I the ideal of R generated by $[U, U]$. Then by [3, proof of Lemma 1.3, p. 4] we have $I \subseteq U + U^2$ and $[I, R] \subseteq U$.

From now on we always assume that $a \neq 0$. We want to prove that $d = 0$. Let $\lambda = \{t \in R \mid td(x)^{n(t,x)} = 0 \text{ for all } x \in U\}$. Then λ is a nonzero left ideal of R since $0 \neq a \in \lambda$.

Lemma 1. *If $ty = 0$, where $t \in \lambda$ and $y \in R$, then $td(y) = 0 = d(t)y$.*

Proof. For $x \in I$ we have $[y, xt] \in [R, I] \subseteq U$. Thus there exists an integer $n = n(t, x, y)$ such that $td([y, xt])^n = 0$. Since $ty = 0$, we have

$$0 = td(yxt)^n = t\big(yd(xt) + d(y)xt\big)^n = t\big(d(y)xt\big)^n.$$

So $\big(td(y)x\big)^{n+1} = 0$. This implies that $td(y)I$ is a nil right ideal of R. Since R has no nonzero nil right ideals, $td(y) = 0$ follows. So $d(t)y = 0$ since $ty = 0$. □

Lemma 2. $\lambda d(\lambda) = 0$.

Proof. Let $t \in \lambda$ and $x \in U$. Then there exists an integer $n = n(t, x)$ such that $td(x)^n = 0$. By Lemma 1, $d(t)d(x)^n = 0$. This implies $d(t) \in \lambda$. Therefore $d(\lambda) \subseteq \lambda$. Let $\overline{\lambda} = \lambda/\big(\lambda \cap r_R(\lambda)\big)$. Then $\overline{\lambda}$ is a prime ring without nonzero nil right ideals since R is. Define the mapping $\delta : \overline{\lambda} \to \overline{\lambda}$ by $\delta(\overline{x}) = \overline{d(x)}$ for all $x \in \lambda$. Then δ is a well-defined derivation of $\overline{\lambda}$. By our assumption we have $\delta(\overline{x})^{n(\overline{x})} = 0$ for all $\overline{x} \in [\overline{I\lambda}, \overline{I\lambda}]$. Now $[\overline{I\lambda}, \overline{I\lambda}]$ is a Lie ideal of the prime ring $\overline{\lambda}$. If $[\overline{I\lambda}, \overline{I\lambda}]$ is a noncommutative Lie ideal of $\overline{\lambda}$, then by [6; Theorem 2] $\delta = 0$, that is, $\lambda d(\lambda) = 0$ as desired. Suppose next that $[\overline{I\lambda}, \overline{I\lambda}]$ is a commutative Lie ideal of $\overline{\lambda}$. It is easy to check that $\lambda[\lambda, \lambda] = 0$.

Let $t \in \lambda,\ x_i \in I$ and $y_i \in [I\lambda, I\lambda] \subseteq I,\ 1 \le i \le k$. Then by the definition of λ there exists an integer $n = n(t, x_i, y_i)$ such that

$$\begin{aligned} 0 &= td\Big(\sum_{i=1}^{k}[x_i, y_i]\Big)^n \\ &= t\Big(\sum_{i=1}^{k}\big([d(x_i), y_i] + [x_i, d(y_i)]\big)\Big)^n \\ &= t\Big(\sum_{i=1}^{k}\big(d(x_i)y_i + x_i d(y_i)\big)\Big)\Big(\sum_{i=1}^{k}\big([d(x_i), y_i] + [x_i, d(y_i)]\big)\Big)^{n-1} \end{aligned} \tag{1}$$

since $ty_i \in \lambda[\lambda, \lambda] = 0$ and $td(y_i) \in \lambda[\lambda, \lambda] = 0$ by $d(\lambda) \subseteq \lambda$. Note that $y_i \in \lambda$ and $d(y_i) \in \lambda$. By (1) the same process implies that $t\left(\sum_{i=1}^{k} \left(d(x_i)y_i + x_i d(y_i)\right)\right)^n = 0$, that is, $td\left(\sum_{i=1}^{k} x_i y_i\right)^n = 0$. But $\sum_{i=1}^{k} x_i y_i \in \lambda$, we have $\delta\left(\sum_{i=1}^{k} \overline{x_i y_i}\right)^n = 0$ in $\overline{\lambda}$. Since $\overline{\lambda}$ is now a commutative domain, this implies that $\delta\left(\sum_{i=1}^{k} \overline{x_i y_i}\right) = 0$. So $\delta\left(\overline{I[I\lambda, I\lambda]}\right) = 0$. Also, $\overline{I[I\lambda, I\lambda]}$ is a nonzero left ideal of $\overline{\lambda}$. This implies that $\delta(\overline{\lambda}) = 0$, that is, $\lambda d(\lambda) = 0$. This completes the proof of the lemma.

With Lemma 2 in hand we claim that d must be Q-inner. Indeed, by Lemma 2 we have $ad(xa) = 0$ for all $x \in R$. That is, $ad(x)a + axd(a) = 0$. If d is not Q-inner, then by Kharchenko's theorem [4] $aya + axd(a) = 0$ for all $x, y \in R$. This implies $a = 0$, a contradiction. □

So we assume from now on that d is a Q-inner derivation. Therefore there exists $b \in Q$ such that $d(x) = [b, x]$ for all $x \in R$. Applying Lemma 2 again we have $a[b, xa] = 0$ for all $x \in R$. That is, $abxa = axab$ for all $x \in R$. Now Martindale's lemma [8] forces $a(b - \mu) = 0$ for some $\mu \in C$. Since $b - \mu$ and b induce the same inner derivation of R, replacing b by $b - \mu$ we may assume from the start that $ab = 0$. Now the aim is to prove $b = 0$. Suppose on the contrary that $b \neq 0$. By the primeness of R there exists $u \in R$ such that $bu \in R$, $bua \in I$, but $bua \neq 0$. Since $bua \in \lambda$, replacing a by bua we may assume further that $a^2 = 0$ and $a \in I$.

The next step is to handle the situation when R is a left primitive ring. We need the following lemma proved by E. A. Whelan [9]. Here we present a short and direct proof.

Lemma 3. *Let R be a left primitive ring with V a faithful irreducible left R-module. Then V can be uniquely defined as a faithful irreducible left Q-module extending the module structure of ${}_RV$. Moreover, $\mathrm{End}({}_QV) = \mathrm{End}({}_RV)$.*

Proof. Let $q \in Q$. Then there exists a nonzero ideal J of R such that $qJ + Jq \subseteq R$. For any $m \neq 0$ in V we have $Jm = V$. Define $q \cdot (xm) = (qx)m$ for all $x \in J$. We claim that it is well-defined. Indeed, let J' be an arbitrary nonzero ideal of R such that $qJ' + J'q \subseteq R$ and let $m' \neq 0$ in V. Then if $xm = x'm'$, where $x \in J$ and $x' \in J'$, then for any $y \in J \cap J'$ we have $(yq)(xm) = (yq)(x'm')$ and hence $y\big((qx)m - (qx')m'\big) = 0$. So $(J \cap J')\big((qx)m - (qx')m'\big) = 0$, which implies that $(qx)m = (qx')m'$ as desired. Now it is easy to check that V is defined as a left Q-module extending the module structure of ${}_RV$. The rest is obvious. We omit details. □

We are now in a position to handle the primitive case.

Lemma 4. *The Main Theorem holds when R is a left primitive ring.*

Proof. Let ${}_RV$ be a faithful irreducible left R-module. By Lemma 3, V can be uniquely extended to a faithful irreducible left Q-module. Let $D = \mathrm{End}({}_QV) = \mathrm{End}({}_RV)$, a division ring. Recall that I is the ideal of R generated by $[U, U] \neq 0$. Thus ${}_IV$ is also a faithful irreducible left I-module with $\mathrm{End}({}_IV) = D$. The rest of the proof is implicit in the argument given in [7]. For completeness we give its argument here.

We may assume that $\dim V_D > 1$. Otherwise, R is a division ring and hence we are done. By the density theorem, I acts densely on V_D. For any given $0 \neq v \in V$ we claim that v and bv are D-dependent. Assume first that $av \neq 0$. Suppose on the contrary that v and bv are D-independent. By the density of I in $\mathrm{End}(V_D)$ there exist two elements $x, y \in I$ such that

$$xv = 0, \quad xbv = bv \quad \text{and} \quad yv = 0, \quad ybv = v.$$

Then there exists an integer n such that $a\big(b[x, y] - [x, y]b\big)^n = 0$ and hence $0 = a\big(b[x, y] - [x, y]b\big)^n v = av$, a contradiction. Thus v and bv are D-dependent. Assume next that $av = 0$. Since $a \neq 0$, we have $aw \neq 0$ for some $w \in V$. Then $a(v + w) = aw \neq 0$. Applying the first case we have

$$bw = w\alpha \quad \text{and} \quad b(v + w) = (v + w)\beta$$

for some $\alpha, \beta \in D$. But v and w are clearly D-independent, there exist two elements x, y in I such that

$$xw = 0, \quad xv = v + w \quad \text{and} \quad yw = v + w, \quad yv = v.$$

Then $[x, y]w = v + w$ and $[x, y](v + w) = w$. Hence for some positive integer n we have

$$0 = a\big(b[x, y] - [x, y]b\big)^n w = \pm aw(\beta - \alpha)^n,$$

which implies $\alpha = \beta$ and hence $bv = v\alpha$ as claimed.

From the above we have proved that $bv = v\alpha(v)$ for all $v \in V$, where $\alpha(v) \in D$ depends on v. By an easy argument as given in [7] there exists $\delta \in \mathcal{Z}(D)$, the center of D, such that $bv = v\delta$ for all $v \in V$. So $d = 0$. This completes the proof of the lemma. □

For further discussion we need one more lemma.

Lemma 5. *Let S be a prime ring without nonzero nil right ideals, L a noncommutative Lie ideal of S and $t \in S$. Suppose that $tx^{n(x)} = 0$ for all $x \in L$, where $n(x)$ depends on x. Then $t = 0$.*

Proof. Denote by J the ideal of S generated by $[L, L] \neq 0$. Then $[J, J] \subseteq L$ and hence $t(xy - yx)^{n(x,y)} = 0$, where $n(x, y)$ depends on x and y, for all $x, y \in J$. Suppose on the contrary that $t \neq 0$. Set $\rho = r_J(t)$. For $z \in \rho$ and $w \in J$ we have $t[zt, w]^{n(z,w)} = 0$. Therefore $(ztw)^{n(z,w)+1} = 0$. Therefore $ztJ = 0$. It follows from the primeness of S that $zt = 0$. That is, $\rho t = 0$ and hence $\rho = 0$. So $(xy - xy)^{n(x,y)} = 0$ for all $x, y \in J$. Now if $x \in J$ satisfies $x^2 = 0$, then $x(xy - yx)^{n(x,y)} = 0$ for all $y \in J$. Expanding it we get $(xy)^{n(x,y)+1} = 0$. So $xJ = 0$ and hence $x = 0$. Therefore J has no nonzero nilpotent elements. This implies that $[J, J] = 0$ and so $[S, S] = 0$, a contradiction. This completes the proof of the lemma. □

Denote by $\mathcal{J}(R)$ the Jacobson radical of R. Then by Lemmas 4 and 5 we have the following result.

Lemma 6. $ad(R)[U, U] \subseteq \mathcal{J}(R)$.

Proof. Let P be a left primitive ideal of R and set $\overline{R} = R/P$ and $\overline{U} = (P + U)/P$. Then $\overline{U}$ is a Lie ideal of $\overline{R}$. If $[\overline{U}, \overline{U}] = 0$, then $[U, U] \subseteq P$. Suppose that $[\overline{U}, \overline{U}] \neq 0$. If $d(P) \subseteq P$, then d canonically induces a derivation $\overline{d}$ on $\overline{R}$ such that $\overline{a}\overline{d}(\overline{u})^{n(\overline{u})} = 0$ for all $\overline{u} \in \overline{U}$. By Lemma 4, $\overline{a}\overline{d}(\overline{R}) = 0$. Thus $ad(R) \subseteq P$. Assume next that $d(P) \not\subseteq P$. Set $W = \big([I, P] + d([I, P])\big)/P$, a Lie ideal of $\overline{R}$. Since $[I, P] \subseteq [I, R] \subseteq U$, we have that $\overline{a}x^{n(x)} = 0$ for all $x \in W$. If $[W, W] \neq 0$, then Lemma 5 forces $\overline{a} = 0$, that is, $a \in P$. If $[W, W] = 0$, then this just implies that $\big[[I, d(P)], [I, d(P)]\big] \subseteq P$. Set $A = (I + P)/P$ and $B = \big(P + d(P)\big)/P$. Then $\big[[A, B], [A, B]\big] = 0$ in $\overline{R}$. Also, B is a nonzero ideal of $\overline{R}$. Now, if $A \neq 0$ in $\overline{R}$, then this implies that $[\overline{R}, \overline{R}] = 0$, i.e., $[R, R] \subseteq P$. Recall that I is the ideal of R generated by $[U, U]$. So $I \subseteq P$, that is, $A = 0$, a contradiction. Thus we have $A = 0$, i.e., $[U, U] \subseteq P$.

At any rate, for any left primitive ideal P of R either $ad(R) \subseteq P$, $a \in P$ or $[U, U] \subseteq P$. So $ad(R)[U, U] \subseteq P$ always holds. But $\mathcal{J}(R)$ is the intersection of all left primitive ideals of R, we have $ad(R)[U, U] \subseteq \mathcal{J}(R)$. This completes the proof. □

Proof of the Main Theorem. We divide the proof into two cases.

Case 1. Suppose that $\big[[U, U], [U, U]\big] = 0$. Since $[U, U] \neq 0$, this implies that char $R = 2$ and $\dim_C RC = 4$ by [5, Lemma 8]. Note in this case $Q = RC$. If RC is a division C-algebra, then we are done. Therefore we may assume that $RC \cong M_2(C)$, the ring of 2 by 2 matrices over C. Note that $[x, y]^2 \in C$ for all $x, y \in RC$. Thus by the assumption we have $[b, u]^2 = 0$ for all $u \in U$. Now [6, Theorem 2] forces $b \in C$. So $b = 0$ since $ab = 0$ and $a \neq 0$.

Case 2. Suppose that $\big[[U, U], [U, U]\big] \neq 0$. Denote by M the ideal of R generated by $\big[[U, U], [U, U]\big]$. Then $M \subseteq [U, U] + [U, U]^2$. Therefore by Lemma 6, $ad(R)M \subseteq \mathcal{J}(R)$. Choose a nonzero ideal N of R such that $bN + Nb \subseteq R$ and $N \subseteq M \cap I$. Let $T = \{t \in I \mid xty = 0 \text{ whenever } xy = 0 \text{ with } x, y \in N\}$. We claim that for any $x \in N$

$$\text{either } ar_N(x) = 0 \text{ or } xbr_N(x) = 0. \tag{1}$$

Given $x, y, r \in N$ with $xy = 0$ we have $(yrx)^2 = 0$ and $yrx = [y, rx] \in [I, I]$. So there exists $n = n(r, x, y)$ such that

$$0 = a[b, yrx]^n yrx = a(-yrxb)^n yrx.$$

Thus $ay(rxby)^{n+1} = 0$. By [2, Theorem 2], $(ay)N(xby) = 0$. So either $ay = 0$ or $xby = 0$. This implies that $r_N(x)$ is the union of its additive subgroups $\{y \in r_N(x) \mid ay = 0\}$ and $\{y \in r_N(x) \mid xby = 0\}$. As a group cannot be the union of its two proper subgroups, this implies that either $ar_N(x) = 0$ or $xbr_N(x) = 0$ as claimed.

The next step is to prove that for any $x \in N$

$$\text{either } ar_N(x) = 0 \text{ or } xb^2 = 0. \tag{2}$$

Indeed, let $x \in N$ be such that $ar_N(x) \neq 0$. For $t \in N^2$ we have $(1+bta)x(1-bta) \in N$ and $(bta)^2 = 0$ since $ab = 0$. Moreover,

$$\begin{aligned} &ar_N\big((1 + bta)x(1 + bta)^{-1}\big) \\ &\qquad = a(1 + bta)r_N(x)(1 + bta)^{-1} = ar_N(x)(1 - bta) \neq 0. \end{aligned}$$

By (1), $(1 + bta)x(1 - bta)b(1 + bta)r_N(x)(1 - bta) = 0$. Using $ab = 0$ and $xbr_N(x) = 0$ we have $xb^2tar_N(x) = 0$. That is, $(xb^2)N^2\big(ar_N(x)\big) = 0$. Since $ar_N(x) \neq 0$, this implies $xb^2 = 0$ as desired.

Now we claim that $a \in T$. Let $x \in N$. Suppose that $ar_N(x) \neq 0$. For any $r \in R$ and $m \in N$ we have $ad(r)m \in \mathcal{J}(R)$. Denote by $\big(ad(r)m\big)'$ the quasi-inverse of $ad(r)m$. Note that $\big(ad(r)m\big)' \in aR$. Then

$$\begin{aligned} ar_N\Big(\Big(1 + \big(ad(r)m\big)'\Big)x\big(1 + ad(r)m\big)\Big) &= a\Big(1 + \big(ad(r)m\big)'\Big)r_N(x)\big(1 + ad(r)m\big) \\ &= ar_N(x)\big(1 + ad(r)m\big) \neq 0 \end{aligned}$$

since $a^2 = 0$. By (2) we have $\Big(1 + \big(ad(r)m\big)'\Big)x\big(1 + ad(r)m\big)b^2 = 0$. So $xad(r)mb^2 = 0$ since $xb^2 = 0$. That is, $xad(R)Nb^2 = 0$. Then either $b^2 = 0$, $xa = 0$ or $d = 0$. If $d = 0$, then we are done. Next, if $b^2 = 0$, then for any $u \in N \subseteq I$ there exists $n = n(u)$ such that $a\Big(\big[b, [ub, u]\big]\Big)^n = 0$. Using $ab = 0 = b^2$ to expand this we have $a(ub)^{2n} = 0$, which implies $aNb = 0$ by [2; Theorem 2]. So $b = 0$ follows. Hence we may assume that $xa = 0$. At any rate, for any $x \in N$ we have $xar_N(x) = 0$. So $a \in T$ as claimed.

Suppose on the contrary that $d \neq 0$. Let $x_1, x_2, x_3 \in N^2$; then $bx_1ad(x_2)x_3 \in \mathcal{J}(R) \cap N$. By the definition of T we have

$$\big(1 + bx_1ad(x_2)x_3\big)a\Big(1 + \big(bx_1ad(x_2)x_3\big)'\Big) \in T,$$

where $\big(bx_1ad(x_2)x_3\big)'$ denotes the quasi-inverse of $bx_1ad(x_2)x_3$. Note that $\big(bx_1ad(x_2)x_3\big)' \in bR$. Therefore this implies that $bx_1ad(x_2)x_3a \in T$, i.e., $bKa \subseteq T$, where $K = N^2ad(N^2)N^2$ is a nonzero ideal of R.

For any $x \in [N, N] \subseteq [I, I]$ and $y \in K$ it follows from our assumption that $axb\big(bya[b, x]\big)^n = 0$, where $n = n(x)$ depends on x. So $(axb^2y)^{n+1} = 0$, which implies that $a[N, N]b^2 = 0$. Then either $a = 0$ or $b^2 = 0$. Since we assume $a \neq 0$, $b^2 = 0$ follows. As before, $b^2 = 0$ implies $b = 0$. This finishes the proof of the Main Theorem. □

References

[1] M. Brešar, A note on derivations, Math. J. Okayama Univ. 32 (1990), 83–88.

[2] B. Felzenszwalb, On a result of Levitzki, Canad. Math. Bull. 21 (1978), 241–242.

[3] I. N. Herstein, Topics in ring theory, University of Chicago Press, Chicago 1969.

[4] V. K. Kharchenko, Differential identities of prime rings, Algebra and Logic 18 (1979), 86–119.

[5] C. Lanski and S. Montgomery, Lie structure of prime rings of characteristic 2, Pacific J. Math. 42 (1972), 117–136.

[6] C. Lanski, Derivations with nilpotent values on Lie ideals, Proc. Amer. Math. Soc. 108 (1990), 31–37.

[7] T.K. Lee and J.S. Lin, A result on derivations, Proc. Amer. Math. Soc. to appear.

[8] W. S. Martindale III, Prime rings satisfying a generalized polynomial identity, J. Algebra 12 (1969), 576–584.

[9] E. A. Whelan, The symmetric ring of quotients of a primitive ring is primitive, Commun. Algebra 18 (3) (1990), 615–633.

Department of Mathematics
National Taiwan University
Taipei
Taiwan 10764, R.O.C.

Department of Business Mathematics
Soochow University
Taipei
Taiwan 10001, R.O.C.

Modules with Krull and Gabriel dimensions over rings and algebras with polynomial identities

*V. T. Markov**

Abstract. The existence of a faithful module which has Krull (or Gabriel) dimension implies some constraints on the ring itself, especially if the latter is a polynomial identity ring. For example, if a *PI*-ring R has a faithful left module M with Krull dimension then its prime radical $\operatorname{rad}(R)$ is nilpotent and the quotient ring $R/\operatorname{rad}(R)$ is Goldie. On the other hand, there are some classes of rings that in general have no Krull dimension but have faithful modules with Krull dimension. We give some examples confirming that the conditions defining these classes cannot be omitted. We show that for every given integers m, r such that $2 \leq m \leq r$ there exist a finitely generated *PI*-algebra R and a cyclic left R-module M such that $\operatorname{Gdim}({}_R M) = m$, $\operatorname{Gdim}({}_R R) = r$, thus answering two questions of Amitsur and Small in the negative.

1. Introduction

In this paper we study some connections between the Krull and Gabriel dimensions of a ring and a faithful module over it. It was proved in [11] that if a *PI*-ring R has a faithful Noetherian left module M then its prime radical $\operatorname{rad}(R)$ is nilpotent and the ring $R/\operatorname{rad}(R)$ is Noetherian. Independently the case of semiprime *PI*-ring was considered in [1], where it was observed that $\operatorname{Kdim}({}_R R) = \operatorname{Kdim}(M)$ (we denote by $\operatorname{Kdim}(M)$ the Krull dimension and by $\operatorname{Gdim}(M)$ the Gabriel dimension of a module M; the definition of Krull and Gabriel dimensions and all the necessary properties one can find in [7]). It is well-known that any left Noetherian module has Krull dimension and that the prime radical of a ring with Krull dimension is nilpotent (cf. [7]). Then this result was generalized by the author and the following two statements were proved.

1.1. Theorem. *Let R be a PI-ring and M be a faithful left R-module. Then*

(a) $\operatorname{rad}(R)$ *is nilpotent;*

(b) *if moreover the left modules ${}_R M$ and ${}_R(\operatorname{rad}(R))$ are finitely generated then R has left Krull dimension and* $\operatorname{l.Kdim}(R) = \operatorname{Kdim}(M)$.

* Work was supported by Russian Basic Research Foundation grant N 26-93-011-1544 and International Science Foundation grant JBX100.

1.2. Theorem. *If a semiprime ring has a faithful module with Krull dimension then it is a finite subdirect sum of prime rings.*

These results were announced in [12], their proof will appear in [13]. Two classes of rings are pointed out in [13] that do not have Krull dimension in general but have a faithful Artinian (i.e., with zero Krull dimension) left module:

(i) a right Artinian ring such that $\operatorname{rad}(R)^2 = 0$ has a faithful Artinian cyclic left module;

(ii) a finitely generated semiprime PI-algebra has a faithful Artinian left module.

We show here that the condition $\operatorname{rad}(R)^2 = 0$ is essential in (i) (see Section 2) and the polynomial identity is essential in (ii) (see Section 3). Moreover we give an example of a finitely generated algebra having no faithful module with Gabriel dimension.

It is evident that the existence of Gabriel dimension of a ring or module is far less restrictive than the existence of Krull dimension. In particular, any finitely generated PI-algebra over a field has Gabriel dimension [8]. Let M be a faithful finitely generated left module over such an algebra R. A question was asked in [8]: is it true in general that $\operatorname{Gdim}(M) = \operatorname{Gdim}({}_R R)$? We answer this question in negative. Moreover we prove the following

1.3. Theorem. *For every given integers m, n such that $2 \leq m \leq n$ there exist a finitely generated PI-algebra R and a cyclic left R-module M such that* $\operatorname{Gdim}({}_R M) = m$, $\operatorname{Gdim}({}_R R) = n$.

The proof of Theorem 1.3 gives also a negative answer to another question asked in [8]: does affine algebra have Krull dimension if its every homomorphic image has finite Goldie dimension?

We mean by PI-ring a ring that satisfies a polynomial identity with integer coefficients one of which is equal to 1 (cf. [14]). It is well-known that any PI-algebra over a field is a PI-ring in this sense.

All the rings considered in this note are associative and unitary, all the modules are left unitary modules, unless the contrary is stated. The sets of all positive integers and of all rationals are denoted by $\mathbb{N}$ and $\mathbb{Q}$ respectively.

2. Modules with Krull dimension

Part (a) of Theorem 1.1 is a consequence of the following formally more general statement:

2.1. Proposition. *Let N be a nil-ideal of a ring R and M be an R-module with Krull dimension. Suppose that there exists a polynomial $f(x_1, \dots, x_d)$ with integer coefficients one of which is equal to 1, such that $f(a_1, \dots, a_d)m = 0$ for any $a_1, \dots, a_d \in R$ and $m \in M$. Then there exists $n \in \mathbb{N}$ such that $N^n M = 0$.*

We shall use induction on the degree d of the polynomial f to prove Proposition 2.1. The case $d = 1$ is trivial. Suppose that $d \geq 2$. Then the polynomial $f(x_1, \dots, x_d)$ without loss of generality may be expanded as

$$f(x_1, \dots, x_d) = \varphi(x_1, \dots, x_{d-1}) + g(x_1, \dots, x_d)$$

and also as

$$f(x_1, \dots, x_d) = x_1\psi(x_2, \dots, x_d) + h(x_1, \dots, x_d)$$

where the polynomial g (polynomial h) contains no non zero terms ending in x_d (starting in x_1) and φ and ψ have degree $d - 1$. These representations are used in the proofs of the following two lemmas, respectively:

2.2. Lemma. *Under the conditions of Proposition 2.1 let $I \subseteq N$ be such an ideal of R that $NM = IM$. Then for any $k \in \mathbb{N}$ there exists $m \in \mathbb{N}$ such that $N^m M \subseteq I^k M$.*

Proof. This lemma is proved by induction on k and application of Proposition 2.1 to the module $I^{k-1}M/I^k M$. □

2.3. Lemma. *Under the conditions of Proposition 2.1, let $I \subseteq N$ be an ideal in R such that $NIM = 0$. Then there exists $n \in \mathbb{N}$ such that $IN^n M = 0$.*

Proof. Application of Proposition 2.1 to the module $M/\mathrm{Ann}_M(I)$. □

2.4. Lemma. *Under the conditions of Proposition 2.1, if $NM \neq 0$ then there exists an ideal $I \subseteq N$ such that $0 \neq IM \neq M$.*

Proof. The lemma follows from the well-known fact that a nil PI-ring (without 1) $A = N/\mathrm{Ann}_R(M) \cap N$ contains a non-zero nilpotent ideal and so does R (cf. [2], 2.2.5 and [14], 1.1.5) □

The proof of the proposition is finished by induction on dual Krull dimension of the module M (cf. [4]) and application of Lenagan's technique (cf. [9]).

2.5. Lemma. *If M is a faithful module with Krull dimension over a PI-ring R then there exists a faithful module M' over the ring $R/\mathrm{rad}(R)$ having the same Krull dimension as M. If moreover M and ${}_R\mathrm{rad}(R)$ are finitely generated modules then one can assume M' to be also finitely generated.*

Proof. Let $N = \mathrm{rad}(R)$, $R' = R/N$. Then $N^n = 0$ for some $n \in \mathbb{N}$. Let

$$M' = (M/NM) \oplus (NM/N^2M) \oplus (N^2M/N^3M) \oplus \cdots \oplus N^{n-1}M.$$

It is easy to check that M' enjoys the required properties. □

2.6. Lemma. *If R is a prime left bounded left Goldie ring and M is a faithful finitely generated R-module then ${}_RR$ isomorphically embeds into a finite direct sum of copies of M.*

Proof. The lemma follows from the proof of Proposition 3 of [1]. □

Proof of Theorem 1.2. It suffices to prove that R contains no infinite direct sums of non-zero two-sided ideals (cf. [2], Chapter 4). Assume that, on the contrary, $I = \oplus_{k=1}^{\infty} I_k$ is such a direct sum. Consider an infinite descending chain of subsets $S_i \subset \mathbb{N}$ such that $S_i \backslash S_{i+1}$ is infinite for every $I \in \mathbb{N}$ (e.g. take $S_i = 2^i\mathbb{N}$). Let $M_i = (\oplus_{k \in S_i} I_k)M$, $B_j = (\oplus_{k \leq j} I_k)M$, $i, j \in \mathbb{N}$. Then by [9] there are some $i, j \in \mathbb{N}$ such that $M_i \subseteq M_{i+1} + B_j$ which is clearly impossible. □

Remark. For semiprime PI-rings a stronger version of Theorem 1.2 was obtained in [3].

Proof of Theorem 1.1(b). Lemma 2.5 reduces the problem to the semiprime ring case and Theorem 1.2 to the prime ring case, in which it is settled by Lemma 2.6. □

We conclude this section with the following

2.7. Example. A right Artinian ring which has no faithful modules with Krull dimension.

This example is inspired by an example in [6]. Let $D \subset E$ be skew fields such that $\dim({}_DE) = \infty$ and $\dim(E_D) < \infty$ (such skew fields were constructed by P. M. Cohn [5]). Consider the set

$$R = D\begin{pmatrix} 1 & 0 & 0 \\ 0 & 1 & 0 \\ 0 & 0 & 1 \end{pmatrix} + \begin{pmatrix} 0 & E & E \\ 0 & 0 & E \\ 0 & 0 & 0 \end{pmatrix}.$$

Here and below the matrix with sets as elements denotes the set of all matrices whose elements belong to the corresponding sets. It is evident that R is a subring the 3×3 matrices over E. It is also clear that R is right Artinian. Let

$$I = \begin{pmatrix} 0 & 0 & E \\ 0 & 0 & E \\ 0 & 0 & 0 \end{pmatrix}, \quad L = \begin{pmatrix} 0 & 0 & E \\ 0 & 0 & 0 \\ 0 & 0 & 0 \end{pmatrix} \subset I.$$

Then I, L are two-sided ideals in R. Let M be an arbitrary R-module having Krull dimension, $m \in M$ and $K = \mathrm{Ann}_R(m)$. Suppose that $K \cap I \subseteq L$. Then we have an epimorphism of R-modules

$$I/(K \cap I) \to I/L \to 0.$$

By definition the R-submodule lattice of I/L coincides with that of ${}_DE$, so I/L has infinite Goldie dimension and $I/(K \cap I)$ has no Krull dimension. On the other hand, $I/(K \cap I) \cong \mathrm{Im}$ and so the left R-module $I/(K \cap I)$ has to have Krull dimension. The contradiction shows that K contains an element

$$u = \begin{pmatrix} 0 & 0 & \beta \\ 0 & 0 & \alpha \\ 0 & 0 & 0 \end{pmatrix},$$

where $\alpha, \beta \in E$ and $\alpha \neq 0$. But then for every $\gamma \in E$ we have

$$\begin{pmatrix} 0 & 0 & \gamma \\ 0 & 0 & 0 \\ 0 & 0 & 0 \end{pmatrix} = \begin{pmatrix} 0 & \gamma\alpha^{-1} & 0 \\ 0 & 0 & 0 \\ 0 & 0 & 0 \end{pmatrix} \times \begin{pmatrix} 0 & 0 & \beta \\ 0 & 0 & \alpha \\ 0 & 0 & 0 \end{pmatrix},$$

so $L \subset Ru \subset K$ and $LM = 0$, so M cannot be faithful. The author doesn't know if such examples exist in the class of *PI*-rings.

3. Affine algebras and Gabriel dimension

3.1. Lemma. *Let A be a commutative ring, $B = A[x, x^{-1}]$ (a group ring of the infinite cyclic group), $C = A[x] \subset B$, $M_0 \supset M_1 \supset \cdots$ a descending chain of ideals in A such that A/M_i is Artinian A-module for every $i > 0$ and $\cap_{i=0}^{\infty} M_i = 0$ and $L = C + \sum_{i=0}^{\infty} M_i x_{-i} \subset B$. Then L is a C-submodule in B and $N = B/L$ is a semi-Artinian C-module, i.e.,* $\mathrm{Gdim}({}_CN) = 1$.

Proof. Note that $M_0 \supset M_1 \supset \cdots$ implies $xL \subset L$, so $CL = L$. Let K be a C-submodule in B such that $L \subseteq K \subset B$ and $U_i = \{a \in A | ax^{-i} \in K\}$ for every $i \geq 0$. Choose the least integer i such that $U_i \neq A$ (it exists since $K \neq B$). Then $Ax^{-i}/U_i x^{-i}$ is a C-submodule in B/K and its submodule lattice coincides with that of A-module A/U_i. The latter is Artinian since it is a homomorphic image of the Artinian module A/M_i hence every non-zero homomorphic image of N has non-zero socle. □

3.2. Lemma. *Under the conditions of Lemma* 3.10 *suppose moreover that A is an affine algebra over a field F. Consider a matrix ring $R = \begin{pmatrix} C & 0 \\ B & D \end{pmatrix}$ where $D = F[x, x^{-1}]$*

and let $L' = \begin{pmatrix} C & 0 \\ L & 0 \end{pmatrix}$. *Then R is an affine F-algebra, $M = R/L'$ is a faithful R-module and* $\mathrm{Gdim}(M) = 2$.

Proof. It is easy to check that if $a_1, \dots, a_p$ generate A over F then the matrices $\begin{pmatrix} a_1 & 0 \\ 0 & 0 \end{pmatrix}, \dots, \begin{pmatrix} a_p & 0 \\ 0 & 0 \end{pmatrix}, \begin{pmatrix} 0 & 0 \\ 0 & x \end{pmatrix}, \begin{pmatrix} 0 & 0 \\ 0 & x^{-1} \end{pmatrix}, \begin{pmatrix} 0 & 0 \\ 1 & 0 \end{pmatrix}$ generate the algebra R. Let $0 \neq r \in L'$ and show that $rR \not\subseteq L'$. Write down $r = \begin{pmatrix} c & 0 \\ u & 0 \end{pmatrix}$ where $c \in C$, $u \in L$. If $c \neq 0$ then $\begin{pmatrix} 0 & 0 \\ c & 0 \end{pmatrix} = r \begin{pmatrix} 0 & 0 \\ 1 & 0 \end{pmatrix} \in rR$ so without loss of generality one can assume that $c = 0$, $u \neq 0$. There is the unique representation $u = x^q(a + xb)$ where $0 \neq a \in A$, $b \in B$ and q is an integer. Choose i such that $a \notin M_i$ (it is possible since $\bigcap_{i=0}^{\infty} M_i = 0$). Then

$$r\begin{pmatrix} 0 & x^{-i-q} \\ 0 & 0 \end{pmatrix} = \begin{pmatrix} 0 & ax^{-i} + cx^{-i+1} \\ 0 & 0 \end{pmatrix} \notin L'.$$

Let $K = \begin{pmatrix} C & 0 \\ B & 0 \end{pmatrix}$. Then there is an exact sequence

$$0 \to K/L' \to M \to R/K \to 0.$$

One has $\mathrm{Gdim}(K/L') = \mathrm{Gdim}({}_C N) = 1$ and $\mathrm{Gdim}(R/K) = \mathrm{Gdim}({}_D D) = 2$, so $\mathrm{Gdim}(M) = \max(\mathrm{Gdim}(K/L'), \mathrm{Gdim}({}_C N)) = 2$. □

Proof of Theorem 1.3. Consider the algebra R of the preceding lemma. It is clear that $\mathrm{Gdim}(R) = \max(\mathrm{Gdim}(C), \mathrm{Gdim}(D)) = \mathrm{Gdim}(C)$ (since A is not an Artinian ring), so it is sufficient to take the polynomial algebra in $(n-2)$ variables as A and the chain of powers of some maximal ideal as $\{M_i\}$ to get the affine algebra R and the faithful cyclic R-module M such that $\mathrm{Gdim}(R) = n$ and $\mathrm{Gdim}(M) = 2$. To complete the proof consider the direct sum $R' = R \oplus S$ of the algebra R and any affine algebra S with $\mathrm{Gdim}(S) = m$ and the faithful cyclic R'-module $M' = (R \oplus S)/(L' \oplus 0)$. □

There was also a conjecture in [8] that an affine algebra has Krull dimension if its every homomorphic image has finite Goldie dimension. Note that the algebra R of Lemma 3.2 gives the counterexample to this conjecture. Indeed consider the descending chain $M_0B \supset M_1B \supset \dots$ and the ascending chain $C \subset Cx^{-1} \subset Cx^{-2} \dots$ of C-submodules in B. If there exist some integers i and j such that $M_iB \subseteq M_{i+1}B + Cx^{-j}$ then the comparison of the components of degree $-j-1$ in x gives $M_i = M_i + 1$ which is impossible. So by [9] the module ${}_CB$ as well as the module ${}_RR$ has no Krull dimension. On the other hand let I be a two-sided ideal in R and $U = \begin{pmatrix} 0 & 0 \\ B & 0 \end{pmatrix}$. One can suppose without loss of generality that $I \subseteq U$

since R/U is a commutative affine (and hence Noetherian) algebra. So $I = \begin{pmatrix} 0 & 0 \\ V & 0 \end{pmatrix}$ where V is an ideal in B. Then $C/(C \cap V)$ is an essential Noetherian C-submodule in B/V so C-module B/V and hence R-module R/I have finite Goldie dimension.

3.3. Example. A finitely generated prime *PI*-algebra (without a polynomial identity of course) which has no faithful modules with Gabriel dimension.

First we consider an idempotent prime countable dimensional Jacobson radical algebra (without 1) U over a field F. It may be constructed in various ways, for example as follows. Let F be a countable field, S be a semigroup of all non-negative rationals with addition as operation. Let $A = FS$ and $M = \sum_{q>0} Fq$. Then M is a prime ideal in A and $M^2 = M$. Consider then the localization $B = A_M$ and its Jacobson radical $U = J(B)$. It is evident that U enjoys the desired properties. Then we apply the construction of [10] giving an F-algebra R with two generators having a left ideal L such that a surjective ring homomorphism $L \to U$ exists and its kernel K is nilpotent of index 2, i.e., $K^2 = 0$. Then of course $L \subseteq J(R)$ and $L \not\subseteq \text{rad}(R)$ since U contains non-nilpotent elements. So it is possible to assume without loss of generality that R is a prime algebra. Now we show that $J(R)$ contains a non-zero idempotent left ideal. If $a, b \in L$ then $a = \sum_{i=1}^{n} x_i y_i + z, b = \sum_{i=1}^{m} u_i v_i + w$ for some $x_1, \ldots, x_n, y_1, \ldots, y_n, u_1, \ldots, u_m, v_1, \ldots, v_m \in L, z, w \in K$. Since $zw = 0$ one has

$$ab = a(\sum_{i=1}^{m} u_i v_i) + (\sum_{i=1}^{n} x_i y_i)b - (\sum_{i=1}^{n} x_i y_i)(\sum_{i=1}^{m} u_i v_i) \in L^3,$$

so $L^2 = L^3$ and $V = L^2$ is an idempotent left ideal. The same proof as in [9] shows that every Gabriel simple module and so every module with Gabriel dimension is annihilated by V.

The author is grateful to K. I. Beidar for helpful discussions.

References

[1] S. A. Amitsur and L. W. Small, Finite-dimensional representations of *PI*-algebras, J. Algebra 133 (1990), 244–248.

[2] V. A. Andrunakievich and Yu. M. Rjabukhin, Radicals of algebras and structure theory, Nauka, Moscow 1979 (in Russian).

[3] K. I. Beidar and V. T. Markov, A semiprime *PI*-ring having a faithful module with Krull dimension is Goldie, Uspekhi Mat. Nauk 48 (1993), 141–142 (in Russian).

[4] D. Chambless, *N*-dimension and *N*-critical modules, Comm. Algebra 8 (1980), 1561–1592.

[5] P. M. Cohn, Quadratic extensions of skew fields, Proc. London Math. Soc. 3 (1951), 187–198.

[6] B. Cortzen and L. W. Small, Finite extensions of rings, Proc. Amer. Math. Soc. 103 (1988), 1058–1062.

[7] R. Gordon and J. C. Robson, Krull dimension, Mem. Amer. Math. Soc. 133 (1973).

[8] R. Gordon and L. W. Small, Affine *PI*-rings have Gabriel dimension, Comm. Algebra 12 (1984), 1291–1300.

[9] T. H. Lenagan, Reduced rank in rings with Krull dimension, in: Ring Theory (Antwerp 1978), Lect. Notes in Pure and Appl. Math. 51, Marcel Dekker, New York 1979, 123–131.

[10] V. T. Markov, Some examples of finitely generated algebras, Uspekhi Mat. Nauk 36 (1981), 185–186 (in Russian).

[11] V. T. Markov, Faithful Noetherian modules over *PI*-rings, in: Abelian Groups and Modules 8, 99–103, Tomsk, 1989 (in Russian).

[12] V. T. Markov, On *PI*-rings having a faithful module with Krull dimension, in: Proceedings of the third international conference on algebra (Krasnoiarsk 1993 August 23–28), Krasnoiarsk 1993, 414–415 (in Russian).

[13] V. T. Markov, On *PI*-rings and semiprime rings having a faithful module with Krull dimension, to appear in Algebra i Logika (in Russian).

[14] C. Procesi, Rings with polynomial identities, Marcel Dekker, New York 1973.

Department of Mechanics and Mathematics
Chair of Higher Algebra
Moscow State University
Moscow 119899, Russia

A Lie algebra with cohomological dimension one over a field of prime characteristic is not necessarily free

A. A. Mikhalev, U. U. Umirbaev, and A. A. Zolotykh**

Abstract. We give an example of a Lie algebra over a field of prime characteristic with cohomological dimension one which is not a free Lie algebra. This gives a counterexample to a hypothesis that the analog of the Stallings–Swan theorem takes place in the case of Lie algebras. The problem in the case of the field of zero characteristic is still open. We formulate also some other related conjectures.

Introduction

J. R. Stallings [St] and R. G. Swan [Sw] proved that any group of cohomological dimension one is free. It was quite natural to conjecture that this assertion takes place also for Lie algebras. Moreover, G. P. Kukin in [Ku2, Ku3] claimed that Lie algebras of cohomological dimension one are free over an arbitrary field. G. L. Feldman proved in [Fe] that any two-generated subalgebra of a Lie algebra with cohomological dimension one is free.

The main result of this article is an example of a nonfree Lie algebra with cohomological dimension one over a field of prime characteristic (Theorem 3).

The article is organized as follows. In Section 1 we collect necessary definitions and preliminary results. In Section 2 we prove the main theorem. In Section 3 we give another proof of the main result using Fox differential calculus.

The results of this article were announced in [MUZ].

*Partially supported by the International Science Foundation, grant M22000, and by the Russian Foundation for Fundamental Research, grant 93-011-1543.

1. Definitions and preliminary results

Let K be a field, L a Lie algebra over K, M an L-module. Consider the set $C^n(L, M)$ of all n-dimensional cochains (i.e., of all K-multilinear skew symmetric mappings of the direct product of n copies of L into M).

The operator of taking the coboundary

$$\partial_n: C^n(L, M) \to C^{n+1}(L, M)$$

is given by

$$(\partial_n f)(v_1, \dots, v_{n+1}) = \sum_{i=1}^{n} v_i f(v_1, \dots, \widehat{v_i}, \dots, v_n) + \sum_{i<j} (-1)^{i+j} ([v_i, v_j], v_1, \dots, \widehat{v_i}, \dots, \widehat{v_j}, \dots, v_{n+1}).$$

The set $Z^n(L, M)$ of all n-dimensional cocycles is the set of all $f \in C^n(L, M)$ such that $\partial_n f = 0$. The set $B^n(L, M)$ of all n-dimensional coboundaries is the set of all $f \in C^n(L, M)$ such that there exists $h \in C^{n-1}(L, M)$ with $f = \partial_{n-1} h$.

One can see that $\partial_{n+1}\partial_n = 0$. The space (abelian group) of n-dimensional cohomologies of L relatively M is given by

$$H^n(L, M) = Z^n(L, M)/B^n(L, M).$$

This space can be calculated in the following way. Consider K as a trivial L-module. Let $U(L)$ be the universal enveloping algebra of L. We can consider all L-modules as right $U(L)$-modules. Let

$$\cdots \longrightarrow P_n \xrightarrow{d_n} P_{n-1} \longrightarrow \cdots \longrightarrow P_2 \xrightarrow{d_2} P_1 \xrightarrow{d_1} P_0 \xrightarrow{\varepsilon} K \longrightarrow 0$$

be a projective resolution of the module K (i.e. an exact sequence of right $U(L)$-modules with projective P_i, $i \geq 0$). Then we have the induced sequence

$$\mathrm{Hom}_{U(L)}(P_{n-1}, M) \xrightarrow{d_n^*} \mathrm{Hom}_{U(L)}(P_n, M) \xrightarrow{d_{n+1}^*} \mathrm{Hom}_{U(L)}(P_{n+1}, M)$$

with

$$d_{n+1}^* \cdot d_n^* = 0,$$

and

$$H^n(L, M) = \mathrm{Ker}\, d_{n+1}^* / \mathrm{Im}\, d_n^*.$$

Definition 1. A Lie algebra L over K *has the cohomological dimension* s (notation: $\mathrm{cd}(L) = s$) if for any L-module M we have that $H^{s+1}(L, M) = 0$, and at the same time there exists an L-module M_0 such that $H^s(L, M_0) \neq 0$.

For a Lie algebra L we denote by $I(L)$ the right ideal $L \cdot U(L)$ of $U(L)$ generated by L. By the Poincaré–Birkhoff–Witt theorem, $U(L)/I(L) \cong K$.

Remark 1. If $I(L)$ is a projective right $U(L)$-module, then $\mathrm{cd}(L) \leq 1$. Indeed, for any L-module M we have the resolution

$$0 \longrightarrow 0 \xrightarrow{d_3} 0 \xrightarrow{d_2} I(L) \xrightarrow{\varepsilon} K \longrightarrow 0,$$

and therefore

$$H^2(L, M) = \operatorname{Ker} d_3^* / \operatorname{Im} d_2^* = 0.$$

As for more details in cohomology theory of Lie algebras see for example [Ja, GG, Fu].

Definition 2. Let X be a set, $L(X)$ the free Lie algebra over K on X. In a natural way one can define the length $\ell(a)$ of any element $a \in L(X)$. We say that a subset S of $L(X)$ is *reduced* if for any $s \in S$ the leading part relative to the length s^0 of s does not depend to the subalgebra of $L(X)$ generated by the subset $\{a^0 \in S \mid a \neq s\}$. A subset S of $L(X)$ is called *independent* if S is a set of free generators of the subalgebra which it generates.

Theorem 1 (A. I. Shirshov, [Sh]). *If K is a field, then any reduced subset of $L(X)$ is an independent subset.*

Definition 3. An element h of $L(X)$ is called *primitive* if there is a subset Y of $L(X)$ such that the subset $\{h\} \cup Y$ is a set of free generators of $L(X)$. Analogously we define primitive elements of the free associative algebra $A(X)$ over K on X.

Theorem 2 (G. P. Kukin, [Ku1]). *Let K be a field, $h \in L(X)$, $\mathrm{id}(h)$ the ideal of $L(X)$ generated by h. Then h is a primitive element of $L(X)$ if and only if the algebra $L(X)/\mathrm{id}(h)$ is a free Lie algebra.*

2. Main theorem

In this section we give an example of a nonfree Lie algebra L over a field of prime characteristic which has cohomological dimension one (moreover, in this example L is not free, but its universal enveloping algebra $U(L)$ is a free associative algebra).

Theorem 3. *Let K be a field, $\mathrm{char}\, K = p > 2$, $X = \{x, y, z\}$, $L(X)$ the free Lie algebra over K on X,*

$$h = x + [y, z] + (\mathrm{ad}\, x)^p(z),$$

where $(\mathrm{ad}\, a)(b) = [a, b]$ for $a, b \in L(X)$. Denote by $\mathrm{id}(h)$ the ideal of $L(X)$ generated by h. Then the Lie algebra $L = L(X)/\mathrm{id}(h)$ is not a free Lie algebra, but at the same time $\mathrm{cd}(L) = 1$ and the universal enveloping algebra $U(L)$ of L is a free associative algebra.

Proof. Firstly, we show that h is not a primitive element. Assume to the contrary that h is a primitive element. Then there exist elements f_1, f_2 of $L(X)$ such that $\{h, f_1, f_2\}$ is a set of free generators of $L(X)$. By Theorem 1 we may assume that $\ell(f_1) \leq \ell(f_2) \leq \ell(h)$, and that the leading part h^0 of h is a Lie polynomial on leading parts f_1^0, f_2^0 of elements f_1, f_2, respectively. Since $h^0 = (\operatorname{ad} x)^p(z)$, we may assume that $f_1^0 = \alpha x$, $0 \neq \alpha \in K$. Thus, $f_1 = \alpha x$ and

$$h - \frac{1}{\alpha} f_1 = [y, z] + (\operatorname{ad} x)^p(z)$$

is a primitive element. It gives a contradiction with the fact that any set of free generators of $L(X)$ forms a basis of $L(X)$ modulo $[L(X), L(X)]$. Therefore, by Theorem 2, the Lie algebra $L(X)/\operatorname{id}(h)$ is not free.

Consider now the free associative algebra $A(X)$. Clearly, $U(L) \cong A(X)/\operatorname{Id}(h)$, where $\operatorname{Id}(h)$ is the ideal of $A(X)$ generated by h. Consider the elements $u = y + x^p$ and $v = z$. One can easily see that the set $\{h, u, v\}$ is a set of free generators of $A(X)$. Therefore, $U(L)$ is a free associative algebra with the set $\{\overline{u}, \overline{v}\}$ of free generators.

As above, we consider the right ideal $I(L)$ of $U(L)$ generated by L. Then

$$I(L) = \overline{u}U(L) \oplus \overline{v}U(L).$$

Since $U(L)$ is the free associative algebra on $\{\overline{u}, \overline{v}\}$, $I(L)$ is a free right $U(L)$-module. Therefore, by Remark 1 we get $\operatorname{cd}(L) \leq 1$. Consider the group $H^1(L, K)$, where

$$H^1(L, K) = \operatorname{Ker} d_2^* / \operatorname{Im} d_1^*,$$

$$\operatorname{Hom}_{U(L)}(U(L), K) \xrightarrow{d_1^*} \operatorname{Hom}_{U(L)}(I(L), K) \xrightarrow{d_2^*} \operatorname{Hom}_{U(L)}(0, K) = 0.$$

Since $I(L)$ is a free $U(L)$-module of the rank 2, we get

$$\dim_K \operatorname{Hom}_{U(L)}(I(L), K) = 2.$$

Taking into account that

$$\operatorname{Ker} d_2^* = \operatorname{Hom}_{U(L)}(I(L), K)$$

and

$$\dim_K \operatorname{Hom}_{U(L)}(U(L), K) = 1,$$

we see that $H^1(L, K) \neq 0$. Thus, $\operatorname{cd}(L) = 1$. This completes the proof. □

At the end of this section we formulate some conjectures in this area.

Conjecture 1. If $\operatorname{char} K = 0$, L is a Lie algebra over K, $\operatorname{cd}(L) = 1$, then L is a free Lie algebra (the first nontrivial case is a Lie algebra given by one defining relation).

Conjecture 2. If $\operatorname{char} K = p > 2$, L is a Lie p-algebra over K, $\operatorname{cd}(L) = 1$, then L is a free Lie p-algebra.

Conjecture 3. If L is a Lie algebra over a field, $\operatorname{cd}(L) = 1$, then its universal enveloping algebra $U(L)$ is a free associative algebra.

It would be also interesting to consider (restricted) Lie algebras of cohomological dimension two.

3. Another proof of the main theorem using Fox differential calculus

Let $X = \{x_1, \dots, x_n\}$, $n \geq 2$, $L(X)$ the free Lie algebra on X over a field K. We consider $L(X)$ as a subset of $A(X) = U(L(X))$. Any element $f \in L(X)$ can be uniquely written in the form

$$f = x_1 f_1 + \dots + x_n f_n,$$

where $f_i = \frac{\partial f}{\partial x_i} \in A(X)$ are called Fox partial derivatives of f.

Theorem 4 (A. A. Mikhalev, A. A. Zolotykh, [MZ1, MZ2]). *Let L_X be the free Lie algebra $L(X)$ in the case when* char $K = 0$ *and L_X the free Lie p-algebra $L^p(X)$ in the case when* char $K = p > 2$, *h an element of L_X. Then the following conditions are equivalent:*

i) *the element h is primitive;*

ii) *there exist elements $m_1, \dots, m_n \in A(X)$ such that*

$$\sum_{i=1}^{n} m_i \frac{\partial h}{\partial x_i} = 1.$$

Theorem 5. *Let $f \in L(X)$. Suppose that*

$$m_1 \frac{\partial f}{\partial x_1} + \dots + m_n \frac{\partial f}{\partial x_n} = 1$$

for some $m_1, \dots, m_n \in A(X)$. Then $\operatorname{cd}(L(X)/\operatorname{id}(f)) = 1$.

Proof. Let $L = L(X)/\operatorname{id}(f)$. Since

$$I(L(X)) = x_1 U(L(X)) \oplus \dots \oplus x_n U(L(X)),$$

we may define the homomorphism of right $U(L(X))$-modules

$$\varphi\colon I(L(X)) \to U(L(X))$$

setting $\varphi(x_i) = m_i$, $i = 1, \dots, n$.

Let now

$$P = \psi\varphi\colon I(L(X)) \to I(L(X)),$$

where the homomorphism

$$\psi: U(L(X)) \to I(L(X))$$

of right $U(L(X))$-modules is defined by $\varphi(1) = f$. Then $P^2 = P$ and

$$I(L(X)) = \operatorname{Ker} P \oplus \operatorname{Im} P,$$
$$\operatorname{Im} P = fU(L(X)).$$

Since $U(L(X)) = A(X)$ is a free ideal ring (see [Co], Chapter 2), there exists a basis $\{g_2, \ldots, g_n\}$ of the free right $U(L(X))$-module $\operatorname{Ker} P$,

$$\operatorname{Ker} P = g_2U(L(X)) \oplus \cdots \oplus g_nU(L(X)).$$

Thus, $\{g_1 = f, g_2, \ldots, g_n\}$ is a free basis of $I(L(X))$.

We show that

$$I(L) = \overline{g_2}U(L) \oplus \cdots \oplus \overline{g_n}U(L),$$

where $\overline{g_i} = g_i + \operatorname{id}(f)$, $g_i \in L(X)$. Indeed, let

$$\overline{g_2}\,\overline{u_2} + \cdots + \overline{g_n}\,\overline{u_n} = \overline{0},$$

where $\overline{u_i} = u_i + \operatorname{Id}(f)$, $u_i \in U(L(X))$. Then

$$u = g_2u_2 + \cdots + g_nu_n \in \operatorname{Id}(f),$$

and therefore

$$u = fs_0 + x_1s_1 + \cdots + x_ns_n, \qquad s_0, s_1, \ldots, s_n \in \operatorname{Id}(f).$$

Using aboveconstructed basis $\{g_1 = f, g_2, \ldots, g_n\}$ of $I(L(X))$, we may write down

$$x_1s_1 + \cdots + x_ns_n = ft_1 + g_2t_2 + \cdots + g_nt_n,$$

where $t_i \in \operatorname{Id}(f)$. Since

$$u = g_2u_2 + \cdots + g_nu_n = f(s_0 + t_1) + g_2t_2 + \cdots + g_nt_n,$$

we get $s_0 + t_1 = 0, u_2 = t_2, \ldots, u_n = t_n$. Thus $\overline{u_i} = \overline{0}$, $i \geq 2$. By Remark 1, $\operatorname{cd}(L) \leq 1$.

If char $K = 0$, then by Theorem 4 f is a primitive element of $L(X)$. If char $K = p > 2$, then by Theorem 4 f is a primitive element of the free Lie p-algebra $L^p(X)$. Therefore in all cases f is a primitive element of $A(X)$. As in Section 2, we have that $\operatorname{cd}(L) = 1$. It completes the proof. □

Second proof of Theorem 3. We now apply Theorem 5 to our element

$$h = x + [y, z] + (\operatorname{ad} x)^p(z).$$

Indeed, since char $K = p$,

$$\frac{\partial h}{\partial x} = 1 + x^{p-1}z, \qquad \frac{\partial h}{\partial y} = z, \qquad \frac{\partial h}{\partial z} = -y - x^p.$$

Consider

$$m_1 = 1, \qquad m_2 = -x^{p-1}, \qquad m_3 = 0.$$

It is clear that

$$m_1 \frac{\partial h}{\partial x} + m_2 \frac{\partial h}{\partial y} + m_3 \frac{\partial h}{\partial z} = 1.$$

Thus $\mathrm{cd}(L(X)/\,\mathrm{id}(h)) = 1$. All other statements can be proved as in Section 2. □

References

[Co] P. M. Cohn, Free rings and their relations, 2nd Ed., Academic Press, London–New York 1985.

[Fe] G. L. Feldman, Ends of Lie algebras, Uspekhi Mat. Nauk 38 (1983), no. 1, 199–200.

[Fu] D. B. Fuchs, Cohomology of Infinite Dimensional Lie Algebras, Nauka, Moscow 1984.

[GG] M. Goto and F. D. Grosshans, Semisimple Lie Algebras, Marcel Dekker, New York–Basel 1978.

[Ja] N. Jacobson, Lie Algebras, Wiley-Interscience, New York 1962.

[Ku1] G. P. Kukin, Primitive elements of free Lie algebras, Algebra i Logica 9 (1970), no. 4, 458–472.

[Ku2] G. P. Kukin, On Lie algebras of cohomology dimension one, 5-th All-Union Symposium on the Theory of Rings, Algebras and Modules, Abstracts of reports, Novosibirsk 1982, p. 86.

[Ku3] G. P. Kukin, Lie algebras of cohomology dimension one, Plenary talk on the IV All-Union School Lie algebras and their applications in mathematics and physics, Kazan 1990, unpublished.

[MUZ] A. A. Mikhalev, U. U. Umirbaev, and A. A. Zolotykh, An example of a nonfree Lie algebra of cohomological dimension one, Uspekhi Mat. Nauk 49 (1994), no. 1, 203–204.

[MZ1] A. A. Mikhalev and A. A. Zolotykh, Rank of an element of the free Lie p-superalgebra, Doklady Akad. Nauk 334 (1994), no. 6, 690–693; English translation: Russian Acad. Sci. Dokl. Math. 49 (1994), no. 1, 189–193.

[MZ2] A. A. Mikhalev and A. A. Zolotykh, Rank and primitivity of elements of free color Lie p-superalgebras, Internat. J. Algebra Comput. 4 (1994), 617–656.

[Sh] A. I. Shirshov, Subalgebras of free Lie algebras, Mat. Sb. 33 (1953), 441–452.

[St] J. R. Stallings, On torsion-free groups with infinitely many ends, Ann. Math. 88 (1968), no. 2, 312–334.

[Sw] R. G. Swan, Groups of cohomological dimension one, J. Algebra 12 (1969), 585–610.

A. A. Mikhalev and A. A. Zolotykh
Department of Mechanics and Mathematics
Chair of Higher Algebra
Moscow State University
Moscow 119899, Russia

U. U. Umirbaev
Department of Mathematics
Kazakhstan State University
Almaty, Kazakhstan

Extended centroid of Ore extensions and injective ring endomorphisms

*V. A. Mushrub**

1. Introduction

Throughout this paper all rings are associative. All rings are assumed to have unity, and it will be assumed that all subrings and homomorphisms are unitary.

The notion of extended centroid was first developed for prime rings in [7]. The aim of this paper is to describe the structure of the extended centroid of Ore extensions of the form $R[x, F]$, where R is an F-prime ring and F is an injective endomorphism of R. Notice that results on extended centroids of skew polynomial rings $R[x, F]$ have been obtained to the present day only under the assumption that F is an automorphism. In particular, in [8] Matczuk proves the following.

Theorem. *Let R be a ring, Q be its left Martindale ring of quotients and f be an automorphism of R. Then the extended centroid of $R[x, f]$ is isomorphic to $Z Z^{-1}$, where Z is the center of $Q[x, f]$.*

In § 3 we study some properties of the Martindale f-ring of quotients and, in particular, prove a generalization of Montgomery's and Passman's lemma ([10]) on inner automorphisms.

In § 4 we extend Matczuk's result mentioned above to f-prime rings.

In § 5, the situation we shall be concerned with is that of a ring R, with an injective ring endomorphism $F: R \longrightarrow R$, which will not be assumed to be surjective. Here we impose a condition on R, called the F-stability condition for left annihilators, which allows us to describe the extended centroid of $R[x, F]$.

In the case under consideration we face some difficulties. For instance, there is no reasonable way of extending F to the maximal ring of quotients of R and the skew

* Partially supported by the Russian Foundation for Fundamental Research, grant 95-01-01185, and by the International Science Foundation and Russian Government, grant TBX100.

Laurent polynomial ring $R[x, x^{-1}, F]$ does not exist. In order to surmount these difficulties we will use the construction of a minimal overring $A(R, F)$ to which F extends as an automorphism. This fruitful construction was first introduced in [5] by D. A. Jordan.

The main results of the paper are Theorem 4.13 and Theorem 5.13.

2. Prerequisites

Let F be an injective endomorphism of the ring R. Then it is possible to define a new ring $R[x, F]$, whose underlying additive group coincides with that of $R[x]$. Multiplication is determined by the distributive law and the rule

$$xr = f(r)x$$

for all $r \in R$. The ring $R[x, F]$ is called the skew polynomial ring over R or the Ore extension of R with respect to F.

An element $b \in R$ is said to be F-invariant if $F(b) = b$. A subset I of R is called F-invariant if $F(I) \subseteq I$. An ideal (a subset) I of R is said to be an F-ideal (an F-subset, resp.) if $F^{-1}(I) = I$.

2.1. Definition. The ring R is called *F-prime* if the product of any two nonzero F-ideals of R is nonzero.

The center of a ring B we will denote by $Z(B)$. If B is a ring then the maximal left ring of quotients of B and the maximal right ring of quotients of B are denoted by $Q_m(B)$ and $Q_m^r(B)$, respectively. $C(B)$ and $C^r(B)$ will stand for the center of $Q_m(B)$ and the center of $Q_m^r(B)$, respectively. For an information on maximal rings of quotients and dense ideals we refer to [6, chapter 4]. All required information on classical rings of quotients can be found in [12] and [3].

The following result is standard.

2.2. Lemma. *Let B be an arbitrary ring. Then $C(B) = \{c \in Q_m(B) : cb = bc$ for all $b \in B\}$.*

For a ring B and a subset M of B, let $l_B(M) = \{b \in B : bM = 0\}$ and $r_B(M) = \{b \in B : Mb = 0\}$. If an element c belongs either $C(B)$ or $C^r(B)$, then we put $D(C) = \{b \in B : cb \in B\}$.

Because in § 4 we study extended centroids not only for prime rings, we must first explain what we mean by the extended centroid of the ring B. It is convenient to define the extended centroid as follows. The center $C(B)$ of $Q_m(B)$ is, by definition, the extended centroid of B if the following condition holds

$$r_B\big(D(c)\big) = 0 \quad \text{for all} \quad c \in C^r(B) \quad \text{and}$$
$$l_B\big(D(c)\big) = 0 \quad \text{for all} \quad c \in C(B) \qquad (*)$$

Otherwise, we mean that the extended centroid of B does not exist.

We remark that if B satisfies condition $(*)$ then $C(B)$ is isomorphic to $C^r(B)$ in a natural manner. Indeed, in this situation both rings $C(B)$ and $C^r(B)$ are isomorphic to the direct limit

$$\varinjlim\Big(\mathrm{Hom}_B({}_BI_B,\ {}_BB_B), I \in \Psi\Big)$$

where Ψ is the filter of all two-sided ideals I of B such that $r_B(I) = 0$ and $\mathbf{l}_B(I) = 0$. If we assume B is a semiprime ring, then the extended centroid of B, as defined above, is just the center of the left Martindale ring of quotients of B.

3. Some properties of the f-ring of quotients

Let B be a ring and f be an automorphism of B. Throughout this section B will be f-prime. In what follows, let $\Phi(B)$ be the set of all nonzero f-ideals of B.

In next definition we define the left Martindale f ring of quotients $Q_f(B)$ as

$$\varinjlim\Big(\mathrm{Hom}_B({}_BI,\ {}_BB) : I \in \Phi(B)\Big)$$

3.1. Let M be the set of all pairs (I, α) where $I \in \Phi(B)$ and $\alpha : I \longrightarrow B$ is a homomorphism of left B-modules. Since B is f-prime, we have that $0 \neq IJ \subseteq I \cap J$ for all $I, J \in \Phi(B)$. Therefore, if $I, J \in \Phi(B)$ then $IJ \in \Phi(B)$ and $I \cap J \in \Phi(B)$. Define the equivalence θ on M by the following way: $(I\alpha)\theta(J, \beta)$ if there exists $K \in \Phi(B)$, $K \subseteq I \cap J$, such that $\alpha(a) = \beta(a)$ for all $a \in K$. Let us denote the equivalence class containing (I, α) by $[I, \alpha]$ and the set M/θ of all equivalence classes by $Q_f(B)$. Then the set $Q_f(B)$ becomes a ring if we put

$$\big[I\alpha\big] + \big[J, \beta\big] = \big[I \cap J, \alpha + \beta\big],$$
$$-\big[I, \alpha\big] = \big[I, -\alpha\big] \quad \text{and}$$
$$\big[I, \alpha\big]\big[J, \beta\big] = \big[IJ, \alpha\beta\big].$$

Definition ([1], [9]). The ring $Q_f(B)$ is called the *left (Martindale) f-ring of quotients* of B. The right (Martindale) f-ring of quotients

$$Q_f^r(B) = \varinjlim\big(\mathrm{Hom}_B(I_B, B_B) : I \in \Phi(B)\big)$$

of B is defined analogously.

Remark. If every ideal of B is an f-ideal, then $Q_f(B)$ is just the left Martindale ring of quotients of B.

3.2. Lemma. *Let A be a ring and f be an automorphism of A. The following conditions (1)–(5) are equivalent:*

(1) *A is f-prime;*

(2) *for any two left f-ideals I and J of A the equality $IJ = 0$ implies that either $I = 0$ or $J = 0$;*

(3) *$a \sum_{i \in \mathbb{Z}} Af^i(b) \neq 0$ for any two nonzero elements $a, b \in A$;*

(4) *$r_A(I) = 0$ for all nonzero f-ideals I of A;*

(5) *$l_A(I) = 0$ for all nonzero f-ideals I of A.*

3.3. Lemma. *Let $I \in \Phi(B)$, J be a left ideal of B and $\alpha, \beta : J \longrightarrow B$ be homomorphism of left B-modules. If $\alpha(a) = \beta(a)$ for all $a \in I \cap J$, then $\alpha \equiv \beta$.*

Proof. Let $b \in J$. If $a \in I$, then $ab \in I$ and $\alpha(ab) = \beta(ab)$. Hence $a\big(\alpha(b) - \beta(b)\big) = 0$ for all $a \in I$. Since $r_B(I) = 0$ by Lemma 3.2, it implies $\alpha(b) = \beta(b)$. □

The ring $Q_f(B)$ inherits some well-known properties of the Martindale ring of quotients of a prime ring. In particular, we have the following.

3.4. Lemma. *The ring $Q_f(B)$ can be embedded in $Q_m(B)$ and $B \subseteq Q_f(B)$. Analogously, $B \subseteq Q_f^r(B) \subseteq Q_m^r(B)$.*

Proof. If $I \in \Phi(B)$ then $r_B(I) = 0$ and, consequently, I is a dense left ideal of B (see [6, § 4.3, Corollary]). Therefore, any element $q = [I, \alpha] \in Q_f(B)$ can be matched up with the element $\varphi(q) \in Q_m(B)$, such that $a\varphi(q) = \alpha(a)$ for all $a \in I$. One can verify that the correspondence $q \longmapsto \varphi(q)$ is a well-defined injective ring homomorphism.

Every element $a \in B$ can be interpretted as the homomorphism of left B-modules $a : B \longrightarrow B$, $b \longmapsto ba$. Since $B \in \Phi(B)$, the ring is embedded in $Q_f(B)$ in the natural way. □

In what follows, we will identify $Q_f(B)$ with the subring $\varphi\big(Q_f(B)\big) = \{q \in Q_m(B)$: there exists $I \in \Phi(B)$ such that $Iq \subseteq B\}$ of $Q_f(B)$. Analogously, the ring $Q_f^r(B)$ may be identified with a subring of $Q_m^r(B)$.

3.5. Let $q \in Q_f(B)$. Denote by $J(q)$ the sum of all f-ideals L of B with $Lq \subseteq B$. Observe that $J(q)$ is an f-ideal of B and $J(q)q \subseteq B$. It follows that $J(q)$ is the largest f-ideal of B with respect to the condition $J(q)q \subseteq B$.

For $q \in Q_m(B)$, let $D(q) = \{a \in Q_m(B) : aq \in B\}$. It is well-known that the automorphism f can be extended to a unique automorphism of $Q_m(B)$, which we will denote by f again. If $q \in Q_m(B)$ and $D = D(q)$, then $f(a)f(q) = f(aq)$ for all $a \in D$.

If $q \in Q_f(B)$, then $af(q) = f\big(f^{-1}(a)q\big) \in B$ and $af^{-1}(q) = f^{-1}\big(f(a)q\big) \in B$ for all $a \in J(q)$. Hence $J(q)f(q) \subseteq B$ and $J(q)f^{-1}(q) \subseteq B$. Thus $f(Q_f(B)) \subseteq Q_f(B)$ and $f^{-1}(Q_f(B)) \subseteq Q_f(B)$ and therefore f can be considered as an automorphism of $Q_f(B)$.

The following lemmas were quoted in [1] without proof.

3.6. Lemma.

(a) *For any $q_1, \dots, q_n \in Q_f(B)$ there exists $I \in \Phi(B)$ such that $Iq_i \subseteq B$ for $i = 1, 2, \dots, n$.*

(b) *If $Iq = 0$ for some $q \in Q_f(B)$ and $I \in \Phi(B)$, then $q = 0$.*

(c) *If $qI = 0$ for some $q \in Q_f(B)$ and $I \in \Phi(B)$, then $q = 0$.*

(d) *If $I \in \Phi(B)$ and $\gamma : I \longrightarrow B$ is a homomorphism of left B-modules, then there exists $q \in Q_f(B)$ such that $\gamma(a) = aq$ for all $a \in I$.*

(e) *$Q_f(B)$ is f-prime.*

Proof. (a) We can put $I = J(q_1) \cap J(q_2) \cap \cdots \cap J(q_n)$. It works.

(b) This follows from 3.1 and Lemma 3.3.

(c) It is evident that $P = \sum_{i \in \mathbb{Z}} J(q)f^i(q)$ is a right f-ideal of B and $PI = \sum J(q)f^i(qI) = 0$. Since B is f-prime, $P = 0$. Hence $J(q)q = 0$ and $q = 0$.

(d) This follows from Definition 3.1.

(e) Let a and b be nonzero elements of $Q_f(B)$. Then it follows from (b) that $\sum_{i \in \mathbb{Z}} J(a)f^i(a)$ and $\sum_{i \in \mathbb{Z}} J(b)f^i(b)$ are nonzero right f-ideals of B. Since B is f-prime, we get $\big(\sum_i J(a)f^i(a)\big)\big(\sum_j J(b)f^j(b)\big) \neq 0$. Thus $a \cdot \sum_{j \in \mathbb{Z}} Q_f(B)f^j(b) \neq 0$ and $Q_f(B)$ is f-prime. □

Remark. There is, of course, a similar result for the right f-ring of quotients of B.

3.7. We put $C_f(B) = \{c \in C(B) : f(c) = c\}$ and $C_f^r(B) = \{c \in C^r(B) : f(c) = c\}$.

3.8. Lemma.

(1) *If $q \in C_f(B)$ then $D(q) = \{b \in B \mid bq \in B\}$ is an f-ideal.*

(2) $C_f(B) = Z(Q_f(B))$.

(3) *Let $q = [I, \alpha] \in Q_f(B)$. Then $q \in Z\big(Q_f(B)\big)$ if and only if α is an B-bimodule homomorphism.*

(4) *The ring $C_f(B)$ is isomorphic to the ring $C_f^r(B)$.*

Proof. (1) Let $a \in B$. Then $qa \in B$ if and only if $f(aq) = f(a)q \in B$. It follows that $D(q) \in \Phi(B)$.

(2) This follows from (1) and Lemma 2.2.

(3) If α is an B-bimodule homomorphism, then (as is well-known and easily seen) $q \in C(B)$ and so $q \in C_f(B)$. The converse follows from (1).

(4) It is easily seen from statement (3) that both $C_f(B)$ and $C_f^r(B)$ are isomorphic to the ring $\{c \in \varinjlim(\mathrm{Hom}_B({}_BI_B, {}_BB_B): I \in \Phi(B)) \mid f(c) = c\}$. □

3.9. Lemma ([2, Lemma 1.2]). *Assume that $q \in Q_f(B)$ verifies $qB = Bq$ and $f(q) = q$. Then q is invertible in $Q_f(B)$. In particular, $C_f(B)$ is a field.*

In what follows, we will use the following generalization of the lemma of Montgomery and Passman ([10]).

3.10. Lemma. *Let $Q = Q_f(B)$ and $0 \neq q \in Q$. If $qf(a) = aq$ for all $a \in B$, then q is an invertible element in Q and f is an inner automorphism of Q determined by q (i.e., $f(a) = q^{-1}aq$ for all $a \in Q$).*

Proof. Let p be an arbitrary element of Q and k be an integer. Then $f^k(q)f(a) = f^k(qf^{1-k}(a)) = f^k(f^{-k}(a)q) = af^k(q)$ for all $a \in B$. It follows that $af^k(q)f(p) = f^k(q)f(a)f(p) = f^k(q)f(ap) = apf^k(q)$ for all $a \in J(p)$. So $J(p)(f^k(q)f(p) - pf^k(q)) = 0$. Thus $f^k(q)f(p) = pf^k(q)$ for all integers k and all $p \in Q$. It implies that $Qf^k(q) = f^k(q)Q$.

We first show that $qf^k(q) \neq 0$ for some integer k. Indeed, if it is not true, then

$$q\sum_{k\in\mathbf{Z}} Qf^k(q) = q\sum_{k\in\mathbf{Z}} f^k(q)Q = 0,$$

contrary to the f-primeness of Q.

Now let $u = qf^k(q) \neq 0$. Then $au = aqf^k(q) = qf(a)f^k(q) = uf^2(a)$ for all $a \in B$. Moreover $u = qf^k(q) = f^k(q)f(q) = f(q)f(f^k(q)) = f(u)$. Hence $f(u) = u$ and $Bu = uB$. It follows then from Lemma 3.9 that u is invertible in Q. Observe that $q(f^k(q)u^{-1}) = uu^{-1} = 1$ and $(u^{-1}f^{k-1}(q))q = u^{-1}qf^k(q) = u^{-1}u = 1$. Thus q is invertible in Q.

As we have seen, $qf(p) = pq$ for all $p \in Q$. Since q is invertible, we have $f(p) = q^{-1}pq$ for all $p \in Q$. □

3.11. Corollary. *Let $0 \neq q \in Q_f^r(B)$. If $qf(a) = aq$ for all $a \in B$ then q is an invertible element in $Q_f^r(B)$ and f is an inner automorphism determined by q.*

Proof. We have $f^{-1}(a)q = qa$ for all $a \in B$. Using the right-sided version of Lemma 3.10, one can prove that q is invertible in $Q_f^r(B)$ and $f^{-1}(a) = qaq^{-1}$ for all $a \in Q_f^r(B)$. □

3.12. Corollary. *If B contains an element x such that $xa = f(a)x$ for all $a \in B$ then $C(B) = C_f(B)$.*

Proof. By Lemma 3.10 f^{-1} is an inner automorphism of $Q_f(B) \subseteq Q_m(B)$ determined by x. It implies that $f(c) = xcx^{-1} = xx^{-1}c = c$ for all $c \in C(B)$. □

4. The case of automorphism

Let B be a ring and f be an automorphism of B. Then by $B\langle x, f\rangle$ we denote the skew Laurent polynomial ring, where multiplication is determined by the rule $x^n b = f^n(b)x^n$ for all $b \in B$ and all integer n. $B\langle x\rangle$ will stand for the Laurent polynomial ring $B[x, x^{-1}]$.

Throughout the sequel let A be an f-prime ring and f be an automorphism A.

It is convenient to fix some notation for this and the succeeding section.

We set $A^* = A\langle x, f\rangle$, $Q = Q_f(A)$, $C_f = C_f(A)$. The unique extension of f to Q will be also denoted by f. Set $Q^* = Q\langle x, f\rangle$ and $Z = Z(Q^*)$. V will stand for the subring of $Q_m(Q^*)$ generated by Q^* and $C(Q^*)$ (i.e. the central closure of Q^*). $\Phi(A)$, $\Phi(A^*)$ and $\Phi(Q^*)$ will denote the sets of all f-ideals of A, A^* and Q^*, respectively. Notice that one can extend f to an automorphism of A^* and to an automorphism of Q^* by putting $f(x) = x$. As we have seen in 3.5, f can be extended in a unique way from Q^* to $Q_m(Q^*)$. This extension will be also denoted by f. The inner automorphism of Q induced by an invertible element $u \in Q$ (i.e. given by $q \longmapsto u^{-1}qu$) will be denoted by σ_u.

Observe that every ideal of A^* (of Q^*) is an f-ideal.

4.1. Lemma. *The rings A^* and Q^* are prime.*

Proof. This can be deduced from Lemma 3.2 and Lemma 3.4. □

Let us identify A^* with a subring of Q^* in a natural way.

4.2. Lemma. *The ring Q^* embeds (over A^*) in $Q_f(A^*)$.*

Proof. Let $p = \sum_{i=m}^{n} q_i x^i \in Q^*$, where $q_i \in Q$. Then by Lemma 3.6 (a) there exists $I \in \Phi(A)$ such that $I_{q_i} \subseteq A$ for $m \leq i \leq n$. Observe that $If^z(q_i) \subseteq A$ for all integers z and all i. Set $I^* = \sum_{z\in\mathbf{Z}} Ix^z$. Clearly, $I^* \in \Phi(A^*)$ and $I^*p \subseteq A^*$. Therefore we can define the homomorphism (of left A^*-module) η_p, corresponding to p, by the rule $\eta_p(a) = ap$ for all $a \in T^*$. This yields an embedding η: $Q^* \longrightarrow Q_f(A^*)$ such that $\eta(p) = [I^*, \eta_p]$. □

It is not difficult to verify that $\eta f = f\eta$. Hence, it is convenient to consider Q^* as a subring of $Q_f(A^*)$, and this does not lead to confusion.

4.3. Lemma. *If B and B_1 are rings and $B \subseteq B_1 \subseteq Q_m(B)$, then $Q_m(B) = Q_m(B_1)$ and $C(B) = C(B_1)$.*

Proof. The first equality can be proved by arguing as in the proof of Proposition 4.3.2 [6]. The second equality follows from Lemma 2.2. □

4.4. Lemma.

(1) $C(A^*) = C_f(A^*)$ *and* $C(Q^*) = C_f(Q^*)$.

(2) $Q_m(Q^*) = Q_m(A^*)$.

(3) $C(A^*) = C(Q^*)$.

(4) $C(A^*) \subseteq V$.

(5) $Z = C(A^*) \cap Q^* = \{z \in Q^*: za = az \text{ for all } a \in A^*\}$.

Proof. (1) follows from Corollary 3.12; (2) and (3) result from Lemma 4.2 and Lemma 4.3. Statement (4) is a consequence of (3). Statement (5) can be obtained from Lemma 2.2 and statement (3). □

4.5. Lemma. *If $I \in \Phi(A)$, $v \in V$ and $Iv = 0$, then $v = 0$.*

Proof. Since $Q^* \subset Q_f(A^*)$ by Lemma 4.2 and $C(Q^*) = C_f(A^*) \subseteq Q_f(A^*)$ by Lemma 4.4, we have that $V \subseteq Q_f(A^*)$. Let $I^* = \sum_{z \in \mathbf{Z}} x^z I$. Then $I^* \in \Phi(A^*)$ and $I^* v = 0$. Hence Lemma 3.6 (b) gives $v = 0$. □

The skew polynomial ring $A[x, f]$ can be considered as the subring of A^* generated by A and x. Let $q = \sum_{i=k}^{n} q_i x^i \in Q^*$ with $q_k \neq 0$, $q_n \neq 0$ and $k \leq n$. Then the constant term q_0 is denoted by $\mathrm{ct}(q)$, the support of q is $\mathrm{supp}(q) = \{k, k+1, \ldots, n\}$. In the case that $h \in A[x, f]$, the degree of h is defined in the obvious manner and denoted by $\deg h$.

Let I be a nonzero ideal of A^*. Let us denote by n the smallest non-negative number such that there exists $0 \neq h \in I \cap A[x, f]$ with $\deg h = n$. We put

$$L^* = \left\{u \in I \cap A[x, f] \mid \deg u \leq n\right\} \quad \text{and} \quad L = \left\{\mathrm{ct}(u) \mid u \in L^*\right\}.$$

Clearly, $L^* \neq 0$ and $L \neq 0$. Since every ideal of A^* is an f-ideal, L^* is an f-set and L is an f-ideal of A.

4.6. Lemma. *There exists $z \in Z$ such that* $\mathrm{ct}(z) = 1$ *and* $p = \mathrm{ct}(p) z$ *for any $p \in L^*$.*

Proof. Let $h \in L^*$ and $\mathrm{ct}(h) = 0$. Then $hx^{-1} \in L^*$ and $\deg hx^{-1} = n - 1$. By the minimality of n we get $hx^{-1} = 0$ and, consequently, $h = 0$. Thus for any $h \in L^*$, $\mathrm{ct}(h) = 0$ implies $h = 0$. Therefore, if $u, v \in L^*$ and $\mathrm{ct}(u) = \mathrm{ct}(v)$, then $u = v$.

Thus, if $a \in L$ then there exists a unique

$$p = a + a_1 x + a_2 x^2 + \cdots + a_n x^n \in L.$$

It is then straightforward to see that the maps $d_i: L \longrightarrow A$ given by $\alpha_i(a) = a_i$ are well-defined homomorphisms of left A-modules for all $1 \leq i \leq n$. Since $L \in \Phi(A)$, there exist elements $q_1, \ldots, q_n$ in Q such that $\alpha_i(a) = aq_i$ for every $a \in L$ and for all i. Now it is enough to take

$$z = 1 + q_1 x + \cdots + q_n x^n.$$

Now we will show that $z \in Z$. Let $a \in L$. Then $az \in L^*$ and $f(az) \in L^*$. Since $f(a) \in L$, we get $f(a)z \in L^*$. However, $\mathrm{ct}\big(f(az)\big) = \mathrm{ct}\big(f(a)z\big) = f(a)$. It follows that $f(a)f(z) = f(a)z$ for all $a \in L$ and, consequently, $L\big(f(z) - z\big) = 0$. Lemma 4.5 gives $z = f(z)$. Therefore $xz = zx$.

Now let $a \in L$, $b \in A$. Then $ab \in L$ and $az, abz \in L^*$. Since $\mathrm{ct}(abz) = \mathrm{ct}(azb) = ab$, we have that $abz = azb$. Hence $L(bz - zb) = 0$ for all $b \in A$. Lemma 4.5 shows that $bz = zb$ for all $b \in A$. Now it is clear that $zh = hz$ for all $h \in A^*$. Using Lemma 4.4(5), we get $z \in Z$. □

4.7. Theorem. *The extended centroid $C(A^*)$ is isomorphic to $Z^{-1}Z$.*

Proof. Lemma 4.4(5) shows that $Z \subseteq C(A^*)$. Since $C(A^*)$ is a field, we obtain that $Z^{-1}Z \subseteq C(A^*)$.

Now we are going to prove the converse inclusion. For let us fix a nonzero element $s \in C(A^*)$. Then Lemma 4.4 yields $s \in C_f(A^*)$. Put $I = \{h \in A^*: hs \in A^*\}$. By Lemma 3.8(1) I is a nonzero f-ideal of A^*. Now let L^*, L and z be as above. Then $Lzs \subseteq L^*s \subseteq A^*$. Take a nonzero element $a \in L$. Since $C(A^*) = C_f(Q^*)$ is a field and $Z \subseteq C(A^*)$, we have that zs is an invertible element of $Q_f(A^*)$, $f(zs) = zs$ and $azs \neq 0$. We put $K = \mathrm{supp}(azs)$ and $J = \{b \in L \mid \mathrm{supp}(bzs) \subseteq \mathrm{supp}(azs)\}$. Then $a \in J$ and hence J is a nonzero f-ideal of A^*.

If $b \in J$ then $bzs = \sum_{k \in K} \alpha_k(b)x^k$ for some $\alpha_k(b) \in A$. It is easily seen that all maps $\alpha_i: J \longrightarrow A$ are homomorphisms of left A-modules. By Lemma 3.6 (d) there are elements q_k in Q (where $k \in K$) such that $\alpha_k(b) = bq_k$ for all $b \in J$ and all $k \in K$. Therefore the equality $J(zs - \sum_{k \in K} q_k x^k) = 0$ holds in V. By Lemma 4.5 we get $zs = \sum_{k \in K} q_k x^k$. This implies that $zs \in Q^*$. Hence $zs \in Q^* \cap C(Q^*) = Z$. Recall that z is a nonzero element of Z. This implies that $s = z^{-1}(zs) \in Z^{-1}Z$. □

4.8. Proposition. *Let B be a ring, D be a left divisor set in B and $D^{-1}B$ be the classical left ring of fractions of B with respect to D. Then there exists an injective ring homomorphism $\varepsilon: D^{-1}B \longrightarrow Q_m(B)$ such that $\varepsilon(b) = b$ for all $b \in B$ and $\varepsilon(D^{-1}B) = \{d^{-1}b \mid d \in D, b \in B\}$.*

The proof is omitted since it is similar to those of Proposition 4.3.8 of [6] and Proposition 4.6.1 of [6].

By pullback along ε we may identify $D^{-1}B$ with a subring of $Q_m(B)$.

4.9. Proposition. *The extended centroid of $A[x, f]$ coincides with that of A^* and is equal to $Z^{-1}Z$.*

Proof. It is easily seen that $X = \{x^n : n \geq 0\}$ is a left divisor set in $A[x, f]$ and the ring A^* together with the natural embedding $A[x, f] \longrightarrow A^*$ is the classical left ring of quotients of $A[x, f]$ with respect to X. Applying Proposition 4.8 and Lemma 4.3, we obtain that $C(A[x, f]) = C(A^*)$.

Using the fact that every element $h = \sum_i a_i x^i \in A^*$ can be re-expressed as $h = \sum_i x^i f^{-i}(a_i)$, one can verify that X is a right divisor set in $A[x, f]$ and that A^* is the classical right ring of fraction of $A[x, f]$ with respect to X. Since A^* is prime, $C(A^*) = C^r(A^*)$ is an extended centroid of A^*. Applying right-sided versions of Lemma 4.3 and Proposition 4.8 we get that $C^r(A[x, f]) = C(A^*)$. Thus, both $C(A[x, f])$ and $C^r(A[x, f])$ coincide with $C(A^*)$. It follows then from Theorem 4.7 that $Z^{-1}Z$ is the extended centroid of $A[x, f]$. □

4.10. Lemma. *Let $h = \sum_{i=m}^{n} a_i t^i \in Z$. Then for each i, $m \leq i \leq n$, we have:* (1) *$f(a_i) = a_i$;* (2) *if $a_i \neq 0$ then a_i is invertible in Q and $f^i = \sigma_{a_i}$;* (3) *$a_i x^i \in Z$.*

Proof. The equality $hx = xh$ gives (1). Since $ba_i = a_i f^i(b)$ for all $b \in Q$, Lemma 3.10 gives (2). Statements (1) and (2) result in (3). □

4.11. There are two mutually exclusive cases to consider:

Case 1. f^k is not inner automorphism of Q for all $k \neq 0$;

Case 2. f^k is an inner automorphism of Q for some $k \neq 0$.

We claim that in Case 2 there exists the smallest positive natural number m such that f^m is an inner automorphism of Q determined by an f-invariant element $u \in Q$.

For let $k \neq 0$ and $f^k = \sigma_q$, $q \in Q$, and let $q_i = f^i(q)$ for $i \neq 0$. Without loss of generality we can assume that $k > 0$. Then $f^k(a) = f^i f^k f^{-i}(a) = f^i(q^{-1} f^k(a) q) = \sigma_{q_i}(a)$ for all $a \in Q$. Therefore $f^k = \sigma_{q_i}$ for any integer i. Set $d = q q_1 \ldots q_{k-1}$. Since $f^k(q) = q$ and $f^k(q_i) = q_i$, we have $f(d) = q_1 \ldots q_{k-1} q = q^{-1} d q = f^k(d) = d$. Thus f^{k^2} is an inner automorphism of Q induced by the element $d \in Q$ with $f(d) = d$.

4.12. Proposition. (1) *In Case* 1, $Z = C_f$. (2) *In Case* 2, $Z = C_f\langle ux^m \rangle$, *where m is as above and u is an f-invariant element determining f^m.*

Proof. (1) This follows from Lemma 4.10.

(2) By Lemma 4.10 ux^m is a polynomial of minimal positive degree in Z. Let $h = \sum_{i=j}^{n} a_i x^i \in Z$. Lemma 4.10 show that $a_i x^i \in Z$ for each $i = j, j+1, \ldots, n$. Let us fix $i = j, j+1, \ldots, n$ and let t and r be integers such that $i = tm + r$

where $0 \le r < m$. Then $a_i u^{-t} x^r = a_i x^i (ux^m)^{-t} \in Z$. By the minimality of m, we have $r = 0$. Lemma 4.10 implies that $a_i u^{-t} \in C_f$ and, consequently, $a_i x^i = a_i u^{-t}(ux^m) \in C_f\langle ux^m \rangle$. □

4.13. Theorem. *Let A be an f-prime ring. Then the extended centroid of $A[x, f]$ exists, coincides with that of A^* up to isomorphism and is equal to $Z^{-1}Z$, where $Z = Z(Q\langle x, f\rangle)$. Furthermore*

1) *in Case* 1, $C(A[x, f]) = C_f$;
2) *in Case* 2, $C(A[x, f]) = C_f(y)$, *the field of rational functions over C_f, where $y = ux^m$, u and m are as in* 4.11.

Proof. This theorem immediately follows from Proposition 4.9 and Proposition 4.12.

5. The case of injective endomorphism

Throughout this section let R be a ring and F be an injective endomorphism of R. All introduced in §§ 2–4 designations remain as these are above. For the sake of brevity let us denote the skew polynomial ring $R[x, F]$ by S.

Given an arbitrary ring R and an injective ring endorphism $F: R \longrightarrow R$, it possible to construct a minimal overring $A = A(R, F)$ of R to which F extends as an automorphism f say.

5.1. Definition ([5], [2, Chapter 7, Theorem 3.4]). Let A be a ring and f be an automorphism of A. The pair (A, f) is called the Cohn-Jordan extension of the pair (R, F) if the following conditions (i) and (ii) hold:

(i) the ring R is a subring of A and the automorphism f is an extension of F;

(ii) $A = \bigcup_{n\ge 0} f^{-n}(R)$.

In what follows, by the symbols A and f we always mean the ring and the automorphism which form the Cohn–Jordan extension of the pair (R, F). Following Jordan, we will also denote the ring A by $A(R, F)$. The ring $A(R, F)$ is called the Cohn–Jordan extension of R with respect to F. This ring is unique up to isomorphism over R.

5.2. Remark. To construct $A(R, F)$ it is enough to put

$$A(R, F) = \varinjlim(R_i, F^{j-i} \mid j \ge i \ge 0)$$

where $R_i = R$ for each non-negative integer i and $F^{j-i}: R_i \longrightarrow R_j$. It is shown in [5] that $A(R, F)$ may be identified with the subring $\{x^{-i}rx^i : r \in R, i \ge 0\}$ of the classical left ring of quotients $X^{-1}S$ of $S = R[x, F]$ with respect to $X = \{x^i : i \ge 0\}$.

Recall that a subset M of F is said to be F-invariant if $F(M) \subseteq M$. In what follows, $\Phi(A)$ and $\Gamma(R)$ will denote respectively the set of all nonzero f-ideals of A and the set of all F-invariant ideals of R.

The following lemma can be easily deduced from theorem 4.7 [5].

5.3. Lemma. *If $I \in \Gamma(R)$ then $\bigcup_{n\geq 0} f^{-n}(I) \in \Phi(A)$.*

5.4. Lemma. *Let I be a right F-invariant ideal of R and $J = \bigcup_{n\geq 0} f^{-n}(I)$. Then for any $a \in J$ anf $b \in A$ there exist a non-negative integer n and elements $r \in J$ and $s \in R$ such that $a = f^{-n}(r)$ and $b = f^{-n}(s)$.*

In what follows, Lemma 5.4 will be used without any indicating.

Note that F has a unique extension to S such that $F(x) = x$. Using Lemma 5.3, one can prove following.

5.5. Lemma. *The following conditions* 1)–4) *are equivalent:*

1) *R is F-prime;*
2) *A is F-prime;*
3) *S is F-prime;*
4) *for any two F-invariant ideals I and J of R the equality $IJ = 0$ implies that either $I = 0$ or $J = 0$.*

Notice that the Ore extension S may be identified with the subring of the skew Laurent polynomial ring A^* generated by R and x.

5.6. Theorem. *$C(S)$ is isomorphism to $C(A^*)$.*

Proof. It is a matter of direct verification to prove that the set $X = \{x^n : n \geq 0\}$ is a left divisor set in S. Identifying S with the subring of A^* generated by x and R, one can show that A^* is the classical left ring of quotients of S with respect to X (see Remark 5.2, a more detailed proof of this fact may be found in [5]). The rest follows from Lemma 4.5 and Proposition 4.8. □

In the remaining of the paper R will always be an F-prime ring.

In this circumstance A is f-prime by Lemma 5.5. This fact allows us to apply results of § 4.

We say that the ring R satisfies the F-stability condition for left annihilators if $l_R(I) = l_R(F^n(I))$ for all F-invariant ideals I of R and for all positive integers n.

Next we state without proof the following.

5.7. Theorem. *If R satisfies the F-stability condition for left annihilators, then S is a prime ring and $l_R(I) = 0$ for all $I \in \Gamma(R)$.*

5.8. Theorem. *Let m be a positive integer. Then the following conditions* 1) *and* 2) *are equivalent:*

1) *f^m is an inner automorphism of $Q_f(A)$ determined by some f-invariant element $u \in Q_f(A)$;*
2) *f^m is an inner automorphism of $Q_f^r(A)$ determined by some f-invariant element $v \in Q_f^r(A)$.*

Proof. First we show that (2)$\Longrightarrow$(1). Suppose that condition (2) holds. Then $v^{-1}av = f^m(a)$ and consequently $av = vf^m(a)$ for all $a \in A$. Let I be a nonzero f-ideal of A such that $vI \subseteq A$. Then $Iv = vf^m(I) = vI \subseteq A$. Lemma 3.6 shows that there exists an element $u \in Q_f(A)$ such that $au = av$ for all $a \in I$. Observe that if $b \in A$ and $a \in I$, then $abu = abv = avf^m(b) = auf^m(b)$. Therefore $I\big(bu - uf^m(b)\big) = 0$ for all $b \in A$. It follows then from Lemma 3.6 that $bu = uf^m(b)$ for all $b \in A$. Using Lemma 3.10 we get (1).

Now the statement of the theorem follows from the reasons of symmetry. □

For $q \in Q_f^r(A)$, let $I(q) = \{r \in R \mid qr \in R\}$.

5.9. Lemma. *Let $q \in Q_f^r(A)$, $f(q) = q$ and $I = I(q)$. Then I is a nonzero right F-invariant ideal of R.*

Proof. Let a be a nonzero element of A such that $qa \in A$. Then by Lemma 5.4 $qa \in f^{-n}(R)$ and $a = f^{-n}(s)$ for some non-negative integer n and some nonzero element $s \in R$. Then $qs = f^n(qa) \in R$ and, consequently, $s \in I$. Hence $I \neq 0$.

Finally, if $r \in I$ then $qF(r) = f(q)f(r) = f(qr) \in R$. This show that I a right F-invariant ideal. □

5.10. Proposition. *Let $T = \{q \in Q_f^r(A) \mid$ there exists $I \in \Gamma(R)$ such that $qI \subseteq R\}$ and $C_F = \{p \in C^r(R) \mid$ there exists $I \in \Gamma(R)$ such that $pI \subseteq R$ and $F(pr) = pF(r)$ for all $r \in I.\}$. Suppose that $l_R(I) = 0$ for all $I \in \Gamma(R)$. Then*

1) *$R \subseteq T$ and T is a subring of $Q_f^r(A)$;*
2) *$C_f^r(A) \subseteq T$;*
3) *there exists a ring homomorphism $\varphi\colon T \longrightarrow Q_m^r(R)$ such that $\varphi(r) = r$ for all $r \in R$ and $C_f^r(A) \cap \operatorname{Ker}\varphi = 0$;*
4) *$\varphi\big(C_f^r(A)\big) = C_F$.*

Proof. Statement 1) is evident.

2) It easily seen from the previous lemma that $I(q)$ is a nonzero F-ideal for each $q \in C_f^r(A)$.

3) As show in [6, § 4.3], an ideal I of R is a dense right ideal if and only if $l_R(I) = 0$. Hence every $I \in \Gamma(R)$ is a dense right ideal.

Let $q \in T$. Then from the construction of T it follows that the map $\alpha: I \longrightarrow R$ given by $\alpha(r) = qr$ for all $r \in I$ (where I is a nonzero F-ideal of R such that $qI \subseteq R$) is a well-defined homomorphism of right R-modules. As shown in Proposition 4.3.6. [6], there exists $\varphi(q) \in Q^r_m(R)$ such that $\varphi(q)r = qr$ for all $r \in R$. Consider the map $\varphi: q \longmapsto \varphi(q)$. We claim that φ is well defined. For let $q \in T$; $I, J \in \Gamma(R)$; $qI \subseteq R$, $qJ \subseteq R$ and $p_1, p_2 \in Q^r_m(R)$ be elements such that $p_1 r = qr$ for all $r \in I$ and $p_2 r = qr$ for all $r \in J$. Then $I \cap J \in \Gamma(R)$ and $p_1 r = p_2 r$ for all $r \in I \cap J$. This shows that $p_1 = p_2$ as required.

If $q, p \in T$, $I, J \in \Gamma(R)$ and $qI \subseteq R$, $pJ \subseteq R$, then $IJ \in \Gamma(R)$ and $\varphi(q)\varphi(p)r = qpr == \varphi(qp)r$ for all $r \in IJ$. It follows that $\varphi(q)\varphi(p) = \varphi(qp)$. Analogously, $\big(\varphi(q \pm p)\big) - \big(\varphi(q) \pm \varphi(p)\big)\ (I \cap J) = 0$ and, consequently, $\varphi(q) \pm \varphi(p) = \varphi(q \pm p)$. Thus we have proved that φ is a homomorphism of rings. Clearly, $\varphi(r) = r$ for all $r \in R$.

Let $c \in C^r_f(A) \cap \operatorname{Ker}\varphi$. Then $cI = 0$ for some $I \in \Gamma(R)$. By Lemma 5.3 $J = \bigcup_{n \geq 0} f^{-n}(I) \in \Phi(A)$. Observe that $cf^{-n}(I) = f^{-n}(cI) = 0$. It follows that $cJ = 0$ and, consequently, $c = 0$.

4) If $c \in C^r_f(A)$, $I \in \Gamma(R)$ and $cI \subseteq R$, then $\varphi(c)F(r) = \varphi\big(cF(r)\big) = f(cr) = F\big(\varphi(c)r\big)$ for all $r \in I$. This shows that $\varphi\big(C^r_f(A)\big) \subseteq C_F$.

On the other hand, let $s \in C_F$ and I be a nonzero F-ideal of R such that $sF(r) = F(sr) \in R$ for all $r \in I$. Then by Lemma 5.3 $J = \bigcup_{n \geq 0} f^{-n}(I) \in \Phi(A)$. It is easy to prove that the map $\gamma: J \longrightarrow A$, such that $\gamma(f^{-n}(r)) = f^{-n}(sr)$ for all $r \in I$ and all $n > 0$, is a well-defined homomorphism of A-bimodules. It follows from lemma 3.8 (3) that there exists $c \in C^r_f(A)$ such that $ca = \gamma(a)$ for all $a \in J$. Now it is straightforward to check that $cr = sr$ for all $r \in I$. Therefore $\varphi(c) = s$. □

5.11. Corollary. *If $l_R l(I) = 0$ for all $I \in \Gamma(R)$, then the ring C_F is isomorphic to $C^r_f(A)$.*

Proof. It follows from Proposition 5.10 (3),(4). □

5.12. Theorem. *Let m be a positive integer. Suppose that $\mathbf{l}_R(I) = 0$ for all $I \in \Gamma(R)$. Then the following conditions* (1) *and* (2) *are equivalent:*

(1) *f^m is an inner automorphism of $Q^r_f(A)$ determined by some f-invariant element $v \in Q^r_f(A)$;*

(2) *there exist an element $w \in Q^r_m(R)$ and an ideal $I \in \Gamma(R)$ such that $wr = F^m(r)w$ for all $r \in R$ and $F(wr) = wF(r)$ for all $r \in I$.*

Proof. (1)$\Longrightarrow$(2). Let v be an f-invariant element of $Q^r_f(A)$ such that $f^m(q) = v^{-1}qv$ for all $q \in Q^r_f(A)$. It is claimed that $v^{-1} \in T$. Indeed, let $I = \{r \in R: v^{-1}r \in R\}$. Then by Lemma 5.9 I is a nonzero right F-ideal of R. One can verify by a direct calculation that $v^{-1}q = f^m(q)v^{-1}$ for all $q \in Q^r_f(A)$. Hence

$v^{-1}RI = f^m(R)v^{-1}I \subseteq R$ and, consequently, $RI \subseteq I$. Thus $I \in \Gamma(R)$ and $v^{-1} \in T$ as claimed.

Put $w = \varphi(v^{-1})$. Then $wr = \varphi(v^{-1}r) = \varphi(f^m(r)v^{-1}) = F^m(r)w$. Using the definition of φ, we see that $F^m(wr) = f^m(v^{-1}r) = v^{-1}f^m(r) = wF^m(r)$ for all $r \in I$.

(2) $\Longrightarrow$(1). Let w and I be as in condition (2). Then by Lemma 5.3 $J = \bigcup_{n\geq 0} f^{-n}(I) \in \Phi(A)$. Consider the map $\alpha: J \longrightarrow A$ defined by $\alpha(f^{-n}(r)) = f^{-n}(wr)$ for all $r \in I$ and all $n \geq 0$. The map α is well-defined since

$$f^{-n-k}(wf^k(r)) = f^{-n-k}(F^k(wr)) = f^{-n}(wr) \text{ for any } n \geq 0.$$

We claim that α is a homomorphism of right A-modules. For let $a = f^{-n}(r) \in J$ where $r \in I$ and $b = f^{-n}(s) \in A$ where $s \in R$. Then $\alpha(ab) = f^{-n}(wrs) = f^{-n}(wr)b = \alpha(a)b$, and the claim is proved. Thus by lemma 3.6 (d) there is an element d in $Q_f^r(A)$ such that $da = \alpha(a)$ for all $a \in J$.

Let $a = f^{-n}(r) \in J$, where $r \in I$. Then $df(a) = f^{-n}(wf(r)) = f^n(f(wr)) = f(da) = f(d)f(a)$. Since $f(J) = J$, we have $(d - f(d))J = 0$ and, consequently, $f(d) = d$.

Now let $a = f^{-n}(r) \in J$ and $b = f^{-n}(s) \in A$ where $r \in I$ and $s \in R$. Then $f^n(dba) = dsr = wsr = F^m(s)wr = F^m(s)dr$. Hence $dba = f^m(b)da$. It implies that $(db - f^m(b)d)J = 0$, for all $b \in A$. It follows from Lemma 3.6 that $db = f^m(b)d$ for all $b \in A$. Using the right-sided version of Lemma 3.10, we obtain that d is invertible in $Q_f^r(A)$ and $f^m(q) = dqd^{-1}$ for all $q \in Q_f^r(A)$. Now it is enough to take $v = d^{-1}$. It works. □

We have the following two mutually exclusive possibilities for the endomorphism F:

Case (i). For some positive integer m there exist an element $w \in Q_m^r(R)$ and a nonzero F-invariant ideal I of R such that $wr = F^m(r)w$ for all $r \in R$ and $F(wr) = wF(r)$ for all $r \in I$.

Case (ii). Case (i) does not hold.

5.13. Theorem. *Let R be an F-prime ring, $C^r(R)$ be the center of maximal right ring of quotients of R and $C_F = \{p \in C^r(R) \mid$ there exists a nonzero F-invariant ideal I of R such that $pI \subseteq R$ and $F(pr) = pF(r)$ for all $r \in I\}$. Suppose that R satisfies the F-stability condition for left annihilators. Then $R[x, f]$ is a prime ring and C_F is a field. Furthermore in Case* (i) *the extended centroid $C(R[x, F])$ of $R[x, F]$ is isomorphic to $C_F(y)$, the field of rational functions over C_F; in Case* (ii) *the extended centroid $C(R[x, F])$ of $R[x, f]$ is isomorphic to C_F.*

Proof. By Theorem 5.7 $S = R[x, F]$ is prime and consequently $C(S)$ is the extended centroid of S. Theorem 5.6 yields the isomorphism $C(S) \simeq C(A^*)$. This shows that while describing the extended centroid of S it is enough to describe that of A^*. A description of $C(A^*)$ is provided by Theorem 4.13. By Lemma 3.9 and Corollary 5.11 $C_f \simeq C_F$ is a field.

Using Theorem 5.7, we get that $l_R(I) = 0$ for all $I \in \Gamma(R)$. Then Theorem 5.8 and Theorem 5.12 show that Case (i) holds if and only if Case 2 (see 4.11) holds. Since Case 1 and Case 2 are mutually exclusive, we obtain that F fits into Case (ii) if and only if f fits into Case 1. Therefore, in virtue of Theorem 4.13 and Corollary 5.11, we have that:

$C(S) \simeq C(A^*) \simeq C_f(y) \simeq C_F(y)$ in Case (i);

$C(S) \simeq C(A^*) \simeq C_f \simeq C_F$ in Case (ii).

Thus we deduce. □

Remark. One can prove that if R satisfies the ascending chain condition for left annihilators then R satisfies the F-stability condition for left annihilators.

Acknowledgments. The author would like to thank Professors A. V. Mikhalev and K. I. Beidar for their helpful suggestions and useful conversations about the content of this paper.

References

[1] Cisneros E., Ferrero M. and Gonzalez M. I., Prime ideals of skew polynomial rings and skew Laurent polynomial rings, Math, J. Okayama Univ. 32 (1990), 61–72.

[2] Cohn P. M., Universal algebra, Harper and Row, New York 1965.

[3] Jategaonkar A. V., Left principal ideal rings, Lecture Notes in Mathematics 123, Springer-Verlag 1970.

[4] Jordan D. A., Ore extensions and Jacobson rings, Ph. D. Thesis, Leeds 1975.

[5] Jordan D. A., Bijective extensions of injective ring endomorphisms, J. London Math. Soc. 35(2) (1982), 435–448.

[6] Lambek J., Lectures on ring and modules, Blaisdell Publ. Company, Waltham 1966.

[7] Martindale W. S., Prime ring satisfying a generalized polynomial identity, J. Algebra 12 (1969), 576–584.

[8] Matczuk J., Extended centroids of skew polynomial rings, Math. J. Okayama Univ. 30 (1988), 13–20.

[9] Montgomery S., Fixed rings of finite automorphism groups of associative rings, Lecture Notes in Mathematics 818, Springer-Verlag 1980.

[10] Montgomery S. and Passman D. S., Outer Galois theory of prime rings, Rocky Mountain J. Math. 14 (1984), 305–317.

[11] Rosen J. D. and Rosen M. P., Extended centroids of skew polynomial rings, Canad. Math. Bull. 28(1) (1985), 67–76.

[12] Stenström B., Rings of quotients, Grundlehren Math. Wiss. 217, Springer-Verlag, Berlin 1975.

Faculty of Mathematics
Moscow Pedagogical State University
Krasnoprudnaja Street 14
107104 Moscow, Russia

Polylinear recurring sequences over modules and quasi-Frobinus modules

*A. A. Nechaev**

1. Definitions and statement of the main results

In the following R is a commutative Artinian ring with identity e, M is a left R-module. For any subsets $I \subseteq R$, $N \subseteq M$ we define annihilators

$$\mathrm{Ann}_M(I) = \{\alpha \in M : I\alpha = 0\}, \quad \mathrm{Ann}_R(N) = \{r \in R : rN = 0\}.$$

We call a module ${}_RM$ *quasi-Frobenius* (a *QF-module*) if for each ideal $I \lhd R$ and for each submodule $N <_R M$ the following equalities hold:

$$\mathrm{Ann}_R(\mathrm{Ann}_M(I)) = I, \quad \mathrm{Ann}_M(\mathrm{Ann}_R(N)) = N.$$

This definition is equivalent to the definition of QF-modules in [2, 3].

This paper display the connections between QF-modules and families of polylinear recurring sequences (LRS-families) over modules. These connections are of two kinds. Firstly, the well-known relations between LRS-families over a field may be generalized to LRS-families over a module ${}_RM$ if and only if it is quasi-Frobenius (see Theorems 1.4, 1.6). Secondly, the theory of LRS over QF-modules enables us to build very constructively a QF-module over any commutative Artinian ring as a k-LRS-family over a principal ideal ring.

Let us introduce some definitions. Let $\mathcal{P}_k = R[x_1, \ldots, x_k]$ be the ring of polynomials in k variables $x_1, \ldots, x_k$. We call an ideal $I \lhd \mathcal{P}_k$ a *monic* (or a *cofinite*) *ideal* if there exist monic polynomials $F_1(x), \ldots, F_k(x)$ in the ring $R[x]$ (polynomials of one variable) such that

$$F_1(x_1), \ldots, F_k(x_k) \in I. \tag{1.1}$$

This condition is equivalent to the condition that $\mathcal{P}_k/I$ is an Artinian ring (§3).

By a *k-sequence* over the module M we mean any function $\mu : \mathbb{N}_0^k \to M$ (note that if $k = 1$, then μ is a usual sequence over M). This definition is equivalent to

*Partially supported by the Russian Foundation for Fundamental Research.

the definition of a k-sequence in [6]. The set $M^{\langle k\rangle}$ of all k-sequences over M is an R-module relative to the usual operation on functions and constants. We denote by $z_1, \dots, z_k$ the free variables on N_0 and in some cases we write $\mu = \mu(z_1, \dots, z_k)$ for $\mu \in M^{\langle k\rangle}$.

Let us define the structure of $\mathcal{P}_k$-modules on $M^{\langle k\rangle}$ in the following way. If $\mu \in M^{\langle k\rangle}$ and

$$F = F(x_1, \dots, x_k) = \sum_{s_1,\dots,s_k \in \mathrm{N}_0} f_{s_1,\dots,s_k} x_1^{s_1} \cdots x_k^{s_k} \in \mathcal{P}_k,$$

then $F\mu = \nu$, where $\nu \in M^{\langle k\rangle}$ and

$$\nu(z_1, \dots, z_k) = \sum_{s_1,\dots,s_k \in \mathrm{N}_0} f_{s_1,\dots,s_k} \mu(z_1 + s_1, \dots, z_k + s_k).$$

For short, the annihilator $\mathrm{Ann}_{\mathcal{P}_k}(N)$ of a subset $N \subset M^{\langle k\rangle}$ in the ring $\mathcal{P}_k$ is denoted by $\mathrm{Ann}(N)$. The sequence $\mu \in M^{\langle k\rangle}$ is called a *k-linear recurring sequence (k-LRS) over M* if $\mathrm{Ann}(\mu)$ is a monic ideal of $\mathcal{P}_k$. Note that if $M = R$, $k = 1$, then this definition is the usual definition of LRS over the ring R (see for example, [4]).

Proposition 1.1. *The collection $\mathcal{L}M^{\langle k\rangle}$ of all k-LRS over M is a submodule of the $\mathcal{P}_k$-module $M^{\langle k\rangle}$* (§3).

If $\mathcal{M}$ is a finitely generated R-module, then we write $\mathcal{M}$ is an *f.g.-R-module.*

Proposition 1.2. *If $\mathcal{M}$ is an f.g.-R-submodule of the $\mathcal{P}_k$-module $M^{\langle k\rangle}$, then* $\mathrm{Ann}(\mathcal{M})$ *is a monic ideal of $\mathcal{P}_k$, i.e., each sequence $\mu \in \mathcal{M}$ is a k-LRS* (§3).

We denote the annihilator $\mathrm{Ann}_{M^{\langle k\rangle}}(I)$ of a subset $I \subset \mathcal{P}_k$ in the $\mathcal{P}_k$-module $M^{\langle k\rangle}$ for short by $L_M(I)$.

Proposition 1.3. *If I is a monic ideal of $\mathcal{P}_k$, then $L_M(I)$ is an f.g.-R-submodule of the $\mathcal{P}_k$-module $\mathcal{L}M^{\langle k\rangle}$* (§3).

Under the hypotheses of Proposition 1.3, we call $L_M(I)$ a *k-LRS-family over M*.

Let $\mathfrak{A}_k$ be the set of all monic ideals of $\mathcal{P}_k$ and $\mathfrak{M}_k$ the set of all f.g.-R-submodules of the $\mathcal{P}_k$-module $M^{\langle k\rangle}$. Propositions 1.2, 1.3 give a pair of Galois correspondence:

$$L_M : \mathfrak{A}_k \to \mathfrak{M}_k, \quad \mathrm{Ann} : \mathfrak{M}_k \to \mathfrak{A}_k, \tag{1.2}$$

i.e., for each $I \in \mathfrak{A}_k$, $\mathcal{M} \in \mathfrak{M}_k$, the following relations hold

$$\mathrm{Ann}(L_M(I)) \supseteq I, \quad L_M(\mathrm{Ann}(\mathcal{M})) \supseteq \mathcal{M}. \tag{1.3}$$

Note that if $I \in \mathfrak{A}_k$, then $S = \mathcal{P}_k/I$ is an Artinian ring and $S = R[\theta_1, \dots, \theta_k]$, where $\theta_s = x_s + I$ for $s \in \{1, \dots, k\}$. Moreover, the $\mathcal{P}_k$-module $L_M(I)$ can be viewed as an S-module if we put

$$F(\theta_1, \dots, \theta_k)\mu = F(x_1, \dots, x_k)\mu \tag{1.4}$$

for any $\mu \in L_M(I)$, $F \in \mathcal{P}_k$.

It is not difficult to see that for any ideals $I_1, I_2 \in \mathfrak{A}_k$ and modules $\mathcal{M}_1, \mathcal{M}_2 \in \mathfrak{M}_k$,the following relations are true:

$$\begin{aligned} L_M(I_1+I_2) &= L_M(I_1) \cap L_M(I_2), \\ L_M(I_1 \cap I_2) &\supseteqq L_M(I_1) + L_M(I_2), \\ L_M(\mathcal{M}_1+\mathcal{M}_2) &= \mathrm{Ann}(\mathcal{M}_1) \cap \mathrm{Ann}(\mathcal{M}_2), \\ L_M(\mathcal{M}_1 \cap \mathcal{M}_2) &\supseteqq \mathrm{Ann}(\mathcal{M}_1) + \mathrm{Ann}(\mathcal{M}_2); \end{aligned}$$

and if $\mu, \nu \in \mathcal{L}M^{\langle k \rangle}$, then $\nu \in \mathcal{P}_k\mu \Rightarrow \mathrm{Ann}(\mu) \subset \mathrm{Ann}(\nu)$.

The main results of this paper which where announced in [10] are the following:

Theorem 1.4. *Let M be an f.g.-R-module. Then the following statements are equivalent.*

(a) *$_RM$ is a QF-module.*

(b) *The Galois correspondences* (1.2) *are inverses of each other* (*i.e., the inclusions* (1.3) *are eqalities*).

(c) *For every ideal $I \in \mathfrak{A}_k$ the family $L_M(I)$ is a QF-module over the ring $S = \mathcal{P}_k/I$ and every module $\mathcal{M} \in \mathfrak{M}_k$ is a QF-module over the ring $S = \mathcal{P}_k/\mathrm{Ann}(\mathcal{M})$.*

(d) *For any ideals $I_1, I_2 \in \mathfrak{A}_k$ and for any modules $\mathcal{M}_1, \mathcal{M}_2 \in \mathfrak{M}_k$ the following equalities hold:*

$$L_M(I_1 \cap I_2) = L_M(I_1) + L_M(I_2),$$

$$L_M(\mathcal{M}_1 \cap \mathcal{M}_2) = \mathrm{Ann}(\mathcal{M}_1) + \mathrm{Ann}(\mathcal{M}_2).$$

(e) *For any k-LRS $\mu, \nu \in \mathcal{L}M^{\langle k \rangle}$ the implications*

$$\nu \in \mathcal{P}_k\mu \Leftrightarrow \mathrm{Ann}(\mu) \subseteq \mathrm{Ann}(\nu)$$

are true (§4).

Note that property (b) is the generalization of the following property of LRS over a field F: each nonzero ideal of $F[x]$ is the annihilator of some LRS-family over F. Properties (d), (e) generalize well-known relations between LRS-families over a field.

The equivalence of statements (a) and (c) of Theorem 1.4 allow us to describe constructively a QF-module over any Artinian commutative ring S. Such a module looks like a k-LRS-family over some principal ideal ring R. As S is a direct sum of a local rings, it is sufficient to propose that S is a local ring. In this case, let $\mathfrak{N}(S)$ be a maximal ideal of S. Then there exists a subring R of S such that R is a local principal ideal ring and $R/\mathfrak{N}(R) \cong S/\mathfrak{N}(S)$ (see for example [8]).

Hence R is a QF-ring (i.e., $_RR$ is a QF-module) and for some $k \in \mathbb{N}_0, \theta_1, \dots, \theta_k \in \mathfrak{N}(S)$, the equality $S = R[\theta_1, \dots, \theta_k]$ holds. It follows that $S \cong \mathcal{P}_k/I$ for some monic ideal $I \lhd \mathcal{P}_k = R[x_1 \dots x_k]$ and according to Theorem 1.4(c), $L_R(I)$ is the desired quasi-Frobenius S-module.

We remind that if F is a field, then each monic ideal $I \lhd F[x]$ is the annihilator of some LRS over F and $L_F(I)$ is a cyclic $F[x]$-module. There are the following generalizations of this result.

Proposition 1.5. *For any monic ideal $I \lhd \mathcal{P}_k$ there exists an f.g.-R-module M such that $I = \mathrm{Ann}(\mu)$ for some $\mu \in \mathcal{L}M^{(k)}$.*

Theorem 1.6. *If ${}_RQ$ is a QF-module, then for a monic ideal $I \lhd \mathcal{P}_k$ the following statements are equivalent:*

(a) $I = \mathrm{Ann}(\mu)$ *for some* $\mu \in Q^{(k)}$;

(b) $L_Q(I)$ *is a cyclic* $\mathcal{P}_k$*-module;*

(c) $\mathcal{P}_k/I$ *is a* QF*-ring.*

Corollary 1.7. *Let $F_1(x_1), \ldots, F_k(x_k)$ be monic polynomials over the ring R, and $S = \mathcal{P}_k/(F_1(x_1), \ldots, F_k(x_k))$. Then S is a QF-ring if and only if R is a QF-ring.*

For the proof of the above results (in §§3, 4) we need some preliminary results about QF-modules.

2. Properties of QF-modules

To complete the account, we state at first some known basic properties of QF-modules. The necessary definition and proofs can be found in [1–5]. Below we consider any R-module ${}_RM$ as an R-R-bimodule ${}_RM_R$ such that $\alpha a = a\alpha$ for all $a \in R, \alpha \in M$.

Theorem 2.1. *There exists a unique QF-module ${}_RM$ up to isomorphism. For any module ${}_RM$ the following conditions are equivalent:*

(a) ${}_RM$ *is a* QF*-module;*

(b) ${}_RM$ *is finitely generated minimal injective cogenerator in the catogory of* R*-modules;*

(c) *the bimodule* ${}_RM_R$ *determines a Morita-duality on category of* R*-modules;*

(d) ${}_RM$ *is a faithful module and for every maximal ideal* I *of* R *the submodule* $\mathrm{Ann}_M(I)$ *is irreducible.*

Note that the ring R is a direct sum $R = R_1 \dotplus R_2 \dotplus \cdots \dotplus R_t$ of local rings. Let e_s be the identity of R_s. Then any R-module M is a direct sum of R-modules $M_s = e_sM$, $s \in \{1, 2, \ldots, t\}$. Moreover, M_s is an R_s-module and ${}_RM$ is a QF-module if and only if ${}_{R_s}M_s$ is a QF-module for $s \in \{1, 2, \ldots, t\}$.

Proposition 2.2. *If R is a local ring, then for a faithful module ${}_RM$ the following conditions are equivalent:*

(a) ${}_RM$ *is a* QF*-module;*

(b) $\mathrm{Ann}_M(\mathfrak{N}(R))$ *is a minimal submodule of* ${}_RM$*;*

(c) $\mathrm{Ann}_M(\mathfrak{N}(R))$ *is a cyclic* R*-module;*

(d) ${}_RM$ *contains the only irreducible submodule.*

Proposition 2.3. *Let* ${}_RQ$ *be a* QF*-module. Then for any ideal* $I \lhd R$ *the submodule* $\mathrm{Ann}_M(I)$ *of* ${}_RQ$ *is a* QF*-module over the ring* R/I*. Any submodule* $N < {}_RQ$ *is a* QF*-module over the ring* $R/\mathrm{Ann}_R(N)$*.*

Let K, N, Q be any R-modules and let $\varphi : K \times N \to Q$ be an R-bilinear map such that each of the equations $\varphi(x, N) = 0$, $\varphi(K, y) = c$ has only the zero solution (correspondingly, in K and in N). Then we say that φ is an *orthogonal pairing* of K and N with respect to Q [2]. For any subsets $K_1 \subset K$, $N_1 \subset N$, we put

$$\begin{aligned}\mathrm{Ann}_N^{\varphi}(K_1) &= \{\beta \in N : \varphi(K_1, \beta) = 0\},\\ \mathrm{Ann}_K^{\varphi}(N_1) &= \{\alpha \in K : \varphi(\alpha, N_1) = 0\}.\end{aligned}$$

Proposition 2.4 ([2]). *Let* ${}_RQ$ *be a* QF*-module and let* φ *be an orthogonal pairing of* ${}_RK$ *and* ${}_RN$ *with respect to* Q*. Then for any submodules* $K_1 < {}_RK$, $N_1 < {}_RN$ *the following equalities hold:*

$$\mathrm{Ann}_K^{\varphi}\mathrm{Ann}_N^{\varphi}(K_1) = K_1, \quad \mathrm{Ann}_N^{\varphi}\mathrm{Ann}_K^{\varphi}(N_1) = N_1.$$

We will use this result in the following special case. For any $n \in \mathbb{N}$, we denote by Q^n ($Q^{(n)}$) the R-module of all rows (columns) of length n over Q. For any $m \times n$-matrix $A_{m\times n}$ over the ring R we denote by $\mathrm{Ann}_{Q^n}^{\{r\}}(A)$ the right annihilator of A in Q^n, i.e., the R-submodule of solutions in $Q^{(n)}$ of the linear equation system

$$A\begin{pmatrix} y_1 \\ \vdots \\ y_m \end{pmatrix} = 0. \tag{2.1}$$

Dually, for any $n \times t$-matrix $B_{n\times t}$ over the module Q we denote by $\mathrm{Ann}_R^{\{l\}}(B)$ the R-module of solutions in R^n of the linear equation system

$$(x_1, \dots, x_n)B = 0. \tag{2.2}$$

Proposition 2.5. *Let* ${}_RQ$ *be a* QF*-module and* $A_{m\times n}$ *be a matrix over the ring* R*. Suppose that* $B_1^1, \dots, B_t^1 \in Q^{(n)}$ *is a generating set of the solution module of the system* (2.1)*. Let* $B = \left(B_1^1, \dots, B_t^1\right)$*. Then the set* $\{A_1, \dots, A_m\}$ *of the rows vectors of the matrix A is the generating set of the solution module in* R^n *of the linear equation system* (2.2)*. Dually, let* $B_{n\times t}$ *be a matrix over* Q *and let* $A = \left(\frac{A_1}{A_m}\right)$*, where* $A_1, \dots, A_m \in R^n$ *is a generating set of the solution module of the system* (2.2)*. Then the set* $\{B_1^1, \dots, B_t^1\}$ *of the column vectors of the matrix B is the generating set of the solution module in* $Q^{(n)}$ *of the system* (2.2)*.*

Proof. We consider the map $\varphi : R^n \times Q^{(n)} \to Q$ such that

$$\varphi\left((a_1, \dots, a_n), (\alpha_1, \dots, \alpha_n)^{\mathrm{T}}\right) = a_1\alpha_1 + \dots + a_n\alpha_n$$

for all $(a_1, \dots, a_n) \in R^n$, $(\alpha_1, \dots, \alpha_n)^{\mathrm{T}} \in Q^{(n)}$. As ${}_RQ$ is an QF-module, φ is an orthogonal pairing. Now Proposition 2.5 follows from Proposition 2.4. □

In conclusion we state an interesting criterion for the solvability of any system of linear equations

$$(x_1, \dots, x_m)A_{m\times n} = (a_1, \dots, a_m), \tag{2.3}$$

where $a_1, \dots, a_n \in R$ and $x_1, \dots, x_m$ are free variables on R. We consider the system of linear equations

$$A_{m\times n}\begin{pmatrix} y_1 \\ \vdots \\ y_n \end{pmatrix} = 0, \tag{2.4}$$

where $y_1, \dots, y_n$ are free variables on the QF-module ${}_RQ$. Let $C_1^1, \dots, C_t^1$ be a generating set of the solution module of the system (2.4) and let $C = \left(C_1^1, \dots, C_t^1\right)$.

Proposition 2.6. *The system of linear equalities* (2.3) *is solvable over the ring R if and only if the equality*

$$(a_1, \dots, a_n)C = 0 \tag{2.5}$$

holds.

Proof. Let N_A be the submodule of ${}_RR^{(r)}$ generated by the system of rows of the matrix A. Then the system (2.3) is solvable if and only if $(a_1, \dots, a_n) \in N_A$. According to the definition of the matrix C, the set of its columns generates the R-module $\mathrm{Ann}_Q^{\{r\}}(A)$. As ${}_RQ$ is a QF-module, in view of Proposition 2.5, the R-module N_A satisfies the equality

$$N_A = \mathrm{Ann}_R^{\{l\}}(C).$$

Hence the inclusion $(a_1, \dots, a_n) \in N_A$ is equivalent to the condition (2.5). □

Note that if R is a finite ring, then we may regard the submodule N_A of ${}_RR^{(r)}$ as a linear code of length n over the ring R with generating matrix A (see for example [9]). Then the matrix C is a checking matrix of the code N_A. So we have

Corollary 2.7. *Any linear code over a finite commutative ring R has a checking matrix over the QF-module ${}_RQ$. Any linear code over R has a checking matrix over R if and only if R is a QF-ring.*

3. Properties of k-LRS families

Below we prove Propositions 1.1–1.3, 1.5 and state some additional results.

Proof of Proposition 1.1. It is enough to prove that if $\mu, \nu \in \mathcal{L}M^{\langle k\rangle}$ and $A(\vec{x}), B(\vec{x}) \in \mathcal{P}_k$, then $A(\vec{x})\mu + B(\vec{x})\nu \in \mathcal{L}M^{\langle k\rangle}$. According to the definition of k-LRS, there exists monic polynomials $F_1(x_1), \ldots, F_1(x_k) \in \mathrm{Ann}(\mu)$ and $G_1(x_1), \ldots, G_1(x_k) \in \mathrm{Ann}(\nu)$. Then $H_s(x_s) = F_s(x_s)G_s(x_s)$ is a monic polynomial and $H_s(x_s)(A(\vec{x})\mu + B(\vec{x})\nu) = 0$ for $s \in \{1, \ldots, k\}$. Hence $H_1(x_1), \ldots, H_{,}(x_k) \in \mathrm{Ann}(A(\vec{x})\mu + B(\vec{x})\nu)$ and $A(\vec{x})\mu + B(\vec{x})\nu \in \mathcal{L}M^{\langle k\rangle}$. □

Proof of Proposition 1.2. Let $\mu^{(1)}, \ldots, \mu^{(r)}$ be a generating system of ${}_R\mathcal{M}$. Then for any $s \in \{1, \ldots, k\}$, $t \in \{1, \ldots, r\}$ a submodule ${}_R(\mu^{(t)}, x_s\mu^{(t)}, \ldots, x_s^i\mu^{(t)}, \ldots)$ of ${}_R\mathcal{M}$ is an f.g.-R-module. Hence there exists an $m \in \mathbb{N}$ such that $x_s^m\mu^{(t)} = c_{m-1}x_s^{m-1}\mu^{(t)} + \cdots + c_1x_s\mu^{(t)} + c_0\mu^{(t)}$ (evidently m and $c_0, \ldots, c_{m-1} \in R$ are functions of s and t). This means that the monic polynomial $F_s^{(t)}(x_s) = c_m x_x^m - c_{m-1}x_s^{m-1} - \cdots - c_1x_s - c_0$ annihilates the sequence $\mu^{(t)}$, and the polynomial $F_s(x_s) = F_s^{(1)}(x_s) \cdot \cdots \cdot F_s^{(r)}(x_s)$ annihilates all the sequences of the module $\mathcal{M}$. Consequently, $F_s(x_s) \in \mathrm{Ann}(\mathcal{M})$ for $s \in \{1, \ldots, k\}$, i.e., $\mathrm{Ann}(\mathcal{M})$ is a monic ideal. □

To prove Proposition 1.3, we suppose that I is a monic ideal of $\mathcal{P}_k$, satisfying the following conditions:

$$F_1(x_1), \ldots, F_k(x_k) \in I, \tag{3.1}$$

$$F_s(x_s) = x_x^{ms} - f_{m_s-1}^{(s)}x_s^{m_s-1} - \cdots - f_0^{(s)} \in R[x_s],\ s \in \{1, \ldots, k\}. \tag{3.2}$$

The usual order on $\mathbb{N}_0$, as well as the partial order it induces on the $\mathbb{N}_0^k$, will be denoted by $\leq$. Moreover, we write $\preceq$ for the well-ordering on $\mathbb{N}_0^k$, having $\vec{0} \in \mathbb{N}^k$ as the minimal element and such that for every $\vec{i} = (i_1, \ldots, i_k)$, $\vec{j} = (j_1, \ldots, j_k)$ from $\mathbb{N}_0^k$ the relation $\vec{i} \preceq \vec{j}$ holds if and only if the first nonzero number in the sequence $(j_1 + \cdots + j_k) - (i_1 + \cdots + i_k), j_1 - i_1, \ldots, j_k - i_k$ is positive.

We put $\vec{m} = (m_1, \ldots, m_k)$ (see (3.2)), $\vec{1} = (1, \ldots, 1)$ and consider the polyhedron

$$\begin{aligned}\Pi = \Pi(\vec{m}) &= \{\vec{i} \in \mathbb{N}_0^k : \vec{i} \preceq \vec{m} - \vec{i}\} \\ &= \{0, \ldots, m_1 - 1\} \times \cdots \times \{0, \ldots, m_k - 1\}.\end{aligned}$$

To each function $\mu \in M^{\langle k\rangle}$, we associate the diagram of values on Π:

$$\mu[\Pi] = \{\mu(\vec{i}) : \vec{i} \in \Pi\}.$$

Let M^{Π} be the set of all such diagrams. Evidently M^{Π} is an R-module relative to usual operations.

It may be assumed that all the terms in the polyhedron Π are well-ordered relative to $\preceq$ such that

$$\vec{i}_0 = \vec{0} \prec \vec{i}_1 \prec \vec{i}_2 \prec \cdots \prec \vec{i}_{m-1}, \quad m = m_1 \ldots m_k, \tag{3.3}$$

and the terms $\mu(\vec{c})$ of each diagram $\mu[\Pi]$ are ordered correspondingly. So, this diagram may be associated with the following row of length m:

$$(\mu(\vec{0}), \mu(\vec{i}_1), \ldots, \mu(\vec{i}_{m-1})) \in M^m,$$

and the module ${}_RM^\Pi$ may be regarded as ${}_RM^m$.

Let us denote $\vec{x} = (x_1, \ldots, x_k)$ and $\vec{x}^{\vec{j}} = x_1^{j_1} \ldots x_k^{j_k}$ for any $\vec{j} \in \mathbb{N}_0^k$. Let $\operatorname{Res}(x_s^j / F_s(x_s))$ be the reminder of x_s^j modulo $F_s(x_s)$. For each $\vec{j} \in \mathbb{N}_0^k$ we put

$$H^{\vec{j}}(\vec{x}) = \sqcap_{s=1}^k \operatorname{Res}(x_s^{j_s} / F_s(x_s)). \tag{3.4}$$

Then the polynomial $H^{\vec{j}}(\vec{x})$ has a form

$$H^{\vec{j}}(\vec{x}) = \sum_{\vec{i} \in \Pi} h_{\vec{i}}^{\vec{j}} \vec{x}^{\vec{i}}. \tag{3.5}$$

Lemma 3.1. *All the terms $\mu(\vec{\delta})$, $\vec{\delta} \in \mathbb{N}_0^k$, of any k-LRS*

$$\mu \in L_M(F_1(x_1), \ldots, F_k(x_k))$$

are uniquely determined by the diagram $\mu[\Pi]$ of the initial values of μ for the following relations:

$$\mu(\vec{j}) = \sum_{\vec{i} \in \Pi} h_{\vec{i}}^{\vec{j}} \mu(\vec{i}), \quad \vec{j} \in \mathbb{N}_0^k. \tag{3.6}$$

Proof. Note that $\mu(\vec{j}) = \nu(\vec{0})$, where $\nu = x^{\vec{j}}\mu$. Moreover, $\nu = H^{\vec{j}}(\vec{x})\mu$, as $F_s(x_s)\mu = 0$ for $s \in \{1, \ldots, k\}$. Now, (3.6) follows from (3.5). □

Let $\vec{F} = (F_1(x_1), \ldots, F_k(x_k))$ and for each $\vec{j} \in \Pi$ the k-sequence $e_{\vec{j}}^{\vec{F}}$ over the ring R be such a k-LRS $u \in L_R(F_1(x_1), \ldots, F_k(x_k))$ that $u(\vec{j}) = e$ and all the others initial values of $u[\prod]$ are zero, and let

$$e^{\vec{F}} = e_{\vec{m}-\vec{1}}^{\vec{F}}.$$

Proposition 3.2. *Let $M = R\alpha_1 + \cdots + R\alpha_n$. Then the family $L_M(F_1(x_1), \ldots, F_k(x_k))$ is generated as R-module by the system of sequences*

$$e_{\vec{j}}^{\vec{F}} \alpha_r, \quad r \in \{1, \ldots, n\}, \vec{j} \in \Pi. \tag{3.7}$$

The family $L_M(F_1(x_1), \ldots, F_k(x_k))$ is generated as $\mathcal{P}_k$-module by the system of sequences

$$e^{\vec{F}} \alpha_1, \ldots, e^{\vec{F}} \alpha_r. \tag{3.8}$$

In particular, $L_M(F_1(x_1), \dots, F_k(x_k))$ is a free R-module if and only if M is a free R-module and in this case

$$\operatorname{rank}_R L_M(F_1(x_1), \dots, F_k(x_k)) = m \cdot \operatorname{rank}_R M.$$

If ${}_R M$ is a cyclic R-module, then $L_M(F_1(x_1), \dots, F_k(x_k))$ is a cyclic $\mathcal{P}_k$-module.

Proof. Let $\mu \in L_M(F_1(x_1), \dots, F_k(x_k))$ and $\mu(\vec{i}) = \sum_{r=1}^{n} c_{\vec{i}}^{(r)} \alpha_r$, where $c_{\vec{i}}^{(r)} \in R$ for $\vec{i} \in \Pi$, $r \in \{1, \dots, n\}$. Then $\nu = \sum_{r=1}^{n} \sum_{\vec{i} \in \Pi} c_{\vec{i}}^{(r)} e_{\vec{i}}^{\vec{F}} \alpha_r$ is a k-LRS from $L_M(F_1(x_1), \dots, F_k(x_k))$ and the diagram $\mu[\Pi]$ is equal to $\nu[\Pi]$. Hence, according to Lemma 3.1 $\mu = \nu$ and (3.1) is a generating system of the R-module $L_M(F_1(x_1), \dots, F_k(x_k))$.

Now, according to the ordering (3.3), we build the sequences $\mu_0, \dots, \mu_{m-1} \in L_M(F_1(x_1), \dots, F_k(x_k))$ by the following rule: $\mu_0 = \mu - \vec{x}^{\vec{m}-\vec{1}} e^{\vec{F}} \mu(\vec{c}_0)$ and

$$\mu_t = \mu_{t-1} - \vec{x}^{\vec{m}-\vec{1}-\vec{i}_t} e^{\vec{F}} \mu_{t-1}(\vec{i}_t) \quad \text{for } t \in \{1, \dots, m-1\}.$$

It is not difficult to see that if $\vec{j} \in \Pi$ and $\vec{j} \preceq \vec{i}_t$, then $\mu_t(\vec{j}) = 0$. Hence all the initial terms of the LRS $\mu_{m-1} \in L_M(F_1(x_1), \dots, F_k(x_k))$ are zero and according to Lemma 3.1 $\mu_{m-1} = 0$. Consequently, the system (3.8) generates $L_M(F_1(x_1), \dots, F_k(x_k))$ over $\mathcal{P}_k$.

The last two statements of Proposition 3.2 are obvious. □

Proof of Proposition 2.3. The inclusion $L_M(I) \subset L_M(F_1(x_1), \dots, F_k(x_k))$ follows from condition (3.1). Hence $L_M(I)$ is an f.g.-R -submodule as $L_M(F_1(x_1), \dots, F_k(x_k))$ is an f.g.-R -module according to Proposition 3.1 and R is an Artinian ring.□

Now we may describe the generating system of the R-module $L_M(I)$. Since R is an Artinian ring, I is a finitely generated ideal, and under the conditions (3.1), (3.2) we may assume that I has a generating system of the form

$$F_1(x_1), \dots, F_k(x_k), G_1(\vec{x}), \dots, G_n(\vec{x}), \tag{3.9}$$

where $\deg_{x_s} G_r(\vec{x}) < m_s$ for $s \in \{1, \dots, k\}$, $r \in \{1, \dots, n\}$, i.e.,

$$G_r(\vec{x}) = \sum_{\vec{j} \in \Pi} g_{r,\vec{j}} \vec{x}^{\vec{j}}, \quad r \in \{1, \dots, n\}. \tag{3.10}$$

For any polynomial $H(\vec{x}) \in \mathcal{P}_k$ we denote the restriction of $H(\vec{x})$ modulo the ideal $(F_1(x_1), \dots, F_k(x_k))$ as $\operatorname{Res}(H(\vec{x})/F_1(x_1), \dots, F_k(x_k))$ (the result of the consecutive dividings with reminder of $H(\vec{x})$ on $(F_1(x_1), \dots, F_k(x_k))$). Then under the designations (3.4) the equality $H^{\vec{j}}(\vec{x}) = \operatorname{Res}(\vec{x}^{\vec{j}}/F_1(x_1), \dots, F_k(x_k))$ holds.

Let $\operatorname{Res}(I/F_1, \dots, F_k) = \{\operatorname{Res}(H/F_1, \dots, F_k) : H \in I\}$. Under the conditions (3.1) and (3.2), the ideal I is well-defined by its R-submodule $\operatorname{Res}(I/F_1, \dots, F_k)$.

Any polynomial $G(\vec{x}) \in \text{Res}(I/F_1, \dots, F_k)$ has the form

$$G(\vec{x}) = \sum_{\vec{i} \in \Pi} g_{\vec{i}} \vec{x}^{\vec{i}},$$

and it is uniquely determined by its diagram of coefficients $G_\Pi = \{g_{\vec{j}} : \vec{j} \in \Pi\}$. Let

$$I_\Pi = \{G_\Pi : G(\vec{x}) \in \text{Res}(I/F_1, \dots, F_k)\}.$$

Then I_Π is a submodule of R^Π isomorphic to $\text{Res}(I/F_1, \dots, F_k)$.

It is not difficult to see that if the ideal I has a generating set (3.9), then the R-module $\text{Res}(I/F_1, \dots, F_k)$ is generated by a set of polynomials

$$G_r^{\vec{u}}(\vec{x}) = \text{Res}(G_r(\vec{x})\vec{x}^{\vec{u}}/F_1, \dots, F_k), \ r \in \{1, \dots, k\}, \vec{u} \in \Pi, \tag{3.11}$$

where

$$G_r^{\vec{u}}(\vec{x}) = \sum_{\vec{i} \in \Pi} g_{r,\vec{i}}^{\vec{u}} \vec{x}^{\vec{i}}. \tag{3.12}$$

Correspondingly, the submodule I_Π of ${}_R R^\Pi$ is generated by the set of diagrams $\{G_{r,\Pi}^{\vec{u}} : r \in \{1, \dots, k\}, \vec{u} \in \Pi\}$. Note that in view of (3.4), (3.5) and (3.10), the coefficients of $G_r^{\vec{u}}(\vec{x})$ have the form

$$g_{r,\vec{i}}^{\vec{u}} = \sum_{\vec{j} \in \Pi} g_{r,\vec{j}} h_{\vec{i}}^{\vec{u}+\vec{j}}. \tag{3.13}$$

According to Lemma 3.1, to describe $L_M(I)$ it is enough to find conditions on the diagram $\mu[\Pi]$ of the LRS $\mu \in L_M(F_1, \dots, F_k)$, which are equivalent to $\mu \in L_M(I)$.

Proposition 3.3. *Let I be a monic ideal of $\mathcal{P}_K$ with the generating set* (3.9) *and let*

$$H_1(\vec{x}), \dots, H_w(\vec{x}) \tag{3.14}$$

be any generating set of the R-module $\text{Res}(I/F_1, \dots, F_k)$. *Then a k-LRS $\mu \in L_M(F_1, \dots, F_k)$ belongs to $L_M(I)$ if and only if the diagram $\mu[\Pi]$ is the solution of the system of linear equations*

$$\sum_{\vec{i} \in \Pi} h_{v,\vec{i}} x(\vec{i}) = 0, \quad v \in \{1, \dots, w\}, \tag{3.15}$$

where $\{x(\vec{i}) : \vec{i} \in \Pi\}$ is the system of independent variables on M.

Proof. Since (3.14) and (3.11) are generating sets of $\text{Res}(I/F_1, \dots, F_k)$, the system of equations (3.15) is equivalent to the system of equations

$$\sum_{\vec{i} \in \Pi} g_{r,\vec{i}}^{\vec{u}} x(\vec{i}) = 0, \quad \vec{u} \in \Pi, r \in \{1, \dots, n\}. \tag{3.16}$$

Hence it is sufficient to prove the proposition for the system (3.16).

Let $\delta_r = G_r(\vec{x})\mu$ for $r \in \{1, \dots, n\}$. Then μ belongs to $L_M(I)$ if and only if $\delta_1 = \dots = \delta_r = 0$. As $\delta_r \in L_M(F_1, \dots, F_k)$, the last condition is equivalent to the condition

$$\delta_r[\Pi] = 0, \quad r \in \{1, \dots, n\}. \tag{3.17}$$

Let $\delta_r^{\vec{u}} = G_r^{\vec{u}}(\vec{x})\mu$. Then in view of (3.11), $\delta_r^{\vec{u}} = \vec{x}^{\vec{u}} G_r(\vec{x})\mu = \vec{x}^{\vec{u}}\delta_r$ and the condition (3.12) is equivalent to the condition

$$\delta_r^{\vec{u}} = 0 \quad \text{for } r \in \{1, \dots, n\}, \vec{u} \in \Pi.$$

It remains to note that according to (3.12)

$$\delta_r^{\vec{u}}(0) = \sum_{\vec{i} \in \Pi} g_{r,\vec{i}}^{\vec{u}} \mu(\vec{i}).$$

□

Corollary 3.4. *Let $\mu'_1, \dots, \mu'_l \in M^{\Pi}$ be a generating set of the solution module of the system* (3.14). *Then the set of k-LRS $\mu_1, \dots, \mu_l \in L_M(F_1, \dots, F_k)$ satisfies the conditions $\mu_t[\Pi] = \mu'_t$, $t \in \{1, \dots, l\}$ and generates the R-module $L_M(I)$.*

Let us now describe the annihilator $\mathrm{Ann}(\mathcal{M}) \lhd \mathcal{P}_k$ of any k-LRS family

$$\mathcal{M} = \mathcal{P}_k \nu_1 + \dots + \mathcal{P}_k \nu_d \tag{3.18}$$

such that $\mathcal{M} < L_M(F_1(x_1), \dots, F_k(x_k))$ and (3.2) holds. As R is an Artinian ring, the ideal $\mathrm{Ann}(\mathcal{M})$ of $\mathcal{P}_k$ has a finite generating system. In view of the condition

$$F_1, \dots, F_k \in \mathrm{Ann}(\mathcal{M}), \tag{3.19}$$

a generating set of $\mathrm{Ann}(\mathcal{M})$ may be found in the form

$$F_1(x_1), \dots, F_k(x_k), M_1(\vec{x}), \dots, M_w(\vec{x}), \tag{3.20}$$

where $M_1(\vec{x}), \dots, M_w(\vec{x})$ is any generating set of $\mathrm{Res}(\mathrm{Ann}(\mathcal{M})/F_1, \dots, F_k)$.

Note that according to (3.18) and (3.19), $\mathcal{M}$ is an f.g.-R-module, generated by the system of sequences

$$\nu_s^{\vec{u}} = \vec{x}^{\vec{u}} \nu_s, \quad \vec{u} \in \Pi, s \in \{1, \dots, d\}, \tag{3.21}$$

and according to (3.10) and (3.4), the relations

$$\nu_s^{\vec{u}}(\vec{j}) = \sum_{\vec{i} \in \Pi} h_{\vec{i}}^{\vec{u}+\vec{j}} \nu_s(\vec{i}), \quad j \in \mathbb{N}_0^k, s \in \{1, \dots, d\} \tag{3.22}$$

hold.

Proposition 3.5. *Let $\mu_1, \dots, \mu_l \in \mathcal{M}$ be any generating set of the R-module ${}_R\mathcal{M}$. Then the polynomial*

$$H(\vec{x}) = \sum_{\vec{j} \in \Pi} h_{\vec{j}} \vec{x}^{\vec{j}} \in \mathcal{P}_k$$

belongs to Ann($\mathcal{M}$) *if and only if the diagram* H_Π *of its coefficients is the solution of the system of linear equations*

$$\sum_{\vec{j}\in\Pi} y_{\vec{j}}\mu_t(\vec{j}) = 0, \quad t \in \{1, \dots, l\}, \tag{3.23}$$

where $\{y_{\vec{j}} : \vec{j} \in \Pi\}$ *is the system of independent variables on* R.

Proof. Let $\delta_s = H(\vec{x})\nu_s$, $s \in \{1, \dots, d\}$. In view of (3.18), $H(\vec{x})$ belongs to Ann($\mathcal{M}$) if and only if $\delta_s = 0$ for $s \in \{1, \dots, d\}$. As $\delta_1, \dots, \delta_d \in L_M(F_1, \dots, F_k)$, the last condition is equivalent to the condition

$$\delta_s(\vec{u}) = 0, s \in \{1, \dots, d\}, \vec{u} \in \Pi. \tag{3.24}$$

Let $\delta_s^{\vec{u}} = \vec{x}^{\vec{u}} H(\vec{x})\nu_s$. Then (3.21) is equivalent to the condition

$$\delta_s^{\vec{u}}(\vec{0}) = 0, \quad s \in \{1, \dots, d\}, \vec{u} \in \Pi.$$

In view of (3.21), $\delta_s^{\vec{u}} = H(\vec{x})\nu_s^{\vec{u}}$ and

$$\delta_s^{\vec{u}}(\vec{0}) = \sum_{\vec{j}\in\Pi} h_{\vec{j}}\nu_s^{\vec{u}}(\vec{j}).$$

Hence $H(\vec{x})$ belongs to Ann($\mathcal{M}$) if and only if the diagram H_Π is the solution of the following linear equation system

$$\sum_{\vec{j}\in\Pi} y_{\vec{j}}\nu_s^{\vec{u}}(\vec{j}) = 0, \quad s \in \{1, \dots, d\}, \vec{u} \in \Pi. \tag{3.25}$$

It remains to note that the system (3.25) is equivalent to the system (3.23), as (3.21) is the generating set of the R-module $\mathcal{M} = R\mu_1 + \dots + R\mu_l$. □

Corollary 3.6. *Let* $H_{1\Pi}, \dots, H_{w\Pi}$ *be the generating set of the solution module of the system* (3.23). *Then the set of polynomials* (3.20) *generate the ideal* Ann($\mathcal{M}$).

Proof of Proposition 1.5. Let $I \lhd \mathcal{P}_k$ be a monic ideal, $S = \mathcal{P}_k/I$, $\theta_s = x_s + I \in S$ for $s \in \{l, \dots, k\}$. Then the sequence μ over the module ${}_RS$ of the form $\mu(\vec{z}) = \theta_1^{z_1} \cdot \dots \cdot \theta_k^{z_k}$ is a k-LRS and Ann(μ) $= I$. In fact, let $H(\vec{x}) = \mathcal{P}_k$ and $\nu = H(\vec{x})\mu$. Then $\nu(\vec{z}) = H(\theta_1 \cdot \dots \cdot \theta_k)\theta_1^{z_1} \cdot \dots \cdot \theta_k^{z_k}$. Therefore

$$\nu = 0 \Leftrightarrow H(\theta_1 \cdot \dots \cdot \theta_k) = 0 \Leftrightarrow H(\vec{x}) \in I.$$

□

In conclusion we note that if $R = R_0 \dotplus R_1$ is a direct sum of ideals and e_s is the unit of R_s, $s \in \{0, 1\}$, then the module ${}_RM$ is a direct sum of submodules: $M = M_0 \dotplus M_1$, where $M_s = e_sM$ is an R-module and R_s-module; any monic ideal $I \lhd \mathcal{P}_k$ is a direct sum of ideals: $I = I_0 \dotplus I_1$, where $I_s = e_sI$ is a monic ideal of the ring $\mathcal{P}_k^{(s)} = R_s[\vec{x}]$ of the polynomials over the ring R_s. In this case, any k-LRS-family $L_M(I)$ is a direct sum $L_M(I) = e_0 L_M(I) \dotplus e_1 L_M(I)$ of $\mathcal{P}_k$-submodules. Moreover, $e_s L_M(I)$ is a k-LRS-family over M, as $e_s L_M(I) \subset L_M(I + e_{1-s}\mathcal{P}_k)$, and $e_s L_M(I) = L_{M_s}(I_s)$ is a k-LRS-family over M_s. Any f.g.-R-submodule $\mathcal{M}$ of $\mathcal{P}_k$-module $\mathcal{L}M^{\langle k\rangle}$ is a direct sum $\mathcal{M} = e_0\mathcal{M} \dotplus e_1\mathcal{M}$, where $e_s\mathcal{M} = \mathcal{M}_s$ is an f.g.-R_s-submodule of $\mathcal{P}_k^{(s)}$-module $\mathcal{L}M_s^{\langle k\rangle}$, and the following equalities hold:

$$\begin{aligned}
\mathrm{Ann}(\mathcal{M}) &= \mathrm{Ann}_{\mathcal{P}_k^{(0)}}(\mathcal{M}_0) \dotplus \mathrm{Ann}_{\mathcal{P}_k^{(1)}}(\mathcal{M}_1),\\
\mathrm{Ann}(L_M(I)) &= \mathrm{Ann}_{\mathcal{P}_k^{(0)}}(L_{M_0}(I_0)) \dotplus \mathrm{Ann}_{\mathcal{P}_k^{(1)}}(L_{M_1}(I_1)),\\
L_M(\mathrm{Ann}(\mathcal{M})) &= L_{M_0}(\mathrm{Ann}_{\mathcal{P}_k^{(0)}}(\mathcal{M}_0)) \dotplus L_{M_1}(\mathrm{Ann}_{\mathcal{P}_k^{(1)}}(\mathcal{M}_1)).
\end{aligned}$$

It is not difficult to see that under above assumption, any of the statements of Theorems 1.4 and 1.6 is true if and only if it is true for each of the modules ${}_{R_s}M_s$, $s \in \{0, 1\}$. Hence it is enough to prove these theorems for a local Artinian ring R, as any Artinian ring is a direct sum of a local rings.

4. Proofs of the main results

Below we prove Theorems 1.4 and 1.6.

Proof of Theorem 1.4. We will prove the implications

$$\text{(a)} \Rightarrow \text{(b)} \Rightarrow \text{(c)} \Rightarrow \text{(d)} \Rightarrow \text{(a)} \Leftrightarrow \text{(e)}.$$

(a) $\Rightarrow$ (b). Let $F_1(x_1), \ldots, F_k(x_k)$ be fixed elementary monic polynomials from $\mathcal{P}_k$. It is enough to prove that if ${}_RM$ is a QF-module, then

$$\mathrm{Ann}(L_M(I)) = I \tag{4.1}$$

for any monic ideal $I \lhd \mathcal{P}_k$ such that $F_1, \ldots, F_k \in I$, and

$$L_M(\mathrm{Ann}(\mathcal{M})) = \mathcal{M} \tag{4.2}$$

for any $\mathcal{P}_k$-submodule $\mathcal{M} < L_M(F_1, \ldots, F_k)$.

Let $H_1(\vec{x}), \ldots, H_w(\vec{x})$ be a generating set of $\mathrm{Res}(I/F_1, \ldots, F_k)$. Then according to Corollary 3.4 the R-module $\mathcal{M} = L_M(I)$ generated by a system of LRS $\mu_1, \ldots, \mu_l \in L_M(F_1, \ldots, F_k)$ such that $\mu_1(\Pi), \ldots, \mu_l(\Pi)$ is the generating set of the R-module of the solutions of the system (3.14). As ${}_RM$ is a QF-module, according to Proposition 2.4, we may state that the set of diagrams $H_{1\Pi}, \ldots, H_{w\Pi}$ is a generating set of the R -module of the solutions of (3.23). Hence in view of Corollary 3.6, the ideal $\mathrm{Ann}(L_M(I))$ is generated by a set of polynomials (3.20) and the equality (4.1) holds.

The equality (4.2) can be proved analogously.

(b) $\Rightarrow$ (c). Let $\mathcal{M} = L_M(I)$. Then for any submodule $\mathcal{K} <_S \mathcal{M}$ the following equalities hold: $\mathrm{Ann}_{\mathcal{S}}(\mathcal{K}) = \widetilde{\mathcal{T}} = \mathcal{T}/I$, where $\mathcal{T} = \mathrm{Ann}_{\mathcal{S}}(\mathcal{K}) \lhd \mathcal{P}_k$, and

$$\mathrm{Ann}_{\mathcal{M}}(\mathrm{Ann}_{\mathcal{S}}(\mathcal{K})) = \mathrm{Ann}_{\mathcal{M}}(\widetilde{\mathcal{T}}) = L_M(\mathcal{T}).$$

Hence according to (b), $\mathrm{Ann}_{\mathcal{M}}(\mathrm{Ann}_{\mathcal{S}}(\mathcal{K})) = \mathcal{K}$.

For any ideal $\widetilde{\mathcal{T}} \lhd \mathcal{S}$, there exists a unique ideal $\mathcal{T} \lhd \mathcal{P}_k$ such that $I \lhd \mathcal{T}$ and $\widetilde{\mathcal{T}} = \mathcal{T}/I$. Then $\mathrm{Ann}_{\mathcal{M}}(\widetilde{\mathcal{T}}) = L_M(\mathcal{T})$, $\mathrm{Ann}(L_M(\mathcal{T})) \supset I$ and $\mathrm{Ann}_{\mathcal{S}}(\mathrm{Ann}_{\mathcal{M}}(\widetilde{\mathcal{T}})) = \mathrm{Ann}_{\mathcal{S}}(L_M((\mathcal{T})) = \mathrm{Ann}(L_M((\mathcal{T}))/I$. In view of (b), $\mathrm{Ann}(L_M((\mathcal{T})) = \mathcal{T}$, we have $\mathrm{Ann}_{\mathcal{S}}(\mathrm{Ann}_{\mathcal{M}}(\widetilde{\mathcal{T}})) = \mathcal{T}/I = \widetilde{\mathcal{T}}$. Consequently, $L_M(I)$ is a quasi-Frobenius $\mathcal{S}$-module.

Let $\mathcal{M} \in \mathfrak{M}_k$ and $I = \mathrm{Ann}(\mathcal{M})$. Then (b) implies that $\mathcal{M} = L_M(I)$ and as it was proved above, $\mathcal{M}$ is a quasi-Frobenius module over the ring $S = \mathcal{P}_k/\mathrm{Ann}(\mathcal{M}) = \mathcal{P}_k/I$.

(c) $\Rightarrow$ (d). To prove (1.7), we put $I = I_1 \cap I_2$, $S = \mathcal{P}_k/I$, $\widetilde{I}_s = I_s/I$ for $s \in \{1, 2\}$, and $\mathcal{M} = L_M(I)$. Then $\mathrm{Ann}_{\mathcal{M}}(\widetilde{\mathcal{I}}_s) = L_M(I_s)$ for $s \in \{1, 2\}$, and to prove (1.7) it is sufficient to prove the equality

$$\mathcal{M} = \mathrm{Ann}_{\mathcal{M}}(\widetilde{\mathcal{I}}_1) + \mathrm{Ann}_{\mathcal{M}}(\widetilde{\mathcal{I}}_2). \tag{4.3}$$

Note that $\widetilde{\mathcal{I}}_1 \cap \widetilde{\mathcal{I}}_2 = \widetilde{0}$ and the equality

$$\mathrm{Ann}_{\mathcal{S}}(\mathrm{Ann}_{\mathcal{M}}(\widetilde{\mathcal{I}}_1) + \mathrm{Ann}_{\mathcal{M}}(\widetilde{\mathcal{I}}_2)) = \mathrm{Ann}_{\mathcal{S}}(\mathrm{Ann}_{\mathcal{M}}(\widetilde{\mathcal{I}}_1)) + \mathrm{Ann}_{\mathcal{S}}(\mathrm{Ann}_{\mathcal{M}}(\widetilde{\mathcal{I}}_2))$$

is true. Hence, as ${}_{\mathcal{S}}\mathcal{M}$ is a QF-module, the following equalities hold:

$$\mathrm{Ann}_{\mathcal{S}}(\mathrm{Ann}_{\mathcal{M}}(\widetilde{\mathcal{I}}_1) + \mathrm{Ann}_{\mathcal{M}}(\widetilde{\mathcal{I}}_2)) = \widetilde{\mathcal{I}}_1 \cap \widetilde{\mathcal{I}}_2 = \widetilde{0} = \mathrm{Ann}_{\mathcal{S}}(\mathcal{M}),$$

and (4.3) is true.

To prove (1.8), we put $\mathcal{M} = \mathcal{M}_1 + \mathcal{M}_2$ and $I = \mathrm{Ann}(\mathcal{M})$. Then $\mathcal{M}$ is a QF-module over the ring $S = \mathcal{P}_k/I$. Note that $\mathrm{Ann}_{\mathcal{S}}(\mathcal{M}_1 \cap \mathcal{M}_2) = \mathrm{Ann}(\mathcal{M}_1 \cap \mathcal{M}_2)/I$ and $\mathrm{Ann}_{\mathcal{S}}(\mathcal{M}_t) = \mathrm{Ann}(\mathcal{M}_t)/I$ for $t \in \{1, 2\}$. Hence to prove (1.8), it is sufficient to prove the equality

$$\mathrm{Ann}_{\mathcal{S}}(\mathcal{M}_1 \cap \mathcal{M}_2) = \mathrm{Ann}(\mathcal{M}_1) + \mathrm{Ann}(\mathcal{M}_2). \tag{4.4}$$

As the equality

$$\mathrm{Ann}_{\mathcal{M}}(\mathrm{Ann}_{\mathcal{S}}(\mathcal{M}_1) + \mathrm{Ann}_{\mathcal{S}}(\mathcal{M}_2)) = \mathrm{Ann}_{\mathcal{M}}(\mathrm{Ann}_{\mathcal{S}}(\mathcal{M}_1)) \cap \mathrm{Ann}_{\mathcal{M}}(\mathrm{Ann}_{\mathcal{S}}(\mathcal{M}_2))$$

is true and ${}_{\mathcal{S}}\mathcal{M}$ is a QF-module, the equality

$$\mathrm{Ann}_{\mathcal{M}}(\mathrm{Ann}_{\mathcal{S}}(\mathcal{M}_1) + \mathrm{Ann}_{\mathcal{S}}(\mathcal{M}_2)) = \mathcal{M}_1 \cap \mathcal{M}_2$$

holds. This implies (4.4) (as ${}_{\mathcal{S}}\mathcal{M}$ is a QF-module).

(d) $\Rightarrow$ (a). As it was noted in the conclusion of §3, we may suppose that R is a local Artinian ring. Assume that ${}_R M$ is not a QF-module. Then according to Proposition 2.2, $\mathrm{Ann}_M(\mathfrak{N}(R))$ is not a cyclic submodule of ${}_R M$. Hence there exist $\alpha_1, \alpha_2 \in M \setminus \{0\}$ such that $\mathfrak{N}(R)\alpha_1 = \mathfrak{N}(R)\alpha_2 = 0$, $R\alpha_1 \cap R\alpha_2 = 0$.

Let us consider the modules $\mathcal{M}_1, \mathcal{M}_2 \in \mathfrak{M}_k$ such that $\mathcal{M}_s$ (for $s \in \{1, 2\}$) is the set of all k-LRS $\mu \in \mathcal{L}M^{(k)}$ with the property $\mu(\vec{0}) \in R\alpha_s$, $\mu(\vec{i}) = 0$ for all

$\vec{i} > \vec{0}$. Then the following relations hold: $\mathcal{M}_1 \cap \mathcal{M}_2 = 0$, $\text{Ann}(\mathcal{M}_1 \cap \mathcal{M}_2) = \mathcal{P}_k$, $\text{Ann}(\mathcal{M}_s) = (x_1, \dots, x_k, \mathfrak{N}(R))$ for $s \in \{1, 2\}$. Hence (d) is not true as (1.8) is not true. It follows that $_RM$ is a QF-module.

(a) $\Rightarrow$ (e). The implication $\nu \in \mathcal{P}_k\mu \Rightarrow \text{Ann}(\mu) \subseteq \text{Ann}(\nu)$ is evident. Let us suppose that $\text{Ann}(\mu) \subseteq \text{Ann}(\nu)$. Note that in view of (a), the condition (e) is true. Then $L_M(\text{Ann}(\mu)) \subseteq L_M(\text{Ann}(\nu))$ and in view of (b) we have $L_M(\text{Ann}(\mu)) = \mathcal{P}_k\mu$ and $L_M(\text{Ann}(\nu)) = \mathcal{P}_k\nu$. Hence (e) is true.

(e) $\Rightarrow$ (a). We may assume that R is a local ring. Let us suppose that $_RM$ is not a QF-module and α_1, α_2 are the same as in the proof of (d) $\Rightarrow$ (a). We consider the sequence $\mu_1, \mu_2 \in \mathcal{L}M^{(k)}$ such that $\mu_s(\vec{0}) = \alpha_s$, $\mu(\vec{i}) = 0$ for all $\vec{i} > \vec{0}$, $s \in \{1, 2\}$. Then $\text{Ann}(\mu_1) = \text{Ann}(\mu_2)$, but $\mathcal{P}_k\mu_1 \nsubseteq \mathcal{P}_k\mu_2$ as $R\alpha_1 \nsubseteq R\alpha_2$. This is a contradiction to (e). □

Proof of Theorem 1.6. (a) $\Rightarrow$ (b). According to Theorem 1.4, we have the equalities $L_Q(I) = L_Q(\text{Ann}(\mathcal{P}_k\mu)) = \mathcal{P}_k\mu$, i.e., $L_Q(I)$ is a cyclic $\mathcal{P}_k$-module.

(b) $\Rightarrow$ (c). In view of Theorem 1.4, the family $\mathcal{M} = \mathcal{P}_k\mu = S_\mu$ is a cyclic module, and according to the definition, $\mathcal{M}$ is a faithful S-module. Therefore there exists an isomorphism $_SS \cong {_S\mathcal{M}}$. Hence $_SS$ is a QF-module, i.e., S is a QF-ring.

(c) $\Rightarrow$ (a). As $S = \mathcal{P}_k/I$ is a QF-ring, the module $_SS$ is quasi-Frobenius one. According to Theorem 1.4, $_SL_Q(I)$ is a QF-module too. Hence, in view of Theorem 2.1, $_SS \cong {_SL_Q(I)}$ and $L_Q(I)$ is a cyclic S-module: $L_Q(I) = S_\mu = \mathcal{P}_k\mu$ for some $\mu \in L_Q(I)$. It follows $I - \text{Ann}(L_Q(I)) = \text{Ann}(\mathcal{P}_k\mu) = \text{Ann}(\mu)$. □

Proof of Theorem 1.7. Let $\mathcal{M} = L_R(F_1(x_1), \dots, F_k(x_k))$. Note that in view of Proposition 3.2, $\mathcal{M}$ is a cyclic S-module: $\mathcal{M} = \mathcal{P}_k e^{\vec{F}} = Se^{\vec{F}}$.

If R is a QF-ring, then according to Theorem 1.6, $_S\mathcal{M}$ is a cyclic QF-module. Hence in view of Theorem 1.6, S is a QF-ring.

Let S be a QF-ring, but R is not a QF-ring. As it was noted in the conclusion of §3, we may assume that R is a local ring. Then it follows from Proposition 2.2 that there exist $a_1, a_2 \in \text{Ann}_R(\mathfrak{N}(R))\backslash\{0\}$ such that $Ra_1 \cap Ra_2 = 0$. In this case $\text{Ann}_R(a_k) = \mathfrak{N}(R)$ for $t \in \{1, 2\}$.

Let us consider the submodules

$$\mathcal{M}_t = \mathcal{P}_k a_t e^{\vec{F}} = Sa_t e^{\vec{F}} \quad (t \in \{1, 2\})$$

of the S-module $\mathcal{M}$. Evidently $\mathcal{M}_1 \neq \mathcal{M}_2$. It is not difficult to see that $\text{Ann}_R(\mathcal{M}_t) = (F_1(x_1), \dots, F_k(x_k), \mathfrak{N}(R))$ for $t \in \{1, 2\}$. Hence $\text{Ann}_S(\mathcal{M}_1) = \text{Ann}_S(\mathcal{M}_2) = \mathfrak{N}(R)S$ and $\mathcal{M}_1 = \text{Ann}_{\mathcal{M}}(\text{Ann}_S(\mathcal{M}_1)) = \text{Ann}_{\mathcal{M}}(\text{Ann}_S(\mathcal{M}_2)) = \mathcal{M}_2$, as $_S\mathcal{M}$ is a QF-module. This is a contradiction to the definitions of $\mathcal{M}_1$ and $\mathcal{M}_2$. So R is a QF-ring. □

Acknowledgement. The author is grateful to Prof. A. V. Mikhalev and Prof. V. T. Markov for a discussion of the results and for their support.

References

[1] G. Azumaya, A duality theory for injective modules (Theory of quasi-Frobenius modules), Amer. J. Math. 81 (1959), 249–278.

[2] L. Cerlienco and F. Piras, On the continuous dual of a polynomial bialgebra, Comm. Algebra 19 (1991), 2707–2727.

[3] C. Faith, Algebra II: Ring Theory, Grundlehren Math. Wiss. 191, Springer-Verlag 1976.

[4] F. Kasch, Module und Ringe, B. G. Teubner, Stuttgart 1977.

[5] V. L. Kurakin, A. S. Kuzmin, A. V. Mikhalev and A. A. Nechaev, Linear recurring sequences over rings and modules, J. Math. Sci. 1994, Thematic Surveys, 130pp.

[6] F. Y. MacWilliams and N. J. A. Sloane, The Theory of Error-correcting Codes, North-Holland Publ. Co., 1977.

[7] K. Morita, Duality for modules and its applications to the theory of rings with minimum condition, Sci. Rep. Tokyo Kyoika Daigaku 6 (1958), 83–142.

[8] M. Nagata, Local Rings, Int. Publ., New York 1962.

[9] A. A. Nechaev, Linear recurring sequences over commutative rings, Discrete Math. Appl. 2 (1992), 659–683, (in Russian: Diskret. Math. 3 (1991)).

[10] A. A. Nechaev, Linear recurring sequences over quasi-Frobenius modules, Russian Mathematical Surveys 48 (1993).

[11] R. Wisbauer, Grundlagen der Modul- und Ringtheorie, Verlag Reinhard Fischer, München, 1982.

Department of Mechanics and Mathematics
Chair of Higher Algebra
Moscow State University
Moscow 119899, Russia

On semi-endomorphism near-rings of non-abelian groups of order pq

L. S. Shiao and C. S. Wang

Abstract. The notion of a semi-homomorphism of groups was introduced by I. N. Herstein in 1968 while the notion of a semi-endomorphism near-ring $S(G)$ of a group G was introduced by Y. Fong and G. Pilz in [4]. Since 1991 the structure of $S(G)$ was investigated in a number of papers. In the present article it is shown that every semi-endomorphism of a nonabelian group G of order pq, where $p, q > 2$ are prime numbers, either is an endomorphism or is an antiautomorphism. Further it is proved that the semi-endomorphism near-ring of the above group G is equal to the endomorphism near-ring and the elementwise description of the endomorphism near-ring of G is presented. Finally an explicit formula for its order is obtained.

1991 Mathematics Subject Classification: 20K30, 16Y30

1. Introduction

An additively written group $(N, +)$ equipped with a binary operation $\cdot : N \times N \to N$, $(x, y) \mapsto xy$, such that $(xy)z = x(yz)$ and $(y+z)x = yx + zx$ for all $x, y, z \in N$ is called a (right) near-ring. We refer the reader to the books of Clay [1], Meldrum [8] and Pilz [9] for the basic results of near-ring theory and its applications.

Given an additively written group G, the mapping $\alpha : G \to G$ is called a semi-endomorphism if $\alpha(x + y + x) = \alpha(x) + \alpha(y) + \alpha(x)$ for all $x, y \in G$. We denote by $S(G)$ $(E(G))$ the subnearring of the transformation near-ring $M(G)$ generated by all semi-endomorphisms (respectively, by all endomorphisms) of G. The notion of semi-endomorphism was introduced and investigated by I. N. Herstein [5] (see also [6]) which, in particular, it was proved that every semi-automorphism of a simple non-abelian group with elements of order two is either an automorphism or an anti-automorphism. Here we also note the following result of Y. Fong [2] which states that every nonzero semi-endomorphism of a finite simple group of order bigger than two is either an automorphism or an antiautomorphism. Our first main result is the following theorem.

Theorem 1.1. *Let p, q be prime numbers and let k be an integer such that $k \not\equiv 1 \pmod q$ but $k^p \equiv 1 \pmod q$. Further let*

$$G = \langle x, y \mid px = 0 = qy, \ -x + y + x = ky \rangle$$

be a group given by generators and relations. Suppose that $p, q > 2$. Then any semi-endomorphism χ of G is either an endomorphism or antiautomorphism of G.

We note that for $p = 2$ and $q > 2$, Y. Fong [2] had constructed an example of a group G described in Theorem 1.1 and a semi-endomorphism α of G which is neither an endomorphism nor an antiautomorphism of G. Therefore Theorem 1.1 is no longer true if $p = 2$.

Let G be the group defined as Theorem 1.1. We denote by Z_p the prime field of order p. Given two polynomials $u(t), v(t) \in Z_q[t]$ and $a \in Z_p$, we denote by $[a, u, v]$ the mapping $G \to G$ given by the rule:

$$[a, u, v](mx + ny) = \begin{cases} v(1)ny & \text{if } m = 0 \\ amx + (u(k^m) + v(k^m)n)y & \text{if } 0 < m < p \end{cases}$$

for all $m, n \in Z$. Further let

$$\mathcal{P}_{p,q} = \{[a, u(t), v(t)] \mid u(1) = 0, \ \deg(u), \deg(v) < p\}.$$

The notion of semi-endomorphism near-ring $S(G)$ of a group G was introduced by Y. Fong and G. Pilz [4] and were investigated in a number of papers [4], [3] and [2]. Our second main result is the following theorem.

Theorem 1.2. *Let $p, q > 2$ be prime numbers and let k be an integer such that $k \not\equiv 1 \pmod q$ but $k^p \equiv 1 \pmod q$. Further let $G = \langle x, y \mid px = 0 = qy, \ -x + y + x = ky \rangle$ be a group given by generators and relations. Then:*

(i) $S(G) = E(G)$;

(ii) $S(G) = \mathcal{P}_{p,q}$ *and* $|S(G)| = pq^{2p-1}$.

Given a group G and $x \in G$, we denote by $O(x)$ the order of the element x. For $x, y \in G$ we set $[x, y] = x + y - x - y$. Here we close this section with the following lemmas which will be used frequently in the sequel.

Lemma 1.3 ([4, Lemma 1.1]). *Let f be a semi-homomorphism of a group G. Then $f(-x) = -f(x)$ for all $x \in G$.*

Lemma 1.4 ([4, Lemma 1.2]). *Let f be a semi-homomorphism of a group G. Then $O(f(0)) \leq 2$ and $[f(x), f(0)] = 0$ for all $x \in G$.*

Lemma 1.5 ([4, Lemma 1.3]). *Let f be a semi-homomorphism of a group G, $x \in G$ and n a natural number. Then*

$$f(nx) = \begin{cases} nf(x) & \text{if } n \text{ is odd,} \\ nf(x) + f(0) & \text{if } n \text{ is even} \end{cases}$$

Corollary 1.6. *Let f be a semi-homomorphism of a group G. Suppose that G has no elements of order 2. Then $f(0) = 0$ and $f(nx) = nx$ for all $x \in G$ and natural number n.*

Lemma 1.7 ([4, Lemma 1.5]). *Let f be a semi-endomorphism of a group G. If $x \in G$ has even order n then the order of $f(x)$ divides n. If $f(0) = 0$, this holds for odd n, too.*

2. Proofs of the main theorems

We start with the following notations and observations. In what follows p, q, and G be as in Theorem 1.1. Further, let a and b be integers. We define endomorphisms α_a and β_b by the rules $\alpha_a(x) = x$, $\alpha_a(y) = ay$, $\beta_b(x) = x + by$, and $\beta_b(y) = y$. We note that β_b is in fact an automorphism of G and α_a is an automorphism of G if, and only if, $a \not\equiv 0 \pmod q$. We denote by τ the antiautomorphism of G given by the rule $\tau(z) = -z$ for all $z \in G$. Let $0 < m < p$ and $0 \leq n < q$. Then

$$\beta_b(mx + ny) = mx + \{b(k-1)^{-1}(k^m - 1) + n\}y \tag{1}$$

Indeed,

$$\begin{aligned}\beta_b(mx + ny) &= m\beta_b(x) + n\beta_b(y) = m(x + by) + ny \\ &= mx + \left(\sum_{i=0}^{m-1} bk^i + n\right) y \\ &= mx + \{b(k-1)^{-1}(k^m - 1) + n\}y.\end{aligned}$$

Remark 2.1. Let a, b be natural numbers. Then

$$O(ax + by) = \begin{cases} p & \text{if } a \not\equiv 0 \pmod p, \\ q & \text{if } a \equiv 0 \pmod p \text{ but } b \not\equiv 0 \pmod q \end{cases}$$

Proof. If $a \equiv 0 \pmod p$ but $b \not\equiv 0 \pmod q$, then clearly $O(by) = q$. Now let $a \not\equiv 0 \pmod p$. Then we have

$$p \cdot (ax + by) = pax + \left(b \cdot \sum_{t=0}^{p-1} k^{at}\right) y = b\left(\frac{k^{ap} - 1}{k^a - 1}\right) y = 0$$

because $k^{ap} = (k^p)^a \equiv 1 \pmod q$. Therefore $O(ax + by) = p$. □

Theorem 2.2. *Let $p, q > 2$ be prime numbers and let k be an integer such that $k \not\equiv 1 \pmod q$ but $k^p \equiv 1 \pmod q$. Further let*

$$G = \langle x, y \mid px = 0 = qy,\ -x + y + x = ky \rangle$$

be a group given by generators and relations and let χ be a semi-endomorphism of the group G such that $\chi(y) \neq 0$. Then there exist natural numbers b, d with $0 \leq b < p$, $0 < d < q$ such that:

(i) *$\chi(y) = dy$ and either $\chi(x) = x + by$, or $\chi(x) = -x + by$;*

(ii) *If $\chi(x) = x + by$, then $\chi = \beta_b \alpha_d$;*

(iii) *If $\chi(x) = -x + by$, then $\chi = \beta_{-bk} \tau \alpha_{-d}$.*

Proof. Clearly $\chi(y) = cx + dy$ for some natural numbers c and d. By Lemma 1.7 we have that $O(\chi(y)) = q$. It follows from Remark 2.1 that $\chi(y) = dy$. Since $\chi(y) \neq 0$, $0 < d < q$. Obviously $\sigma = \chi \alpha_d^{-1}$ is a semi-endomorphism of G and $\sigma(y) = y$. Clearly $\sigma(x) = ax + by$ for some natural numbers a, b. We claim that either $a \equiv -1 \pmod p$ or $a \equiv 1 \pmod p$. Indeed, we have that

$$\begin{aligned} \sigma(y + 2x + y) &= \sigma(y) + 2\sigma(x) + \sigma(y) = y + 2(ax + by) + y \\ &= 2ax + (k^{2a} + bk^a + b + 1)y \end{aligned} \tag{2}$$

On the other hand

$$y + 2x + y = 2x + (k^2 + 1)y = x + (k^2 + 1)ry + x$$

where $kr \equiv 1 \pmod q$. Therefore

$$\begin{aligned} \sigma(y + 2x + y) &= \sigma(x + (k^2 + 1)ry + x) = \sigma(x) + (k^2 + 1)r\sigma(y) + \sigma(x) \\ &= ax + by + (k^2 + 1)ry + ax + by \\ &= 2ax + (bk^a + (k^2 + 1)rk^a + b)y \\ &= 2ax + (bk^a + k^{a+1} + k^{a-1} + b)y \end{aligned} \tag{3}$$

From (2.2) and (2.3) we see that

$$k^{2a} - k^{a+1} - k^{a-1} + 1 = (k^{a+1} - 1)(k^{a-1} - 1) \equiv 0 \pmod q$$

Therefore either $k^{a+1} - 1 \equiv 0 \pmod q$ or $k^{a-1} - 1 \equiv 0 \pmod q$. Since k is an element of order p of the multiplicative group of q element field, we infer that either $a \equiv -1 \pmod p$ or $a \equiv 1 \pmod p$ and so our claim is established. Since $\sigma(x) = \chi\alpha_d(x) = \chi(x)$, (i) is proved.

Suppose now that $a \equiv 1 \pmod p$. Then $\sigma(x) = x + by$ and $\sigma(y) = y$. We claim that $\sigma = \beta_b$. Indeed, it is enough to show that $\theta = \beta_b^{-1}\sigma = \mathrm{id}_G$ the identity map of G. We have that $\theta(x) = x$, $\theta(y) = y$ and θ is a semi-endomorphism of G. Let $u = mx + ny$ for some natural numbers m, n. We have that $u = mlx + nr^{ml}y + mlx$ where $2 \cdot l \equiv 1 \pmod p$ and $kr \equiv 1 \pmod q$. It follows from Corollary 1.6 that

$$\theta(u) = \theta(mlx + nr^{ml}y + mlx) = mlx + nr^{ml}y + mlx = u$$

and so $\theta = \mathrm{id}_G$ and our claim is established. Therefore $\chi\alpha_d^{-1} = \beta_b$ and $\chi = \beta_b\alpha_d$ which proves (ii).

Finally we suppose that $a \equiv -1 \pmod p$. Then $\sigma(x) = -x + by$ and $\sigma(y) = y$. We claim that σ is an antiautomorphism of G. Indeed, setting $d = bk$ and $\zeta = \beta_d\sigma\alpha_{-1}$, we note that ζ is a semi-endomorphism of G, $\zeta(y) = -y$ and

$$\begin{aligned}\zeta(x) &= \beta_d\sigma\alpha_{-1}(x) = \beta_d\sigma(x) = \beta_d(-x+by)\\ &= -\beta_d(x) + \beta_d(by) = -(x+bky) + by\\ &= -bky - x + by = -bry + (-x+by+x) - x = -bky + bky - x = -x\end{aligned}$$

Therefore $\zeta(x) = -x$, $\zeta(y) = -y$ and so $\tau\zeta(x) = x$, $\tau\zeta(y) = y$. Since $\tau\zeta$ is a semi-endomorphism, by the above result we have that $\tau\zeta = \mathrm{id}_G$ and so

$$\tau\beta_{bk}\chi\alpha_d^{-1}\alpha_{-1} = \mathrm{id}_G.$$

Since $(\alpha_d^{-1}\alpha_{-1})^{-1} = \alpha_{-d}$, $\tau^{-1} = \tau$ and $\beta_{bk}^{-1} = \beta_{-bk}$, we see that $\chi = \beta_{-bk}\tau\alpha_{-d}$ which completes the proof. □

Corollary 2.3. *Let χ be a semi-endomorphism of G such that $\chi(y) \neq 0$. Then χ is either an automorphism or an antiautomorphism of G.*

Proof of Theorem 1.1. According to Corallary 2.3 it is enough to consider the case when $\chi(y) = 0$. We claim that χ is an endomorphism of G. Indeed, let $z \in G$. Then there exist natural numbers m and n such that $z = mx + ny$. Since $p > 2$, there exists a natural number l such that $2 \cdot l \equiv 1 \pmod p$. Choose a natural number r such that $kr \equiv 1 \pmod q$. Then we have $mx + ny = mlx + nr^{ml}y + mlx$ and so from Corollary 1.6 we see that

$$\begin{aligned}\chi(z) &= \chi(mx+ny) = \chi(mlx + nr^{ml}y + mlx) = \chi(mlx) + \chi(nr^{ml}y) + \chi(mlx)\\ &= ml\chi(x) + nr^{ml}\chi(y) + ml\chi(x) = (2\cdot l)m\chi(x) = m\chi(x).\end{aligned}$$

Let now $u = ax + by$ where a, b are natural numbers. Then $z + u = (m+a)x + (nk^a + b)y$ and hence

$$\chi(z+u) = (m+a)\chi(x) = \chi(z) + \chi(u).$$

Therefore χ is an endomorphism of the group G. □

Proof of Theorem 1.2. Clearly $-\mathrm{id}_G = \tau$ and so $\tau \in E(G)$. Applying Theorem 2.2 and Theorem 1.1 we conclude that $E(G)$ contains all the semi-endomorphisms of G. Hence $S(G) = E(G)$ and (i) is proved.

Let $0 \leq a < p$ and $u(t), v(t) \in Z_q[t]$. Denote by $u^*(t)$ and $v^*(t)$ the remainders of $u(t)$ and $v(t)$ modulo $t^p - 1$. Since $k^p \equiv 1 \pmod q$, it follows that

$$[a, u(t), v(t)] = [a, u^*(t), v^*(t)].$$

Note that if $u(1) = 0$, then $u^*(1) = 0$ as well.

We claim that $\beta_b = [1, b(k-1)^{-1}(t-1), 1]$ and $\alpha_d = [1, 0, d]$. Indeed, setting $a = 1$, $u(t) = b(k-1)^{-1}(t-1)$ and $v(t) = 1$ we see that $\beta_b(ny) = ny = v(1)ny$.

Further if $0 < m < p$, then by (2.1)

$$\begin{aligned}\beta_b(mx+ny) &= m\beta(x)+n\beta(y) = mx+\{b(k-1)^{-1}(k^m-1)+n\}y \\ &= amx+\{u(k^m)+v(k^m)n\}y.\end{aligned}$$

Therefore $\beta_b = [a, u(t), v(t)] = [1, b(k-1)^{-1}(t-1), 1]$. Next we set $a=1$, $u(t)=0$ and $v(t)=d$. Then we have $\alpha_d(ny) = dny = v(1)ny$. If $0 < m < p$, then

$$\begin{aligned}\alpha_d(mx+ny) &= m\alpha_d(x)+n\alpha_d(y) = mx+dny \\ &= amx+\{u(k^m)+v(k^m)n\}y\end{aligned}$$

which proves our claim. Therefore $\alpha_d, \beta_b \in \mathcal{P}_{p,q}$.

Given $[a,u,v], [a',u',v'] \in \mathcal{P}_{p,q}$, one can readily checks that

$$[a,u,v]+[a',u',v'] = [a+a', u'', v''],$$

where $u'' = (t^{a'}u(t)+u'(t))^*$ and $v'' = (t^{a'}v(t)+v')^*$, and

$$[a,u,v][a',u',v'] = [aa', u''', v'''],$$

where $u''' = (u(t^{a'})+u'(t)v(t^{a'}))^*$ and $v''' = (v(t^{a'})v'(t))^*$. Therefore $\mathcal{P}_{p,q}$ is a subnearring of the transformation nearring $M_0(G)$. It follows from Theorem 2.2, that $\operatorname{Aut}(G) \subseteq \mathcal{P}_{p,q}$ and so $A(G) \subseteq \mathcal{P}_{p,q}$.

Next we claim that $A(G) = \mathcal{P}_{p,q}$. Indeed, let $[a,u,v] \in \mathcal{P}_{p,q}$. It follows that $\beta_b\alpha_d = [1, b(k-1)^{-1}(t-1), d]$ and

$$\begin{aligned}&[1, b_1(k-1)^{-1}(t-1), d_1]+\cdots+[1, b_n(k-1)^{-1}(t-1), d_n] \\ &= \left[n, \left(\sum_{i=0}^{n-1} t^{n-1-i} b_i(k-1)^{-1}(t-1)\right)^*, \left(\sum_{i=0}^{n-1} t^{n-1-i} d_i\right)^*\right]\end{aligned}$$

setting $n = p+a$, we can find $b_1, \ldots, b_n, d_1, \ldots, d_n$ such that

$$\left(\sum_{i=0}^{n-1} t^{n-1-i} b_i(k-1)^{-1}(t-1)\right)^* = u(t)$$

and

$$\left(\sum_{i=0}^{n-1} t^{n-1-i} d_i\right)^* = v(t)$$

which implies that $A(G) = \mathcal{P}_{p,q}$. According to [7], $A(G) = E(G)$. Therefore $S(G) = E(G) \subseteq \mathcal{P}_{p,q}$. Since the number of polynomials in $Z_q[t]$ of degree $\leq p-1$ having root 1 is equal to q^{p-1}, we see that $|\mathcal{P}_{p,q}| = pq^{p-1}q^p = pq^{2p-1}$ which completes the proof. □

We would like to express our deepest gratitude to Y. Fong for his numerous useful comments and suggestions.

References

[1] Clay J. R., Near-rings: Geneses and Applications, Oxford Univ. Press, Oxford 1992.

[2] Fong Y., On semi-endomorphisms of groups, to appear.

[3] Fong Y., Huang F. K., Ke W. F., Yeh Y. N., On Semi-endomorphisms of Finite Abelian Groups and Transformation Near-rings, to appear.

[4] Fong Y., Pilz G., Near-ring generated by semi-endomorphisms of groups, in: Contribution to General Algebra 7, Verlag Hölder-Pichler-Temsky, Wien 1991; Verlag B. G. Teubner, Stuttgart, 159–168.

[5] Herstein I. N., Semi-homomorphisms of groups, Canad. J. Math. 20 (1968) 384–388.

[6] Herstein I. N., Ruchte M. F., Semi-automorphisms of groups, Proc. Amer. Math. Soc. 9 (1958), 145–150.

[7] Malone J. J., Mason G., ZS-metacyclic groups and their Endomorphism Near-rings, to appear in Monatsh. Math.

[8] Meldrum J. D. P., Near-rings and Their Links with Groups, Pitman, Marshfield, MA, 1985.

[9] Pilz G., Near-rings, 2nd ed., 23, North Holland/American Elsevier, Amsterdam, 1983.

[10] Thomsen G. S. M. J., Syskin S. A., Endomorphism Near-rings of Finite Groups, a Report, to appear.

Graduate Institute of Mathematics Education
National Kaohsiung Normal University
Kaohsiung, Taiwan, ROC

Chinese Air Force Academy
Kang San
Taiwan, ROC

Local structure of linearly compact algebras and local finiteness of coalgebras

Arkadii Slinko

The theory of coalgebras was initially developed as a part of the structure theory of Hopf algebras which have appeared in topology in various ways. Hopf algebras were the first examples of bialgebras studied by algebraists and they were essentially associative objects with associative multiplication and comultiplication (see [1] and the recent monograph [2] devoted to this theory). Lie coalgebras having associative coenveloping coalgebras were considered in [3, 4, 5, 6]. Michaelis [7, 8] considered Lie coalgebras from the abstract point of view, and he discovered that not all of them have associative coenveloping coalgebras, but only locally finite ones. Lie bialgebras were introduced by V. Drinfel'd [9] for the sake of quantum groups.

Since then it became clear that every variety of algebras has a corresponding category of coalgebras. For coalgebras of an arbitrary variety all necessary definitions were first given by J. Anquela, T. Cortés and F. Montaner [10], where basic results of the associative theory were generalized to coalgebras of an arbitrary variety. But one of the most important results, the so called "Fundamental Theorem on Coalgebras", which states that every associative coalgebra is locally finite, cannot be generalized to coalgebras of an arbitrary variety. In fact, Michaelis showed that it is not true for Lie coalgebras. Nevertheless, J. Anquela, T. Cortés and F. Montaner proved that "Fundamental Theorem on Coalgebras" is true for alternative and Jordan coalgebras. After this result were delivered at the 3rd International Conference on Nonassociative Algebras and Their Applications, E. Taft, in a private conversation with the author expressed the opinion that it would be interesting to find some necessary and sufficient conditions on the variety under which all coalgebras of this variety are locally finite. I. Shestakov suggests that these conditions must be similar to those (see [11]) under which the locally nilpotent radical exists. Partially motivated by this question the author suggested in [12] a necessary and sufficient condition for the Lie coalgebra M to be locally finite. It was given in terms of its dual Lie algebra L, which is a linearly compact topological algebra with respect to the linear weak topology. It was shown that local finiteness of M is equivalent to the existence of a base of ideal neighborhoods of zero of L which is, in turn, equivalent to topological algebraicity of L.

In this paper we shall specify a broad class of varieties of algebras for which necessary and sufficient condition can be obtained, similar to the condition for Lie algebras in [12]. As varieties of Jordan and alternative algebras are contained in the specified class of varieties and since the necessary and sufficient conditions mentioned become trivial for them, the result can be considered as a generalization of the result of

J. Anquela, T. Cortés and F. Montaner [10]. On the other hand the results on the local structure of linearly compact alternative and Jordan algebras are also new. Due to this connection with coalgebras the results on the structure of linearly compact algebras are getting more and more important. The reader may notice their similarity to the structure of compact rings [13, 14].

1. Coalgebras and their dual algebras

Let k be a field. A coalgebra over k is a vector space C over k equipped with an arbitrary k-linear map $\Delta : C \to C \otimes C$. This mapping Δ is called the diagonal mapping or comultiplication of the coalgebra C. In practice this comultiplication obeys certain "identities", so that we can speak of associative, Lie, and other coalgebras. Although it is possible to give these definitions directly in each particular case the general rule will be hidden. It comes to the surface with the definition of the dual algebra of a coalgebra.

Let us consider the dual space $A = C^*$ of C and define a multiplication on it by the formula

$$(f \cdot g)(c) = (f \otimes g)(\Delta c), \tag{1}$$

or in Sweedler's notations [1], if $\Delta c = \sum_c c_1 \otimes c_2$, then

$$(f \cdot g)(c) = \sum_c f(c_1)g(c_2). \tag{2}$$

Definition 1.1. Let $\mathcal{M}$ be an arbitrary variety of algebras over k. A coalgebra C is called a coalgebra of the variety $\mathcal{M}$ if its dual algebra $A = C^*$ is an algebra of $\mathcal{M}$.

As it is well-known [15], A as a vector space has a natural topology on it, which is defined by the neighborhood base of zero consisting of the subspaces

$$V^\perp = \{f \in A \mid f(V) = 0\}, \quad \dim V < \infty.$$

It is linear in a sense that it possesses a neighborhood base of zero consisting of subspaces and it is called the linear weak topology or the weak-* topology. Let us recall also [16] that a linear topology of a vector space E is said to be linearly compact if any filter base on E consisting of closed linear varieties has a non-empty intersection.

Proposition 1.2. *With the linear weak topology A is a linearly compact topological algebra.*

Proof. Given that $\Delta c = \sum_c c_1 \otimes c_2$, we define the subspaces $L(c)$ and $R(c)$ to be the spans of all c_1's and c_2's, respectively.

Suppose that $V^\perp$ for $V = \{v_1, \dots, v_k\}$ is one of the neighborhoods from the base. Set $U = L(v_1) + \cdots + L(v_k)$ and $W = R(v_1) + \cdots + R(v_k)$. Then (2) implies that

$$U^\perp A \subseteq V^\perp, \quad AW^\perp \subseteq V^\perp.$$

It shows the continuity of the multiplication of A. The proof of linear compactness of A stems from the isomorphism of the vector space of A with the direct topological product k^d, where d is a dimension of A (for more details see [15, p. 97]). □

Definition 1.3. A subspace $B \subseteq C$ is said to be a subcoalgebra of C if $\Delta(B) \subseteq B \otimes B$. A coalgebra C is locally finite if any $c \in C$ lies in some finite dimensional subcoalgebra. As the sum of subcoalgebras is a subcoalgebra, it is equivalent to the condition that any finite number of elements $c_1, \ldots, c_k \in C$ lie in some finite dimensional subcoalgebra.

It is known that all associative coalgebras are locally finite [1]. This theorem is of a great importance for the theory of associative coalgebras and it is called the "Fundamental Theorem on Coalgebras". Local finiteness is also an important property for Lie coalgebras (although not always fulfilled) as it is exactly for the class of locally finite Lie coalgebras that a dual Poincaré–Birkhoff–Witt Theorem holds [8].

If $F \subseteq A$, we also denote $F^{\perp} = \{c \in C \mid f(c) = 0 \text{ for all } f \in F\}$. Note that the sign $\perp$ here has a slightly different meaning than before.

Proposition 1.4. *If S is a finite dimensional subcoalgebra in C, then $S^{\perp}$ is an ideal of A of a finite codimension. If I is an ideal of A of a finite codimension, then $I^{\perp}$ is a finite dimensional subcoalgebra of C.*

The proof is straightforward and immediately follows from (2).

Proposition 1.5. *C is locally finite if and only if A has a basis of neighborhoods of zero consisting of ideals of finite codimension.*

Proof. Suppose that C is locally finite. Consider a neighborhood of zero $V^{\perp}$, $\dim V < \infty$, from the base of neighborhoods of zero in A. Then the finite basis of V—and hence V as a whole—is contained in a finite dimensional subcoalgebra W of C and therefore $W^{\perp} \subseteq V^{\perp}$ is an ideal neighborhood of A, which lies in $V^{\perp}$.

Suppose now that A has a basis of neighborhoods of zero consisting of ideals. As these ideals are open, they all have finite codimensions. For any $x \in C$ the set $x^{\perp}$ is open and there exists an ideal I from this basis such that $I \subseteq x^{\perp}$. As $\operatorname{codim} I < \infty$, subcoalgebra $I^{\perp}$ is finite dimensional and $x \in I^{\perp}$. □

2. Admissible varieties of algebras; topological restrictedness

Let A be an arbitrary algebra over k, which we assume to be of characteristic $\neq 2$. For every element $a \in A$ two mappings from the algebra to itself are defined:

$$R_a : x \mapsto xa, \quad L_a : x \mapsto ax.$$

These endomorphisms of the vector space of the algebra A are called the operators of right and left multiplication by the element a, respectively. The subalgebra of the algebra of endomorphisms $\mathrm{End}_k(A^\sharp)$ of the vector space $A^\sharp$ of the algebra A generated by all possible operators R_a and L_a, where $a \in A$, is called the multiplication algebra of the algebra A and is denoted by $M(A)$. Elements of $M(A)$ will be considered as acting on elements of A from the left, for example:

$$R_a L_b R_c x = (b(xc))a.$$

The identities satisfied by algebras imply a series of relations for the multiplication operators. It is possible also to define identities by means of these relations. We shall list some of them which we shall refer to later. The general reference for them will be [17, 18].

In any Jordan algebra A for arbitrary $x, y \in A$ the following relations are valid:

$$R_x = L_x, \tag{3}$$
$$2R_x R_y R_x = -R_{x^2 y} + 2R_x R_{xy} + R_y R_{x^2}, \tag{4}$$

or

$$R_x \equiv L_x, \tag{5}$$
$$R_x R_y R_x \equiv 0, \tag{6}$$

where the equivalence means equality modulo the linear combination of operator words of smaller operator length, i.e., containing fewer operators of right and left multiplications.

In any Mal'tsev algebra A for arbitrary $x, y \in A$ the following relations are valid:

$$R_x = -L_x, \tag{7}$$
$$R_x R_y R_x = R_x R_x R_y + R_{(xy)x} + R_{xy} R_x, \tag{8}$$

or

$$R_x \equiv -L_x, \tag{9}$$
$$R_x R_y R_x \equiv R_x R_x R_y, \tag{10}$$

For an alternative algebra A for arbitrary $a, b \in A$

$$R_a R_b = -R_b R_a + R_{ab+ba},$$
$$L_a L_b = -L_b L_a + L_{ab+ba},$$
$$L_a R_b = R_b L_a + R_b R_a - R_{ba},$$
$$R_a L_b = L_b R_a - R_b R_a + R_{ba},$$

from which the following relations can be derived:

$$R_x R_y R_x \equiv R_x L_y R_x \equiv L_x L_y L_x \equiv L_x R_y L_x \equiv 0. \tag{11}$$

$$R_x R_y L_x \equiv R_x L_x R_y, \tag{12}$$

$$
\begin{aligned}
R_x L_y L_x &\equiv -R_x L_x L_y, & (13)\\
L_x L_y R_x &\equiv L_x R_x L_y, & (14)\\
L_x R_y R_x &\equiv -L_x R_x R_y. & (15)
\end{aligned}
$$

Generalizing we can proceed with the folowing definition which will embrace these three important cases. For convenience of notation we will denote also $M_x^+ = R_x$ and $M_x^- = L_x$, so that an arbitrary operator can be denoted as M_x^ϵ, where $\epsilon \in \{+,-\}$ or $M^\epsilon(x)$, especially if x itself is an expression containing indices.

Definition 2.1. We shall say that a homogeneous variety of algebras $\mathcal{M}$ is admissible if for an arbitrary algebra A of $\mathcal{M}$ and for all $x, y \in A$ all eight operator products of the type $M_x^{\epsilon_1} M_y^{\epsilon_2} M_x^{\epsilon_3}$, where $\epsilon_i \in \{+,-\}$, are expressible in the following two ways:

1. as linear combinations of operator products of the type $M_x^{\tau_1} M_x^{\tau_2} M_y^{\tau_3}$, $\tau_i \in \{+,-\}$, and products of smaller operator length:
$$M_x^{\epsilon_1} M_y^{\epsilon_2} M_x^{\epsilon_3} \equiv \sum_\tau \alpha_\tau M_x^{\tau_1} M_x^{\tau_2} M_y^{\tau_3}, \qquad (16)$$
where $\alpha_\tau \in k$,
2. as linear combinations of operator products of the type $M_y^{\tau_1} M_x^{\tau_2} M_x^{\tau_3}$, $\tau_i \in \{+,-\}$, and products of smaller operator length:
$$M_x^{\epsilon_1} M_y^{\epsilon_2} M_x^{\epsilon_3} \equiv \sum_\tau \beta_\tau M_x^{\tau_1} M_x^{\tau_2} M_y^{\tau_3}, \qquad (17)$$
where $\beta_\tau \in k$.

Relations (5)–(6), (9)–(10), (11)–(15) mean that varieties of Jordan, Mal'tsev and alternative algebras are admissible. Note that (16) and (17) are, indeed, the same as in [11].

Suppose now that A is a topological algebra. In this case the multiplication algebra can be also endowed with a topology by the base of neighborhoods of zero consisting of the subsets

$$\Psi(U) = \{\omega \in M(A) \mid \omega A \subseteq U\},$$

where U is a neighborhood of zero in A. This topology clearly converts the multiplication algebra into a topological algebra, and we assume further that the multiplication algebra is always topologized in this manner. Of course, this topology can be discrete for some algebras but not for linearly compact ones as we shall see from the following lemma. We can prove it without any use of identities i.e., for arbitrary nonassociative algebras.

Lemma 2.2. *Let A be an arbitrary linearly compact algebra over a field k and let U be an arbitrary neighborhood of zero in A. Then there exists a neighborhood of zero W such that $WA \subseteq U$ and $AW \subseteq U$.*

Proof. It is sufficient to prove only the first inclusion. As the multiplication in A is continuous there exists a neighborhood of zero V such that $V^2 \subseteq U$. Since V has a finite codimension we can write $A = ka_1 + \cdots + ka_m + V$. There exist also neighborhoods $V_1, \ldots, V_m$ such that $V_i a_i \subseteq U$ for $i = 1, \ldots, m$. It is clear now that for the neighborhood $W = V \cap V_1 \cap \cdots \cap V_m$

$$WA = Wa_1 + \cdots + Wa_m + WV \subseteq V_1 a_1 + \cdots + V_m a_m + V^2 \subseteq U,$$

and the lemma is proved. □

Let A be an algebra and $M(A)$ be its multiplicative algebra. By $M_n(A)$ we shall denote the subspace of $M(A)$ spanned by all operator products of operator length less than or equal to n.

Corollary 2.3. *Let A be an arbitrary linearly compact algebra, let U be a neighborhood of zero in A and n be an integer. Then there exists a neighborhood $W = W(n, U)$ such that*

$$M_n(A)W \subseteq U.$$

Proof. The result follows from the lemma by an easy induction. □

This lemma also implies that for an arbitrary linearly compact algebra A the topology of its multiplication algebra $M(A)$ is not discrete and the regular right and left representations

$$R : x \mapsto R_x, \quad L : x \mapsto L_x$$

are continuous.

Definition 2.4. We shall say that a topological algebra A is topologically restricted if for an arbitrary element $x \in A$ and an arbitrary neighborhood of zero W of $M(A)$ there exists an integer $n = n_{x,W}$ such that for all $\epsilon_1, \ldots, \epsilon_n \in \{+, -\}$

$$M_x^{\epsilon_1} \ldots M_x^{\epsilon_n} \equiv \omega \in W. \tag{18}$$

Note that topological restrictedness is not a restriction at all for Jordan and alternative algebras. For Jordan algebras it follows from (4) that

$$R_x R_x R_x \equiv 0$$

One should also remember that Jordan algebras are commutative ones. Therefore $n = 3$ will always do the job. The same is true for alternative algebras. It follows from relations (11)–(15) since

$$R_x R_x \equiv 0, \quad L_x L_x \equiv 0.$$

But, of course, it is a restriction, say for Lie algebras.

Proposition 2.5. *If a topological algebra A has a neighborhood base of zero consisting of ideals of finite codimension, then it is topologically restricted.*

Proof. Let $\Psi(U)$ be a neighborhood of zero from the base of $M(A)$. We can assume that U is an ideal neighborhood of A. Consider $\bar{A} = A/U$. Let $x \in A$ be an arbitrary and $\bar{x} = x + U$. Then, as $\bar{A}$ is finite dimensional, $M(\bar{A})$ is also finite dimensional and there exists an integer n such that

$$M_{\bar{x}}^{\epsilon_1} \dots M_{\bar{x}}^{\epsilon_n} \equiv \bar{0},$$

or, what is the same,

$$M_x^{\epsilon_1} \dots M_x^{\epsilon_n} \equiv \omega \in \Psi(U),$$

and A is topologically restricted. □

This proposition explains our interest towards topological restrictedness since it is a necessary condition for having a neighborhood base of ideals of finite codimension, which is important to us in view of Proposition 1.5. We shall prove that for linearly compact algebras of admissible variety this condition is also sufficient.

3. The local structure of linearly compact algebras of admissible varieties

We shall start with the following combinatorial lemma. For brevity we shall refer to algebras of admissible varieties as to admissible algebras.

Lemma 3.1. *Let A be an admissible linearly compact topologically restricted algebra over a field k and U be a neighborhood of zero in A. Then there exists an integer $n = n(U)$ such that*

$$M(A) = M_n(A) + \Psi(U). \tag{19}$$

Proof. Let W be a neighborhood, which by Lemma 2.2 satisfies $WA \subseteq U$ and $AW \subseteq U$. Intersecting W with U, if necessary, we can assume that $W \subseteq U$. As A is linearly compact, this neighborhood is a linear subspace of a finite codimension. Thus $A/W = \{\bar{a}_1, \dots, \bar{a}_m\}$, $\bar{a}_i = a_i + W$, for elements of some finite subset $B = \{a_1, \dots, a_m\}$ of A. Suppose that p_i is an integer such that for arbitrary $\epsilon_1, \dots, \epsilon_{p_i} \in \{+, -\}$

$$M^{\epsilon_1}(a_i) \dots M^{\epsilon_{p_i}}(a_i) \equiv \omega_i \in \Psi(W). \tag{20}$$

Such integers p_i exist since A is topologically restricted. Set now $n = 2m(p-1)$, where p is the maximal integer among $p_1, \dots, p_m$. We shall prove that this integer satisfies the desired property.

It is sufficient to show that an arbitrary operator product

$$\eta = M^{\epsilon_1}(x_1) \dots M^{\epsilon_t}(x_t) \tag{21}$$

is contained in the set $M_n(A) + \Psi(U)$. Since it is clear for $t \leq n$ we can proceed by induction on t.

For any $j = 1, \ldots, t$ we can write $x_j = a_{i_j} + v_j$, where $a_{i_j} \in B$, $v_j \in W$. Thereby we can suppose in (21) without loss of generality that $x_j \in B \cup W$.

Linearizing (16) gives us

$$M_x^{\epsilon_1} M_y^{\epsilon_2} M_z^{\epsilon_3} \equiv -M_z^{\epsilon_1} M_y^{\epsilon_2} M_x^{\epsilon_3} + \sum_\tau \alpha_\tau M_x^{\tau_1} M_z^{\tau_2} M_y^{\tau_3} + \sum_\tau \alpha_\tau M_z^{\tau_1} M_x^{\tau_2} M_y^{\tau_3}. \quad (22)$$

This means that we can move operators L_z and R_z to the left modulo words of smaller operator length.

Suppose that for some i the total amount of operators R_{a_i} and L_{a_i} in the operator word η is p or more. Using (22) let us move them all to the left until all subwords $M^{\epsilon_1}(x)M^{\epsilon_2}(y)M^{\epsilon_3}(a_i)$, where $x \neq a_i$ and $y \neq a_i$, will be eliminated. Then there might be some subwords of the type $M^{\epsilon_1}(a_i)M^{\epsilon_2}(y)M^{\epsilon_3}(a_i)$, which can be eliminated by (16) and again (22). At the end of this process we will gather all R_{a_i} and L_{a_i} on the left, starting at the first or the second position of the operator word. Therefore we shall represent η in one of the two forms:

$$\eta \equiv M^{\epsilon_1}(a_i) \ldots M^{\epsilon_p}(a_i) M^{\tau_1}(x_1) \ldots M^{\tau_q}(x_q), \quad (23)$$

or

$$\eta \equiv M^{\tau_1}(x_1) M^{\epsilon_1}(a_i) \ldots M^{\epsilon_p}(a_i) M^{\tau_2}(x_2) \ldots M^{\tau_q}(x_q), \quad (24)$$

where $x_i \in B \cup W$ and $p + q = t$. Due to (20) in the first case

$$\eta \equiv \omega_i M^{\tau_1}(x_1) \ldots M^{\tau_q}(x_q) \in \Psi(W) \subseteq \Psi(U), \quad (25)$$

and in the second

$$\eta \equiv M^{\tau_1}(x_1) \omega_i M^{\tau_2}(x_2) \ldots M^{\tau_q}(x_q) \in M^{\tau_1}(x_1)\Psi(W) \subseteq \Psi(U). \quad (26)$$

Suppose now that for all i the total amount of operators R_{a_i} and L_{a_i} in the operator word η is less than p. Linearizing (17) we obtain

$$M_x^{\epsilon_1} M_y^{\epsilon_2} M_z^{\epsilon_3} \equiv -M_z^{\epsilon_1} M_y^{\epsilon_2} M_x^{\epsilon_3} + \sum_\tau \beta_\tau M_y^{\tau_1} M_x^{\tau_2} M_z^{\tau_3} + \sum_\tau \beta_\tau M_y^{\tau_1} M_z^{\tau_2} M_x^{\tau_3}. \quad (27)$$

Using (27) we can eliminate from η all subwords $M^{\epsilon_1}(a_i)M^{\epsilon_2}(w_1)M^{\epsilon_3}(w_2)$ moving now a_i's to the right. After elimination of these subwords η will be equivalent either to an operator word starting with an operator of right or left multiplication by an element of W or to the word containing of no more than $2m(p-1) = n$ operators. In the first case η is equivalent to an operator word from $\Psi(W) \subseteq \Psi(U)$, while in the second case it is congruent to the word from $M_n(A)$. The lemma is proved. □

Theorem 3.2. *Every admissible linearly compact topologically restricted algebra A has a basis of neighborhoods consisting of ideals (of finite codimension).*

Proof. Let U be an arbitrary neighborhood of zero of A. Then by Lemma 3.1 there exists an integer n for which

$$M(A) = M_n(A) + \Psi(U).$$

By Corollary 2.2 there exists also a neighborhood V, which is contained in U and such that

$$M_n(A)V \subseteq U.$$

The ideal of A generated by V is given by

$$\langle V \rangle = V + M(A)V = V + M_n(A)V + \Psi(U)V \subseteq U,$$

and it is open since it contains V. This proves the theorem. □

Corollary 3.3. *Every linearly compact simple alternative, Jordan or algebraic Mal'tsev algebra is finite dimensional and discrete.*

Corollary 3.4. *Let A be a linearly compact alternative or Jordan algebra. Then its (quasiregular) radical $J(A)$ is topologically nilpotent, i.e., for an arbitrary neighborhood of zero U there exists an integer $n = n(U)$ such that*

$$J(A)^n \subseteq U.$$

Proof. We can find an ideal neighborhood W inside every neighborhood of zero U. Since every open subset of A is a subspace of finite codimension, the quotient algebra $\bar{A} = A/W$ is finite dimensional, and its radical is nilpotent: $J(\bar{A})^n = \bar{0}$. As the radical $J(A)$ is in the inverse image of $J(\bar{A})$ we have $J(A)^n \subseteq U$. □

4. Local finiteness of coalgebras

In this short section we can finally formulate our main result concerning coalgebras and formulate a necessary and sufficient condition for the coalgebras of admissible variety to be locally finite.

Theorem 4.1. *A coalgebra C of an admissible variety is locally finite if and only if its dual algebra $A = C^*$ is topologically restricted.*

The proof immediately follows from Theorem 3.2 and Propositions 1.5 and 2.4.

Corollary 4.2 (Anquela–Cortés–Montaner). *Every alternative or Jordan coalgebra is locally finite.*

References

[1] M. E. Sweedler, Hopf Algebras, W. A. Benjamin, New York 1969, pp.1–336.

[2] S. Montgomery, Hopf Algebras and Their Actions on Rings, CBMS 82, Amer. Math. Soc., Providence, R.I., 1993, pp. 238.

[3] M. André, L'algébre de Lie d'un anneaux local, Simposia Mathematica 4, Istituto Nazionale di Alta Matematica Roma, Academic Press, London–New York 1970, 337–375.

[4] M. André, Hopf algebras with divided powers, J. Algebra 18 (1971), 19–50.

[5] W. D. Nichols, Pointed irreducible bialgebras, J.Algebra 57 (1979), 64–76.

[6] R. E. Block, Commutative Hopf Algebras, Lie Coalgebras, and Divided Powers, J. Algebra 96 (1985), 275–306.

[7] W. Michaelis, Lie Coalgebras, Adv. Math. 38 (1980), 1–54.

[8] W. Michaelis, The Dual Poincaré–Birkhoff–Witt Theorem, Adv. Math. 57 (1985), 93–162.

[9] V. G. Drinfel'd, Quantum Groups, Proc Internat. Congr. Math. 1, Berkeley 1986, 789–820.

[10] J. A. Anquela, T. Cortés, F. Montaner, Nonassociative Coalgebras, Comm. Algebra 22(12) (1994), 4693–4716.

[11] I. P. Shestakov, K. A. Zhevlakov, On local finiteness in the sense of Shirshov, Algebra i Logika 12 (1973), 41–73.

[12] A. Slinko, Local finiteness of coalgebraic Lie coalgebras, Comm. Algebra 23(3) (1995), 1165–1170.

[13] A. Slinko, Nonassociative topological bimodules, Math. USSR Sbornik 61 (1988), 259–270.

[14] A. Slinko, Compact Lie Rings, Comm. Algebra 20(2) (1992), 387–408.

[15] G. Köthe, Topological Vector Spaces I, Springer-Verlag, Berlin–Heidelberg–New York 1969, pp. 456.

[16] S. Lefschetz, Algebraic Topology, Amer. Math. Soc. Colloq. Publ. XXVII, New York 1942.

[17] K. A. Zhevlakov, A. M. Slin'ko, I. P. Shestakov, A. I. Shirshov, Rings that are nearly associative, Academic Press, New York–London 1982, pp. 371.

[18] E. N. Kuz'min, I. P. Shestakov, Nonassociative structures, in: Encyclopedia of Mathematical Sciences 57, A. I. Kostrikin, I. R. Shafarevich (eds.), Algebra–6, Springer-Verlag, 1994.

Department of Mathematics
The University of Auckland
Private Bag 92019
Auckland, New Zealand

Azumaya algebras in skew group rings

George Szeto and Yuen-Fat Wong

Abstract. Let R be a ring with 1, G a finite automorphism group of R, R^G the subring of the elements fixed under G, $R * G$ the skew group ring and V the commutator subring of R^G in $R * G$. Conditions are given under which that any two of the rings R^G, V and $R * G$ are Azumaya algebras implies that the other one is also an Azumaya algebra.

1991 Mathematics Subject Classification: Primary 16H05, Secondary 16W20.

1. Introduction

Let R be a ring with 1, C the center of R and G a group. Then a relation among the group ring RG, the group algebra CG and R was given by F. R. DeMeyer and G. J. Janusz ([5]). That is, RG is an Azumaya algebra if and only if so are R and CG. A similar relation was also obtained for projective group rings by the present authors ([7, Theorem 3.5]). Noting that CG is the commutator subring of R in RG, we shall investigate such a relation among the skew group ring $R * G$ where G is a finite automorphism group of R, the subring R^G (the set of elements in R fixed under G) and the commutator subring V of R^G in $R * G$. We shall give conditions under which that any two of $R * G$, R^G and V are Azumaya algebras implies that the other one is also an Azumaya algebra.

2. Basic definitions

Throughout, R denotes a ring with 1, G a finite automorphism group of R, $R^G = \{r \in R \mid g(r) = r \text{ for all } g \in G\}$, C the center of R, $V = \{r \in R * G \mid tr = rt$ for all $t \in R^G\}$ and $R * G$ the skew group ring. That is, the ring with elements of G as a free basis over R such that $gr = g(r)g$ for all g in G and r in R. As defined in [2] and [4], let $S \subseteq T$ be a ring extension (that is, S is a subring of T with the same 1). Then T is called a *separable* extension of S if there exist elements $\{a_i, b_i \in T \mid \sum ta_i \otimes b_i = \sum a_i \otimes b_i t$ for all $t \in T$ where $i = 1, 2, \ldots, n$ for some integer n and $\otimes$ is over $S\}$, and $\sum a_i b_i = 1$. If S is contained in the center of T, then the separable extension T of S is called a *separable algebra*; and in particular, the separable algebra T over S is called an *Azumaya algebra* if S is the center of T. A special separable extension T of S is called an *H-separable extension* (Hirata's separable) if $T \otimes T$ is a direct summand of m-copies of T as $(T$-$T)$-bimodules for some integer m where $\otimes$ is over S (see [1] and [6]). Let T be a ring extension of S with a finite automorphism group G. Then T is called a *G-Galois extension* of S if $T^G = S$ and there exist elements $\{c_i, d_i \in T \mid \sum c_i g(d_i) = 0$ for each $g \neq 1$ in G and $\sum c_i d_i = 1$ where $i = 1, 2, \ldots, k$ for some integer $k\}$. The set $\{c_i, d_i\}$ is called a *G-Galois set* for T. It is well-known that T is a G-Galois extension of S if and only if T is finitely generated and projective as a right S-module and $T * G \cong \mathrm{Hom}_S(T, T)$ ([2, Theorem 1]).

3. Azumaya algebras in skew group rings

Keeping the notations given in Section 2, we shall study when any two of $R * G$, R^G and V are Azumaya algebras implies that the other one is also an Azumaya algebra. We begin with the well-known commutator theorem for Azumaya algebras ([4, Theorem 4.3, p. 57]). In this section, $\otimes$ is over C^G unless stated otherwise.

Proposition 1. *Let A be an Azumaya algebra over its center Z. (1) If B is a separable subalgebra, then $V_A(B)$ (the commutator subring of B in A) is also a separable subalgebra. (2) If B is an Azumaya subalgebra, then so is $V_A(B)$ such that $A \cong B \otimes_Z V_A(B)$ by the multiplication map. (3) If $A \cong X \otimes_Z Y$ for some Z-subalgebras X and Y, then X and Y are Azumaya subalgebras.*

Noting that $V_R(R^G)$ is a G-invariant subring of R, we have our first relation.

Theorem 2. *Let Z be the center of $R * G$. If $R * G$ is an Azumaya algebra and R^G is an Azumaya algebra over C^G, then so is V and $R \cong R^G \otimes V_R(R^G)$ where V is the commutator of R^G in $R * G$.*

Proof. Since Z is the center of $R * G$, C^G is a C^G-direct summand of Z by a direct computation. Since R^G is an Azumaya C^G-algebra, $R^G \otimes Z$ is an Azumaya Z-algebra ([4, Corollary 1.10, p. 45]). It is easy to see that $R^G \otimes Z \cong R^G Z$ by the multiplication map, so $V_{R*G}(R^G Z)$ is an Azumaya Z-algebra by Proposition 1. Noting that $V_{R*G}(R^G Z) = V_{R*G}(R^G) = V_R(R^G) * G$, we conclude that V $(= V_{R*G}(R^G))$ is an Azumaya Z-algebra. Moreover, by Proposition 1 again, $R * G \cong (R^G \otimes Z) \otimes_Z (V_R(R^G) * G) \cong R^G \otimes (V_R(R^G) * G)$, so $R \cong R^G \otimes V_R(R^G)$. □

The second relation of the Azumaya algebras is the following.

Theorem 3. *Assume that $V_R(R^G) * G$ is an Azumaya algebra over its center Z and that R^G is an Azumaya algebra over C^G. Then, $R * G$ is an Azumaya algebra if and only if $R \cong R^G \otimes V_R(R^G)$.*

Proof. Assume $R \cong R^G \otimes V_R(R^G)$. Then $R * G \cong R^G \otimes (V_R(R^G) * G) \cong (R^G \otimes Z) \otimes_Z (V_R(R^G) * G)$ as Z-algebras. Since $(R^G \otimes Z) \otimes_Z (V_R(R^G) * G)$ is an Azumaya Z-algebra, so is $R * G$. The converse is a consequence of Theorem 2. □

Next is another relation of the Azumaya algebras.

Theorem 4. *Assume that $R * G$ is an Azumaya algebra over its center Z. Then, R^G is an Azumaya C^G-algebra if and only if $R \cong R^G \otimes V_R(R^G)$. When the equivalence conditions hold, V is an Azumaya Z*

Proof. Assume $R \cong R^G \otimes V_R(R^G)$. Then

$$R * G \cong R^G \otimes (V_R(R^G) * G) \cong (R^G \otimes Z) \otimes_Z V$$

for $Z \subset V = V_R(R^G) * G$. Since $R * G$ is an Azumaya Z-algebra, so are $R^G \otimes Z$ and V by Proposition 1. But C^G is a C^G-direct summand of Z, so R^G is an Azumaya C^G-algebra ([4, Corollary 1.10, p. 45]). The converse and the rest are consequences of Theorem 2. □

From Theorem 2, there are rings R such that $R \cong R^G \otimes V_R(R^G)$ which is also a necessary and sufficient condition in Theorems 3 and 4. In the following, we give another class of rings R satisfying the condition that $R \cong R^G \otimes V_R(R^G)$.

Lemma 5. *If $V_R(R^G)$ is a G-Galois algebra over C^G, then R is a G-Galois extension such that $R \cong R^G V_R(R^G)$.*

Proof. Let $\{a_i, b_i \in V_R(R^G), i = 1, 2, \ldots, n$ for some integer $n\}$ be a G-Galois system for $V_R(R^G)$. That is, $\sum a_i b_i = 1$ and $\sum a_i g(b_i) = 0$ for each $g \neq 1$ in G. Then $\{\mathrm{tr}_G(a_i \_), b_i\}$ is a dual basis for the finitely generated and projective right module over C^G where tr_G is the trace map induced by G ([2, Proof of Theorem 1, p. 118]). But then $R^G V_R(R^G)$ and R are both G-Galois extensions over R^G with the same G-Galois system $\{a_i, b_i\}$. Thus $r = \sum \mathrm{tr}_G(a_i r) b_i$ for each r in R. Therefore $R = R^G V_R(R^G)$. □

Theorem 6. *Assume that R is an Azumaya C^G-algebra. If $V_R(R^G)$ is a G-Galois algebra over C^G, then R is a G-Galois extension over R^G such that $R \cong R^G \otimes V_R(R^G)$ as Azumaya algebras.*

Proof. By Lemma 5, R is a G-Galois extension over R^G such that $R = R^G V_R(R^G)$, so the center of $V_R(R^G)$ is C^G (for the center of R is C^G by hypothesis). But $V_R(R^G)$ is a G-Galois algebra over C^G, so it is a separable C^G-algebra. Hence $V_R(R^G)$ is an Azumaya C^G-algebra. Since $V_R(R^G)$ is a G-Galois algebra over C^G, $V_R(R^G) * G = \mathrm{Hom}_{C^G}(V_R(R^G), V_R(R^G))$ and $V_R(R^G)$ is finitely generated and projective over C^G. We have that $V_R(R^G)$ is a progenerator over C^G. Thus $\mathrm{Hom}_{C^G}(V_R(R^G), V_R(R^G))$ is an Azumaya C^G-algebra ([4, Proposition 4.1, p. 56]); and so V $(= V_R(R^G) * G)$

is also an Azumaya C^G-algebra. Thus there exists an element d in the center of $V_R(R^G)$ such that $\mathrm{tr}_G(d) = 1$ ([1, Proposition 1.4, p.3]). This gives a R^G-bimodule homomorphism $\mathrm{tr}_G(d\_)$ from R $(= R^G V_R(R^G))$ onto R^G. Since $\mathrm{tr}_G(d\_)I$ is equal to the identity of R^G where I is the inclusion map from R^G to R, thus, R^G is a direct summand of R as a bimodule over R^G. Moreover, R is an Azumaya C^G-algebra by hypothesis, so it is an H-separable extension over C^G. Thus $V_R(V_R(R^G)) = R^G$ ([6, (5), p. 296]). Therefore, by Proposition 1, $R \cong R^G \otimes V_R(R^G)$ as Azumaya C^G-algebras. □

As given in the proof of Theorem 6 for $C = C^G$, we have:

Corollary 7. *If R is a central G-Galois algebra, then the order of G is a unit in R.*

Proof. Since R is a G-Galois over its center C, $R * G \cong \mathrm{Hom}_C(R, R)$, $R * G$ is a separable algebra over C. Hence $R * G$ is a separable extension of R. Thus there exists an element d in C such that $\mathrm{tr}_G(d) = 1$ ([1, Proposition 1.4, p. 3]). That is $nd = 1$ where n is the order of G. □

4. Examples

In this section, we give two examples of rings to demonstrate our results in Section 3. One is an Azumaya skew group ring $R * G$, and the other is a ring R with a finite automorphism group G such that R is not isomorphic to $R^G \otimes V_R(R^G)$.

1. Let S be the real field, $Q = S[i, j, k]$, the real quaternion algebra, $R = Q \oplus Q$, the direct sum of Q, and $G = \{g \mid g(a, b) = (b, a) \text{ for } (a, b) \in R\}$. Then we have the following properties:

(a) G is an automorphism group of R of order 2.

(b) $R^G = \{(a, a) \mid a \in Q\} \cong Q$.

(c) The center C of R is $S \oplus S$.

(d) Q is an Azumaya algebra over S.

(e) R^G is an Azumaya algebra over S by (b) and (d).

(f) $V_R(R^G) = S \oplus S = C$ by (c).

(g) $C^G = \{(b, b) \mid b \in S\}$.

(h) $R \cong R^G \otimes V_R(R^G) = Q \otimes_S (S \oplus S) \cong Q \oplus Q$.

(i) The commutator subring of Q^G in $Q * G$ is $(S \oplus S) * G$ which is $V_R(R^G) * G$ by (f).

(j) $R * G \cong R^G \otimes (V_R(R^G) * G) \cong R^G \otimes (C * G)$.

(k) The center of $R * G = C^G = S$ by (g).

(l) $R * G$ is an Azumaya C^G-algebra.

(m) V = the commutator of R^G in $R * G = C * G = (S \oplus S) * G$ which is an Azumaya C^G-algebra. Thus Theorems 2, 3 and 4 hold for $R * G$.

2. Let Q be the real quaternion algebra as given in Example 1. $G = \langle g_i \mid g_i$ is the inner automorphism of Q induced by $i\rangle$, the cyclic group generated by g_i. Then the following hold:

(a) The order of G is 2.

(b) $Q^G = S[i]$.

(c) $V_Q(Q^G) = S[i]$ which is a commutative ring.

(d) The center of Q is S and $C^G = S$.

(e) Q is not isomorphic to $Q^G \otimes V_Q(Q^G)$ $(= S[i] \otimes S[i])$ where $\otimes$ is over C^G $(= S)$.

(f) Q^G is not an Azumaya algebra over C^G by (b).

(g) $Q * G$ is an Azumaya algebra (for the order of G is a unit in Q). Thus the sufficient condition of Theorem 2 and the necessary condition of Theorem 4 do not hold for $Q * G$.

References

[1] R. Alfaro, Separability and G-Galois actions, Centre Derecerca Matematica, Institut D'Estudis Catalans, #133, September 1991.

[2] F. R. DeMeyer, Some notes on the general Galois theory of rings, Osaka J. Math. 2 (1965), 117–127.

[3] F. R. DeMeyer, Galois theory in separable algebras over commutative rings, Illinois J. Math. 2 (1966), 287–295.

[4] F. R. DeMeyer and E. Ingraham, Separable Algebras over Commutative Rings, Lecture Notes in Math. 181, Springer-Verlag, Berlin–Heidelberg–New York 1971.

[5] F. R. DeMeyer and G. J. Janusz, Group rings which are Azumaya algebras, Amer. Math. Soc. 279 (1983), 389–395.

[6] K. Sugano, On a special type of Galois extensions, Hokkaido Math. J. 4 (1975), 123–128.

[7] G. Szeto and Y. F. Wong, On Azumaya projective group rings, Contemporary Math. 124 (1992), 251–256.

Department of Mathematics
Bradley University
Peoria, IL 61625, USA

Department of Mathematics
DePaul University
Chicago, IL 60614, USA

On distributive rings and modules

*A. A. Tuganbaev**

All rings are assumed to be associative and (unless otherwise specified) with nonzero identity. A module is called *distributive* if the lattice of its submodules is distributive. A ring is called *right invariant* if all of its right ideals are ideals. Right distributive right noetherian rings are right invariant [1]. Our first main result is Theorem 1.

Theorem 1. *Let A be a semiprime right distributive ring and let for every minimal prime ideal H of A the quotient ring A/H be right noetherian. Then A is right invariant ring.*

Infinite direct products of skew fields are not right (left) noetherian and satisfy the conditions of Theorem 1.

The Jacobson radical, the set of all nilpotent elements, the group of all invertible elements of a ring A are denoted by $J(A)$, $T(A)$ and $U(A)$ respectively. An endomorphism f of a module M is called locally nilpotent if for every cyclic submodule D of M there exists a natural number $n = n(D)$ such that $f^n(D) = 0$. If B, C are subsets of a ring A, then BC denotes the set of all finite sums $\sum b_i c_i$ with $b_i \in B$, $c_i \in C$. The endomorphism ring, the set of all endomorphisms with essential (in M) kernels, the set of all endomorphisms with superfluous (in M) images, the set of all locally nilpotent endomorphisms of a module M are denoted by $\mathrm{End}(M)$, $F(M)$, $G(M)$ and $H(M)$, respectively. $F(M)$ and $G(M)$ are always ideals of $\mathrm{End}(M)$.

The following result was proved in [2]: Let A be a right distributive ring, $T = T(A)$, $U = U(A)$. Then: (a) T is a nil subring (without identity) of $J(A)$; (b) T is a sum of all nilpotent ideals of a nil ring T; (c) $UT = TU$ and UT is an ideal of $J(A)$; (d) if T is right or left t-nilpotent, then T coincides with the prime radical of A. Our second main result is Theorem 2.

Theorem 2. *Let M_A be a distributive module, $R \equiv \mathrm{End}(M)$, $T \equiv T(R)$, $U \equiv U(R)$ and $H \equiv H(M)$. Then:*

(a) *H is a subring (without identity) of the ideal $F(M)$ of R;*

(b) *T is a nil subring (without identity) of the ideal $F(M) \cap J(R)$ of R;*

(c) *T is a nil ideal of H and coincides with the sum of all nilpotent ideals of H;*

*Partially supported by the Russian Foundation for Fundamental Research.

(d) $UH = HU$, $UT = TU$;

(e) UT *is an ideal of* $J(R)$;

(f) *If* T *is right or left t-nilpotent, then* T *coincides with the prime radical of* R;

(g) *If* M *is a locally noetherian module, then* $F(M) = H \subseteq J(R)$.

The proofs of Theorems 1 and 2 consist of several lemmas. The singular submodule of a module M is denoted by $\mathcal{Z}(M)$. A module M is called *nonsingular* (*singular*) if $\mathcal{Z}(M) = 0$ ($\mathcal{Z}(M) = M$). A module is called *uniserial* if any two of its submodules are comparable with respect to set inclusion. A ring with no nonzero nilpotent elements is called *reduced*. The set of all maximal right ideals of a ring A is denoted by $\max(A)$. A subset T of a ring A is called a *right denominator set* if there exist a ring A_T and a ring homomorphism $f_T \equiv f : A \to A_T$ such that all elements of $f(T)$ are invertible in A_T, $\ker(f) = \{a \in A \mid at = 0 \text{ for some } t \in T\}$, $A_T = \{f(a)f(t)^{-1} \mid a \in A, t \in T\}$. In this situation A_T is called a *right ring of fractions* of A with respect to T, f_T is called a *canonical homomorphism*. If $T = A \setminus M$ where M is a right ideal of A, we will write M instead of T in the index. A ring A is called *right localizable* if A_M exists for every maximal ideal M of A.

Lemma 3. *Let* M *be a module,* $R \equiv \mathrm{End}(M)$. *Then:*

(a) $F(M) \cap G(M) \subseteq J(R)$;

(b) *If* $f \in H(M)$, *then* $1 - f \in U(R)$;

(c) *If* $H(M)$ *is a right (left) ideal of* R, *then* $H(M) \subseteq J(R)$;

(d) *If* M *is locally noetherian, then* $F(M) \subseteq H(M) \cap J(R)$;

(e) *If* M *is nonsingular, then* $F(M) = 0$.

Proof. (a) Let $f \in F(M) \cap G(M)$, $g = 1 - f \in R$. Then $\ker(g) \cap \ker(f) = 0$, $\mathrm{Im}(g) + \mathrm{Im}(f) = M$. Since $f \in F(M)$, $\ker(g) = 0$. Since $f \in G(M)$, $\mathrm{Im}(g) = M$. Therefore, $g \in U(R)$. In addition, $F(M) \cap G(M)$ is an ideal of R.

(b) Since f is locally nilpotent, the rule $g(x) = \sum_{n=0}^{\infty} f^n(x)$ defines an endomorphism $g \in R$. Then $(1 - f)g = g(1 - f) = 1_M$.

(c) follows from (b).

(d) Let D be any cyclic submodule of M, $f \in F(M)$, $N_i \equiv D \cap \ker(f^i)$. Since $N_i \subseteq N_{i+1}$ and D is a noetherian module, $N_n = N_{n+1}$ for some integer n. Then f^n induces an isomorphism $h : D/N_n \to f^n(D)$. Let $E \equiv f^n(D) \cap \ker(f)$, $h^{-1}(E) = B/N_n$ where $N_n \subseteq B \subseteq D$. Suppose $f^n(D) \neq 0$. Since M is an essential extension of $\ker(f)$, N_n is a proper submodule of B. In addition, $f^{n+1}(B) = f(E) = 0$. Then $B \subseteq N_{n+1} = N_n$, $B = N_n$, a contradiction.

(e) Let $f \in F(M)$, $T \equiv \ker(f)$. Then M/T is a singular module and $f(M) \cong M/T$. Therefore $f(M) \subseteq Z(M) = 0$, $f = 0$. □

Lemma 4 ([1]). *If* N *and* T *are submodules of a distributive module and* $N \cap T = 0$, *then* $\hom(N, T) = 0$.

Lemma 5. *Let f be an endomorphism of a distributive module M. Then:*

(a) *If D is a submodule of M and $f^{n+1}(D) \subseteq \sum_{i=1}^{n} f^i(D)$ for some integer $n \geq 0$, then $f(D) \subseteq D$;*

(b) $\ker(f^n)$ *is an essential extension of* $\ker(f)$ *for every natural number n;*

(c) $\cup_{i=1}^{\infty} \ker(f^i)$ *is an essential extension of* $\ker(f)$*;*

(d) *If M is an essential extension of* $\cup_{i=1}^{\infty} \ker(f^i)$*, then M is an essential extension of* $\ker(f)$.

Proof. Part (a) is proved in Theorem 3.2 of [1].

(b) It is sufficient to prove that $\ker(f^{n+1})$ is an essential extension of $\ker(f^n)$ for every natural number n. Let $T \equiv \ker(f^n)$, N is a submodule of $\ker(f^{n+1})$ and let $N \cap T = 0$. Since $f^{n+1}(N) = 0$, $f(N) \subseteq T$. By Lemma 2, $N \subseteq \ker(f) \subseteq T$. Therefore, $N = N \cap T = 0$ and $\ker(f^{n+1})$ is an essential extension of T.

(c), (d). Part (c) follows from part (b) and part (d) follows from part (c). □

Lemma 6. *Let M be a distributive module, $R \equiv \mathrm{End}(M)$, $F \equiv F(M)$ and $G \equiv G(M)$. Then:*

(a) *Quotient ring R/F, R/G and $R/(F \cap G)$ are reduced;*

(b) *All nilpotent elements of R lie in $F \cap G \cap J(R)$;*

(c) *$H(M) \subseteq F$;*

(d) *If M is nonsingular, then R is reduced;*

(e) *If M is locally noetherian, then $F = H(M) \subseteq J(R)$.*

Proof. (a) By Lemma 2(d), R/F is reduced. Let n be a natural number, $f \in R$ and let $f^{n+1}(M)$ be a superfluous submodule of M. Then $f(D) + f^{n+1}(M) = f(M) \supseteq f^n(M)$. Therefore, $M = D + f(D)$. By Lemma 3(a), $f(D) \subseteq D$. Therefore $D = M$, and $f^n(M)$ is a superfluous submodule of M. A similar reasoning shows that $f^{n-1}(M)$ is a superfluous submodules of M, and so. Therefore, R/G is a reduced ring. In addition, R/F is reduced. Therefore, $R/(F \cap G)$ is a reduced ring.

(b), (c), (d), (e). Part (b) follows from part (a) and Lemma 1(a). Part (c) follows from Lemma 2(d). Part (d) follows from part (a) and Lemma 1(e). Part (e) follows from part (c) and Lemma 1(d). □

Lemma 7. *Let M_A be a distributive module and let D be a submodule of M. Then:*

(a) *$f(N) \subseteq N$ for every submodule N of M and for every $f \in H(M)$;*

(b) *If $f_i, g_i \in H(M)$ $(1 \leq i \leq n,\ 1 \leq j \leq n)$ and $(f_1 f_2 \dots f_n)(D) = 0$, then $(f_1 g_1 f_2 g_2 \dots f_n g_n)(D) = 0$;*

(c) *If $f \in H(M)$, $f^n(D) = 0$ and H_1 is a subring of* $\mathrm{End}(M)$ *generated by $H(M)$ and by the identity of* $\mathrm{End}(M)$*, then $(f H_1)^n(D) = 0$;*

(d) *If* $f, g \in H(M)$ *and* $f^m(D) = g^n(D) = 0$, *then* $(f+g)^{m+n}(D) = 0$.

Proof. Part (a) follows from the fact that, by Lemma 3(a), $f(E) \subseteq E$ for every cyclic submodule E of M. Part (b) follows from part (a). Parts (c), (d) follow from parts (a) and (b). □

Lemma 8 ([3]). *Let all nilpotent elements of a ring* A *be central. Then* A *is right distributive if and only if* A *is right localizable and the right ring of fractions* A_M *is right uniserial for every* $M \in \max(A)$.

The right (left) annihilator of a subset B of a ring A is denoted by $r(B)$ $(l(B))$.

Lemma 9. *Let* A *be a reduced ring. Then:*

(a) $r(B) = l(B)$ *for every subset* B *of* A;

(b) *If* $a_1, \ldots, a_n \in A$ *and* $a_1 \ldots a_n = 0$, *then every product of all elements* a_i *(in any order) is equal to zero;*

(c) A/H *is a domain for every minimal prime ideal* H *of* A;

(d) A *is a right (left) nonsingular;*

(e) *If* A *is a prime ring, then* A *is a domain;*

(f) $r(a) = r(a^n)$ *for every* $a \in A$ *and every natural number* n;

(g) *If* T *is a right denominator subset of* A, *then the right ring of fractions* A_T *is a reduced ring.*

Proof. Parts (a), (b), (c) and (d) are proved in [4, pp. 286, 288, 316]. Part (e) follows from part (c).

(f) The inclusion $r(a) \subseteq r(a^n)$ is obvious. Let $b \in r(a^n)$. Then $a^n b^n = 0$. By part (a), $(ab)^n = 0$. Therefore, $ab = 0$, and $r(a^n) \subseteq r(a)$.

(g) Let $f : A \to A_T$ be the canonical homomorphism, $a \in A$, $t \in T$, $g = f(a)f(t)^{-1}$ and let $g^2 = 0$. Since $f(t)^{-1}f(a) \in A_T$, there exist elements $u \in T$, $b \in A$ such that $au = tb$. Then $0 = q^2 f(tu) = f(ab)$, $ab \in \ker(f)$. Therefore, $abs = 0$ for some $s \in T$. By part (a), $r(a)$ is an ideal of A. Therefore $bs \in r(a)$, $0 = atbs = a^2us = a^2x$ where $x \equiv us \in T$. By part (f), $r(a) = r(a^2)$. Therefore $ax = 0$, $a \in \ker(f)$, $q = 0$. □

Lemma 10. *Let* A *be a right localizable ring. Then:*

(a) A *is right distributive* $\Leftrightarrow$ A_M *is a right uniserial ring for every* $M \in \max(A)$;

(b) *If* A *is right noetherian, then* A_M *is a right noetherian ring for every* $M \in \max(A)$;

(c) *If* A_M *is a right invariant ring for every* $M \in \max(A)$, *then* A *is a right invariant ring;*

(d) *If A_M is a right noetherian right uniserial ring for every $M \in \max(A)$, then A is right invariant;*

(e) *A is a reduced ring $\Leftrightarrow$ A_M is a reduced ring for every $M \in \max(A)$.*

Proof. Parts (a) and (c) are proved in [3]. Part (b) is well-known (it follows, for example, from Lemma 8 of [3]). By [1] right distributive right noetherian rings are right invariant. Therefore, part (d) follows from part (c). The implication $\Rightarrow$ of part (e) follows from Lemma 7(g). The implication $\Leftarrow$ of part (e) is proved in [3]. □

Lemma 11. *Let $N(A)$ be the set of all minimal prime ideals of a ring A. Then the following conditions are equivalent:*

(a) *A is a semiprime right distributive ring and for every $H \in N(A)$, the quotient ring A/H is right nonsingular;*

(b) *A is a semiprime right distributive ring and for every $H \in N(A)$ the quotient ring A/H satisfies either the maximum condition for right annihilators or the maximum conditions for left annihilators;*

(c) *A is right distributive, right nonsingular;*

(d) *A is a reduced right distributive ring;*

(e) *A is right localizable and the right ring of fractions A_M is a right uniserial domain for every $M \in \max(A)$.*

Proof. The equivalence of the conditions (c), (d) follows from Lemma 7(d), 4(d). The implication (e) $\Rightarrow$ (d) follows from Lemma 8(a), (e). The implication (d) $\Rightarrow$ (b) follows from the following assertions: (1) A/H is a domain for every $H \in N(A)$ (it follows from Lemma 4(d) and 7(e)); (2) Every semiprime ring A is a subdirect product of quotient rings A/H ($H \in N(A)$); (3) Every subdirect product of reduced rings is a reduced ring. Since every right uniserial right nonsingular ring is a domain, the implication (d) $\Rightarrow$ (e) follows from Lemma 6, 7(b), 8(e). □

Lemma 12 ([4], Lemma 1.1.5). *Let B be an ideal of a ring A and let C be a nilpotent ideal of B. Then the ideal of A generated by C is nilpotent.*

Proof of Theorem 1. By Lemma 9 A is a reduced right localizable ring. Let $M \in \max(A)$, $Q \equiv A_M$ and let $f : A \to Q$ be the canonical homomorphism with the kernel D. Let $h : A \to A/D$ be a natural epimorphism. By Lemma 9, Q is a domain. Therefore D is a prime ideal. Then D contains a minimal prime ideal H of A. By condition, A/H is a right noetherian ring. Therefore $h(A)$ is a right noetherian ring. Since Q is a domain, $h(A)$ is a right distributive domain. Furthermore, $h(M) \in \max(h(A))$. By Lemmas 9 and 8(b) the right ring of fractions $h(A)_{h(M)}$ exists and is right uniserial right noetherian. Since $h(A)_{h(M)} \cong Q$, Q is a right uniserial right noetherian ring. By Lemma 8(d) A is a right invariant ring. □

Proof of Theorem 2. Parts (a), (b) follows from Lemmas 4(b), (c) and 5(c), (d). Part (c) follows from part (b) and from Lemma 5(c). Part (g) is proved in Lemma 4(e).

(d) Let $a \in U$ and $h \in H$, and let D be a cyclic submodule of M. Then $h^n(a^{-1}(D)) = 0$ for some natural number n. Therefore

$$(aha^{-1})(D) = (ah^n a^{-1})(D) = 0,$$

and so $aha^{-1} \in H$. Similarly, $a^{-1}ha \in H$. Therefore $UH = HU$. Similarly, $UT = TU$.

(e) By part (b), $T \in J(R)$. Therefore, $UT \subseteq J(R)$. By part (d), $UT = TU$. Let $a \in U$, $t \in T$, $b \in J(R)$. Then $1 + b \in U$, $bat = (1 + b)at - at \in UT$, $tab = ta(1 + b) - ta \in TU = UT$.

(f) Let $a \in U$, $f \in T$, $f_i \equiv a^i f a^{-i}$ $(i = 1, 2, \ldots)$. Let T be right t-nilpotent. Then $(f_1 \ldots f_n)(M) = 0$ for some natural number n. Therefore $(af)^n(M) = (f_1 \ldots f_n a^n)(M) = 0$. Then $T = UT$. By part (e) T is an ideal of $J(R)$. From part (c) it follows that T is a sum of nilpotent ideals of T. With this fact and with Lemma 10 we can prove that T coincides with the prime radical of R. Similarly, if T is left t-nilpotent, then T coincides with the prime radical of R. □

References

[1] W. Stephenson, Modules whose lattice of submodules is distributive, Proc. London Math. Soc. 28 (1974), 291–310.

[2] R. Mazurek and E. R. Puczylowski, On nilpotent elements of distributive rings, Comm. Algebra 18 (1990), 463–471.

[3] A. Tuganbaev, Hereditary rings, Mat. Zametki 41 (1987), 303–312, (Russian).

[4] V. A. Andrunakievich and Yu. M. Ryabuchin, The radical of algebras and a structural theory, Nauka, Moscow 1979 (Russian).

Department of Mechanics and Mathematics
Chair of Higher Algebra
Moscow State University
Moscow 119899, Russia

Chains of motley gems—and their Wiener indices

*Bo-Yin Yang and Yeong-Nan Yeh**

**In homage of my beloved late Teresa Teng who has supported me spiritually through years*

Abstract. Looking into the problem of determining the Wiener indices of mixed polygonal chains, we find general recursion algorithms applicable to all polygonal chains, yet simple enough for hand calculation. For chains composed of even-sided polygons only, we derive pithy explicit formulas. In the process, we also find simpler proofs for previously known formulas.

Keywords: polygonal chain, Wiener index

1. History

Definition 1. For a connected graph $G = (V, E)$, and its distance function $d : V \times V \to \mathbb{N} = \{0, 1, 2, 3, \ldots\}$, we define the *Wiener index*:

$$\mathcal{W} = \mathcal{W}(G) \equiv \sum_{\{u,v\} \subset V} d(u, v).$$

For any given point u in $V = V(G)$, we define

$$\mathcal{W}(u|G) \equiv \sum_{v \in V} d(u, v),$$

sometimes called the *partial Wiener index* of u (with respect to G).

In short, the Wiener index of a graph is the sum of distances between pairs of its vertices. Obviously we have $\mathcal{W} = \frac{1}{2} \sum_{u \in V} \mathcal{W}(u|G)$.

This quantity was first used in 1947 by Harold Wiener in his seminal article [35] (and used in a later series of articles, see [36]–[38]) where he came up with this surprisingly good approximation for the boiling points of high alkanes:

$$\text{b.p.} \approx \alpha \mathcal{W} + \beta w_3 + \gamma,$$

with α, β, γ being empirical constants, $\mathcal{W}$ the Wiener index of the molecular graph of the alkane, and w_3 the 'path number' (number of vertex pairs 3 apart in the graph).

After Wiener, applications of graph theory to chemistry lay dormant for a decade or so, and not until the concept was rediscovered in 1962 and used to estimate the critical constants of alkanes did the Wiener index really resurface. Wiener's original definition only applied to alkanes (the total number of carbon-carbon bonds between all pairs of carbon atoms), but it is easy to generalize to all molecular graphs. Eventually, Hosoya pointed out [13] in 1971 the usefulness of treating the Wiener index as a quantity associated with a molecular graph only and basically gave the present definition.

In the '70s and '80s, numerous studies of the Wiener index were undertaken, both from the chemical and mathematical angles, and it is surprising therefore that a bewildering array of names for this concept had been used as no one seems to have bothered checking the literature before inventing his own terminology. Those with a graph theory bent called it the *total transmission* like Šoltés (see [33]) or *gross(total) status* like Harary ([11] & [12]); some like Plesnik [25] and Rouvray [27]–[30] used descriptions, (*sum of distances* and *sum of the distance matrix elements* resp.), obviously both being twice the Wiener index; also used are *total distance* (Mohar [22]) and *total weight* (Teh and Shee, [34]). However 'Wiener index' or 'Wiener number' seems to have stuck as the most popular.

Eventually, the Wiener index $\mathcal{W}$ became one of the best (also likely the most often used) descriptors of molecular shape deducible from the molecular graph. We recommend, in addition to those specifically referred to below, [1], [3], [5], [17], [20], [21], [24], and [31], for those interested. Cites to most of the chemistry can also be found in the survey [8].

It should be fairly easy to see that the Wiener index of the molecular graph provides a measure of the compactness and the extent of branching in the molecule. Indeed, those physical and chemical properties that depends primarily on the intermolecular forces[1] (whose primary determining factor is the molecular volume-to-surface ratio) or the extent of branching in the skeleton are usually well linearly correlated with $\mathcal{W}$. For most families of hydrocarbons whether cyclic or acyclic, aromatic or aliphatic, these form an impressive list. As summarized in [2], these include heats of formation, atomization, isomerization, and vaporization; density, boiling point, critical pressure and temperature, refractive index, surface tension, velocity of sound propagation, and viscosity. In polymer chemistry, the Wiener index have been used successfully as predictors for physical properties such as melting points and others such as π-electron energies in conjugated polymers.

Since both pharmacology and material science often involve physical and chemical properties having to do with intermolecular forces, it is not surprising that Wiener

[1] The fact that $\mathcal{W}$ is correlated with so many physico-chemical properties of hydrophobic molecues means that it just has to be a rough measure of intermolecular forces. Recently it was shown that for medium-sized hydrocarbons of various families, both of the most widely used measures of intermolecular forces, Pitzer's acentric factor [26] and the Van der Waals a-coefficient have linear correlation coefficients higher than 90% to the Wiener index.

indices also found applications to both fields. For the former, Lukovits showed that the Wiener index of certain pharmacologically significant families of compounds has very strong correlations and their respectice cytostatic, antihistaminic, and estrogen-bonding activities; recently, he provided a good estimator of the partition coefficient to water for certain compounds (see [18]), a vital parameter in forecasting pharmacological utility. For the latter, $\mathcal{W}$ was recently used to characterize crystal defects and stability of lattices.

Lots of water has passed under the bridge and by now Wiener indices are relatively familiar objects to both graph theorists and theoretical chemists, and connections to other branches of mathematics have been found (e.g. [32]). Algorithms for computing the Wiener index for specific types of graphs are known for some time (see [23] for a good summary) and in some specific instances linear-time algorithms can be found. However, chemists are more interested in methods that lead to relatively simple general expressions for Wiener indices of families of molecules, especially that enables calculation by paper-and-pencil. For the case of a tree $G = (V, E)$, Wiener [35] himself provided this pretty formula:

$$\mathcal{W}(G) = \sum_{e \in E} n_1(e) n_2(e), \tag{1.1}$$

where $n_1(e), n_2(e)$ are the number of vertices on either side of edge e.

Merris and McKay [19] independently came up with another which connects Wiener indices to spectura of Laplacian matrices of trees. Let $N = |V|$ and $\lambda_1 \geq \cdots \geq \lambda_{n-1} \geq \lambda_n = 0$ be the spectrum of the Laplacian matrix $L = D - A$ of the graph G, then

$$\frac{\mathcal{W}(G)}{N} = \sum_j \frac{1}{\lambda_j}. \tag{1.2}$$

Gutman (see [8]) proved yet a third nice formula:

$$\mathcal{W}(G) = \binom{n+1}{3} - \sum_{\substack{\deg x \geq 3 \\ x \in V}} \left(\sum_{i>j>k} n_i(x) n_j(x) n_k(x) \right), \tag{1.3}$$

where $n_1 \geq n_2 \geq \cdots$ are the number of vertices in the components resulting the deletion of x from G.

Other facts about the Wiener indices of trees can be found in [8].

For graphs that are not trees (usually representing cyclic molecules), there isn't nearly so much in the way of general results aside from the formulas for Wiener indices of composite graphs by Gutman and Yeh [10]. Most results have been worked out on a case-by-case basis and open problems abound.

One such had been the computation of Wiener indices of polygonal graphs, especially chains. For a long time, calculations involve long computer runs that sometimes even give the wrong result, and the only noteworthy results were those of Gutman *et al* (see [6] & [7]) on hexagonal chains, which still were not closed-form. Recently, the authors have found closed-form solutions which indeed allow paper-and-pencil calculations for hexagonal [16] and pentagonal chains [39] (the latter an entirely open problem despite much effort), as well as for regular two-dimensional polygons patterns [15]. Here we give a general closed-form formula for all chains of even-sided polygons (Section 2) and demonstrate (via examples) a suitably easy algorithm for 'straight chains' (Section 3) and from it for any polygonal chain (Section 4).

2. Finding the Wiener index of a generic even chain

To begin with, we need to define our terms:

Definition 2. A *motley (polygonal) chain* is a graph of concatenated polygons (cycles sharing an edge) in which (for the moment) no vertex has degree more than 3.

Definition 3 (Representation of motley chains). Given n ordered pairs of non-negative integers $S = (a_1, b_1), \ldots, (a_n, b_n)$, we create a graph as follows: take the graph $P_2 \times P_n$ (noting it to be composed of n adjacent squares joined up side by side) and subdivide the j-th upper and lower edges by inserting a_j and b_j extra vertices respectively. We will call it the *motley chain* associated with S and write it as $M(S)$ or, when context permits, just S. The Wiener index of the graph $M(S)$ will be denoted by $\mathcal{W}(S)$. Obviously, as long as two sequences only differ in the first and last pairs with the sums $a_1 + b_1$ and $a_n + b_n$ identical, the associated graphs will remain the same. Hereafter, we will represent such a equivalence class as:

$$\left\langle (a_1 + b_1) \begin{matrix} a_2 & a_3 & \ldots & a_{n-1} \\ b_2 & b_3 & \ldots & b_{n-1} \end{matrix} (a_n + b_n) \right\rangle .$$

A motley chain clearly determines its own representation as above, uniquely up to the reversal of the order of the sequences, or the exchange of the numbers in each pair (both of these being involutions, i.e., at most 4 distinct representations in a K_4 group).

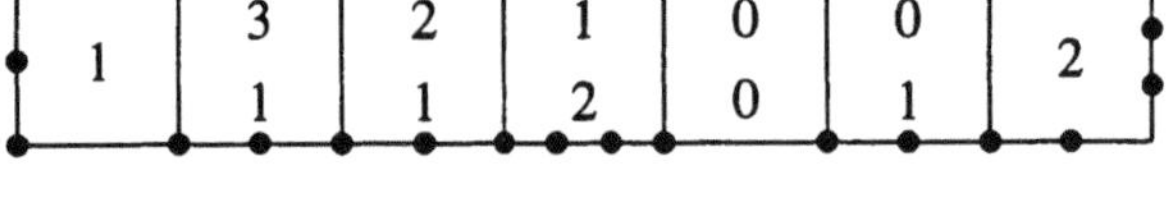

Figure 1. A "chain of motley gems."

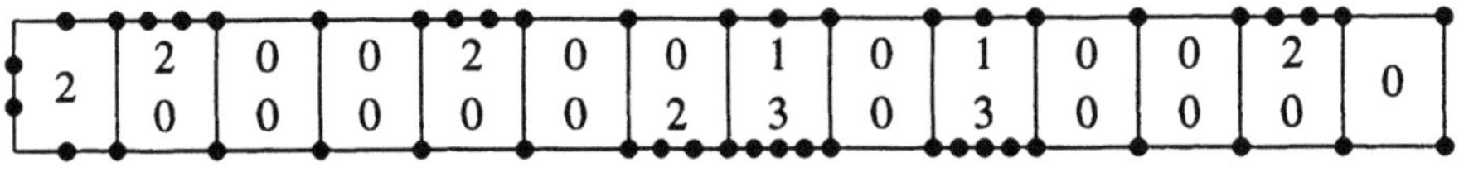

Figure 2. An even motley chain.

Figures 1 and 2 show how to make the construction as defined above. Thus, the work in [16] and [39] handle respectively the cases where each polygon is a hexagon and a pentagon (or when each pair of integers (a_i, b_i) adds up to 2 and 1) respectively.

In the special cases where each of the polygons has an even number of sides (or the each pairs of numbers are of two of the same parity—see Figure 2), there is an alternative representation:

Definition 4 (Alternative representation of even motley chains). We correspond each sequence $E \equiv k_0, (k_1, j_1), (k_2, j_2), \ldots, (k_n, j_n), k_{n+1}$, (where for every i we have $k_i \in \mathbb{P}$ and $j_i \in \mathcal{Z}$, $|j_i| < k_i$) with the motley chain $M(S)$, where

$$S \equiv \hat{E} \equiv \left\langle 2(k_0 - 1) \begin{matrix} k_1 - j_1 - 1 & k_2 - j_2 - 1 & \ldots & k_n - j_n - 1 \\ k_1 + j_1 - 1 & k_2 + j_2 - 1 & \ldots & k_n + j_n - 1 \end{matrix} 2(k_{n+1} - 1) \right\rangle .$$

To make the notation more compact, we will also write this as (see Figures 2 and 3)

$$E = k_0 (k_1)_{j_1} (k_2)_{j_2} \ldots (k_n)_{j_n} k_{n+1}$$

and subscript 0's, especially when the corresponding k_i is 1, will often be omitted as well—see Figure 3. We will call $K \equiv \sum_{j=0}^{n+1} k_j$ the *total length* of the even motley chain.

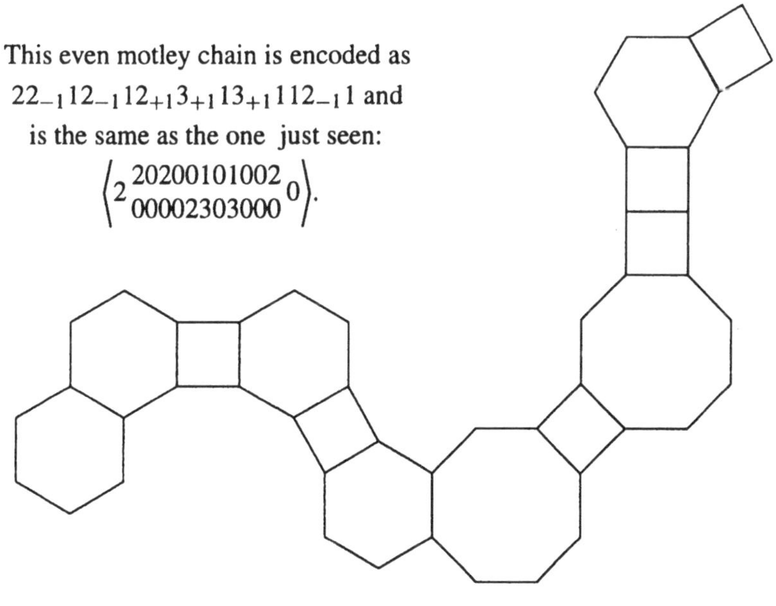

Figure 3. The last-shown motley chain in the eyes of chemists.

Using methods reminiscent of those in [15], we will first establish a base value from which we can evaluate deviations; for even motley chains this base value is obvious—it must be the Wiener index of a straight chain. Before we do that, however, we need a small lemma

Lemma 1. *The Wiener index of a chain of n squares is*

$$\mathcal{W}((1)^n) = \mathcal{W}(P_{n+1} \times P_2) = \frac{1}{3}(n+1)(n+3)(2n+1),$$

which differs from the Wiener index of a $2(n+1)$*-gon (or* $2(n+1)$*-cycle) by* $2\binom{n+1}{3}$.

Proof. Theorem 4 of [8] gives the Wiener index for the Cartesian product. That $\mathcal{W}(C_{2n}) = n^3$ has been known from long ago. □

Theorem 1 (Straight evens). *For a 'straight' even motley chain E, represented as*

$$E \equiv k_0k_1k_2\ldots k_n \equiv \left\langle 2(k_0-1) \begin{matrix} k_1-1 & k_2-1 & \ldots & k_n-1 \\ k_1-1 & k_2-1 & \ldots & k_n-1 \end{matrix} 2(k_{n+1}-1) \right\rangle,$$

and has among its $(n+2)$ *polygons a total of* n_4 *squares,* n_6 *hexes,* n_8 *octagons, etc. and total length* $K = \sum_i (n_{2i+2}\, i)$*, we have as the expression for its Wiener index:*

$$\mathcal{W}(E) = \frac{1}{3}(K+1)(K+3)(2K+1) + \frac{1}{3}\sum_{j=0}^{n+1} k_j(k_j+1)(k_j-1) \qquad (2.4)$$

$$= \frac{1}{3}(K+1)(K+3)(2K+1) + 2\sum_i n_{2i}\binom{i}{3}, \qquad (2.5)$$

which is independent of the order in which the polygons are arranged.

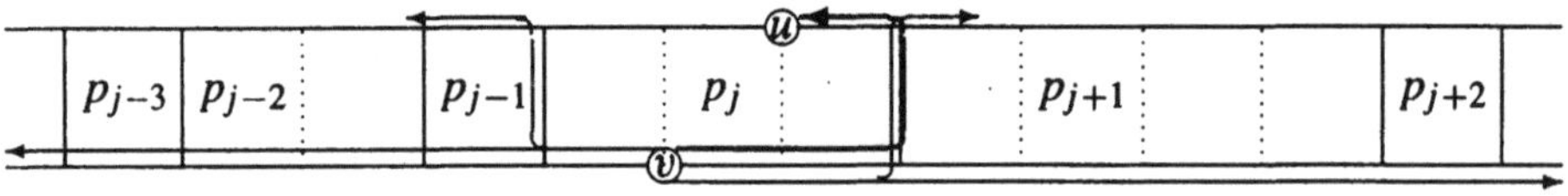

Figure 4. Wiener indices of straight even motley chains.

Proof. We compare the chain E to the even chain 1^K (a chain of squares). See Figure 4: we can see that for any two vertices $u, v \in E$, either $d_E(u,v) = d_{1^K}(u,v)$ or u and v belong to the same polygon p_j in the chain. Ergo:

$$\begin{aligned}\mathcal{W}(E) &= \sum_{\{u,v\}\in\binom{E}{2}} d_E(u,v) \\ &= \sum_{\{u,v\}} d_{(1^K)}(u,v) + \sum_{j=0}^{n+1}\sum_{\{u,v\}\in p_j}\left[d_E(u,v) - d_{(1^K)}(u,v)\right] \\ &= \mathcal{W}(P_{K+1}\times P_2) + \sum_{j=0}^{n+1}\left[\mathcal{W}(C_{2k_j+2} - \mathcal{W}(P_{k_j+1}\times P_2)\right].\end{aligned}$$

and the final substitutions follow from the lemma. □

Having deduced the base value, we can obtain results for *any* even motley chain by *'straightening'* it out in stages! Take any even chain

$$E = k_0(k_1)_{j_1}(k_2)_{j_2}\dots(k_n)_{j_n}k_{n+1}$$

and define the 'intermediate stages of straightening':

$$E_\ell \equiv k_0(k_1)_{j_1}(k_2)_{j_2}\dots(k_\ell)_{j_\ell}(k_{\ell+1})_0(k_{\ell+2})_0\dots(k_n)_0k_{n+1}.$$

Obviously, we have

$$\mathcal{W}(E) = \mathcal{W}(E_0) - \sum_{i=1}^{n}\left[\mathcal{W}(E_{i-1}) - \mathcal{W}(E_i)\right], \tag{2.6}$$

and therefore we need only know each $\mathcal{W}(E_{i-1}) - \mathcal{W}(E_i)$ in order to find out $\mathcal{W}(E)$. For convenience, we will use these notations so as to be parallel to the usage of [16] and [39]: for each stage of the 'straightening' we mark out the vertices u_0 and u_1, and $v_0, v_1, \dots, v_\ell$ where ℓ is the number of sides of p_j, the polygon in question, minus three (see Figure 5). The partial Wiener index $\mathcal{W}(v_i, M(S))$ is written as $\mathcal{X}_i(S)$. T will refer to the portion of straight even chain at the tail.

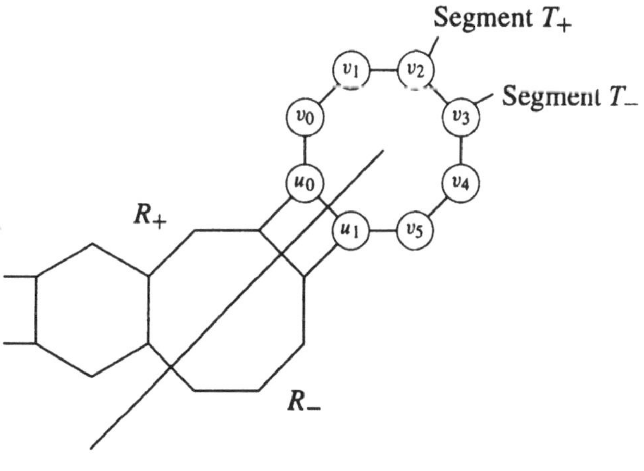

Figure 5. Marking the advance vertices.

It should be clear that some of the vertices in the 'unstraightened' part of S will have shortest paths into T through u_0 and some through u_1; to be quite precise, we can draw a figurative border down the middle of the next polygon and everything on the same side of u_1 will have the minimum distance access route through u_1, and vice versa, and we will call the first part R_+ and the other R_-, just to mark them (in Figure 6, these are colored light gray and black respectively). It should also be clear that there is a total of $2\sum_{m=j+1}^{n+1} k_m$ vertices in T, and we label the 'top' half T_+ and the 'bottom' half T_- as in the Figure 5.

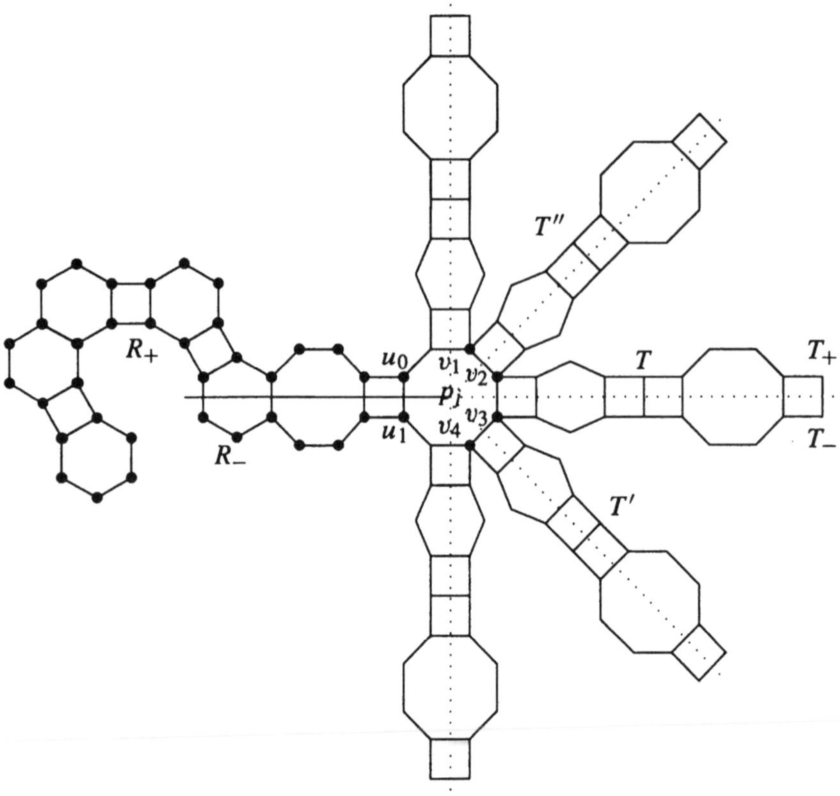

Figure 6. During the process of straightening!

We can see from Figure 6 that in fact as far as Wiener indices are concerned, the polygon p_j itself doesn't matter. Of the portion R remaining unstraightened, we can see that when the chain is 'bent' such that T goes into T', each point in R_+ gets closer to each point in T_- by one, and farther away from each point in T_+ by one— for no net difference; however, each point in R_- gets closer to any point in the tail by one. Hence, we know that the difference in Wiener indices is $|R_-| \times |T|$. Similarly, when we rotate T into T'', the Wiener index decreases by $|R_+| \times |T|$.

What about bigger turns? For each move after the first, we have the *entire* R approaching the whole of T and the net decrease in Wiener index is $|R| \times |T|$. Notice that bigger turns are impossible for chains of just hexagons (and squares)—this last case only become relevant in the case of an octagon or larger.

The above can be summed up succintly thus:

Lemma 2.

$$\mathcal{W}(E_{i-1}) - \mathcal{W}(E_i) = \begin{cases} |T|[j_i|R_+| + (j_i - 1)|R_-|], & \text{if } j_i > 0, \\ |T|[(-j_i - 1)|R_+| - j_i|R_-|], & \text{if } j_i < 0. \end{cases} \tag{2.7}$$

Given an even chain $E = k_0(k_1)_{j_1}(k_2)_{j_2}\dots(k_n)_{j_n}k_{n+1}$ of whose j's precisely those with the indices $i_1, i_2, \dots, i_m$ are non-zero. We define more notations:

$$K_\ell \equiv \sum_{h=0}^{i_\ell - 1} k_h, \text{ (and } K_0 \equiv 0,\text{)}$$

$$K'_\ell \equiv \sum_{h=i_\ell+1}^{n+1} k_h,$$

$$\tau_\ell \equiv (2|j_{i_\ell}| - 1)K_\ell + \operatorname{sgn}(j_{i_\ell} j_{i_{\ell-1}})K_{\ell-1}.$$

Theorem 2 (General evens).

$$\mathcal{W}(E_0) - \mathcal{W}(E) = 2\sum_{h=1}^{m} K'_\ell \tau_\ell, \tag{2.8}$$

where $\mathcal{W}(E_0)$ is given by Equation 2.5.

Proof. In the i_ℓ-th stage of 'straightening out', $|T|$ in Equation 2.7 is obviously $2K'_\ell$. It remains to calculate $|R_+|$ and $|R_-|$. Again from Figure 6 we can see that $|R|$, the sum of the two, is $2K_h$ and the difference is $2K_{\ell-1}$; hence, the smaller of the two is $K_\ell - K_{\ell-1}$ while the larger is $K_\ell + K_{\ell-1}$. So the difference at this polygon adds up to $2K'_\ell \tau_\ell$. □

Example 1. We tabulate in a table the calculations needed to find the Wiener index of the even motley chain shown in Figures 2 and 3.

h	0	1	2	3	4	5	6	7	8	9	10	11	12
Top	1	2	0	2	0	0	1	0	1	0	0	2	0
Bottom	1	0	0	0	0	2	3	0	3	0	0	0	0
k_h	2	2	1	2	1	2	3	1	3	1	1	2	1
j_h	0	-1	0	-1	0	1	1	0	1	0	0	-1	0
K		2		5		8	10		14			19	
K'		18		15		12	9		5			1	
τ		2		7		3	18		24			5	
$2\tau K'$		72		210		72	324		240			10	

Table A: Calculating the Wiener index of motley chain in Figures 2 and 3.

Total length of the chain is $K = 22$, $n_6 = 5$, $n_8 = 2$, so:

$$\mathcal{W}(E_0) = \frac{1}{3}(23 \times 25 \times 45) + (5 \times 2 + 2 \times 8) = 8651,$$

$$\mathcal{W}(E) = \mathcal{W}(E_0) - (72 + 210 + 72 + 324 + 240 + 10) = 7723.$$

Corollary 1. *Assume that the k_i's above comprise n_4 ones (squares) and n_6 twos (hexagons) and define its hex-length to be $N \equiv n_6 + \frac{n_4}{2}$, we assign to each hex the* *sum of half the number of squares and the number of hexs that comes before it in the chain (thus, either integer or half-odd) as demonstrated in Figure 7.*

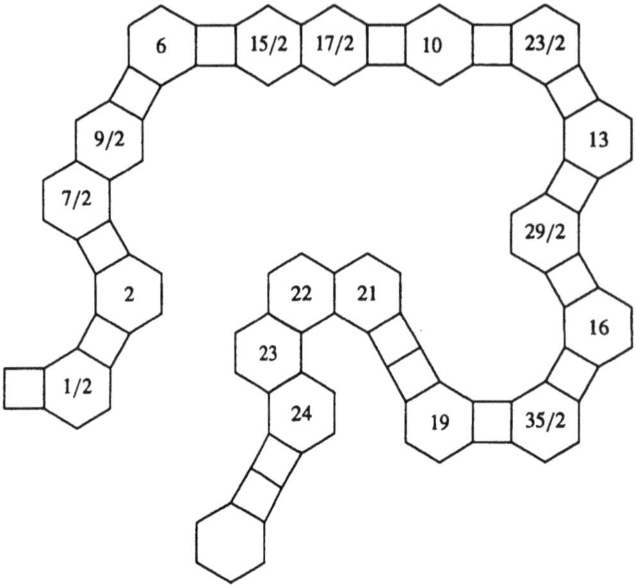

Figure 7. Marking the chain □0□0□21□2□11□1□2□2□0□2□2□□0002□□★.

Assume B to be the set of numbers assigned to hexes with turns (i.e., represented by 2_{+1} or 2_{-1}) plus the number $N-1$. For each $i \in B$, define $\nu(i)$ to be the next larger element in B, and take C to be the elements i of B such that the $j_i\, j_{\nu(i)} > 0$. We further define $\psi(i)$ to be $\nu(i) - i$ for each i that is in B but not in C, and $2(N-1) - \nu(i) - i$ for $i \in C$. Then we have:

$$\mathcal{W}(E) = \frac{1}{3}(2N+1)(2N+3)(4N+1) + 2n_6 - 8\left[\sum_{i \in B} i\psi(i)\right]. \tag{2.9}$$

This is a direct generalization of Theorem 1 in [16], and almost every single one of its consequences hold for square-hex chains without need for extensive modification!

We note that the notation for hex-chains used in [16] can be carried over with these modifications: the numbers 0, 1, 2 will still denote a hex with a left 120 degree turn, a straight move, and a right 120 degree turn respectively; a box will denote a square, and for hexes at the end (not expressly shown in the notation of [16]) will be denoted by a star.

Example 2. The even motley chain $212_0 2_1 12_1 2_1 112_{-1} 12_{-1} 1$ in Figure 8 has 7 hexes and 6 squares, thus is '10 hexes long', and can be denoted ★□12□22□□0□0□. The calculations needed to find its Wiener index (easily manageable by hand–even the monster chain in Figure 7 is quite doable with paper, pencil and a little patience!) is shown below and the reader is invited to compare this with similar calculations in [16].

Hex:i	0	$\frac{3}{2}$	$\frac{5}{2}$	4	5	7	$\frac{17}{2}$
B			•	•	•	•	•
C			•	•		•	
ν			4	5	7	$\frac{17}{2}$	9
ψ			$\frac{23}{2}$	9	2	$\frac{5}{2}$	$\frac{1}{2}$
$i\psi(i)$			$\frac{115}{4}$	36	10	$\frac{35}{2}$	$\frac{17}{4}$

$$\begin{aligned} \mathcal{W} &= \frac{1}{3}(21 \times 23 \times 41) + 2 \times 7 \\ &\quad -8\left\{\frac{5}{2} \times \frac{23}{2} + 4 \times 9 + 5 \times 2 \right. \\ &\quad \left. +7 \times \frac{5}{2} + \frac{17}{2} \times \frac{1}{2}\right\} \\ &= 6615 - 7 \times \frac{193}{2} \\ &= 5843 \end{aligned}$$

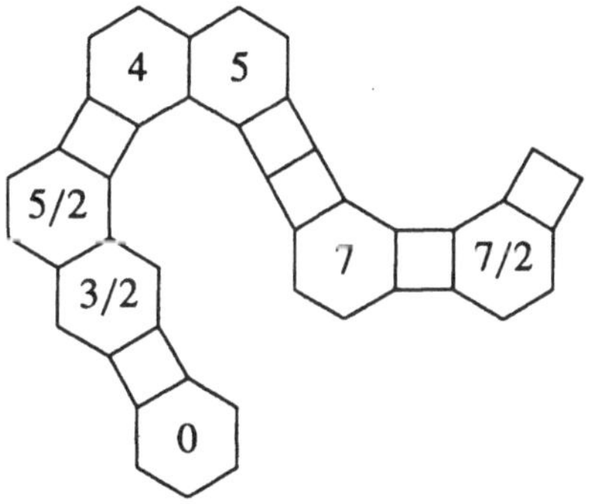

Figure 8. The chain ⋆⊓12□22□□0□0□, otherwise written $212_02_112_12_1112_{-1}12_{-1}1$, and its length is $N = 10$.

3. Zigzagging chains including odd polygons

We have seen that chains with only even-sided polygons can be handled comprehensively via Equation 2.8. When the situation involves odd-sided polygons, the situation is quite a bit stickier. First of all we need a base value against which Wiener indices of chains can be compared, which means a kind of 'standard' chain must be defined—as below.

Definition 5 (Zigzagging chain). A motley chain is said to be *straight* or *zigzagging* if for one of its representations (as in Definition 3) $S = (a_1, b_1), \ldots, (a_n, b_n)$ the set $\{\sum_{i=1}^{j}(a_i - b_j) | 1 \leq j \leq n\}$ is contained in $\{0, 1\}$ or $\{0, -1\}$. We will denote by $\mathcal{Z}_{n_1,n_2,\ldots,n_k}(n)$ the Wiener index of a straight chain consisting of n repetitions of polygons of $n_1, n_2, \ldots, n_k$ sides. A pictorial representation is given in Figure 9:

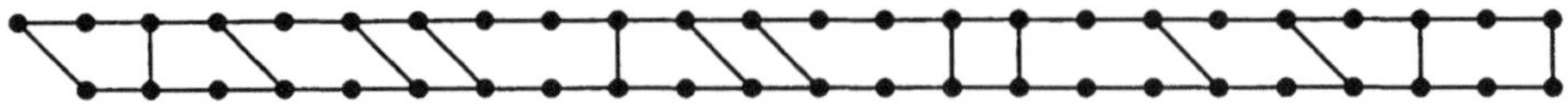

Figure 9. A 'straight' motley chain.

And as the following example will demonstrate, the Wiener index of straight chains composed of just odd-sided polygons is relatively straightforward:

Example 3 (Wiener index of a chain of zigzagging pentagons and septagons). We want to find $\mathcal{Z}_{5,7}(n)$, the Wiener index of a straight motley chain composed of alternating pentagons and septagons. To do this, we start with a straight chain of n decagons, the Wiener index we know from Equation 2.5 to be

$$\mathcal{Z}_{10}(n) = \frac{1}{3}(4n+1)(4n+3)(8n+1) + 20n.$$

Figure 10 should illustrate our method well. We will subdivide each decagon into a pentagon and a septagon starting from one end. Clearly, the distances *between the vertices which comprise each original decagon* changes with the subdivision, as is given by the difference (found from Equation 2.8) between the Wiener number of a $2(m+n)$-gon and a $(2m+1)$-gon fused with a $(2n+1)$-gon:

$$\begin{aligned}\Delta(m,n) &= \mathcal{W}(C_{2m+2n}) - \mathcal{W}(\langle(2m-3),(2n-3)\rangle)\\ &= \mathcal{W}(C_{2(m+n-1)}) - \mathcal{W}(\langle(2m-4),(2n-4)\rangle)\\ &= mn(m+n) - \frac{1}{2}(m^2+n^2+4mn) + \frac{3}{2}(m+n) - 1, \qquad (3.10)\end{aligned}$$

so between the Wiener indices of a Decagon (= 125) and a Septagon-Pentagon pair (= 107) there is a difference of 18. This difference is the same for each subdivision.

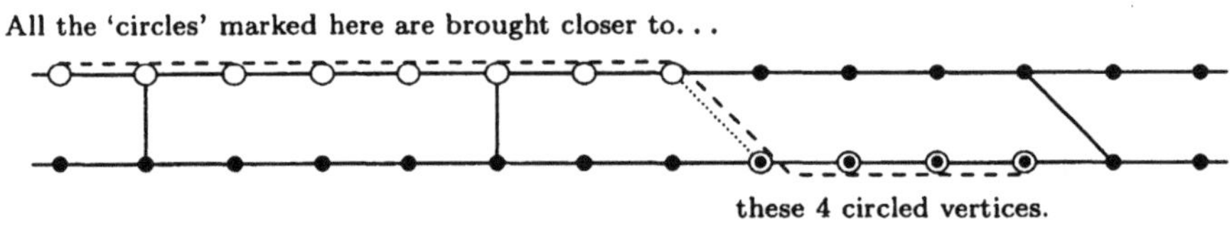

Figure 10. Obtain the Wiener index of a zigzagging pentra-septa chain by subdivision.

However, this aside, certain vertices on either side of the divide get closer to each other as shown in Figure 10. One can observe that the distances between two vertices in the graph decreases by one when they are one circled and one gray marked vertices in the figure, *except* for the already-counted cases *when both of the vertices are in the decagon*–which is two times three, or six cases in all here, for each subdivision. We can observe that

- at the beginning of the chain only *two* 'grays' as shown in the figure exist;
- when the j-th decagon from the end is subdivided there are $4j-1$ 'circles'.

So we get

$$\mathcal{Z}_{10}(n) - \mathcal{Z}_{7,5}(n) = 12n + 4\sum_{j=1}^{n-1}(4j-1) + 2(4n-1) = 2(2n+1)^2. \quad (3.11)$$

Similarly for a chain of zigzagging septagons (given that a dodecagon and two septagons differ by $\Delta(3,3) = 35$ in terms of Wiener indices) we have:

$$\mathcal{Z}_{12}(n) - \mathcal{Z}_7(2n) = (35 - 3\times 3)n + 5\sum_{j=1}^{n-1}(5j-2) + 3(5n-1) = 4 + \frac{37n}{2} + \frac{25n^2}{2}. \quad (3.12)$$

It should be obvious now that any 'straight' (that is, zigzagging) polygonal chain made up of an *even* number of *odd-sided* polygons (or one in which the odd-sided polygons pair together nicely) can be handled in the same manner.

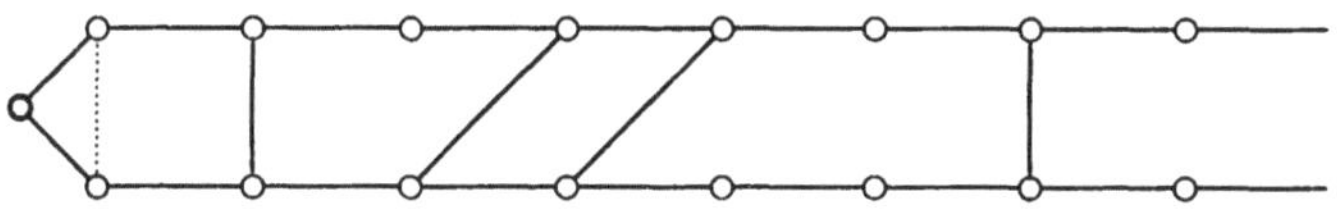

Figure 11. 'Tacking on' an extra at one end of a zigzagging chain.

To see that an *odd* number of odd-siders poses no particular problem either, we show that the Wiener index of a zigzagging septagon-chain can be found painlessly thus: consider the straight even chain $S = \underbrace{5\ldots5}_{n}2$, that is, n dodecagons followed by an hexagon, and consider the increment in Wiener index when the trailing hex is changed to a septagon—that is, when an extra vertex is tacked onto the end of the chain. The difference is clearly $15 + 5n(5n+7)$, the first term being the difference between $\mathcal{W}(C_6) = 27$ and $\mathcal{W}(C_7) = 42$, and the second being the sum of distances between the tacked-on vertex and the vertices not in the original hex. It can then be seen that this tacked-on vertex does not change its distance to any point in the process of subdivision. Therefore we get $\mathcal{Z}_7(2n+1)$ by subdivision of the chain S above and adding in the difference.

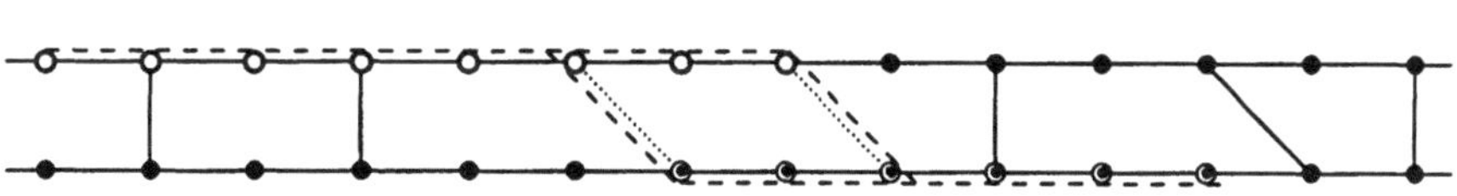

Figure 12. A more general sort of subdivision.

In the previously introduced procedure of subdivision, the most important relation is given by Equation 3.10. To make sure that we can handle any zigzagging chain, in any arrangement of the polygons, we need to handle a split of a $2(m+n+k)$-gon into polygons with $2m+1$, $2(k+1)$, and $2n+1$ sides, in that order. So we need

$$\begin{aligned}\Delta(m,n;k) &\equiv \mathcal{W}(C_{2(m+n+k)}) - \mathcal{W}\left(\left((2m-3)\,{}^{k-1}_{k-1}\,(2n-3)\right)\right)\\ &= -1+2k-k^2+\frac{3m}{2}-2km+k^2m-\frac{m^2}{2}+km^2+\frac{3n}{2}-2kn\\ &\quad +k^2n-2mn+2kmn+m^2n-\frac{n^2}{2}+kn^2+mn^2. \qquad (3.13)\end{aligned}$$

One can see that setting $k=0$ reduces to Equation 3.10, furthermore having more than one even-sided polygon between the two odd-sided ones present no problem because one can subdivide an even-sided polygon quite easily with the method of the previous section.

Example 4 (Repeating straight chain). Suppose that we wish to calculate $\mathcal{Z}_{5,6,7}(n)$. From the above, we start with a chain of n straight quadradecagons which has Wiener index

$$\mathcal{Z}_{14}(n) = (2n+1)(6n+1)(12n+1)+70n.$$

The difference for each subdivision in the 14-gon itself is $\Delta(2,3,2)=68$. For the division with j 14-gons left the 'circles' number $6j-1$, and there are always *six* grays (except for 4 at the beginning). Ergo, we have

$$\begin{aligned}\mathcal{Z}_{14}(n) - \mathcal{Z}_{7,6,5}(n) &= (68-5\times 4)n + 6\sum_{j=1}^{n-1}(6j-1) + 4(6n-1)\\ &= 18n^2+48n+2.\end{aligned}$$

Hence

$$\mathcal{Z}_{7,6,5}(n) = -1+42n+90n^2+144n^3. \qquad (3.14)$$

4. Wiener indices of generic polygonal chains

Having found a standard Wiener index against which to calculate deviations, we will proceed in a similar fashion to Section 2. However there is one essential difference, viz: when handling even motley chains, every vertex in a chain was *either* closer to one or of its end-vertices (marked '+' and '−') *or* the other, as depicted by the gray and black dots in Figure 6. Such is not the case when dealing with odd-sided polygons since some (marked as white dots in following 4 figures) may be equi-distant to both end-vertices. More precisely:

Lemma 3. *Marking the upper and lower end-vertices of a motley chain as u_+ and u_- and represent the chain as* $S = \begin{pmatrix} a_1 & a_2 & \dots & a_n \\ b_1 & b_2 & \dots & b_n \end{pmatrix}$ *(with the a_1, b_1 side next to the end-vertices). Define $d_i \equiv a_i - b_i$, $e_i \equiv a_i + b_i + 2$, and where applicable ℓ the smallest i for which $d_i \neq 0$. Further denote by S_+, S_-, S_O the set of non-end vertices closer to u_+, closer to u_-, and equidistant respectively.. Then when $d_l > 0$ (otherwise just switch the $+$'s and $-$'s):*

- *If $d_\ell \geq 2$, (Figure 13), then*

$$|S_+| = \left\lfloor \frac{1}{2} \sum_{j=1}^{\ell} e_j \right\rfloor, \quad |S_O| = (e_\ell \bmod 2), \quad |S_-| = |S_+| + \sum_{j=\ell+1}^{n} e_j.$$

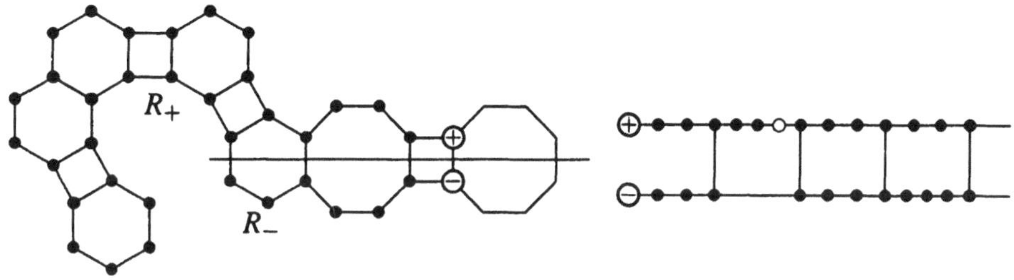

Figure 13. A turn to the right, same as the even-only case.

- *If $d_\ell = 1$, then* $|S_+| = \left\lfloor \left[\left(\sum_{j=1}^{\ell} e_j\right) - 1\right]/2 \right\rfloor$, *and furthermore:*
- *if $d_i = 0$ or 1, $\forall i > \ell$ (Figure 14, a straight chain), then:*

$$|S_O| = \left\lfloor \frac{1}{2}\left(3 + \sum_{j=\ell+1}^{n} e_j\right) \right\rfloor, \quad |S_-| = \left\lfloor \frac{1}{2}\left(\sum_{j=1}^{n} e_j\right) \right\rfloor.$$

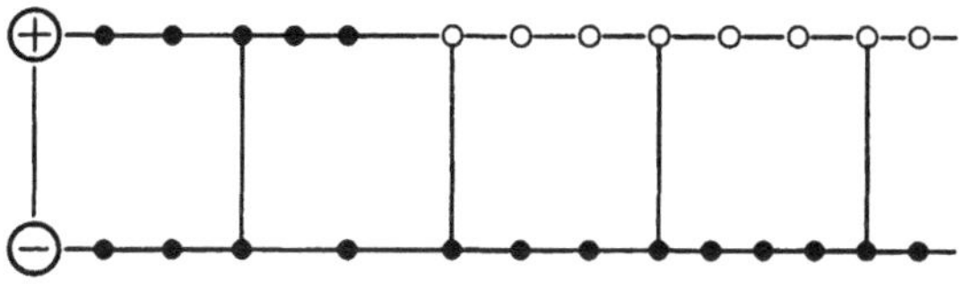

Figure 14. A zigzagging chain and its vertices.

- *Suppose $d_m \geq 2$, where $m = \min\{i \mid d_i \neq 0 \text{ or } 1\}$ (Figure 15), then:*

$$|S_O| = \left\lfloor \frac{1}{2}\left(3 + \sum_{j=\ell+1}^{m} e_j\right) \right\rfloor, \quad |S_-| = \left\lfloor \frac{1}{2}\left(\sum_{j=1}^{m} e_j\right) \right\rfloor + \sum_{j=m+1}^{n} e_j;$$

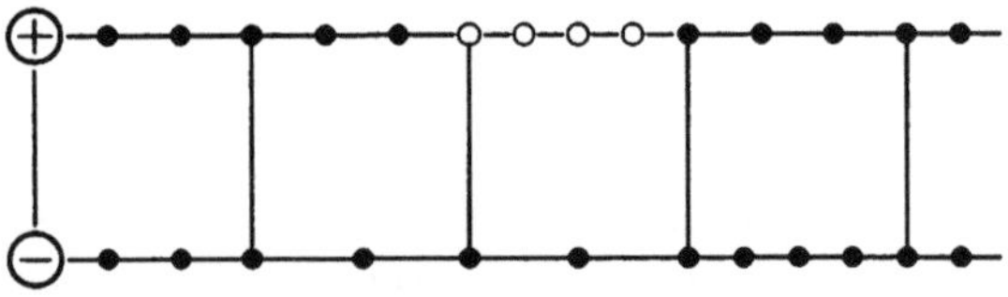

Figure 15. Here the equal-distance portion is transient.

- *as above, but if* $d_m < 0$ *(Figure 16), then:*

$$|S_0| = \left\lfloor \frac{1}{2}\left(3 + \sum_{j=\ell+1}^{m} e_j\right)\right\rfloor + \sum_{j=m+1}^{n} e_j, \quad |S_-| = \left\lfloor \frac{1}{2}\left(\sum_{j=1}^{m} e_j\right)\right\rfloor.$$

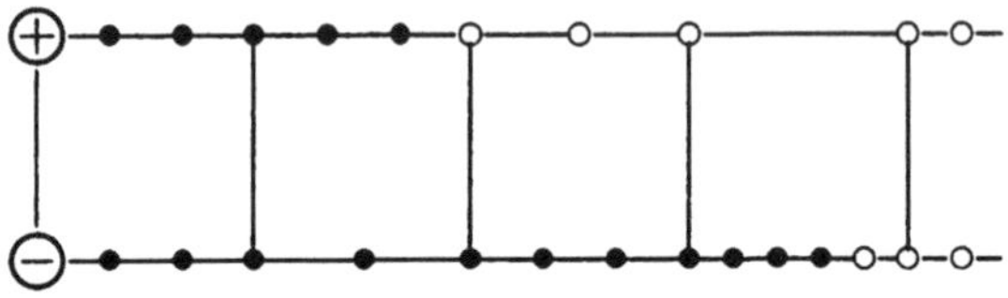

Figure 16. Here most vertices are equidistant to the ends.

- *If there is no* ℓ*, chain is straight even, vertices split evenly between* S_+ *and* S_-.

Proof. Pretty much by inspection. □

With Lemma 3 under our belt, we can proceed to the following lemma which we will need to derive the Wiener index of a generic motley chain from a straight chain. Before that, however, we need to define our notations. Given any motley chain $S = \begin{pmatrix} a_1 & a_2 & \dots & a_n \\ b_1 & b_2 & \dots & b_n \end{pmatrix}$, we can find a chain $S_0 = \begin{pmatrix} x_1 & x_2 & \dots & x_n \\ y_1 & y_2 & \dots & y_n \end{pmatrix}$ which is *straight* or *zigzagging* according to Definition 4, and satisfying $x_i + y_i = a_i + b_i$, $\forall i$. The *stages of straightening* $S_j = \begin{pmatrix} a_1 & a_2 & \dots & a_j & x_{j+1} & \dots & x_n \\ b_1 & b_2 & \dots & b_j & y_{j+1} & \dots & y_n \end{pmatrix}$ will be defined in direct analogy to the notations used in Section 2 (one example shown in Figure 17).

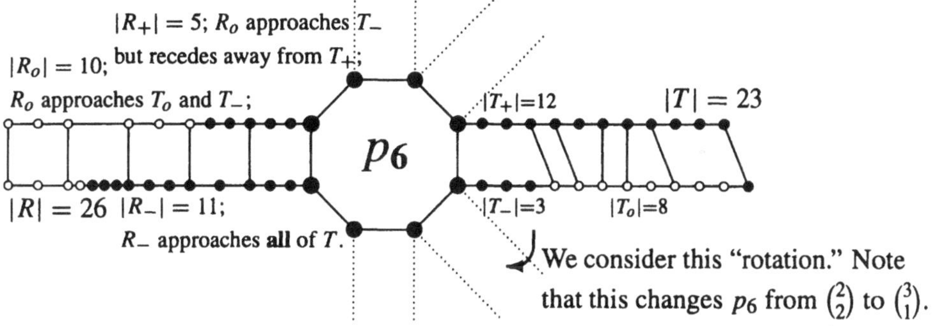

Figure 17. p_6, a 'stage of the straightening' of a generic motley chain!

Lemma 4 (Rotation at one polygon). *If we mark the two remaining chains to be R and T, the latter being straight (zigzagging), then we need only the following to calculate the difference in Wiener indices between straight motley chain and general positions. Partition R into three parts R_+, R_-, and R_O and T ditto as in the previous lemma, and:*

- *If $x_i = y_i$ and $x_i - a_i > 0$ (as demonstrated in Figure 17—switch + and − for the opposite case) then*

$$\mathcal{W}(S_{i-1}) - \mathcal{W}(S_i) = (x_i - a_i)|R||T| - (|R_O||T_-| + |R_-||T_O| + 2|R_-||T_-|).$$

- *If $x_i - y_i = x_i - a_i = 1$ (Figure 18), then*

$$\mathcal{W}(S_{i-1}) - \mathcal{W}(S_i) = (|R_+||T_+| + |R_O||T_+| + |R_+||T_O|) - (|R_-||T_O| + |R_-||T_O| + |R_-||T_-|).$$

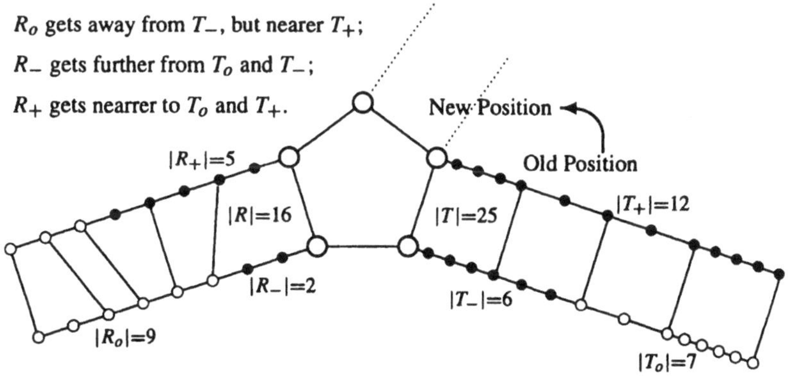

Figure 18. 'Buckling' between bearing up and bearing down!

- *If* $x_i - y_i = 1,\ a_i - x_i > 0$, *then*

$$\mathcal{W}(S_{i-1}) - \mathcal{W}(S_i) = (a_i - x_i)|R||T|.$$

Lemma 3 and Lemma 4 are basically sufficient to calculate, by hand if necessary (although somewhat cumbersome), the Wiener number of any motley chain. For chains in which deviations from the straight is not too numerous, the process is quite straightforward, and we will now give a concrete example.

Example 5. We will do a very simple illustration with the motley chain in Figure 1 and its straightened form, which can be represented as

$$S = \left\langle 1 \begin{matrix} 3\ 2\ 1\ 0\ 0 \\ 1\ 1\ 2\ 0\ 1 \end{matrix} 2 \right\rangle, \quad \text{and } S_0 = \left\langle \begin{matrix} 0\ 2\ 2\ 1\ 0\ 1\ 1 \\ 1\ 2\ 1\ 2\ 0\ 0\ 1 \end{matrix} \right\rangle.$$

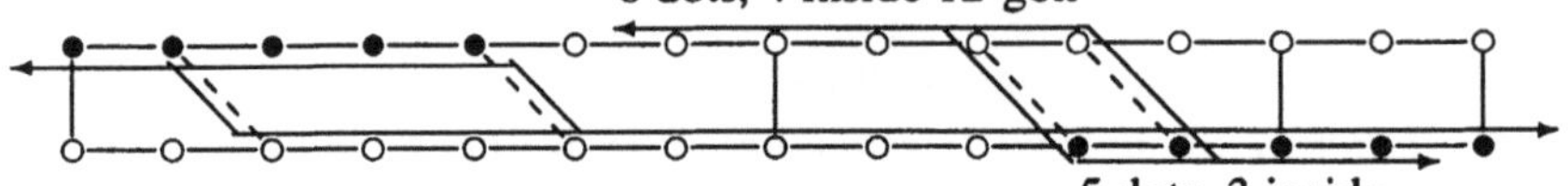

Figure 19. The straightened chain.

So, before the subdivisions are drawn into S_0, we have an even motley chain of length 14, made up of an hexodecagon, a dodecagon, and a hexagon. Equation 2.5 gives its Wiener index as $\frac{1}{3}(15 \times 17 \times 29) + 2(\binom{8}{3} + \binom{6}{3} + \binom{3}{3}) = 2619$.

The hexodecagon subdivided into a pentagon, an octagon, and a septagon. makes a difference of $\Delta(2, 3, 3) = 105$; the extra difference between rows is $5 \times 7 = 35$, for 140 total. The dodecagon subdivided into a septagon, a tetragon, and a pentagon, make a difference of $\Delta(3, 2, 1) = 39$; the extra difference comes to $6 \times 5 - 4 \times 3 = 18$, for a total of 57. Finally $\mathcal{W}(S) = \mathcal{W}(S_6)$ is found via:

i	1	2	3	4	5	6	7
x_i	0	2	2	1	0	1	1
y_i	1	2	1	2	0	0	1
e_i	3	6	5	5	2	3	4
$\lvert T_-\rvert$		9	2	4		2	
$\lvert T_O\rvert$		8	6	3		0	
$\lvert T_+\rvert$		2	6	2		2	
a_i	0	3	2	1	0	0	1
b_i	1	1	1	2	0	1	1
$\lvert R_-\rvert$		1	6	2		3	
$\lvert R_O\rvert$		1	0	4		9	
$\lvert R_+\rvert$		1	3	8		9	

$$\begin{aligned} \mathcal{W}(S_0) &= 2619 - 140 - 57 = 2422, \\ \mathcal{W}(S_2) &= \mathcal{W}(S_0) - 43 = 2379, \\ \mathcal{W}(S) &= \mathcal{W}(S_2) - 12 = 2367. \end{aligned}$$

5. Discussion

So far as we know, this is the first attempt on systemic calculation of Wiener indices of arbitrary polygonal chains. Perceptive readers should have by now realised that even though Sections 3 and 4 are mostly in examples, the procedure thus outlined actually contains a generally applicable algorithm which takes considerably less actual computation time than a brute-force recursion approach—even though a complete description would be tedious and too long in view of space limitations.

The eagle-eyed reader would also have spotted the fact that despite the original 'chemical' definition of 'catacondensed' polygonal chains (as used in [16] and [39]) not allowing vertices of degree 4, but the contents of this article apply to them with equal force. To be quite precise, they correspond to motley chains that have some of the a_i or b_i being -1, but not both in a pair at the same time (since the smallest polygon that is meaningful is the triangle). Most formulas in the text apply with undiminished validity!

Acknowledgements. The authors are greatly indebted to National Science Council of Taiwan (R. O. China) for financial support under Grants NSC84-2121-M-032-006 and NSC84-2121-M-001-009 respectively.

References

[1] D. Bonchev, Information theoretic indices for characterization of chemical structure, Research Studies Press, Chichester 1983.

[2] D. Bonchev and O. Mekenyan, Comparability, Graphs and Molecular Properties, J. Chem. Soc. Faraday II, 80 (1983), 695.

[3] F. Buckley and F. Harary, Distances in Graphs, Addison-Wesley, New York 1990.

[4] J.-C. Chen, I. Gutman and Y. Yeh, On the sum of all distances in graphs, Tamkang J. of Mathematics 25 (1993), 83–86.

[5] I. Gutman and O. E. Polansky, Mathematical concepts in organic chemistry, Springer-Verlag, Berlin 1986.

[6] I. Gutman, Wiener numbers of bezenoid hydrocarbons: Two theorems, Chem. Phys. Letters 136 (1987), 134–136.

[7] I. Gutman, J. W. Kennedy and L. V. Quintas, Wiener numbers of random bezenoid chains, Chemical Physics Letters 173 (1990), 403–408.

[8] I. Gutman, S.-L. Lee, Y.-L. Luo and Y. Yeh, Recent results in the theory of the Wiener number, Indian J. Chem. 32A (1993), 651–661.

[9] I. Gutman and D. H. Rouvray, A new theorem for the Wiener branching index of trees with perfect matchings, to appear in Computers and Chemistry.

[10] I. Gutman and Y. Yeh, On the sum of all distances in composite graphs, to appear in Discrete Math.

[11] F. Harary, Graph Theory, Addison-Wesley, Reading, MA, 1971.

[12] F. Harary, Status and contrastatus, Sociometry 22 (1959), 23–43.

[13] H. Hosoya, Topological index. A newly proposed quantity characterizing the topological nature of structural isomers of saturated hydrocarbons, Bull. Chem. Soc. Japan 44 (1971), 2332–2339.

[14] H. Hosoya, On some counting polynomials in Chemistry, Discrete Appl. Math. 19 (1988), 239–254.

[15] I.-W. Huang, B.-Y. Yang, and Y. Yeh, Wiener Indices of Hex Carpets—from Hexagon Models to Square Grids, preprint.

[16] W.-C. Huang, B.-Y. Yang, and Y. Yeh, From ternary Strings to Wiener Indices of Bezenoid Chains, preprint.

[17] S.-L. Lee and C.-Y. Lin, Graph Theory and Chemistry, Chinese Chem. Soc. (of Taiwan) 49 (1991), 179–187.

[18] I. Lukovits, Correlation between Components of the Wiener Index and Partition Coefficients of Hydrocarbons, Internat. J. Quantum Chem. and Quantum Biol. Symposium 19 (1992), 247.

[19] R. Merris, Laplacian Matrices of Graphs: a Survey, preprint.

[20] R. Merris, An edge version of the matrix-tree theorem and the Wiener Index, Linear and Multilinear Algebra 25 (1989), 291–296.

[21] Z. Mihalić, D. Veljan, D. Amić, S. Nikolić, D. Plavšić, and N. Trinajstić, The Distance Matrix in Chemistry, J. Math. Chem. 11 (1992), 223–258.

[22] B. Mohar, Eigenvalues, Diameter, and Mean Distance in graphs, Graphs Combin. 7 (1991), 53–64.

[23] B. Mohar and T. Pisanski, How to compute the Wiener index of a graph, J. Math. Chem. 2 (1988), 267–277.

[24] I. Motoc and A. Balaban, Topological indices: intercorrelation, physical meaning, correlational ability, Revue Roumaine de Chimie 26 (1981), 593–600.

[25] J. Plesnik, On the sum of all distances in a graph or digraph, J. Graph Theory 8 (1984), 1–21

[26] K. S. Pitzer, The Volumetric and Thermodynamic Properties of Fluids. Theoretical Basis and Virial Coefficients, J. Amer. Chem. Soc. 77 (1955), 3427.

[27] D. H. Rouvray, On the sum of distance matrix elements, in: Chemical Applications of Graph Theory (A. T. Balaban, ed.), Academic Press, New York 1976.

[28] D. H. Rouvray, The role of the topological distance matrix in chemistry, in: Mathematics and Computational Concepts in Chemistry (N. Trinajstić, ed.), Horwood Publishers, Chichester 1986, 295–306.

[29] D. H. Rouvray, The modelling of chemical phenomena using topological indexes, J. Comput. Chem. 8 (1987), 470–480.

[30] D. H. Rouvray, The challenge of characterizing molecular branching in molecular species, Discrete Appl. Math. 19 (1988), 317–338.

[31] E. Ruch and I. Gutman, The Branching Extent of Graphs, J. Combin. Inform. System Sci. 4 (1979), 285–295.

[32] B. E. Sagan and Y. Yeh, The Wiener Polynomial of a Graph, preprint.

[33] L. Šoltés, Transmission in graphs: a bound and vertex removing, Math. Slovaka 41 (1991), 11–16.

[34] H.-H. Teh and S.-C. Shee, Algebraic Theory of Graphs, Lee Kong Chian Inst. of Math. and Comp. Sci. Nanyang Univ. Singapore.

[35] H. Wiener, Structural determination of paraffin boiling points, J. Amer. Chem. Soc. 69 (1947), 17–20.

[36] H. Wiener, Correlation of Heats of Isomerization and Differences in Heats of Vaorization of Isomers, among the Paraffin Hydrocarbons, J. Amer. Chem. Soc. 69 (1947), 2636.

[37] H. Wiener, Vapor Pressure—Temperature Relationships among the Branched Paraffin Hydrocarbons, J. Phys. Chem. 52 (1948), 425.

[38] H. Wiener, Relation of the Physical Properties of the Isomeric Alkanes to Molecular Structure, J. Phys. Chem. 52 (1948), 1082.

[39] B.-Y. Yang, and Y. Yeh, Zigging and Zagging in Pentachains, to appear in Adv. in Appl. Math.

Department of Mathematics
Tamkang University
Tamsui
Taipei County, Taiwan, ROC

Institutc of Mathematics
Academia Sinica
Nankang
Taipei City, Taiwan, ROC

Bei Fragen zur Produktsicherheit wenden Sie sich bitte an:
If you have any questions regarding product safety,
please contact:

Walter de Gruyter GmbH
Genthiner Straße 13
10785 Berlin
productsafety@degruyterbrill.com